Springer Series in
OPTICAL SCIENCES 137

founded by H.K.V. Lotsch

Springer Series in
OPTICAL SCIENCES

The Springer Series in Optical Sciences, under the leadership of Editor-in-Chief *William T. Rhodes*, Georgia Institute of Technology, USA, provides an expanding selection of research monographs in all major areas of optics: lasers and quantum optics, ultrafast phenomena, optical spectroscopy techniques, optoelectronics, quantum information, information optics, applied laser technology, industrial applications, and other topics of contemporary interest.
With this broad coverage of topics, the series is of use to all research scientists and engineers who need up-to-date reference books.

The editors encourage prospective authors to correspond with them in advance of submitting a manuscript. Submission of manuscripts should be made to the Editor-in-Chief or one of the Editors. See also www.springer.com/series/624

A. Erko M. Idir
T. Krist A.G. Michette
(Eds.)

Modern Developments in X-Ray and Neutron Optics

With 299 Figures

Springer

Professor Dr. Alexei Erko
BESSY GmbH
Albert-Einstein-Str. 15, 12489 Berlin, Germany
E-mail: erko@bessy.de

Dr. Mourad Idir
Synchrotron Soleil L'orme des Merisiers Saint Aubin
BP 48, 91192 Gif-sur-Yvette cedex, France
E-mail: mourad.idir@synchrotron-soleil.fr

Dr. Thomas Krist
Hahn-Meitner Institut Berlin GmbH
Glienicker STr. 100, 14109 Berlin, Germany
E-mail: krist@hmi.de

Professor Alan G. Michette
University of London, King's College London, Department of Physics
Centre for X-Ray Science
Strand, London WC2R 2LS, UK
E-mail: alan.michette@kcl.ac.uk

ISSN 0342-4111

ISBN 978-3-540-74560-0 Springer Berlin Heidelberg New York

Library of Congress Control Number: 2007940819

Springer is a part of Springer Science+Business Media.

springer.com

Typesetting: SPi, Pondicherry, India
Cover: eStudio Calamar Steinen

Printed on acid-free paper SPIN: 11901648 56/3180/SPi 5 4 3 2 1 0

Preface

This book is based on the joint research activities of specialists in X-ray and neutron optics from 11 countries, working together under the framework of the European Programme for Cooperation in Science and Technology (COST, Action P7), initiated by Dr. Pierre Dhez in 2002–2006, and describes modern developments in reflective, refractive and diffractive optics for short wavelength radiation as well as recent theoretical approaches to modelling and ray-tracing the X-ray and neutron optical systems. The chapters are written by the leading specialists from European laboratories, universities and large facilities. In addition to new ideas and concepts, the contents provide practical information on recently invented devices and methods.

The main objective of the book is to broaden the knowledge base in the field of X-ray and neutron interactions with solid surfaces and interfaces, by developing modelling, fabrication and characterization methods for advanced innovative optical elements for applications in this wavelength range. This aim follows from the following precepts:

- Increased knowledge is necessary to develop new types of optical elements adapted to the desired energy range, as well as to improve the efficiency and versatility of existing optics.
- Enhanced optical performances will allow a significant increase in the range of applications possible with current and future X-ray and neutron sources.
- Better cooperation between national groups of researchers in the design and application of X-ray and neutron optics will lead to improvements in many key areas fundamental to societal and economic developments.

Behind each of these precepts is the knowledge that similar optical components are required in many X-ray and neutron systems, although the optics may have originally been developed primarily for X-rays (e.g., zone plates) or for neutrons (e.g., multilayer supermirrors). Bringing together expertise from both fields has led to efficient, cost-effective and enhanced solutions to common problems.

The editors are very grateful to Prof. Dr. h.c. Wolfgang Eberhardt, BESSY scientific director, for his continuous support of the COST P7 Action on X-ray and neutron optics and for his great help in the preparation of this book. The editors also wish to thank Prof. Dr. William B. Peatman for his critical analysis of the original manuscripts. Their support has contributed significantly to the publication of this book. Finally, the editors want to express their thanks to BESSY and the Hahn-Meitner-Institute, Berlin (HMI) for financial support, as well as Prof. Dr. Norbert Langhoff and Dr. Reiner Wedell for their help.

Berlin, Paris and London,
February 2008

A. Erko
M. Idir
Th. Krist
A.G. Michette

Contents

Part V Diffraction Optics

Contributors

Vladimír Áč
Alexander Dubček
University of Trenčín
Študentská 2, Trenčín
SK 91150, Slovakia
vac@tnuni.sk

Johannes Bahrdt
BESSY GmbH
Albert-Einstein-Straße 15
12489 Berlin, Germany
Bahrdt@bessy.de

Tilo Baumbach
Forschungszentrum
Karlsruhe GmbH, ISS
Postfach 3640
D-76021 Karlsruhe
Germany
Tilo.Baumbach@iss.fzk.de

Fred Bijkerk
FOM Institute for Plasma
Physics
Rijnhuizen P.O. Box 1207
3430 BE Nieuwegein
The Netherlands
bijkerk@rijnhuizen.nl

Aniouar Bjeoumikhov
IfG-Institute for Scientific
Instruments GmbH
Rudower Chaussee 29/31 (OWZ)
12489 Berlin, Germany
and
Institute for Computer Science
and Problems of Regional
Management (RAS)
Inessa Armand Street 32A
360000 Nalchik, Russia
bjeoumikhov@ifg-adlershof.de

Semfira Bjeoumikhova
Bundesanstalt für Materialforschung
und -prüfung (BAM)
Unter den Eichen 87, 12205 Berlin
Germany
gescheva@ifg-adlershof.de

Christine Borel
Multilayer Laboratory
European Synchrotron
Radiation Facility
6, rue Jules Horowitz
BP220, 38043 Grenoble Cedex
France
Christine.borel@esrf.fr

Marion Bowler
STFC Daresbury Laboratory
Warrington
WA4 4AD UK
m.bowler@dl.ac.uk

Inna Bukreeva
Istituto Fotonica e Nanotecnologie
(IFN) - CNR
V. Cineto Romano 42
00156 Roma, Italy
and
RAS – P.N. Lebedev
Physics Institute
Leninsky pr. 53
119991 Moscow, Russia
bukreeva@ifn.cnr.it

Alessia Cedola
Istituto Fotonica e Nanotecnologie
(IFN) - CNR
V. Cineto Romano 42
00156 Roma, Italy
alessia.cedola@ifn.cnr.it

Oleg Chubar
SOLEIL Synchrotron
L'Orme des Merisiers – Saint Aubin
BP 48, 91192 GIF-sur-YVETTE
CEDEX, France
oleg.chubar@synchrotron-soleil.fr

Daniele Cocco
Sincrotrone Trieste ScpA
S.S. 14, km 163.5 in Area Science
Park, I-34012 Trieste
Italy
Daniele.cocco@elettra.trieste.it

Liberato De Caro
Istituto di Cristallografia-CNR
via Amendola 122/O
70125 Bari, Italy
liberato.decaro@ic.cnr.it

Alexei Erko
BESSY GmbH
Albert Einstein Str. 15
12489 Berlin, Germany
erko@bessy.de

Claudio Ferrari
Institute CNR-IMEM
Parco Area delle Scienze 37/A
I-43010 Fontanini (PR) Italy
ferrari@imem.cnr.it

Alexander Firsov
BESSY GmbH
Albert Einstein Str. 15
12489 Berlin, Germany
firsov@bessy.de

Johan Floriot
Imagine Optic
18 rue Charles de Gaulle
91400 Orsay, France
jfloriot@imagine-optic.com

Rolf Follath
BESSY GmbH
Albert-Einstein-Strasse 15
12489 Berlin, Germany
follath@bessy.de

János Füzi
Research Institute for Solid State
Physics and Optics
Konkoly-Thege út 29-33
H-1121 Budapest, Hungary
fuzi@szfki.hu

Fabrizio Germini
Institute CNR-IMEM
Parco Area delle Scienze 37/A
I-43010 Fontanini (PR) Italy
germini@imem.cnr.it

Thomas Hänsel
Leibniz-Institut für
Oberflächenmodifizierung e.V.-IOM
Permoserstr. 15
04318 Leipzig, Germany
haensel@uni-leipzig.de

Jaromíra Hrdá
Institute of Physics
of the ASCR, v.v.i.
Na Slovance 2, 18221 Praha 8
Czech Republic
and
Charles University in Prague Faculty of Science
Institute of Hydrogeology
Engineering Geology
and Applied Geophysics
Albertov 6, 12843 Praha 2
Czech Republic
hrda@fzu.cz

Jaromír Hrdý
Institute of Physics
of the ASCR, v.v.i.
Na Slovance 2 182 21 Praha 8
Czech Republic
hrdy@fzu.cz

Mourad Idir
Synchrotron SOLEIL
L'Orme des Merisiers –
Saint Aubin, BP 48
91192 Gif- sur-Yvette Cedex
France
Mourad.idir@synchrotron-soleil.fr

Werner Jark
Sincrotrone Trieste S.c.p.A.
S.S. 14 km 163.5 in Area
Science Park
34012 Basovizza (TS)
Italy
werner.jark@elettra.trieste.it

Matej Jergel
Institute of Physics
Slovak Academy of Sciences
Dúbravská 9
845 11 Bratislava, Slovakia
matej.jergel@savba.sk

Dušan Korytár
Institute of Electrical Engineering
Slovak Academy of Sciences
Vrbovská cesta 110
SK-921 01 Piešt'any
Slovak Republic
elekdkor@savba.sk

Rita Kovács-Mezei
MIRROTRON Multilayer
Laboratory Ltd.
Konkoly Thege út 29-33
H-1121 Budapest, Hungary
kovmez@hotmail.com

Thomas Krist
Hahn-Meitner-Institut Berlin
Glienicker Str. 100
D-14109 Berlin
Germany
krist@hmi.de

Michael Krumrey
Physikalisch-Technische
Bundesanstalt
X-ray Radiometry, Abbestraße 2-12
10587 Berlin, Germany
michael.krumrey@ptb.de

Jiří Kulda
Institut Laue-Langevin
6, rue Jules Horowitz
38042 Grenoble Cedex 9
France
kulda@ill.fr

Stefano Lagomarsino
Istituto Fotonica e Nanotecnologie
(IFN) - CNR
V. Cineto Romano 42
00156 Roma, Italy
Stefano.lagomarsino@ifn.cnr.it

Heiner Lammert
BESSY GmbH
Albert-Einstein-Str. 15
12489 Berlin, Germany
lammert@bessy.de

Juan I. Larruquert
Instituto de Física Aplicada. CSIC
C/ Serrano 144
28006 Madrid, Spain
larruquert@ifa.cetef.csic.es

Xavier Levecq
Imagine Optic
18 rue Charles de Gaulle
91400 Orsay, France
xlevecq@imagine-optic.com

Eva Majková
Institute of Physics
Slovak Academy of Sciences
Dúbravská 9
84511 Bratislava, Slovakia
eva.majkova@savba.sk

Igor Matko
Laboratoire des Matériaux
et du Génie Physique
INP Grenoble – Minatec
3, parvis Louis Néel BP 257
38016 Grenoble Cedex
France
igor.matko@inpg.fr

Marco Matteucci
Sincrotrone Trieste S.c.p.A.
S.S. 14 km 163.5
in Area Science Park
34012 Basovizza (TS)
Italy
marco.matteucci@elettra.
trieste.it

Evgeni Meltchakov
CNRS, L2MP, Case 131
Faculté des Sciences de St Jérome
13397 Marseille Cedex 20, France
evgueni.meltchakov@l2mp.fr

Pascal Mercère
Synchrotron SOLEIL
L'Orme des Merisiers –
Saint Aubin, BP 48
91192 Gif- sur-Yvette Cedex
France
pascal.mercere@synchrotron-
soleil.fr

Alan G. Michette
Department of Physics Strand
King's College London
London
WC2R 2LS, UK
alan.michette@kcl.ac.uk

Pavol Mikula
Nuclear Physics Institute
v.v.i. of CAS and Research Centre
Řež Ltd.
250 68 Řež, Czech Republic
mikula@ujf.cas.cz

Petr Mikulík
Department of Condensed
Matter Physics
Masaryk University
Kotlářská 2, CZ-6137 Brno
Czech Republic
mikulik@physics.muni.cz

Christian Morawe
European Synchrotron
Radiation Facility
6, rue Jules Horowitz
BP220, 38043 Grenoble Cedex
France
morawe@esrf.fr

Stephan Müllender
LIT-OCE Carl Zeiss SMT AG
73446 Oberkochen
Germany
muellender@smt.zeiss.com

Andreas Nickel
Leibniz-Institut für
Oberflächenmodifizierung e.V.-IOM
Permoserstr. 15
04318 Leipzig, Germany
nickel@uni-leipzig.de

Frédéric Ott
Laboratoire Léon Brillouin
CEA/CNRS UMR12
Centre d'Etudes de Saclay
91191 Gif sur Yvette
France
Frederic.Ott@cea.fr

Daniele Pelliccia
Institut für Synchrotronstrahlung
– ANKA Forschungszentrum Karlsruhe in der Helmholtz-Gemeinschaft
Herman-von-Helmholtz-Platz 1
76344 Eggenstein-Leopoldshafen
Germany
daniele.pelliccia@iss.fzk.de

Fréderic Pérennès
Sincrotrone Trieste S.c.p.A.
S.S. 14 km 163.5 in Area
Science Park
34012 Basovizza (TS), Italy

and

European Patent Office
PB 5818 Patentlaan 2
2280-HV-Rijswijk (ZH)
The Netherlands
perennesf@gmail.com

Peter Perichta
Alexander Dubček University
of Trenčín
Študentská 2, Trenčín
SK 91150, Slovakia
perichta@tnuni.sk

Luca Peverini
European Synchrotron
Radiation Facility
6 rue J. Horowitz, BP220
38043 Grenoble Cedex, France
peverini@esrf.fr

Slawka J. Pfauntsch
Department of Physics Strand
King's College London
London
WC2R 2LS, UK
slawka.pfauntsch@kcl.ac.uk

Ladislav Pina
Department of Physical Electronics
Faculty of Nuclear Sciences
and Physical Engineering
Czech Technical University in Prague
Brehova 7, 115 19 Prague 1
Czech Republic
ladislav.pina@fjfi.cvut.cz

A. Keith Powell
Department of Physics Strand
King's College London
London
WC2R 2LS, UK
powell.keith@gmail.com

Stefan Rehbein
BESSY GmbH
Albert Einstein Str. 15
12489 Berlin
Germany
rehbein@bessy.de

Dmitry Roshchupkin
Institute of Microelectronics
Technology
Russian Academy of Sciences
142432 Chernogolovka
Moscow District, Russia
rochtch@iptm.ru

Amparo Rommeveaux
European Synchrotron
Radiation Facility
6 rue J. Horowitz, BP220
38043 Grenoble Cedex
France
vivo@esrf.fr

László Rosta
Research Institute of Solid State
Physics and Optics
Konkoly Thege út 29-33
H-1121 Budapest, Hungary
rosta@szfki.hu

Jan Šaroun
Nuclear Physics Institute, v.v.i.
ASCR and Research Center
Řež Ltd. 25068 Řež
Czech Republic
saroun@ujf.cas.cz

Franz Schäfers
BESSY GmbH
Albert Einstein Str. 15
12489 Berlin, Germany
schaefers@bessy.de

Igor Schelokov
Institute of Microelectronics
Technology
Russian Academy of Sciences
142432 Chernogolovka
Moscow District, Russia

Axel Schindler
Leibniz-Institut für
Oberflächenmodifizierung e.V.-IOM
Permoserstr. 15
04318 Leipzig, Germany
axel.schindler@iom-leipzig.de

Gerd Schneider
BESSY GmbH
Albert Einstein Str. 15
12489 Berlin, Germany
schneider@bessy.de

Frank Siewert
BESSY GmbH
Albert-Einstein-Str. 15
12489 Berlin, Germany
siewert@bessy.de

Anatoly Snigirev
European Synchrotron
Radiation Facility
6 rue J. Horowitz, BP220
38043 Grenoble Cedex, France
snigirev@esrf.fr

Irina Snigireva
European Synchrotron
Radiation Facility
6 rue J. Horowitz, BP220
38043 Grenoble Cedex
France
snigireva@esrf.fr

Daniele Spiga
INAF/Osservatorio Astronomico
di Brera Via E. Bianchi 46
23807 Merate (LC) – Italy
daniele.spiga@brera.inaf.it

Alessandro Surpi
Istituto Fotonica e Nanotecnologie
(IFN) - CNR
V. Cineto Romano 42
00156 Roma, Italy
and
Ångströmlaboratoriet
institutionen för teknikvetenskaper
Elektromikroskopi och
Nanoteknologi
Lägerhyddsvägen 1
Box 534 SE-751 21, Upppsala
and
Institutionen för Biologi och
Kemiteknik
Mälardalens Höghskola
Gamla Tullgatan 2
SE-632 20, Eskilstuna
alessandro.surpi@angstrom.uu.se

Anke Teichert
Hahn-Meitner-Institut Berlin
Glienicker Str. 100
D-14109 Berlin, Germany
teichert@hmi.de

Muriel Thomasset
Synchrotron Soleil
L'Orme des Merisiers
Saint Aubin, BP 48
91192 Gif-sur-Yvette Cedex, France
muriel.thomasset@synchrotron-soleil.fr

Patrik Vagovič
Institute of Electrical Engineering
Slovak Academy of Sciences
Vrbovská cesta 110, SK-921 01
Piešt'any, Slovak Republic
elekvago@savba.sk

Bernard Vidal
CNRS, L2MP, Case 131
Faculté des Sciences de St. Jérome
13397 Marseille Cedex 20, France
Bernard.Vidal@l2mp.fr

Vladimir Vidal
CNRS, L2MP, Case 131
Faculté des Sciences de St. Jérome
13397 Marseille Cedex 20, France
vlad_vidal@hotmail.com

Miroslav Vrána
Nuclear Physics Institute ASCR
25068 Rez, Czech Republic
vrana@ujf.cas.cz

Volker Wagner
GKSS Research Center GmbH
Max-Planck Strasse 1
21502 Geesthacht
Germany
and
Physics Technische Bundesanstalt
Bundesallee 100
38116 Braunschweig
Germany
Volker.Wagner@ptb.de

Stephan Werner
BESSY GmbH
Albert Einstein Str. 15
12489 Berlin
Germany
werner@bessy.de

Thomas Zeschke
BESSY GmbH
Albert-Einstein-Str. 15
12489 Berlin, Germany
zeschke@bessy.de

Erwin Zoethout
FOM Institute for Plasma
Physics Rijnhuizen
P.O. Box 1207
3430 BE Nieuwegein
The Netherlands
zoethout@rijnhuizen.nl

1

X-Ray and Neutron Optical Systems

A. Erko, M. Idir, Th. Krist, and A.G. Michette

Abstract. Although X-rays and neutrons can provide different information about samples, there are many similarities in the ways in which beams of them can be manipulated. The rationale behind bringing experts in the two fields together was the desire to find common solutions to common problems. The intention of this brief introduction is to give a flavour of the state-of-the-art in X-ray and neutron optics as well as an indication of future trends.

1.1 X-Ray Optics

There is a growing need for the determination and characterization of elements at trace concentrations that can be well below one part per million by weight. This is true in many fields of human activity, including the environmental sciences and cultural heritage as well as the more obvious physical and biological sciences. Although for quantitative as well as qualitative investigations, X-ray microanalysis is an established method for determining elemental composition, this is now often insufficient, a distribution map of each element being much more useful. However, this can be achieved only with *large flux, optimal excitation energy*, and *high lateral resolution*. For these to be satisfied appropriate optical elements must be developed to transport radiation from source to sample, providing powerful, highly concentrated and possibly monochromatic X-ray beams. As a result *X-ray optics* has grown rapidly in recent years as an important branch of physics and technology.

The phrase "X-ray optics" encompasses a wide range of optical elements exploiting reflection, diffraction, and refraction – or combinations of these – utilizing sub-micrometer and sub-nanometer artificial structures and natural crystals to focus, monochromate or otherwise manipulate X-ray beams. Historically, natural crystals can be regarded as prototypes of many of the artificial structures now in use or proposed. The development of multilayer interference mirrors for the nanometer wavelength range which provides efficient reflection at angles close to normal incidence was a great step forward.

These mirrors can be stable in ultrahigh power X-ray beams and can be also used for broadband high-flux monochromators providing energy resolutions $E/\Delta E \sim 25$ in the range $\sim$100 eV – 15 keV. Developments in the design and manufacture of multilayers have also allowed their capabilities as broadband polarizers and analyzers to be explored. In addition, recent technological progress now allows the manufacture of highly aspheric multilayer-coated optics. These preserve the efficiency, flux, and ease of alignment of spherical or partially aspheric mirrors, while eliminating spherical aberration, thus allowing high collection and convergence angles and small spot sizes.

A compromise between the reflected flux and the necessary energy resolution can be achieved by the choice of a suitable crystal or multilayer monochromator. Low-resolution monochromators can be also built from diffractive elements such as transmission and reflection zone plates. Zone plates as focusing elements and X-ray waveguides to relay sources of nanometer size are recognized as significant optical elements in the nano-world. The great majority of X-ray microscopes and microprobes currently use zone plates, and this has allowed such devices to become available as laboratory instruments and at synchrotron radiation facilities. However, during the last decade conventional zone plate technology has reached the theoretical limit of spatial resolution, with volume diffraction effects in the outer zones (with sizes comparable to X-ray wavelengths) providing the fundamental limitation of zone plate resolution.

Further development of micro and nanofabrication techniques, in particular for planar nanometer-scale structures with sizes of the order of X-ray wavelengths, as well as the deposition and growth of thin films of different materials, has enabled the manufacture of a new generation of diffractive optical elements. In a similar fashion, the fabrication and successful tests of a synthesized X-ray hologram on a crystal have been reported. With such improvements in nanotechnology, mostly for microelectronic applications, methods have been developed to create nanostructures and multilayer films for the effective control of X-rays to provide sub-micrometer spatial resolutions. These include two- and three-dimensional Fresnel and Bragg–Fresnel optical elements based on *zone plates*, with lateral resolutions as good as 15 nm, and diffraction gratings in combination with natural crystals or artificial multilayer structures. The recent development of graded crystals allows simultaneous focusing and enhancement of the spectral flux at the sample by several orders of magnitude. All these optical elements are related via the basic principles of Bragg, Bragg–Laue, or Bragg–Fresnel diffraction on artificially made volume structures and differ from other types of optic through combinations of optical properties. *Refractive/diffractive X-ray optics* were first realized in 1986 and have successfully been used with third-generation synchrotron radiation sources, as they are ideal for high-energy undulator radiation characterized by low divergence in both the vertical and horizontal directions.

Capillary X-ray optics, including microchannel plates, have been successfully used with conventional X-ray sources. Straight glass monocapillaries are efficient in transporting X-rays from the source leading to increased radiation intensity on the sample. Tapered monocapillaries are used in synchrotron beamlines for focusing radiation into micrometer and submicrometer spots. Polycapillary arrays with curved channels can be used for transforming divergent radiation from a point source into a quasiparallel beam or for focusing a divergent beam onto a small spot. Straight polycapillary arrays have been used for X-ray imaging and for beam splitting and filtering. Recent developments have been in making arrays with different geometries to enhance the performances of such optics. Also, of late, elliptically bent *Kirkpatrick–Baez mirrors* have been used to produce submicrometer size X-ray beams. These optics are achromatic and have relatively long focal distances compared to capillaries. This property can be very important for microfluorescence applications in special environments, for example when the sample needs to be contained in a gas-filled temperature-controlled chamber. *Refractive X-ray optics* represent a rapidly emerging option for focusing high energy synchrotron radiation from micrometer to nanometer dimensions. These devices are simple to align, offer a good working distance between the optics and the sample, and are expected to become standard elements in synchrotron beamline instrumentation in general and in high energy X-ray microscopy in particular.

1.2 Metrology

Most synchrotron radiation facilities and large industrial companies have developed their own metrology laboratories to meet the needs of optical characterization in terms of microroughness, radius of curvature, slope errors, and shape errors. The instrumentation used consists mainly of commercial instruments: phase shift interferometers for microroughness characterization, Fizeau interferometers for bidimensional topography, and optical profilometers – for measurements of long optical components – such as the long trace profiler (LTP) or the nanometer optical component measuring machine (NOM). In this book an attempt is made to systemize recent knowledge in ultraprecise surface metrology. This is directly linked to instrument calibration, but up to now there is no standardization of calibration. In round-robin endeavor, typical X-ray mirrors – plane, spherical or toroidal – were examined by the various laboratories using their own instrumentation in order to better understand the accuracy achievable. The ultimate goal of this Round Robin was to create a database of the measurement results in order to provide these references as calibration tools available to the metrology community.

1.3 Neutron Optics

Because of their unique properties, neutrons are used to investigate a growing number of research areas, in both traditional and new fields and from fundamental science to technology; no end to this growth can be foreseen. Currently 4,000–5,000 European researchers are using neutron scattering for their scientific work. This demand drives the improvement of neutron instrumentation which, to a large extent, is related to neutron optics. During the last decade the main advancement has been the introduction of supermirrors (with lateral or transverse grading of the laying thicknesses, either quasiperiodically or aperiodically) for neutron transport in guides, while the next decade will see the increased application of focusing and polarizing devices which will also be based mainly on supermirror coatings. It is in this field of multilayers where there is much similarity between neutron and X-ray optics. Focusing systems for neutrons also have much in common with their X-ray equivalents; in particular, focusing tests using capillary optics and Fresnel zone plates have been performed with neutrons.

An important property of a supermirror is its critical angle, θ_c, the glancing angle up to which it reflects efficiently. By convention θ_c is measured in multiples, m, of the critical angle of nickel, which has the largest critical angle of all naturally occurring elements.

For multilayer production an important advance has been the reduction of stress development during the growth of the film coatings; this is important for the production of X-ray multilayers as well as those designed as neutron reflectors. By varying several parameters during the sputtering process, their influences on the stress development have been determined, leading to an order of magnitude decrease in the stress. Another important step has been progress in the production of polarizing and nonpolarizing supermirrors. For polarizing supermirrors the critical angle up to which they reflect neutrons has been increased and the magnetic field necessary to retain good polarization has been lowered. For nonpolarizing supermirrors the critical angle has also been increased, while the temperature and radiation stability as well as the corresponding crystal structures were characterized and the homogeneity increased for the coating of large areas. Phase space mapping of a neutron beam following neutron optical devices containing supermirrors has also been demonstrated.

Research on bent perfect crystals has been aimed at the development of the technique and demonstrating the properties of systems based on one or two components. One component enables ultrahigh resolution for monochromatizsation or analysis, while two components allow for an adjustable spectral resolution and collimation of $\sim 10^{-3}$–10^{-4}. Such systems have allowed the realization and test of a multianalyzer module for a three-axis spectrometer consisting of an array of 31 individual channels, covering a scattering angle range of 75°. This new device offers improved momentum resolution and enhanced data collection efficiency in experiments aimed at mapping of inelastic response over extended areas in momentum/frequency space and, at the same time, keeps the high incident flux and most of the flexibility of up-to-date triple axis spectrometry using doubly focusing crystal optics.

Part I

Theoretical Approaches and Calculations

2

The BESSY Raytrace Program RAY

F. Schäfers

Abstract. The raytracing program RAY simulates the imaging properties of an optical system. It randomly creates a set of rays within various types of light sources and traces them according to the laws of geometric optics through optical elements onto image planes. The distribution of the rays at the source, optical elements and image planes can be displayed.

A ray is described not only by its coordinates with respect to a suitable coordinate system, but also by its energy and its polarisation determined by the Stokes vector. Different source types are implemented with special emphasis on a realistic simulation of source intensity, volume and emission characteristics, especially for synchrotron radiation including dipole and undulator sources. Optical elements can be reflection mirrors of nearly any figure (plane, cylindrical, spherical, aspherical...), gratings, zone plates, foils or crystals. The absolute transmission of the optics including the effect of optical (multilayered) coatings is calculated according to the reflection/refraction/transmission process from the optical constants of the materials involved. The influence of misalignment of the source and/or the optical elements, slope errors and thermal deformation of the optics can also be taken into account. A graphical display of spot patterns at any position of the beam, intensity and angular distributions, absolute flux, polarisation, energy resolution is possible.

2.1 Introduction

The development of the raytracing program *RAY* was started at BESSY in 1984 for basic raytracing calculations of VUV- and soft X-ray optical schemes [1]. Since that time *RAY* has been in continuous evolution and it has grown into a widely used design tool for synchrotron radiation beamlines as well as for other optical systems. Most of the BESSY I monochromators have been designed using *RAY*. To meet the requirements of the new undulator-based third generation storage ring BESSY II, many new features have been implemented into the code in the last 10 years such that *RAY* now has become an indispensable tool for modern beamline design. Its capabilities are similar to the widely used *SHADOW*–XOP program [2,3]. Considerable effort has been made to ensure that it is a user friendly, easy accessable and easy-to-learn program for everyday use with a minimum effort on data and file handling.

Alternative to these programs based on *intensity* distributions and *geometric optics, wavefront* propagation codes have been developed such as *PHASE* [4], which applies the Stationary Phase Approximation and *SRW* [5] employing Fourier Optics, which on the basis of the complex electric field of the radiation are able to intrinsically take into account interference and coherence effects. These codes are treated separately in this book [6].

This report is intended to be a practical reference and to give an outline of the underlying geometrical, mathematical, physical and optical principles which can be found in textbooks [7–9] or synchrotron radiation handbooks [10]. In particular, Chap. 3.2 of [10] (Ray tracing) is strongly recommended as an introductory guide before calculating a real beamline design. Here the procedure, problems, limitations and the importance of *checking* the raytrace results for the various kinds of errors that can occur are discussed. Various specific *RAY*-features have been described previously: crystal optics in [11] and zoneplate optics employing Fresnel diffraction where the collective effects are treated on a statistical (Monte Carlo) basis [12, 13]. Extended manuals for *RAY* [14] and the reflectivity program *REFLEC* [15] which share the same optics software library are also available. Examples for the use of the program in a variety of synchrotron radiation applications are given in [16]: plane grating monochromator (PGM-) beamlines, [17] IR-beamlines, [18] elliptical undulator beamlines, [19] gradient crystal monochromators, [20] μ-focus X-ray beamline.

Chapter 3 explains the basic statistical treatment to simulate any kind of intensity patterns, while the next chapters describe the simulation of sources (Chap. 4), optical elements (Chap. 5) and of the treatment of absolute reflectivity and polarisation (Chap. 6).

In Chap. 7 crystal diffraction optics employing dynamical theory is described. Looking ahead, in 'Outlook', the time evolution of the rays to describe wave, coherence and interference phenomena is discussed (Chap. 8). This extension of the program and the implementation of the zoneplate optics [12, 13] have been made possible by support through the COST-P7 action and intensive discussions during the COST meetings.

The complete code is available as a PC-Windows version.

2.2 Beamline Design and Modelling

The raytracing program *RAY* simulates the imaging and focussing properties of an optical system. It randomly creates a set of rays within various types of light sources and traces them through one or more optical elements on image planes. The geometric distribution of the rays at the source, at all optical elements and at the image planes can be visualized.

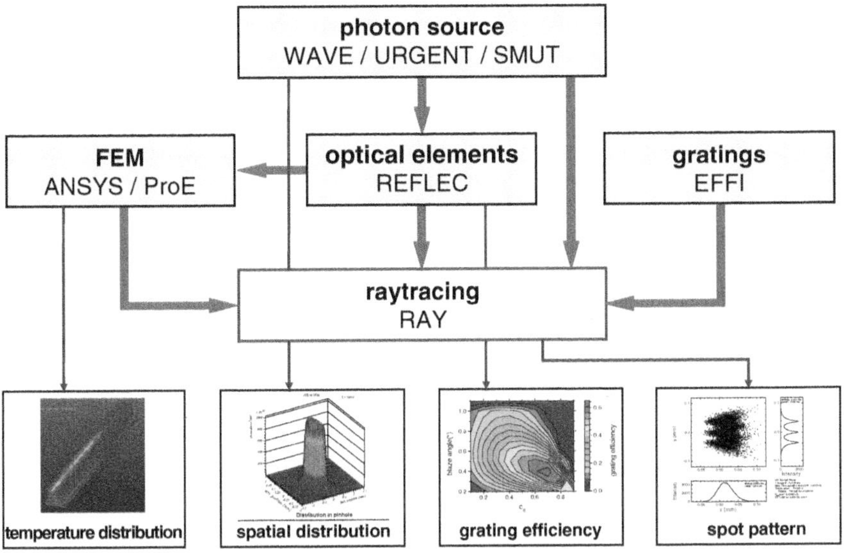

Fig. 2.1. BESSY soft X-ray computational tools and their interplay

Various interesting features like focal properties, power distribution, energy resolution, rocking curves, absolute transmission and polarisation characteristics of an optical setup are simulated. It combines pure geometrical raytracing with calculations of the absolute transmission and is, thus, a central and indispensable part of the BESSY software tools for the design and optimization of new monochromators and beamlines from the infrared spectral region to the hard X-ray range. The interplay of the software tools available at BESSY [21, 22], is demonstrated in Fig. 2.1 as a flowchart.

Special emphasis was put on realistic simulations of beamlines, in particular those employing synchrotron radiation: the path of the photons can be followed from any source, including bending magnets and insertion devices via reflection/diffraction/transmission at optical elements through apertures, entrance and/or exit slits on the sample. The influence of slope errors, surface roughness, thermal bumps, measured or calculated surface profiles as well as a misalignment of the source and optical elements can be studied in a simple way. Thus, it is possible to predict the real performance of the beamline under realistic conditions and to specify the requirements for all the components to be ordered.

In a well defined source volume, rays are created within a given horizontal and vertical divergence. Each ray has the same intrinsic probability. The spatial and angular intensity *distribution* of the source is given by the spatial and angular *density* of the rays (i.e. rays per volume and solid angle). Thus, the outgoing rays simulate the intensity distribution of the corresponding source. The rays are traced according to geometrical optics through one or more optical elements (mirrors, gratings, foils, crystals, slits, zoneplates) of which the surface can have nearly any figure such as plane, cylindrical, spherical, toroidal, paraboloidal or ellipsoidal and can be arranged in any geometry (horizontal, vertical, oblique). The absolute transmission of the optics including the effect of (multilayer-) coatings is calculated according to the reflection/refraction/transmission/diffraction processes from the optical constants of the involved materials. Special monochromator mounts and (coma-corrected) varied line-spacing (VLS-) gratings and (graded) crystals with automatic calculation of structure factors can also be handled.

A ray is determined not only by its coordinates with respect to a suitable coordinate system (e.g. by its starting point) and by its direction, but also by its energy E, its polarisation, described by the Stokes vector $S = (S_0, S_1, S_2, S_3)$, and its pathlength. Thus, a ray is described by 12 parameters, which are traced through the optical setup and for which the geometrical and optical modifications are calculated according to its interaction with the optical coating (reflection/refraction/transmission). Since all rays have equal probability (the intensity of a ray, S_0, is either 1 or 0), the throughput of a beamline is simply given by the number of rays, for SR-sources multiplied with the absolute photon flux as scaling factor.

For a first overview of the focal properties of an optical system, the horizontal and vertical widths of the beam can be visualized along the beam path for the determination of the focus position. At any position along the beam path image planes can be defined. The footprints of the rays on the optical elements and the focal properties of the optical system are analyzed and are visualized graphically as point diagrams, 2D or 3D intensity distributions etc.

The menu-driven program is user friendly and so a first-performance test of an optical design can be gained rapidly without any file handling. Once the beamline has been defined the parameters are stored and can be modified in a subsequent run. The graphics output is directed to monitors, printers, or PS or EPS-files, and alternatively ASCII-data tables of all results can be created for further data evaluation and display.

A flowchart of the program is shown in Fig. 2.2.

2.3 Statistics: Basic Laws of RAY

2.3.1 All Rays have Equal Probability

To simulate realistic intensity patterns on optical elements and image planes (e.g. for heat load studies) it is necessary to create the source points and the rays in such a way that the same intensity is attributed to each ray.

Generally there are two possibilities:

- A *systematic* distribution of the rays within the source so that the real emission characteristic is simulated. For this a large number of rays is required and needs to be calculated before an optical setup is completely described.

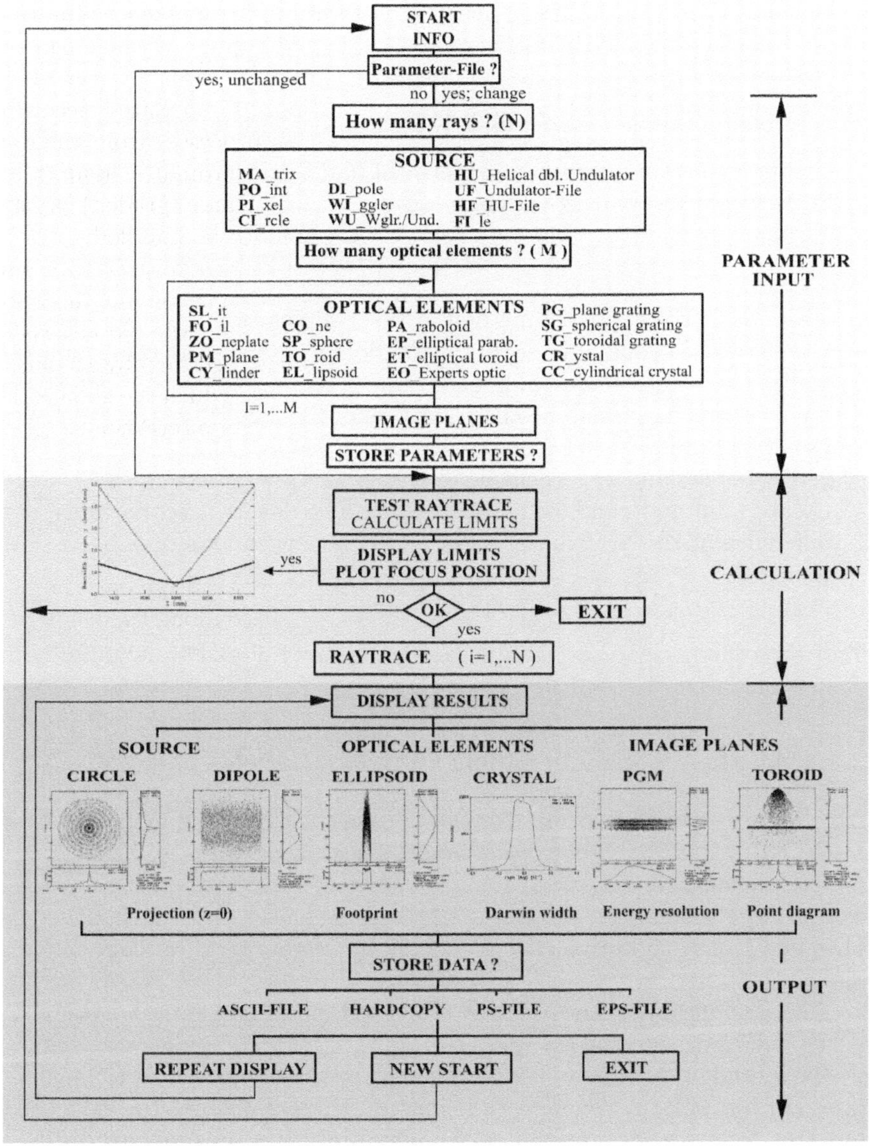

Fig. 2.2. Flow chart of RAY

- The rays are distributed *statistically* within the source so that within the statistical error the real emission characteristic is simulated. The intensity distribution of the source is thus understood as the probability distribution of the necessary parameters, namely position and angle. The main advantages of this Monto–Carlo procedure are its simplicity and the fact that a calculation of relatively few rays already is enough to create a reasonable simulation of the optics. When the statistics and the accuracy seem to be sufficient, the calculation can always be interrupted without making a systematic error.

This second option is realized in *RAY*. The procedure is as follows:

1. Create a random number ran_1 between 0 and 1.
2. Scale the corresponding variable, e.g. the x-coordinate of the source point:

$$x = (ran_1 - 0.5)\,\mathrm{d}x, \tag{2.1}$$

   where $\mathrm{d}x$ is the source-dimension in the x-direction.
3. Calculate the probability, w, of this randomly chosen start value for x (normalized to a maximum value of 1), for example the electron density in a dipole-source (gaussian profile $w(x) = \exp(-x^2/(2\sigma_{\mathrm{y}}^2))$ or the synchrotron radiation intensity for a fixed wavelength at a definite horizontal and vertical emission angle (Schwinger theory [23]).
4. Create a second random number ran_2. The ray is accepted only if the difference of the probability $w(x)$ and this new random number is larger than zero:

$$w\,(x) - ran_2 > 0. \tag{2.2}$$

5. If the difference is less than zero neglect this ray and start again with a new one according to (2.1).

2.3.2 All Rays are Independent, but... (Particles and Waves)

All rays are independent, and so they are considered as individual particles not knowing anything about each other. Thus, *RAY* works exclusively in the particle model. Nevertheless, the statistical method explained above is an elegant way to overcome the particle–wave dualism and to simulate wave phenomena and collective effects such as interference, diffraction, coherence and wave fronts.

This is done by a statistical treatment of an ensemble of individual rays which behave within the statistical errors as a collective unit, as a wavefront.

This random selection of a parameter is used extensively throughout the program not only to simulate the emission characteristics of a light source, but also, for example, to simulate the reflection angle on a mirror to simulate slope errors that are assumed to be gaussian. It is used to simulate reflection losses of rays where $w(x) = R$ with $(0 < R < 1)$ by which the surviving ray is assigned a probability of 1.

Furthermore, it is applied to simulate diffraction effects on slits for which the outgoing beam direction is modulated by a sin v/v term for the case of rectangular slits or by a bessel function for the case of circular slits.

The same diffraction routine is used for zone plate optics to simulate airy patterns at the focus point in first, third and fifth harmonic [12,13].

2.4 Treatment of Light Sources

Various light sources are incorporated in *RAY*. Generally the rays are starting in a defined source volume and are emitted with a defined horizontal and vertical divergence. Either hard (flat-top) edges or a gaussian distribution profile can be simulated. In the latter case, rays are created statistically (see 3.1) within a $\pm 3\sigma$-width of the gaussian profile (i.e. more than 99.9% of the intensity).

For synchrotron radiation beamlines, the polarised emission characteristic of bending magnets, wigglers and undulators is incorporated. For other sources, such as twin or helical undulators, or to take beam emittance effects into account, the input can be given as an ASCII-file taken from programs for undulator radiation: *URGENT* [24], *SMUT* [25] or *WAVE* [26]. In this file the intensity and polarisation patterns of the light source must be described as intensity (photons/seconds) and Stokes parameters at a distance of 10 m from the centre of the source in a suitable x-y mesh.

Each ray is attributed an energy, E, and a polarisation. The energy can be varied continuously within a 'white' hard-edge band of $E_0 \pm \Delta E$, or toggled between three discrete energies E_0, $E_0 + \Delta E$ and $E_0 - \Delta E$. This feature allows one to determine easily the energy dispersion and the spatial separation of discrete energies for monochromator systems, thereby giving a picture of the energy resolution that one can expect.

Table 2.1 lists the main features of the different light sources.

The source coordinate system for the case of bending magnet synchrotron radiation is given in Fig. 2.3. The storage ring is located in the x-z plane,

Table 2.1. Parameters of the *RAY*-sources

	Name	Width x	Height y	Length z	Div. hor. ϕ	Div. vert. ψ	S_0	S_1, S_2, S_3
Matrix	**MA**	Hard	Hard	Hard	Hard	Hard	1	Input
Point	**PO**	Hard soft	Hard soft	Hard	Hard soft	Hard soft	1	Input
Circle	**CI**	Hard	Hard	Hard	Hard	Hard	1	Input
Dipole	**DI**	Soft	Soft	Hard	Hard	Calc.	Flux	Calc.
Wiggler	**WI**	Soft	Soft	$L = n\lambda_u$	Hard	Calc.	Flux	Calc.
Wiggler/Undul.	**WU**	Soft	Soft	0	Calc.	Calc.	Flux	Input
Double–Undul.	**HU**	Soft	Soft	Hard	Soft	Soft	Flux	Input
Undul.–data file	**UF**	Soft	Soft	Hard	File	File	Flux	File
Helical Undul. data file	**HF**	Soft	Soft	Hard	File	File	Flux	File
Source data file	**FI**	Soft	Soft	Hard	File	File	1	Input

hrd, a hard (flat-top) edge; *soft*, soft – a gaussian distribution of the respective variable within a 6σ-width is simulated; *calc*, calculated according to a theoretical model (e.g. Schwinger theory); n, number of wiggler periods; L, length of undulator; λ_u, period length; *file*, parameters taken from data-file; *input*, parameters to be given interactively

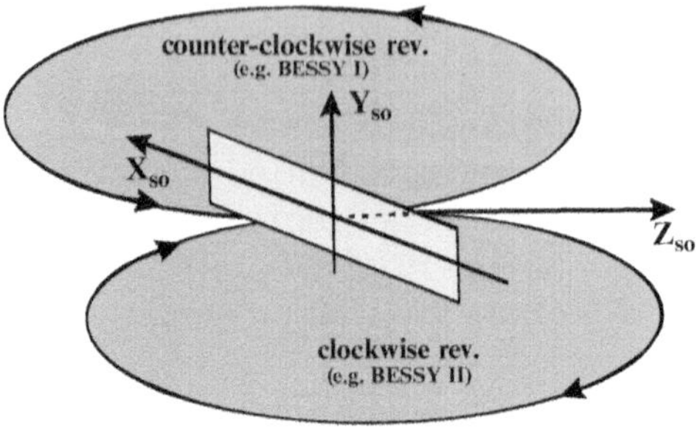

Fig. 2.3. Coordinate system for storage ring-bending magnet sources (*DI_pole*) as viewed from above

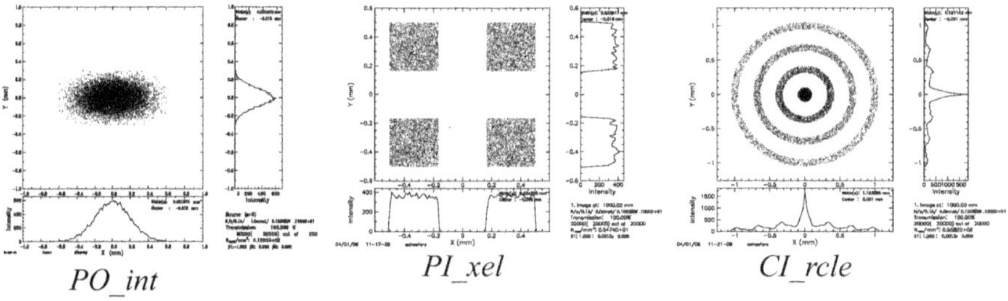

Fig. 2.4. Spot pattern of various source types in x-y plane, projected onto $z = 0$

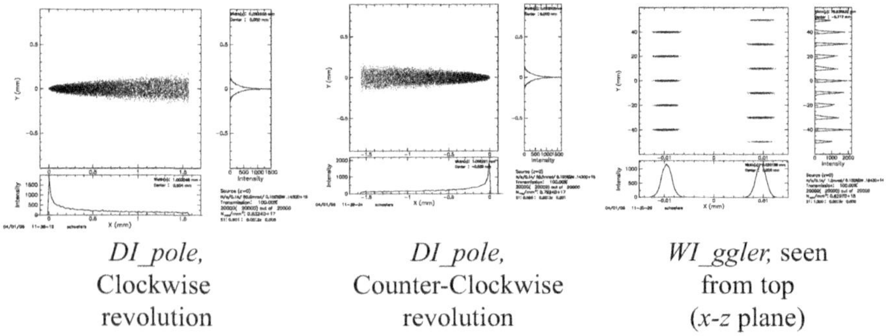

Fig. 2.5. Spot pattern of synchrotron radiation sources in x-y plane, projected onto $z = 0$

for clockwise revolution of the electrons the x-axis is pointing away from the centre, while for counter-clockwise revolution the x-axis is pointing inside the storage ring centre. This is important to be noticed especially for optical systems with large horizontal divergence (e.g. IR-beamlines), where the source cross section is very asymmetric because of the depth-of-field effect (see Fig. 2.5).

Examples of the intensity distribution (footprints) of various sources are given in the Figs. 2.4 and 2.5.

2.5 Interaction of Rays with Optical Elements

2.5.1 Coordinate Systems

The definition of the coordinate system used in *RAY* is shown in Figs. 2.6 and 2.7. Its origin lies in the centre of the source (with the x-axis in general (e.g. SR) being horizontal). The coordinate system is transformed along the optical path from the source to the optical elements and then to the image planes. The z-axis points into the direction of the central ray, the x-axis is perpendicular to the plane of reflection, i.e. horizontal in the case of a vertically deviating optical setup (azimuthal angles 0° or 180°), and it is vertical for horizontal mounts (azimuthal angles 90° (to the right) and 270° (to the left), respectively). The y-axis is *always* the normal in the centre of the optical element. The plane of reflection or dispersion is, thus, *always* the y-z plane and the surface of the optical elements is the x-z-plane, regardless of the azimuthal angle χ chosen. After the optical element the coordinate system

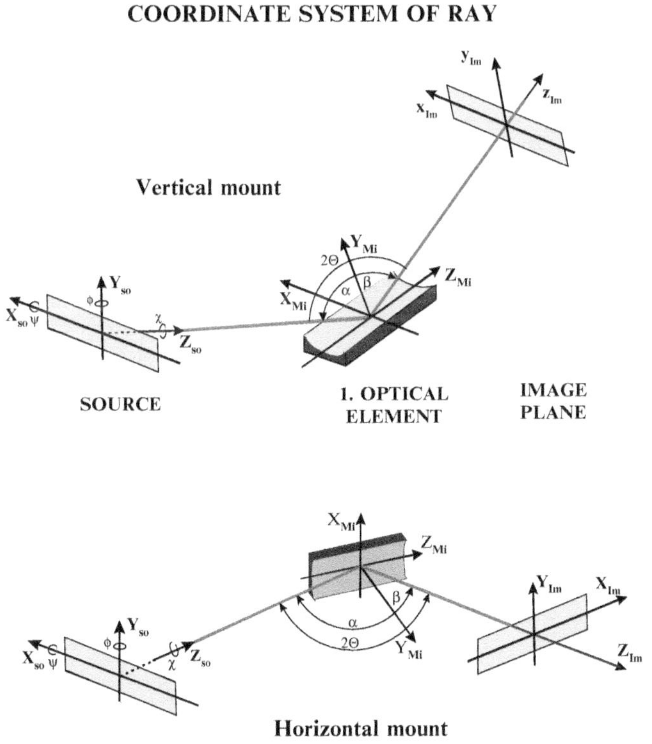

Fig. 2.6. Coordinate system (right-handed screw) and angles used in RAY. (*Top*) Vertical deviation (upwards (downwards)) mount (azimuthal angle $\chi = 0°$ (180°)). (*Bottom*) Horizontal deviation (to the right (left)) (azimuthal angle $\chi = 90°$ (270°)). The optical element is always in the X_M-Z_M-plane

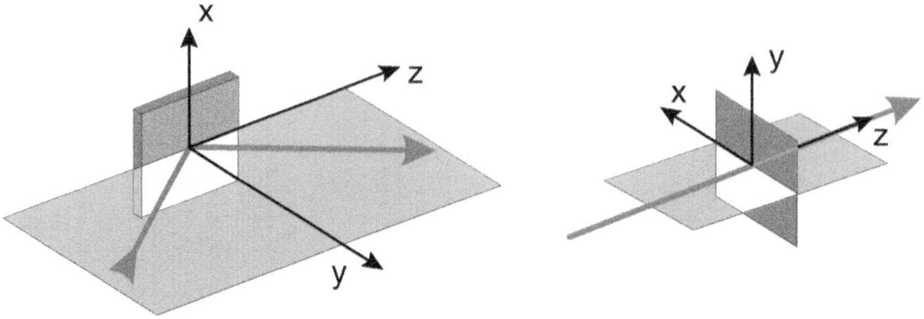

Fig. 2.7. Coordinate systems used in RAY. For optical elements (*left*) the coordinate system is fixed to the optical surface (X-Z plane). Transmission elements, screens and image planes (*right*) are in the X-Y plane, the x-axis is in the horizontal plane. The red line is the light beam

for the outgoing ray is rotated back by $-\chi$, i.e. it has the same orientation as before the optical element. In this way another optical element can be treated in an identical manner.

2.5.2 Geometrical Treatment of Rays

The geometric calculations proceed in the following way:

Statistical creation of a ray within a given source volume and emission cone and within the 'correct' statistics (see Chap. 3). The ray is determined by its source coordinates (x_s, y_s, z_s) and its direction cosines (l_s, m_s, n_s) determined by the horizontal and vertical emission angles φ and ψ (see Fig. 2.8):

$$\vec{\alpha}_S = \begin{pmatrix} l_S \\ m_S \\ n_S \end{pmatrix} = \begin{pmatrix} \sin\varphi\cos\psi \\ \sin\psi \\ \cos\varphi\cos\psi \end{pmatrix} \tag{2.3}$$

The vector equation of the ray is then

$$\vec{x} = \vec{x}_S + t\vec{\alpha}_S \quad \text{with } t\varepsilon\Re_0^+ \tag{2.4}$$

or, in coordinates

$$\begin{pmatrix} x \\ y \\ z \end{pmatrix} = \begin{pmatrix} x_s \\ y_s \\ z_s \end{pmatrix} + t \begin{pmatrix} l_S \\ m_S \\ n_S \end{pmatrix} \tag{2.5}$$

or

$$\frac{x - x_s}{l_S} = \frac{y - y_s}{m_S} = \frac{z - z_s}{n_S} \tag{2.6}$$

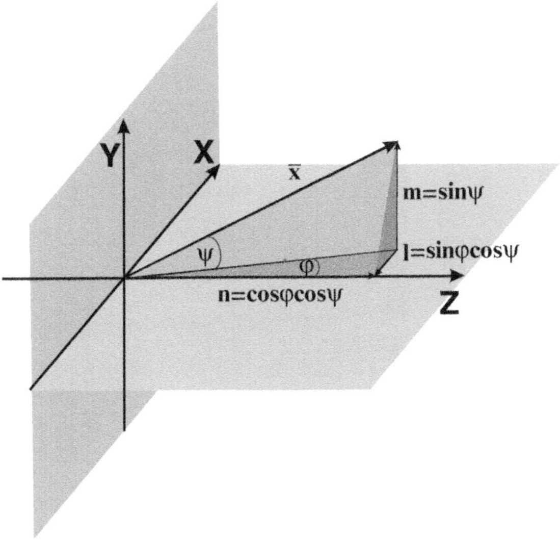

Fig. 2.8. Source coordinate system: Definition of angles and direction cosines in *RAY*

2.5.3 Intersection with Optical Elements

The source coordinate is translated into a new coordinate system with the origin in the centre of the first optical element (hit by the central ray), and the z-axis parallel to a symmetry axis of the optical element (for a simplified equation). The coordinate system is translated by the '*distance from the source*' to the optical element, z_{q}, rotated around z by the azimuthal angle, χ, and around the new $\tilde{x}$-axis by the grazing incidence angle, θ. The transformation to the new-coordinate system is performed by the following matrix operations:

$$\vec{x}_{\mathrm{S}'} = D_{\tilde{x}}\left(\theta\right) D_{z}\left(\chi\right) T_{z}\left(z_{q}\right) \vec{x}_{\mathrm{S}} \tag{2.7}$$

- z_{q} distance source to first optical element or nth to $(n+1)$th element
- θ rotation angle around x (y-z plane)
- χ azimuthal rotation around z (x-y plane) (clockwise),

which corresponds to

$$\begin{pmatrix} x_{\mathrm{S}'} \\ y_{\mathrm{S}'} \\ z_{\mathrm{S}'} \end{pmatrix} = \begin{pmatrix} 1 & 0 & 0 \\ 0 & \cos\theta & -\sin\theta \\ 0 & \sin\theta & \cos\theta \end{pmatrix} \circ \begin{pmatrix} \cos\chi & -\sin\chi & 0 \\ \sin\chi & \cos\chi & 0 \\ 0 & 0 & 1 \end{pmatrix} \circ \left(\begin{pmatrix} x_{\mathrm{S}} \\ y_{\mathrm{S}} \\ z_{\mathrm{S}} \end{pmatrix} - \begin{pmatrix} 0 \\ 0 \\ z_{\mathrm{q}} \end{pmatrix} \right) \tag{2.8}$$

or finally

$$\begin{pmatrix} x_{\mathrm{S}'} \\ y_{\mathrm{S}'} \\ z_{\mathrm{S}'} \end{pmatrix} = \begin{pmatrix} x_{\mathrm{s}} \cos\chi - y_{\mathrm{s}} \sin\chi \\ x_{\mathrm{s}} \sin\chi \cos\theta + y_{\mathrm{s}} \cos\chi \cos\theta - (z_{\mathrm{s}} - z_{\mathrm{q}}) \sin\theta \\ x_{\mathrm{s}} \sin\chi \sin\theta + y_{\mathrm{s}} \cos\chi \sin\theta + (z_{\mathrm{s}} - z_{\mathrm{q}}) \cos\theta \end{pmatrix} \tag{2.9}$$

The direction cosines are transformed correspondingly:

$$\vec{\alpha}_{S'} = D_{\tilde{x}}(\theta)\, D_z(\chi)\, \vec{\alpha}_S, \tag{2.10}$$

and finally
$$\begin{pmatrix} l_{S'} \\ m_{S'} \\ n_{S'} \end{pmatrix} = \begin{pmatrix} l_s \cos\chi - m_s \sin\chi \\ l_s \sin\chi \cos\theta + m_s \cos\chi \cos\theta \\ l_s \sin\chi \sin\theta + m_s \cos\chi \sin\theta \end{pmatrix}. \tag{2.11}$$

In the new coordinate system the ray is described by

$$\begin{pmatrix} x \\ y \\ z \end{pmatrix}(t) = \begin{pmatrix} x_{S'} \\ y_{S'} \\ z_{S'} \end{pmatrix} + t \begin{pmatrix} l_{S'} \\ m_{S'} \\ n_{S'} \end{pmatrix} \tag{2.12}$$

2.5.4 Misalignment

A six-dimensional misalignment of an optical element can be taken into account: three translations of the coordinate system by δx, δy and δz and three rotations by the misorientation angles $\delta\chi$ (x-y plane), $\delta\varphi$ (x-z plane) and $\delta\psi$ (y-z plane). Since the rotations are not commutative, the coordinate system is first rotated by these angles in the given order and then translated. For the outgoing ray to be described in the non-misaligned system, the coordinate system is backtransformed (in reverse order). Thus, the optical axis remains unaffected by the misalignment.

2.5.5 Second-Order Surfaces

Optical elements are described by the general equation for second-order surfaces:

$$\begin{aligned} F(x, y, z) = a_{11}x^2 + a_{22}y^2 + a_{33}z^2 + 2a_{12}xy + 2a_{13}xz \\ + 2a_{23}yz + 2a_{14}x + 2a_{24}y + 2a_{34}z + a_{44} = 0. \end{aligned} \tag{2.13}$$

This description refers to a right-handed coordinate system attached to the centre of the mirror with its surface in x-z plane, and y-axis points to the normal). This coordinate system is used for the optical elements *PL_ane, CO_ne, CY_linder* and *SP_here.*

Note that for the elements *EL_lipsoid* and *PA_raboloid* a coordinate system is used, which again is attached to the centre of the mirror (with x-axis on the surface), but the z-axis is parallel to the symmetry axis of this element for an easier description in terms of the a_{ij} parameters (see Figs. 2.9 and 2.10). The a_{ij}-values of Table 2.2 are given for this system. Thus, the rotation angle of the coordinate system from source to element is here $\theta + \alpha$ (EL) and 2θ (PA), respectively, θ being the grazing incidence angle and α the tangent angle on the ellipse.

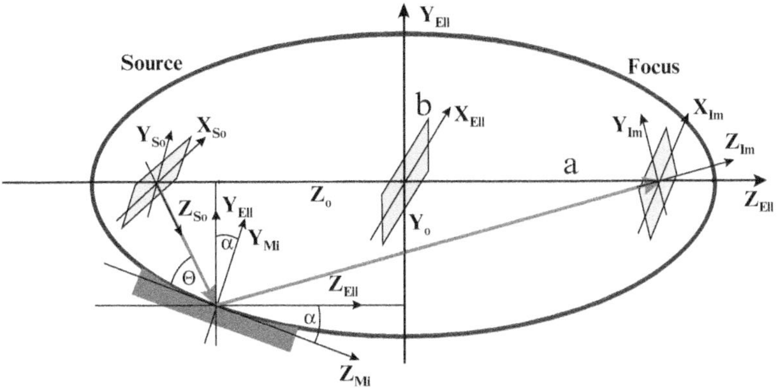

Fig. 2.9. Ellipsoid: definitions and coordinate systems

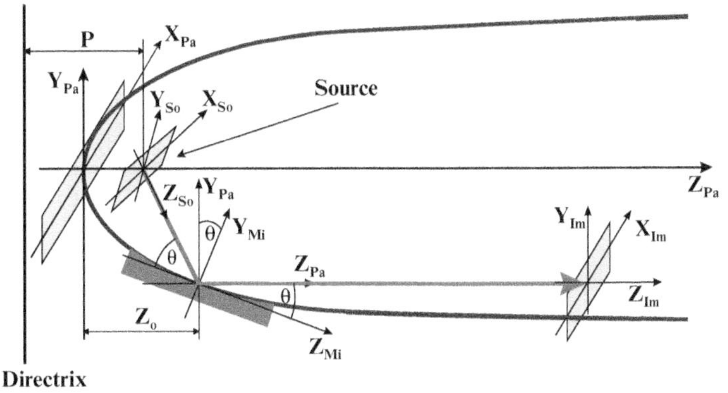

Fig. 2.10. Paraboloid: Definitions and coordinate systems

The individual surfaces are described by the following equations:

- *Plane* $y = 0$
- *Cylinder(in z − dir.)* $x^2 + y^2 = 0$
- *Cylinder(in x − dir.)* $y^2 + z^2 = 0$
- *Sphere* $x^2 + (y - R)^2 + z^2 - R^2 = 0$ (2.14)
- *Ellipsoid* $x^2/C^2 + (y - y_0)^2/B^2 + (z - z_0)^2/A^2 - 1 = 0$
- *Paraboloid* $x^2/C^2 + (y - y_0)^2/B^2 - 2P(z - z_0) = 0$

Alternatively to the input of suitable parameters, such as mirror radii or half axes of ellipses, in an experts modus (EO), the a_{ij} parameters can be directly given, such that any second-order surface, whatever shape it has, can be simulated.

Table 2.2. Parameters of the second-order optical elements

Name	PM	CY	CO	SP	EL	PA
a_{11}	0	1/0	$1-c_{\mathrm{m}}$	1	B^2/C^2	P^2/C^2
a_{22}	0	1	$1-2c_{\mathrm{m}}$	1	1	1
a_{33}	0	0/1	0	1	B^2/A^2	0
a_{12}	0	0	0	0	0	0
a_{13}	0	0	0	0	0	0
a_{23}	0	0	$\sqrt{c_{\mathrm{m}}-c_{\mathrm{m}}^2}$	0	0	0
a_{14}	0	0	0		0	0
a_{24}	-1	$\rho \cdot$ sign	$-a_{23}\frac{R}{\sqrt{c_{\mathrm{m}}}}-\frac{z_{\mathrm{m}}}{2}$	$R \cdot$ sign	$-y_0$	$-y_0$
a_{34}	0		0	0	$z_0 B^2/A^2$	$-P$
a_{44}	0	0	0	0	$y_0^2+z_0^2B^2/A^2-B^2$	$y_0^2-2Pz_0-P^2$
Sign (concave/convex)	0	$1/-1$	$1/-1$	$1/-1$	1	1
$F_x-(a_{11}x+a_{12}y+a_{13}z+a_{14})$	0	$-x/0$		$-x$	$-a_{11}x$	$-a_{11}x$
$F_y-(a_{22}y \cdot \mathrm{sign}+a_{12}x\ a_{23}z+a_{24})$	1	$\rho \cdot$ sign		$R \cdot$ sign	y_0-y	y_0-y
$F_z-(a_{33}z+a_{13}x+a_{23}y+a_{34})$	0	$0/-z$		$-z$	$-(z_0+z)(B/A)^2$	P
					$z_0=A^2/B^2y_0\tan(\alpha)$ $y_0=r_{\mathrm{a}}\sin(\theta-\alpha)$ $\tan(\alpha)=\tan(\theta)$ $(r_{\mathrm{a}}-r_{\mathrm{b}})/(r_{\mathrm{a}}+r_{\mathrm{b}})$	$z_0=f\cos(\alpha,\beta)\cdot sig$ $y_0=f\sin(2\alpha,\beta)$ $P=2f\sin^2(\theta)\mathrm{sig}$
		ρ: radius	R, ρ: radii z_{m}: mirror length $c_{\mathrm{m}}=\left(\frac{(r-\rho)}{z_{\mathrm{m}}}\right)^2$	R: radius	A, B, C half axes in z,y,x-dir; r_{a}, r_{b}: mirror to focus 1,2; θ: grazing angle of central ray; α: tangent angle Plane Ell.: $C=$ infty.; Rotational Ell.: $B=C<>A$; Ellipsoid: $A\neq B\neq C$; Sphere: $A=B=C$	f: mirror–source/focus–dist. C: halfpar. in x; Sig $=\pm 1$; f. collimation/focussing; θ: grazing angle of central ray; $\alpha,\beta=2\theta,0$ (coll); $\alpha,\beta=0,2\theta$ (foc.) Plane $P:C=$ infty.; Rotational $P:C=P$; Elliptical $P:C\neq P$

2.5.6 Higher-Order Surfaces

A similar expert modus is available for surfaces, which cannot be described by the second-order equation. The general equation is the following:

$$\begin{aligned} F(x,y,z) = {} & a_{11}x^2 + \text{sign} a_{22}y^2 + a_{33}z^2 + 2a_{12}xy + 2a_{13}xz + 2a_{23}yz \\ & + 2a_{14}x + 2a_{24}y + 2a_{34}z + a_{44} + b_{12}x^2y + b_{21}xy^2 \\ & + b_{13}x^2z + b_{31}xz^2 + b_{23}y^2z + b_{32}yz^2 = 0 \end{aligned} \tag{2.15}$$

Here, again all a_{ij} and b_{ij} parameters can be given explicitly by the user to describe any geometrical surface.

For special higher order surfaces the surface is described by the following equations.

Toroid

$$F(x,y,z) = \left((R-\rho) + \text{sign}\,(\rho)\,\sqrt{\rho^2 - x^2}\right)^2 - (y-R)^2 - z^2 = 0 \tag{2.16}$$

$$\text{Sign} = \pm 1 \text{ for concave/convex curvature.}$$

The surface normal is calculated according to (see Chap. 5.7)

$$F_x = \frac{-2x\ \text{sign}(\rho)}{\sqrt{\rho^2 - x^2}} \left((R-\rho) + \text{sign}(\rho)\sqrt{\rho^2 - x^2}\right)^2 \tag{2.17}$$

$$F_y = -2(y - R) \tag{2.18}$$

$$F_z = -2z. \tag{2.19}$$

Elliptical Paraboloid

$$F(x,y,z) = \frac{2fx^2}{2f - \frac{z+z_0}{\cos 2\theta}} - 2p(z+z_0) - p^2 = 0. \tag{2.20}$$

Elliptical Toroid

In analogy to a spherical toroid, an elliptical toroid is constructed from an ellipse (instead of a circle) in the (y, z) plane with small circles of fixed radius ρ attached in each point perpendicular to the guiding ellipse.

The mathematical description of the surface is based on the description of a toroid, where in each point of the ellipse a 'local' toroid with radius $R(z)$ and center $(y_c(z),\ z_c(z))$ is approximated (Fig. 2.11).

Following this description the elliptical toroid surface is given by

$$F(x,y,z) = 0 = (z - z_c(z))^2 + (y - y_c(z))^2 - \left(R(z) - \rho + \sqrt{\rho^2 - x^2}\right)^2 \tag{2.21}$$

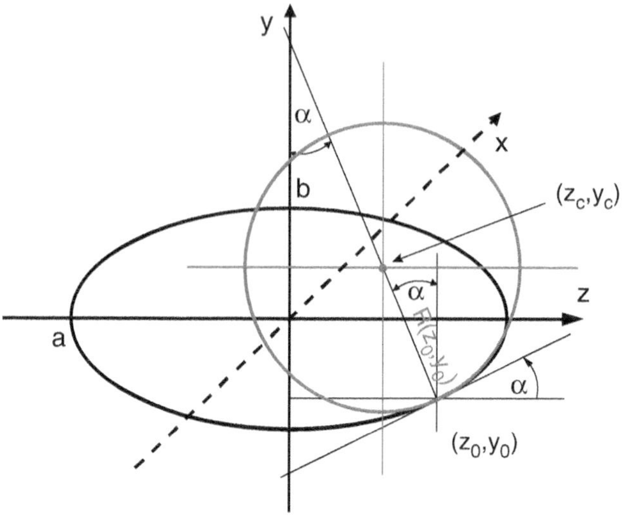

Fig. 2.11. Construction of an elliptical toroid. The ET is locally approximated by a conventional spherical toroid with radius $R(z)$ and center $(z_c(z),\ y_c(z))$

with $R(z) = a^2b^2\left(\frac{z^2}{a^4} + \frac{(a^2-z^2)}{a^2b^2}\right)^{\frac{3}{2}} = \frac{1}{ab}\left(\frac{b^2-a^2}{a^2}z^2 + a^2\right)^{\frac{3}{2}}$

$$\begin{aligned}
z_c(z) &= z - R(z)\sin\alpha(z),\\
y_c(z) &= y(z) + R(z)\cos\alpha(z),\\
z_c' &= 1 - R'\sin\alpha - R\alpha'\cos\alpha,\\
y_c' &= y' + R'\cos\alpha - R\alpha'\sin\alpha,\\
y(z) &= -\frac{b}{a}\sqrt{a^2-z^2},\\
\alpha &= \arctan(y') = \arctan\left(\frac{b}{a}\frac{z}{\sqrt{a^2-z^2}}\right),\\
y' &= \frac{\partial y}{\partial z} = \tan\alpha = \frac{b}{a}\frac{z}{\sqrt{a^2-z^2}}\\
\alpha' &= \frac{\partial \alpha}{\partial z} = \frac{y''}{1+y'^2},\\
y'' &= \frac{\partial^2 y}{\partial z^2} = \frac{ab}{(a^2-z^2)^{\frac{3}{2}}}.
\end{aligned}$$

The surface normal is given by the partial derivatives

$$\frac{\partial F}{\partial x} = 2\frac{x}{\sqrt{\rho^2-x^2}}\left(R - \rho + \sqrt{\rho^2-x^2}\right), \tag{2.22}$$

$$\frac{\partial F}{\partial y} = 2(y - y_c), \tag{2.23}$$

$$\frac{\partial F}{\partial z} = 2(z - z_c(1 - z_c') - 2y_c'(y - y_c) - 2R'\left(R - \rho + \sqrt{\rho^2 - x^2}\right). \tag{2.24}$$

2.5.7 Intersection Point

The intersection point (x_M, y_M, z_M) of the ray with the optical element is determined by solving the quadratic equation in t generated by inserting (2.12) into (2.13) or (2.15). For the special higher-order surfaces (TO, EP, ET) the intersection point is determined iteratively.

Then the local surface normal for this intersection point $\vec{n} = n(x_M, y_M, z_M)$ is found by calculating the partial derivative of $F(x_M, y_M, z_M)$

$$\vec{f} = \nabla F, \tag{2.25}$$

with the components

$$f_x = \frac{\partial F}{\partial x} \qquad f_y = -\frac{\partial F}{\partial y} \qquad f_z = \frac{\partial F}{\partial z}. \tag{2.26}$$

The local surface normal is then given by the unit vector

$$\vec{n} = \begin{pmatrix} n_x \\ n_y \\ n_z \end{pmatrix} = \frac{1}{\sqrt{{f_x}^2 + {f_y}^2 + {f_z}^2}} \begin{pmatrix} f_x \\ f_y \\ f_z \end{pmatrix}. \tag{2.27}$$

Whenever the intersection point found is outside the given dimensions of the optical element, the ray is thrown away as a geometrical loss and the next ray starts within the source according to Chap. 5.2.

2.5.8 Slope Errors, Surface Profiles

Once the intersection point and the local surface normal is found, these are the parameters that are modified to include real surfaces as deviations from the mathematical surface profile, namely figure and finish errors (slope errors, surface roughness), thermal distortion effects or measured surface profiles.

The surface normal is modified incrementally by rotating the normal vector in the y-z (meridional plane) and in the x-y plane (sagittal). The determination of the rotation angles depends on the type of error to be included.

1. Slope errors, surface roughness: the rotation angles are chosen statistically (according to the procedure described in Sect. 2.3.1) within a 6σ-width of the input value for the slope error.
2. Thermal bumps: a gaussian height profile in x- and z-direction with a given amplitude, and σ-width can be put onto the mirror centre.

3. Cylindrical bending: a cylindrical profile in z-direction (dispersion direction) with a given amplitude can be superimposed onto the mirror surface.
4. Measured surface profiles, e.g. by a profilometer.
5. Surface profiles calculated separately, e.g. by a finite element analysis program.

In cases (2–5) the modified mirror is stored in a 251 × 251 surface mesh which contains the amplitudes (y-coordinates). For cases (2) and (3) this mesh is calculated within *RAY*, for the cases (4) and (5) ASCII data files with surface profilometer data (e.g. LTP or ZEISS M400 [27]) or finite-element-analysis data (e.g. ANSYS [28]) can be read in. The new y-coordinate of the intersection point and the local slope are interpolated from such a table accordingly.

2.5.9 Rays Leaving the Optical Element

For those rays that have survived the interaction with the optical element – geometrically and within the reflectivity statistics (Chap. 6) – the direction cosines of the reflected/transmitted/refracted ray $(\vec{\alpha}_2) = (l_2, m_2, n_2)$ are calculated from the incident ray $(\vec{\alpha}_1) = (l_1, m_1, n_1)$ and the local surface normal $\vec{n}$.

Mirrors

For mirrors and crystals the entrance angle, α, is equal to the exit angle, β. In vector notation this means that the cross product is

$$n \times (\vec{\alpha}_2 - \vec{\alpha}_1) = 0, \tag{2.28}$$

since the difference vector is parallel to the normal. For the direction cosines of the reflected ray the result is given by

$$\alpha_2 = \vec{\alpha}_1 - 2(\vec{n} \circ \vec{\alpha}_1)\vec{n} \tag{2.29}$$

or in coordinates

$$l_2 = l_1 - 2n_x \frac{\ln_x + mn_y + nn_z}{{n_x}^2 + {n_y}^2 + {n_z}^2} \tag{2.30}$$

and, correspondingly, for m_2 and n_2.

Gratings

The emission angle β for diffraction gratings is obtained by the grating equation

$$k\lambda = d\,(\sin\alpha + \sin\beta)\,, \tag{2.31}$$

k, diffraction order; λ, wavelength; d, grating constant.

1. The grating is rotated by $\delta\chi = a\tan(n_x/n_y)$ around the z-axis and by $\delta\psi = a\sin(n_z)$ around the x-axis, so that the intersection point is plane (surface normal parallel to the y-axis). The grating lines are parallel to the x-direction.
2. Then the direction cosines of the diffracted beam are determined by

$$\begin{pmatrix} l_2 \\ m_2 \\ n_2 \end{pmatrix} = \begin{pmatrix} l_1 \\ \sqrt{m_1^2 + n_1^2 - (n_1 - a_1)^2} \\ n_1 - a_1 \end{pmatrix}, \tag{2.32}$$

$$a_1 = k\frac{\lambda}{d}\cos\delta\psi.$$

3. The grating is rotated back to the original position by $-\delta\psi$ and $-\delta\chi$.

For varied line spacing (VLS) gratings, the local line density $n = 1/d(l/mm)$ as a function of the (x, z)-position is determined by [29]

$$n = n_0 \cdot \left(1 + 2b_2 z + 3b_3 z^2 + 4b_4 z^3 + 2b_5 x + 3b_6 x^2 + 4b_7 x^3\right). \tag{2.33}$$

Transmitting Optics

For transmitting optics (*SL_it, FO_il*) the direction of the ray is unchanged by geometry. However, diffraction is taken into account for the case of rectangular or circular slits by randomly modifying the direction of each ray according to the probability for a certain direction φ

$$P(\varphi) = \frac{\sin u}{u}, \tag{2.34}$$

$$\text{with } u = \frac{\pi b \sin\varphi}{\lambda} (b,\ \text{slit opening};\ \lambda,\ \text{wavelength}),$$

so that for a statistical ensemble of rays a Fraunhofer (rectangular slits) or bessel pattern (circular slits) appears (see Fig. 2.12). *ZO_neplate* transmitting optics are described in [12,13].

Azimuthal Rotation

After successful interaction with the optical element the surviving ray is described in a coordinate system, which is rotated by the reflection angle θ and the azimuthal angle χ, such that the z-axis follows once again the direction of the outgoing central ray as it was for the incident ray. The old values of the source/mirror points and direction cosines are replaced by these new ones, so that a new optical element can be attached now in similar way.

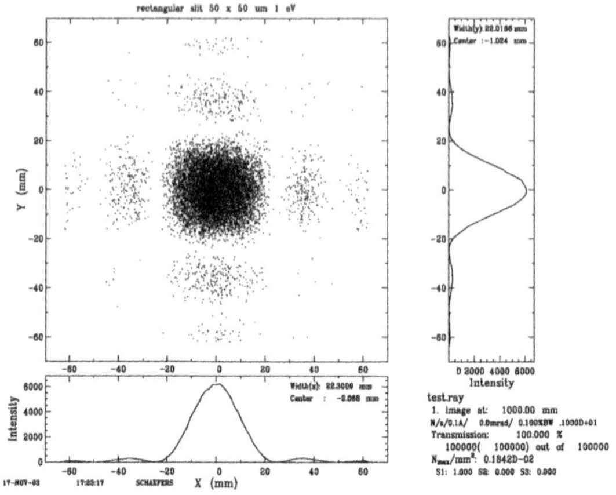

Fig. 2.12. Fraunhofer diffraction pattern on a rectangular slit

2.5.10 Image Planes

If the ray has traversed the entire optical system, the intersection points $(x_I,\ y_I)$ with up to three image planes at the distances $z_{I_{1,2,3}}$ are determined according to

$$\begin{pmatrix} x_I \\ y_I \end{pmatrix} = \begin{pmatrix} x \\ y \end{pmatrix} + \frac{1}{n}\begin{pmatrix} l \\ m \end{pmatrix}(z_{I_{1,2,3}} - z). \tag{2.35}$$

Once a ray reaches the image plane or whenever a ray is lost within the optical system a new ray is created within the source and the procedure starts all over.

2.5.11 Determination of Focus Position

For the case of imaging systems, if the focus position is to be determined, the x- and y-coordinates of that ray which has the largest coordinates are stored along the light beam in the range of the expected focal position (*search in a distance from last OE of ... +/− ...*). The so found cross section of the beam (width and height) is displayed graphically. Since at each position a different ray may be the outermost one, there may be bumps in this focal curve which depend on the quality of the imaging. Especially, for optical systems with large divergences (and thus large optical aberrations) or which include dispersing elements, this curve is only schematic and serves as a quick check of the focal properties of the system.

2.5.12 Data Evaluation, Storage and Display

The x, z-coordinates of the intersection point (x, y for source, slits, foils, zoneplates and image planes) and the angles l, n (l, m, respectively) are stored into 100×100 matrices. These matrices are multichannel arrays, one for the source, for each optical element and for each image plane, whose dimensions (and with it the pixel size) have been fixed before in a 'test-raytrace' run. They represent the illuminated surface in x-z projection. The corresponding surface pixel element that has been hit by a ray is increased by 1, so that intensity profiles and/or heat load can be displayed.

Additionally, the x- and z-coordinates (y, respectively) of the first 10,000 rays are stored in a 10,000x2 ASCII matrix to display footprint patterns of the optical elements, for point diagrams at the image planes or for further evaluation outside the program.

2.6 Reflectivity and Polarisation

Not only the geometrical path of the rays is followed, but also the intensity and polarisation properties of each ray are traced throughout an optical setup. Thus, it is easily possible to preview depolarisation effects throughout the optical path, or to optimize an optical setup for use as, for example, a polarisation monitor. For this, each ray is treated individually with a defined energy and polarisation state.

RAY employs the Stokes formalism for this purpose. The Stokes vector $\vec{S} = (S_0,\ S_1,\ S_2,\ S_3)$ describing the polarisation (S_1, S_2: linear, S_3: circular polarisation) for each ray is given either as free input parameter or, for dipole sources, is calculated according to the Schwinger theory. S_0, the start intensity of the ray from the source $\left(S_0 = \sqrt{S_1{}^2 + S_2{}^2 + S_3{}^2}\right)$, is set to 1 for the artificial sources. It is scaled to a realistic photon flux value for the synchrotron sources Dipole, Wiggler or the Undulator-File.

The Stokes vector is defined by the following equations:

$$\begin{aligned} S_0 &= \left[(E^{\mathrm{o}}{}_{\mathrm{p}})^2 + (E^{\mathrm{o}}{}_{\mathrm{s}})^2\right]\Big/2 = 1, \\ S_1 &= \left[(E^{\mathrm{o}}{}_{\mathrm{p}})^2 - (E^{\mathrm{o}}{}_{\mathrm{s}})^2\right]\Big/2 = P_{\mathrm{l}}\cos(2\delta), \\ S_2 &= E^{\mathrm{o}}{}_{\mathrm{p}}E^{\mathrm{o}}{}_{\mathrm{s}}\cos(\phi_{\mathrm{p}} - \phi_{\mathrm{s}}) = P_{\mathrm{l}}\sin(2\delta), \\ S_3 &= -E^{\mathrm{o}}{}_{\mathrm{p}}E^{\mathrm{o}}{}_{\mathrm{s}}\sin(\phi_{\mathrm{p}} - \phi_{\mathrm{s}}) = P_{\mathrm{c}}, \end{aligned} \tag{2.36}$$

with the two components of the electric field vector defined as

$$E_{\mathrm{p,s}}(z,t) = E^{\mathrm{o}}_{\mathrm{p,s}} \exp\left[\mathrm{i}\left(\omega t - kz + \phi_{\mathrm{p,s}}\right)\right]. \tag{2.37}$$

and P_{l}, P_{c} are the degree of linear and circular polarisation, respectively. δ is the azimuthal angle of the major axis of the polarisation ellipse. Note that

$$\begin{aligned} P_{\mathrm{l}} &= P\cos(2\varepsilon) \\ \text{and} \quad P_{\mathrm{c}} &= P\sin(2\varepsilon), \end{aligned} \tag{2.38}$$

with P being the degree of total polarisation and ε the ellipticity of the polarisation ellipse $(\tan\varepsilon = R_{\mathrm{p}}/R_{\mathrm{s}})$.

Table 2.3. Definition of circular polarisation

Phase ϕ_{p-s}	$90°$, $-270°$ ($\pi/2$, $-3\pi/2$)	$-90°$, $270°$ ($-\pi/2$, $3\pi/2$)
Rotation sense (in time)	Clockwise	Counter-clockwise
Rotation sense (in space)	Counter-clockwise	Clockwise
Polarisation (optical def.)	R(ight) CP	L(eft) CP
Helicity (atomic def.)	Negative ($\sigma-$)	Positive ($\sigma+$)
Stokes vector	Negative	Positive

Table 2.4. Physical interaction for the different optical components

Mirrors	Gratings	Foils	Slits	Zone-plates	Crystals
Fresnel equations	Diffraction	Fresnel equations	–	–	Dynamic theory
Reflectivity R_s, R_p, Δ_{sp}	Efficiency E_s, E_p, Δ_{sp}	Transmission T_s, T_p, Δ_{sp}	Transmission $T_s, T_p = 1$ $\Delta_{sp} = 0$	Transmission $T_s, T_p = 1$, $\Delta_{sp} = 0$	Reflectivity R_s, R_p, Δ_{sp}

Since the SR is linearly polarised within the electron orbital plane ($I_{perp} = 0$), the plane of linear polarisation is coupled to the x-axis (i.e. horizontal). Thus, the Stokes vector for SR is defined in our geometry as (see Chap. 3.4)

$$P_{lin} = S_1 = (I_{perp} - I_{par})/(I_{perp} + I_{par}) = (I_y - I_x)/(I_y + I_x) = -1, \quad (2.39)$$

$S_1 = +1$ would correspond to a vertical polarisation plane.

For the definition of the circular polarisation the nomenclature of Westerfeld et al. [30] and Klein/Furtak [31] has been used. This is summarised in Table 2.3:

For example, for the case of synchrotrons and storage rings, the radiation that is emitted off-plane, upwards, has negative helicity, right-handed CP ($S_3 = -1$), when the electrons are travelling clockwise, as seen from the top.

The modification of the Stokes vector throughout the beamline by interaction of the light with the optical surface is described by the following steps (see e.g. [28]):

(1) Give each ray a start value for the Stokes parameter within the source, S_{ini}, according to input or as calculated for SR sources
(2) Calculate the intensity loss at the first optical element for s- and p-polarisation geometry and the relative phase, $\Delta = \delta_s - \delta_p$, according to the physical process involved (see Table 2.4):
   - Mirrors, Foils

   The optical properties of mirrors, multilayers, filters, gratings and crystals are calculated from the compilation of atomic scattering factors in the spectral range from 30 eV to 30 keV [32]. Another data set covers the X-ray range from 5 up to 50 keV [33]. Additional data for lower energies down to 1 eV are also available for some elements and molecules [34].

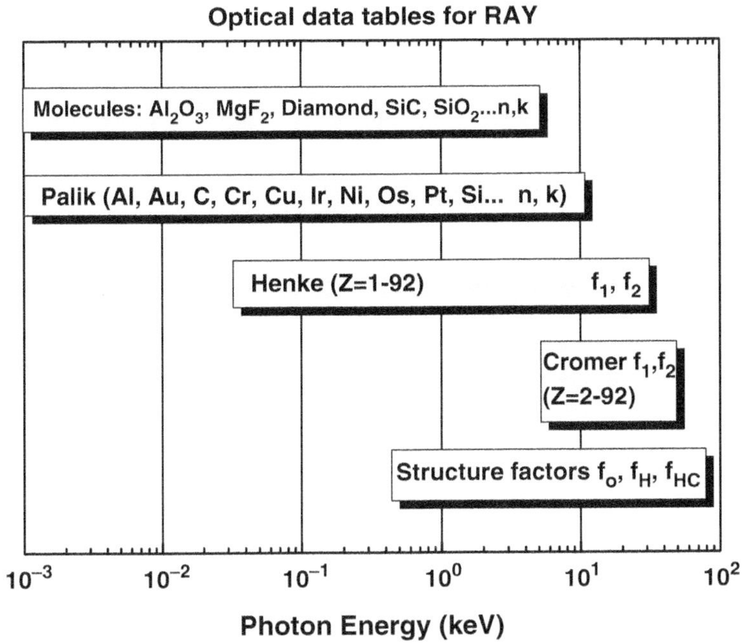

Fig. 2.13. Data bases used for the calculation of optical properties

A summary of the various data tables available within the program is given in Fig. 2.13. For compound materials that can be defined by the case sensitive chemical formula (e.g. MgF_2), the contributions of the chemical elements are weighted according to their stochiometry. A tabulated or, if not available, calculated value for the density is proposed but can be changed. The surface roughness of mirrors or multilayers is taken into account according to the Nevot–Croce formalism [35].

All reflection mirrors and transmission foils in an optical setup can have a multilayer coating (plus an additional top coating). The optical properties of these structures are calculated in transmission and reflection geometry by a recursive application of the Fresnel equations. For periodic multilayers, the layer thickness, the density and the surface roughness must be specified for each type of interface. For aperiodic structures like broad-band or supermirrors, the exact structure has to be provided in a data-file.

- Gratings

For the calculation of (monolayer covered) reflection gratings, a code developed by Neviere is used [36], which allows for the calculation for three different grating profiles (sinusoidal, laminar or blazed). In addition to fixed deviation angle mounts, optionally the incidence angle can be coupled to the photon energy and the c_{ff} factor in the case of a Petersen

SX700 type monochromator (PGM or SGM with a plane pre-mirror which enables the deviation angle across the grating to be varied).

- Crystals

For crystals the diffraction properties are calculated from the dynamical theory using the Darwin–Prins formalism [37]. For all crystals with zinc blende structure such as Si, Ge or InSb as well as for quartz and beryl, the crystal structure factors are determined within the program for any photon energy and the corresponding Bragg angle. For other crystals, the rocking curves can also be evaluated if the structure factors are known from other sources. The calculation is possible for any allowed crystal reflection and asymmetry (see Chap. 2.7).

(3) Transform the incident Stokes vector, $\vec{S}_{\mathrm{ini}}$, into the coordinate system of the optical element $\vec{S}_{\mathrm{M}}$ by rotation around the azimuthal angle χ (R-matrix)

$$\vec{S}_{\mathrm{M}} = R_{\tilde{y}}(\chi)\vec{S}_{\mathrm{ini}}, \tag{2.40}$$

$$\vec{S}_{\mathrm{M}} = \begin{pmatrix} S_{0\mathrm{M}} \\ S_{1\mathrm{M}} \\ S_{2\mathrm{M}} \\ S_{3\mathrm{M}} \end{pmatrix} = \begin{pmatrix} 1 & 0 & 0 & 0 \\ 0 & \cos 2\chi & \sin 2\chi & 0 \\ 0 & -\sin 2\chi & \cos 2\chi & 0 \\ 0 & 0 & 0 & 1 \end{pmatrix} \bullet \begin{pmatrix} S_{0\mathrm{ini}} \\ S_{1\mathrm{ini}} \\ S_{2\mathrm{ini}} \\ S_{3\mathrm{ini}} \end{pmatrix}. \tag{2.41}$$

Thus, the azimuthal angle of an optical element determines the polarisation geometry of the interaction. For instance, for horizontally polarised synchrotron radiation ($S_1 = -1$), an azimuthal angle of $\chi = 0°$ corresponds to an s-polarisation geometry (polarisation plane perpendicular to the reflection plane) with the beam going upwards. Since the coordinate system is right-handed, $\chi = 90°$ corresponds to a deviation to the right, when looking with the beam and a p-polarisation geometry (polarisation plane parallel to reflection plane). Similarly $\chi = 180°$ and $270°$, respectively, determine a beam going down and to the left, respectively. Note that the azimuthal angle is coupled to the coordinate system and not to the polarisation state. $\chi = 0\,°$ always determines a deviation upwards, but this may be an s-polarisation geometry, as in our example above, and can also be a p-geometry (when $S_{1\mathrm{inc}} = +1$).

(4) Calculate the Stokes vector after the optical element $\vec{S}_{\mathrm{final}}$ by applying the Müller matrix, M, onto $\vec{S}_{\mathrm{M}}$

$$\vec{S}_{\mathrm{final}} = M\vec{S}_{\mathrm{M}} \tag{2.42}$$

$$\begin{pmatrix} S_{0\mathrm{final}} \\ S_{1\mathrm{final}} \\ S_{2\mathrm{final}} \\ S_{3\mathrm{final}} \end{pmatrix} = \begin{pmatrix} \frac{R_{\mathrm{s}}+R_{\mathrm{p}}}{2} & \frac{R_{\mathrm{p}}-R_{\mathrm{s}}}{2} & 0 & 0 \\ \frac{R_{\mathrm{p}}-R_{\mathrm{s}}}{2} & \frac{R_{\mathrm{s}}+R_{\mathrm{p}}}{2} & 0 & 0 \\ 0 & 0 & R_{\mathrm{s}}R_{\mathrm{p}}\cos\Delta & R_{\mathrm{s}}R_{\mathrm{p}}\sin\Delta \\ 0 & 0 & -R_{\mathrm{s}}R_{\mathrm{p}}\sin\Delta & R_{\mathrm{s}}R_{\mathrm{p}}\cos\Delta \end{pmatrix} \circ \begin{pmatrix} S_{0\mathrm{M}} \\ S_{1\mathrm{M}} \\ S_{2\mathrm{M}} \\ S_{3\mathrm{M}} \end{pmatrix}. \tag{2.43}$$

(5) Accept this ray only when its intensity ($S_{0,\text{final}}$) is within the 'correct' statistic, i.e. when

$$(S_{0,\text{final}}/S_{0,\text{ini}} - ran\,(z)) > 0. \tag{2.44}$$

(6) Rotate the Stokes vector $\vec{S}_{\text{final}}$ back by $-\chi$ and take this as incident Stokes vector for the next optical element

$$\vec{S}'_{\text{ini}} = R_{\tilde{y}}(-\chi)\vec{S}_{\text{final}}. \tag{2.45}$$

(7) Store the Stokes vector for this optical element, go to the next one (2) or start with the next ray within the source (1).

2.7 Crystal Optics (with M. Krumrey)

For ray tracing, the geometrical point of view is most relevant. In this aspect, the main difference between crystals and mirrors or reflection grating is that the radiation is not reflected at the surface, but at the lattice planes in the material. In contrast to gratings which have already been treated as dispersive elements, reflection for a given incidence angle on the lattice plane occurs only if the well-known Bragg condition is fulfilled:

$$\lambda = 2d\sin\Theta, \tag{2.46}$$

where λ is the wavelength, d is the lattice plane distance and Θ is the incidence angle of the radiation with respect to the lattice plane. The selected lattice planes are not necessarily parallel to the surface, resulting in an asymmetry described by the asymmetry factor b:

$$b = \frac{\sin(\theta_\text{B} - \alpha)}{\sin(\theta_\text{B} + \alpha)}, \tag{2.47}$$

with Θ_B being the Bragg angle for which (2.46) is fulfilled and α the angle between the lattice plane and the crystal surface.

The subroutine package for crystal optics in *RAY* is based on the description of dynamic theory [38–40] as given by Matsushita and Hashizume in [41] and the paper from Batterman and Cole [37]. The reflectance is calculated according to the Darwin–Prins formalism, which requires the knowledge of the crystal structure factors F_o, F_h and F_hc. These factors can be derived for any desired crystal reflection, identified by the Miller indices (*hkl*), if the crystal structure, the chemical elements involved and the lattice constants (or constants for non-cubic crystals) are known. For some crystals with zinc blende structure (e.g. Si, InSb, etc.) or quartz structure, the structure factors are calculated automatically. This calculation combines the geometrical properties, especially the atomic positions in the unit cell which are read from a file, with the element-specific atomic scattering factors. The atomic scattering factor, f, is written here as

$$f = f_0 + \Delta f_1 + \Delta f_2. \tag{2.48}$$

This form allows one to separate the form factor f_0, which is calculated in dependence on $(\sin\theta_B)/\lambda$ based on a table of nine coefficients which are read for every chemical element from a file. The photon energy dependent anomalous dispersion corrections Δf_1 and Δf_2 are calculated from the Henke tables for photon energies up to 30 keV. For higher photon energies, the Cromer tables are directly used up to 50 keV and extrapolated beyond. Both data sets are also stored in files for all chemical elements.

Using the structure factors F_o, F_h and F_{hc}, which can, for other crystals, also be inserted by the user, the reflectance is obtained as

$$R = (\eta \pm \sqrt{\eta^2 - 1})s. \tag{2.49}$$

Here, s is simply defined as

$$s = \sqrt{F_h/F_{hc}} \tag{2.50}$$

while the parameter η is calculated according to

$$\eta = \frac{2b(\alpha - \Theta_B)\sin 2\Theta + \gamma F_o(1-b)}{2\gamma\,|P|\,s\sqrt{|b|}}, \tag{2.51}$$

where γ is defined as

$$\gamma = \frac{r_e\lambda^2}{\pi V_C}. \tag{2.52}$$

Here, r_e is the classical electron radius and V_c is the crystal unit cell volume. The polarisation is taken into account by the factor P, which equals unity for σ-polarisation and $\cos 2\Theta_B$ for π-polarisation.

In addition to the reflectance, the dispersion correction $\Delta\Theta$ for the incident and the outgoing ray at the crystal surface is calculated. For this purpose a crystal reflection curve is calculated according to (2.49) and the difference from its centre to the Bragg angle Θ_B is extracted. Only in the case of symmetrically cut crystals are the dispersion corrections identical:

$$\Delta\Theta_{out} = b\Delta\Theta_{in}. \tag{2.53}$$

At present, plane and cylindrical crystals are treated in reflection geometry (Bragg case). Also crystals with a d-spacing gradient (graded crystals with $d = d(z)$) are taken into account. This versatility enables a realistic simulation to be made of nearly every X-ray-optical arrangement in use with conventional X-ray sources or at synchrotron radiation facilities (double-, four

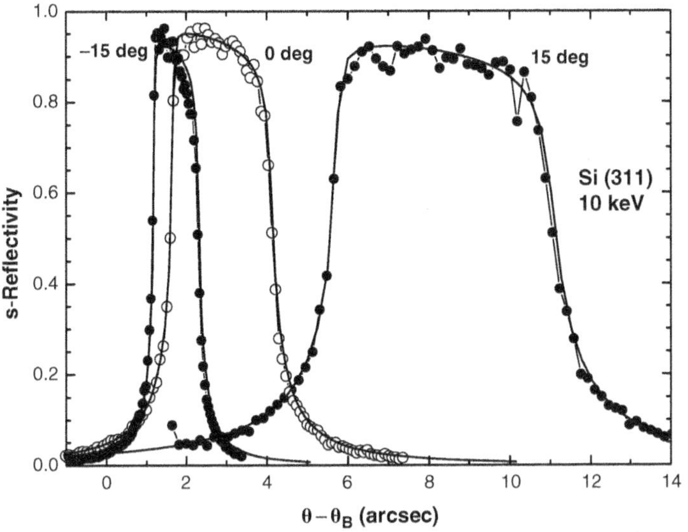

Fig. 2.14. Rocking curves of Si(311) crystal with asymmetric cut (15° and −15°) and symmetric cut (0°) for σ-polarisation at a photon energy of 10 keV

crystal monochromators, 2-bounce, 4-bounce in-line geometries for highest resolution, dispersive or non-dispersive settings, etc. [42, 43]).

Typical X-ray reflectance curves obtained with this subroutine package are shown for illustration. The raytracing code was applied for the calculations of Si(311) asymmetrically and symmetrically cut flat crystals. The angle of asymmetry was chosen to be 15° and −15°. In Fig. 2.14 the comparative results between *RAY* and *REFLEC* [12] codes for the σ-polarisation state are shown. *RAY* results in this figure are represented by the noisy curve. The statistics are determined by the number of rays calculated (10^6 incident rays, distributed into 100 channels).

2.8 Outlook: Time Evolution of Rays (with R. Follath, T. Zeschke)

In this article a program has been described, which is capable of simulating the behaviour of an optical system. Originally the program was designed for the calculation of X-ray optical setups on electron storage rings for synchrotron radiation. Similar programs had been written at most of the facilities for in-house use tailored to their specific applications. Many of them have not survived. Over more than 20 years of use by many people and continuous upgrade, debugging and development, the RAY-program described here has turned into a versatile optics database, by which almost all of the existing synchrotron radiation beamlines from the infrared region to the hard X-ray range can be accessed. In addition, other sources can be modelled since the light sources are described by relatively few parameters.

However, the program has limitations, of course, and it is essential to be aware of them when using it:

- The results are valid only within the mathematical or physical model implemented.
- The program may still have bugs (it has – definitely!!).
- The user may have made typing errors in the input menu.
- The user may have made errors in interpreting unclear or ambiguous input parameters or results.

The program is in continuous development and new ideas about sources or optical elements are implemented relatively fast, so that new demands can be addressed quickly.

One of the latest developments was driven by the advent of the new generation Free Electron Light (FEL) Sources at which the time structure of the radiation in the femto-second regime is of utmost importance. As outlook for the future of raytracing this development, which is still in progress, is discussed here briefly.

To handle the time structure, a ray is not only described by its geometry, energy and polarisation, but also by its geometrical path length or, in other words, by its travel time.

This enables one to follow the time evolution of an ensemble of rays, starting with a well-defined time-structure in the source, through an optical system. By storing the individual path lengths of each ray a pulse-broadening at each element and at the focal plane can be detected.

In the source, each ray is given a start-clock time, t_0, which can be either $t_0 = 0$ for all rays (complete coherence), or have a gaussian or flat-top distribution (less than complete coherence).

The path length of a ray is calculated as difference between the coordinates of the previous optical element (*x_old, y_old, z_old*) (or, for the first optical element, the source coordinates (*x_so, y_so, z_so*)) and the actual coordinates (x,y,z). The path length is measured with respect to the path length of the principal ray, given by the *distance to the preceding element zq*. Only geometrical differences are taken into account, no phase changes on reflection or penetration effects on multilayers are considered.

The path length is given by the equation

$$\mathrm{pl} = \sqrt{((x - x_{\mathrm{old}})^2 + (y - y_{\mathrm{old}})^2 + (z - z_{\mathrm{old}})^2)} - zq. \tag{2.54}$$

The phase of the ray with respect to the central ray and its relative travel time is then

$$\varphi = \frac{2\pi}{\lambda}\mathrm{pl}, \tag{2.55}$$

$$t = \frac{\mathrm{pl}}{c} \qquad c\text{: speed of light (m s}^{-1}\text{)}. \tag{2.56}$$

Assuming pl in millimetre, the travel time is given in nanoseconds.

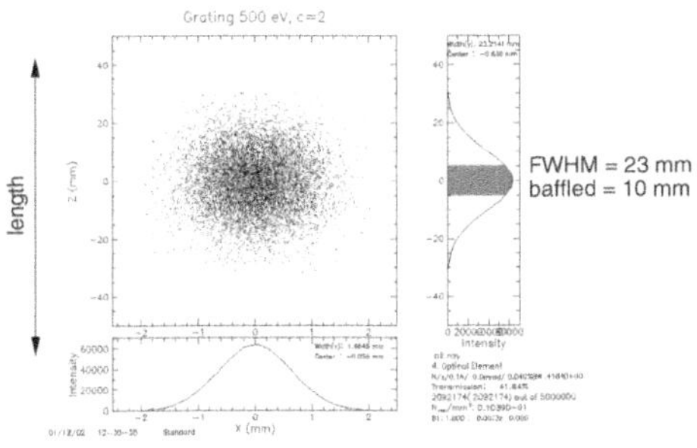

Fig. 2.15. Illumination of a reflection grating and baffling to preserve the time structure of the light beam

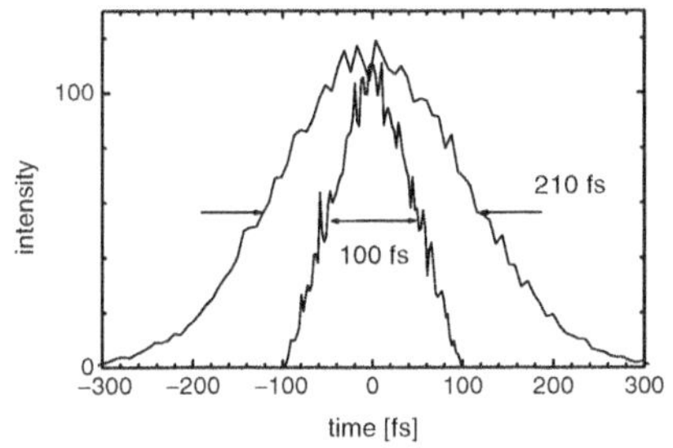

Fig. 2.16. Time structure of the rays after travelling through the beamline; confined–unconfined by the grating of Fig. 2.15

As an example, Fig. 2.15 shows the illumination of a reflection grating, which is part of a soft X-ray plane grating monochromator (PGM-) beamline that has been modelled for the TESLA FEL project [44], in which the conservation of the fs-time structure is essential. By baffling the illuminated grating length down to 10 mm in the dispersion direction the pulse broadening of the monochromatic beam (Fig. 2.16) can be kept well within the required 100 fs, which corresponds to the time structure of a SASE-FEL-source. As a result, the pulse length remains essentially unchanged by the optics.

By combining the path length information of each ray with its spatial information (footprint on an optical element or focus) a three-dimensional space–time picture over an ensemble of rays can be constructed. Such an example is given in Fig. 2.17. Here the focus of a highly demagnifying toroidal mirror (10:1) illuminated at grazing incidence (2.5°) by a diffraction-limited gaussian source with $\sigma = 0.2$ mm cross section and $\sigma = 0.3$ mrad divergence is shown. The illumination is coherent, i.e. all rays have the same start-time within the source. The focus (Fig. 2.17a) shows the typical blurring due to coma and astigmatic coma, and the grey scale colour attributed to each ray (Fig. 2.17b) determines the relative travel time (i.e. phase) with respect to the central ray. This is a snap shop over the focus; rays arrive at the focus in a time indicated by an increasing grey-scale.

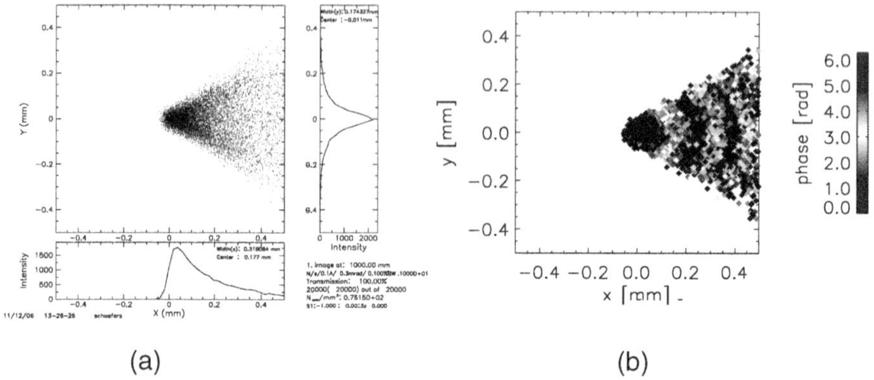

Fig. 2.17. Footprint of rays (**a**) and their individual phases (**b**) arriving at the focus of a toroidal mirror in grazing incidence ($\theta = 2.5°$, 10:1 demagnification)

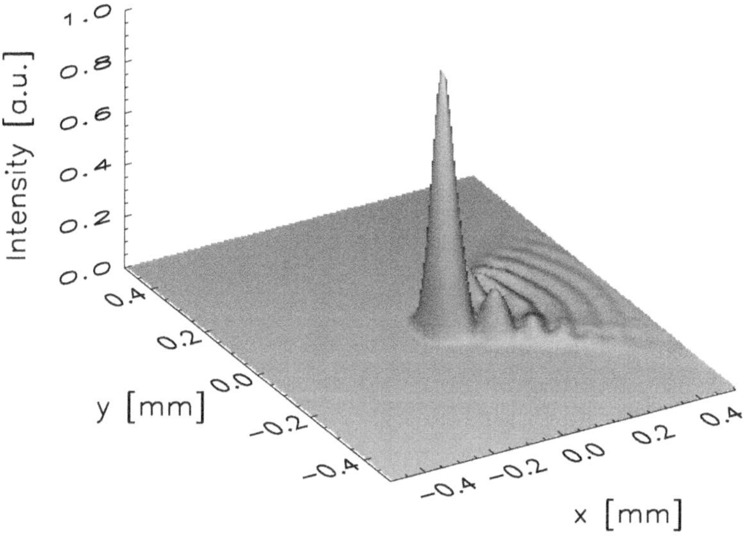

Fig. 2.18. Interference pattern at the focus of a 2.5° incidence toroidal mirror, 10:1 demagnification

In the individual phases an interference pattern in the coma blurred wings becomes visible. After complex addition of all rays within a certain array element according to

$$I = \left| \sum_j e^{i\varphi_j} \right|^2 , \tag{2.57}$$

an interference pattern becomes visible also in the intensity profile (Fig. 2.18). This profile looks very similar to the results obtained with programs on the basis of Fourier-Optics (see this book [6]) and shows the potential of a conventional raytrace program in treating interference effects.

So far in this simple example only the phase and the space coordinates of the rays have been connected to demonstrate the treatment of collective interference effects in the particle model. This model can be extended further to incoherent or partially coherent illumination simply by modifying the incident time-variable of the source suitably. Coherent packages within a total ensemble of rays can be extracted, which are determined by the same wavelength, the same polarisation plane, the same x-y-position (lateral coherence length) or the same path length (transversal coherence). Hence, there is a huge potential for further development of wave-phenomena within the particle model.

Acknowledgements

Thanks are due to hundreds of users of the program over more than 20 years, in particular to all collegues of the BESSY optics group. Without their comments, questions, critics, suggestions, problems and patience over the years the program would not exist.

In particular, Josef Feldhaus as the 'father' of the program, William Peatman for encouragement, support and worldwide advertisement, A.V. Pimpale, K.J.S. Sawhney and M. Krumrey for assistance in implementing essential additional features such as new sources and crystal optics are to be gratefully acknowledged. G. Reichardt implemented the grating calculations and was indispensable in formulating the mathematical aspects of this manuscript. D. Abramsohn managed successfully the adaptation of the FORTRAN source code to any PC-WINDOWS platform and by this he made it accessible world wide. A. Erko is to be thanked for the implementation of zoneplate optics, continuous encouragement and never-ending ideas for implementation of new optical elements.

References

1. J. Feldhaus, *RAY* (unpublished) and personal communication (1984)
2. C. Welnak, G.J. Chen, F. Cerrina, Nucl. Instrum. Methods Phys. Res. A **347**, 344 (1994)
3. T. Yamada, N. Kawada, M. Doi, T. Shoji, N. Tsuruoka, H. Iwasaki, J. Synchrotron Radiat. **8**, 1047 (2001)
4. J. Bahrdt, Appl. Opt. **36**, 4367 (1997)
5. O. Chubar, P. Elleaume, in *Proceedings of 6th European Particle Accelerator Conference EPAC-98*, 1998, pp. 1177–1179
6. M. Bolder, J. Bahrdt, O. Chubar, *Wavefront Propagation* (this book, Chapter 5)
7. P.R. Bevington, *Data Reduction and Error Analysis for the Physical Sciences* (McGraw-Hill, New York, 1969)
8. M. Born, E. Wolf, *Principles of Optics*, 6th edn. (Pergamon Press, New York, 1980)

9. F.A. Jenkins, H.E. White, *Fundamentals of Optics*, 4th edn. (McGraw-Hill, New York, 1981)
10. W.B. Peatman, *Gratings, Mirrors and Slits* (Gordon & Breach, New York, 1997)
11. A. Pimpale, F. Schäfers, A. Erko, Technischer Bericht, BESSY **TB 190**, 1 (1994)
12. A. Erko, F. Schäfers, N. Artemiev, in *Advances in Computational Methods for X-Ray and Neutron Optics SPIE-Proceedings*, vol. 5536, 2004, pp. 61–70
13. A. Erko, in *X-Ray Optics; Raytracing model of a Zoneplate*, ed. by B. Beckhoff et al. Handbook of Practical X-Ray Fluorescence Analysis (Springer, Berlin Heidelberg New York, 2006) pp. 173–179
14. F. Schäfers, Technischer Bericht, BESSY **TB 202**, 1 (1996)
15. F. Schäfers and M. Krumrey, Technischer Bericht, BESSY **TB 201**, 1 (1996)
16. H. Petersen, C. Jung, C. Hellwig, W.B. Peatman, W. Gudat, Rev. Sci. Instrum. **66**, 1 (1995)
17. W.B. Peatman, U. Schade, Rev. Sci. Instrum. **72**, 1620 (2001)
18. M.R. Weiss, et al. Nucl. Instrum. Methods Phys. Res. A **467–468**, 449 (2001)
19. A. Erko, F. Schäfers, W. Gudat, N.V. Abrosimov, S.N. Rossolenko, V. Alex, S. Groth, W. Schröder, Nucl. Instrum. Methods Phys. Res. A **374**, 408 (1996)
20. A. Erko, F. Schäfers, A. Firsov, W.B. Peatman, W. Eberhardt, R. Signorato Spectrochim. Acta B **59**, 1543 (2004)
21. *EFFI: Software Code to Calculate VUV/X-ray Optical Elements*, developed by F. Schäfers, BESSY, Berlin (unpublished)
22. *OPTIMO: Software Code to Optimize VUV/X-ray Optical Elements*, developed by F. Eggenstein, BESSY, Berlin (unpublished)
23. J. Schwinger, Phys. Rev. **75**, 1912 (1949)
24. *URGENT: Software Code for Insertion Devices*, developed by R.P. Walker, B. Diviacco, Sincrotrone Trieste, Italy (1990)
25. C. Jacobson, H. Rarback, in *Insertion Devices for Synchrotron Radiation.* SPIE Proc., vol. 582 (*SMUT: Software code for insertion devices*), 1985, p. 201
26. *WAVE: Software Code for Insertion Devices*, developed by M. Scheer, BESSY, Berlin (unpublished)
27. *M400: Coordinate Measuring Machine*, ZEISS, Oberkochen, Germany
28. *ANSYS: Finite Element Analysis (FEM) Program*, registered trademark of SWANSON Analyzer Systems, Inc., Houston, TX
29. M. Fujisawa, A. Harasawa, A. Agui, M. Watanabe, A. Kakizaki, S. Shin, T. Ishii, T. Kita, T. Harada, Y. Saitoh, S. Suga, Rev. Sci. Instrum. **67**, 345 (1996)
30. W.B. Westerveld, K. Becker, P.W. Zetner, J.J. Corr, J.W. McConkey, Appl. Opt. **24**, 2256 (1985)
31. M.V. Klein, T.E. Furtak, *Optik* (Springer-Lehrbuch, Berlin Heidelberg New York, 1988)
32. B.L. Henke, E.M. Gullikson, J.C. Davis, At. Data Nucl. Data Tables **54**, 181 (1993)
33. D.T. Cromer, J.T. Waber, *Atomic Scattering Factors for X-Rays in International Tables for X-Ray Crystallography*, vol. IV (Kynoch Press, Birmingham, 1974), pp. 71–147
34. E.D. Palik (ed.), *Handbook of Optical Constants of Solids* (Academic Press, New York, 1985); J.H. Weaver et al., Phys. Data **18**, 2 (1981)
35. L. Nevot, P. Croce, Rev. Phys. Appl. **15**, 761 (1980)
36. M. Neviere, J. Flamand, J.M. Lerner, Nucl. Instrum. Methods **195**, 183 (1982); R. Petit (ed.) *Electromagnetic Theory of Gratings*, (Springer Verlag, Berlin Heidelberg New York, 1980) and references therein

37. B.W. Batterman, H. Cole, Rev. Mod. Phys. **36**, 681 (1964)
38. D.W.J. Cruickshank, H.J. Juretsche, N. Kato (eds.), *P.P. Ewald and his Dynamical Theory of X-ray Diffraction* (International Union of Crystallography and Oxford University Press, Oxford, 1992)
39. W.H. Zachariasen, *Theory of X-ray Diffraction in Crystals*, (Wiley, New York, 1945)
40. Z.G. Pinsker, *Dynamical Scattering of X-rays in Crystals*, (Springer Verlag, Berlin Heidelberg New York, 1978)
41. T. Matsushita, H. Hashizume, In: *Handbook of Synchrotron Radiation*, by E.E. Koch (ed.) (North Holland, Amsterdam, 1993)
42. J.W.M. DuMond, Phys. Rev. **52**, 872 (1937)
43. Yu. Shvyd'ko, in *X-Ray Optics, High-Energy-Resolution Applications*, Springer Series in Optical Sciences vol. 98 (Springer, Berlin Heidelberg New York, 2004)
44. R. Follath, AIP Conf. Proc. 879-I, 513 (2007)

3

Neutron Beam Phase Space Mapping

J. Füzi

Abstract. A method based on energy-resolved pinhole camera imaging is proposed and characterized as a tool for neutron-beam phase-space mapping. It relies on time-resolved, two-dimensional position-sensitive neutron detection. Examples of applications in neutron source brightness evaluation, quality assessment of neutron optical components and velocity selector transfer function determination are presented.

The neutron beam phase space in real space is defined as the flux distribution with respect to five parameters: two positional (x and y with respect to the beam axis or a laboratory reference axis), two angular (δ_x and δ_y with respect to the same axis) dimensions and wavelength (velocity, energy). The result of its determination is a five-dimensional array that characterizes the beam in a given cross section.

The knowledge of the neutron beam phase space can serve several purposes:

- Experimental verification of numerical simulations
- Quality assessment of neutron optical components
  - Brightness of moderators and cold neutron sources
  - Average reflectivity, alignment accuracy, throughput of neutron guides
  - Selectivity, transmission of monochromators
  - Transfer functions of velocity selectors, focusing devices
- Information for corrective actions
- Input data for downstream instrument design and optimization

Direct and simultaneous measurement of the whole phase space is not possible because the detectors cannot assess the orientation of the neutron velocity. Pinhole imaging offers the possibility to measure the beam intensity distribution with respect to the neutron velocity direction in real space, for the position in the beam cross section defined by the pinhole.

Moreover, exceedingly intense beams lead to saturation or even damage of the detectors. The use of small pinholes leads to the reduction of the total flux that reaches the detector below its saturation level. The image obtained also becomes clearer as the pinhole diameter is reduced.

A third issue is that small pinholes together with narrow chopper openings allow accurate determination of the flight time, thus improving the energy/wavelength resolution.

The main advantage of pinhole imaging is that the neutron paths can be traced back to the source, including – if necessary – a few reflections.

3.1 Measurement Principle

The energy resolved pinhole imaging technique [1] enables successive measurement of three-dimensional restrictions (to the position of the pinhole in the beam cross section) of the five-dimensional phase space to be made. The neutrons originating from a pinhole situated at the position $(x_0,\ y_0)$ with respect to the reference axis z will reach the detector situated at distance l from the pinhole at a point $(x,\ y)$ with respect to the same axis (Fig. 3.1):

$$\begin{aligned} x &= x_0 + \Delta_x = x_0 + l \tan \delta_x \\ y &= y_0 + \Delta_y = y_0 + l \tan \delta_y . \end{aligned} \tag{3.1}$$

The speed components with respect to the reference axis are connected to the divergence angles by:

$$\begin{aligned} v_x &= v_z \tan \delta_x \\ v_y &= v_z \tan \delta_y \end{aligned} \quad v_z = \frac{v}{\sqrt{1 + \tan^2 \delta_x + \tan^2 \delta_y}} . \tag{3.2}$$

The neutron wavelength is determined by measurement of the flight time, t, required for the neutron to cover the distance between the source and the

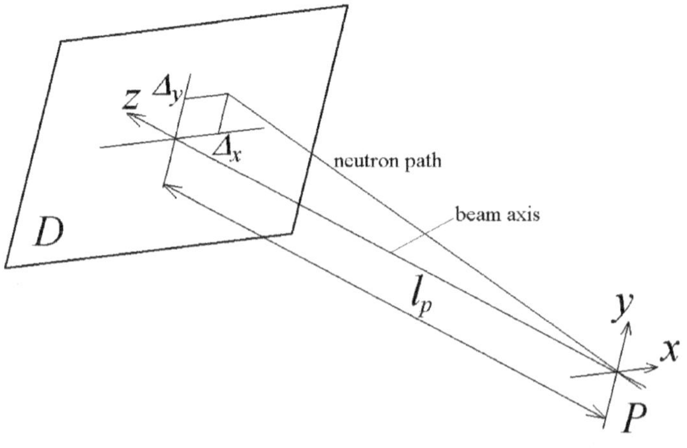

Fig. 3.1. Parameter definition and measurement principle

detector. In case of continuous sources a chopper is required to perform this measurement, and the origin of the flight distance is at the chopper. When the pinhole and the chopper are at the same position,

$$v = \frac{l}{t}\sqrt{1 + \tan^2 \delta_x + \tan^2 \delta_y} \tag{3.3}$$

$$\lambda = \frac{h}{m_\mathrm{n} v} \approx \frac{3{,}956}{v}$$

holds, where h is the Planck constant and m_n the neutron mass. The wavelength units are in Angstrom if the velocity is expressed in m s^{-1}.

The flux distribution is computed according to

$$\Phi_\mathrm{d} = \frac{I}{\eta \mathrm{d}\lambda \mathrm{d}\Omega \mathrm{d}At} k_\mathrm{ch}, \tag{3.4}$$

where η is the detector gas absorption efficiency; I the number of counts measured on $n_x \times n_y$ detector pixels in one time bin; $\mathrm{d}A$ [cm^2] the pinhole area; $\mathrm{d}\Omega = n_x \cdot n_y \cdot d^2/{l_\mathrm{p}}^2$ the solid angle corresponding to the observed area; l_p [mm] the pinhole-detector distance; d [mm] the detector pixel size; t [s] the measurement time; $\mathrm{d}\lambda = 3{,}956 \cdot t_\mathrm{d}/l_\mathrm{f}$ [Å] the incremental wavelength; t_d [ms] the length of a time bin, l_f [mm] the flight length (chopper-detector distance in case of continuous sources, moderator–detector distance in case of pulsed sources); k_ch the chopper ratio (chopper period/chopper open time): $k_\mathrm{ch} = t_\mathrm{ch}/t_\mathrm{o}$. The detector gas absorption efficiency is

$$\eta = 1 - \exp\left(-\frac{g}{\mu}\right) = 1 - \exp\left(-\frac{gp}{k}\lambda\right), \tag{3.5}$$

where p [bar] is the ^{3}He gas pressure, g [cm] its active thickness μ [cm] the gas absorption length. For $p = 2.5$ bar and $g = 3.5$ cm, the constant $k = 12.98$ cm bar Å results. Finally the formula

$$\Phi_\mathrm{d} = I \frac{l_\mathrm{f} l_\mathrm{p}^2}{3{,}956 \eta t_\mathrm{d} n_x n_y d^2 \mathrm{d}At} k_\mathrm{ch} \quad [n\ \mathrm{cm}^{-2}\ \mathrm{s}^{-1}\ \mathrm{sr}^{-1}\ \mathrm{Å}^{-1}] \tag{3.6}$$

holds for the brightness determination of the pinhole area and of the beam cross section, respectively, of the elements viewed through it.

The uncertainty of the wavelength determination is

$$\Delta\lambda = \frac{l_\mathrm{f}}{l_\mathrm{f}^2 - \frac{g^2}{4}} \left[\lambda g + 3{,}956(t_\mathrm{o} + t_\mathrm{d})\right], \tag{3.7}$$

where l_f is expressed in cm.

The angular accuracy is defined by the pinhole radius r, parallax error, the detector pixel size and the quality-filtering limit, ε, defined as the accepted error of the sum of the delay times in the two senses with respect to the delay line length:

$$\Delta\delta = \frac{r + \frac{g}{2}\tan\delta + d + \varepsilon}{l}. \tag{3.8}$$

3.2 Measurement Results

The experimental setup for beam phase space investigations at the exit of a tapered supermirror guide on a beam at the Budapest Neutron Centre is shown in Fig. 3.2. Dimensions are given in mm. CNS stands for the cold neutron source, M is a graphite monochromator tuned to 4.26 Å, G is a tapered neutron guide, Ch, P a chopper with 1 mm wide slit and a 1 mm diameter pinhole at 95 mm from chopper axis ($k_{\mathrm{ch}} = 600$) and D a ^{3}He multiwire gas chamber with 1.3 mm pixel size, $n_x \times n_y \times n_{\mathrm{t}}$ array size: 156 × 156 × 2,000, time bin length: $t_{\mathrm{d}} = 15$ ms.

The data array obtained for one position of the pinhole makes possible the determination of the restriction $f(x_0,\ y_0,\ \delta_x,\ \delta_y,\ \lambda)$ of the five-dimensional phase space $f(x,\ y,\ \delta_x,\ \delta_y,\ \lambda)$.

Figure 3.3 shows beam images for wavelength range selections, according to

$$f_1(x_0, y_0, \delta_x, \delta_y) = \int_{\lambda_1}^{\lambda_2} f(x_0, y_0, \delta_x, \delta_y, \lambda)\mathrm{d}\lambda. \tag{3.9}$$

The wavelength–horizontal divergence distribution

$$f_3\left(\delta_x, \lambda\right) = \int_{\delta y_1}^{\delta y_2} f(x_0, y_0, \delta_x, \delta_y, \lambda)\mathrm{d}\delta_y \tag{3.10}$$

for ±1.5 mrad vertical divergence range and the wavelength–vertical divergence distribution

$$f_2(\delta_y, \lambda) = \int_{\delta x_1}^{\delta x_2} f(x_0, y_0, \delta_x, \delta_y, \lambda)\mathrm{d}\delta_x \tag{3.11}$$

for ±1.5 mrad horizontal divergence range are plotted in Fig. 3.4.

It can be observed how the supermirrors come into effect as the critical angle increases with increasing wavelength. The narrow dark stripes are direct and reflected images of the guide interruption at the location of the monochromator. The large dark stripes on the 4–4.5 Å picture of Fig. 3.3, the wide vertical stripe 4 and the dark rectangles in Fig. 3.5 are due to the

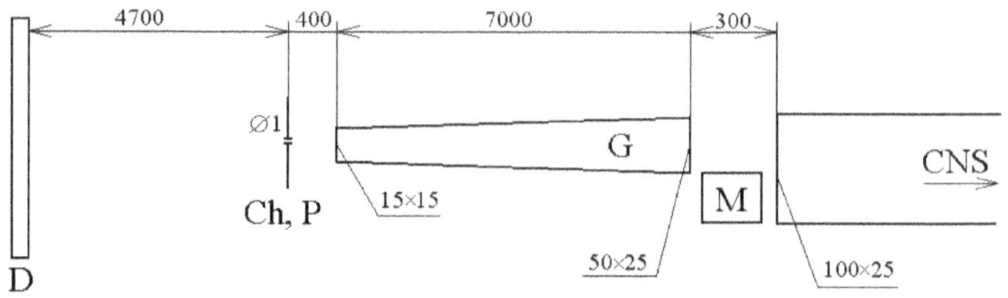

Fig. 3.2. Measurement setup

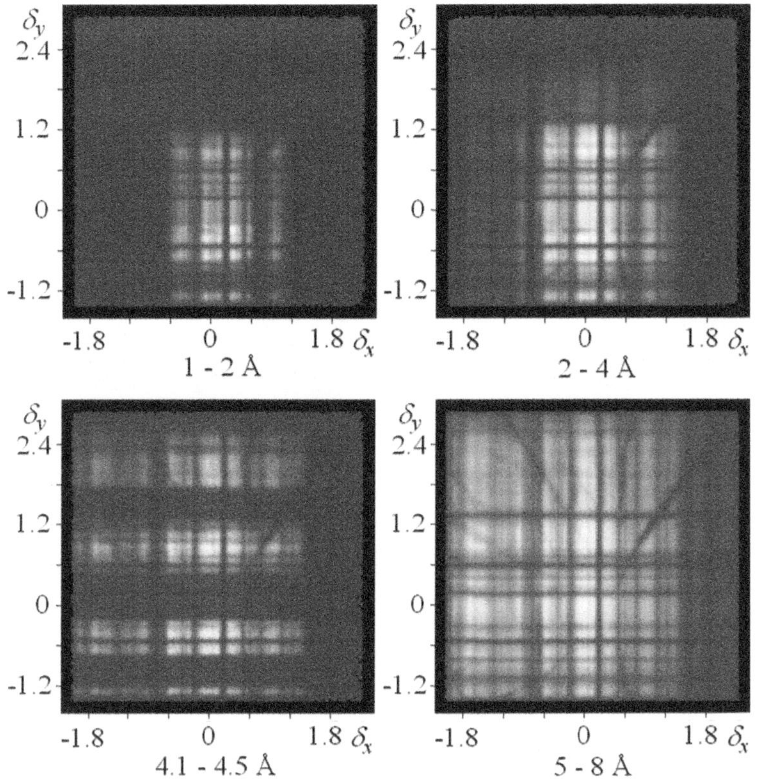

Fig. 3.3. Beam images in several wavelength ranges (angular dimensions in mrad)

monochromator which is opaque at 4.26 Å (the wavelength diverted toward a reflectometer). The $\delta_x(\lambda)$ plot shows a continuous dark stripe since its summation area lies entirely in the monochromator shadow, while the $\delta_y(\lambda)$ plot shows discrete dark rectangles because the monochromator occupies only a part of the beam cross section in vertical direction (as shown in Fig. 3.6). The wide dark horizontal stripe on the $\delta_y(\delta_x)$ plot for 4.1–4.5 Å wavelength range around the zero vertical divergence value is the direct view of the monochromator, the other wide stripes are its reflections on the guide walls.

The spectra

$$f_4(\lambda) = \int_{\delta x_1}^{\delta x_2} \int_{\delta y_1}^{\delta y_2} f(x_0, y_0, \delta_x, \delta_y, \lambda) \mathrm{d}\delta_x \mathrm{d}\delta_y \tag{3.12}$$

are plotted in Fig. 3.5 for the summation areas shown on the insert. The first- and second-order gaps caused by the monochromator are observable in the spectrum of an area lying inside the monochromator shadow (a) but missing from the spectra of areas lying outside the shadows (b and c).

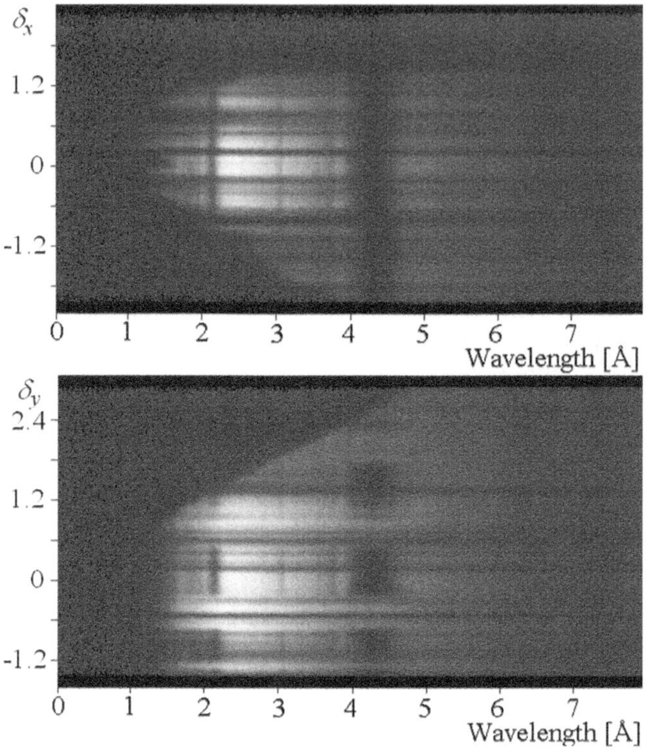

Fig. 3.4. Wavelength–horizontal divergence and wavelength–vertical divergence images

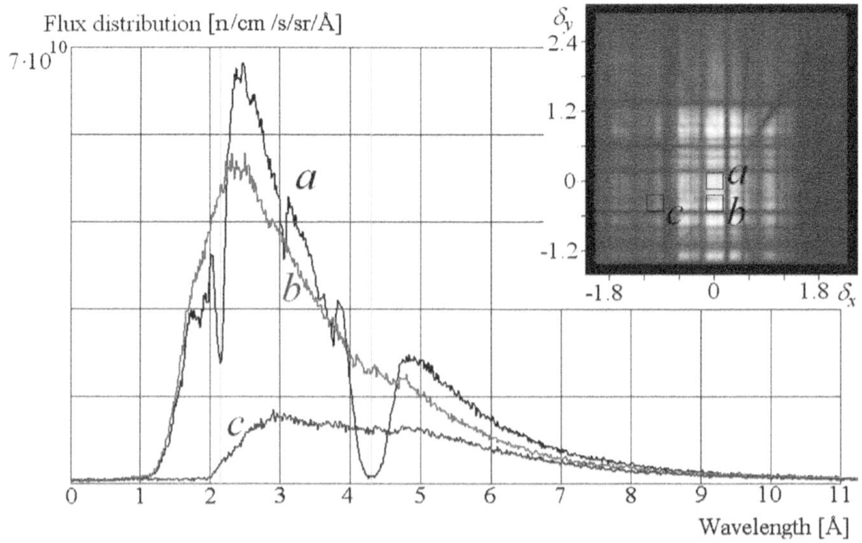

Fig. 3.5. Spectra of selected areas. (*Insert*) beam image for 0–10 Å range showing the summation areas

3.3 Neutron Guide Quality Assessment

The energy resolved pinhole imaging method has been applied for surveying the condition of the neutron beam extraction system and for determining the characteristics of the neutron moderator at the Los Alamos Neutron Scattering Center [2]. Figure 3.6 shows the setup of the experiment [3]. The dimensions are given in mm, the vertical-to-horizontal scale is 50:1. M is the moderator surface, E the FP12 neutron beam extraction guide, including a shutter, D is the position sensitive neutron detector, P a 1 mm diameter pinhole in a Cd plate. Also shown are the limits of the upper single reflection (bb) and the corresponding direct view region (d). It turned out that there was an eccentricity of 3.5 mm in the horizontal and 4.5 mm in the vertical direction of the pinhole with respect to the neutron guide axis. The neutron guide inner cross-sectional area is $9.5 \times 9.5\,\mathrm{cm}^2$.

The experiment can be considered to be an in situ reflectometry measurement. To each pixel (x_1, y_1) of the reflection region corresponds a pixel (x_0, y_0) of the direct view. The corresponding ray tube – not to scale – is plotted in dark gray. The ratio

$$R = \left.\frac{I\,(x_1, y_1, \lambda)}{I\,(x_0, y_0, \lambda)}\right|_{\lambda \in (\lambda_1, \lambda_2)} \tag{3.13}$$

is the reflectivity at the relative momentum transfer ($q/q^{\mathrm{c}}{}_{\mathrm{Ni}}$ where q is the momentum transfer of the mirror, $q^{\mathrm{c}}{}_{\mathrm{Ni}}$ is the critical momentum transfer of natural Ni) given by

$$m = 20\frac{\theta}{\lambda_1 + \lambda_2}. \tag{3.14}$$

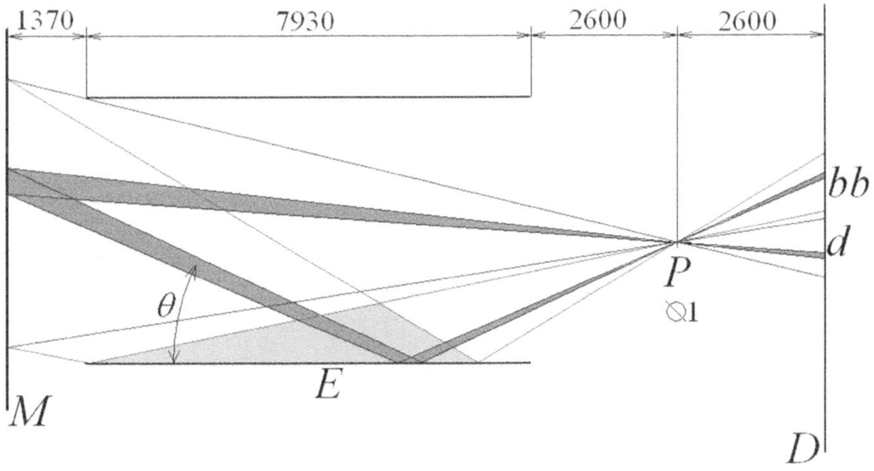

Fig. 3.6. Experimental setup at the Los Alamos Neutron Scattering Center

The wavelength range in (3.14) is selected as small as allowed by the number of neutrons per pixel required to ensure acceptable statistics.

The measurement gives an average reflectivity over the mirror surfaces viewed through the given pinhole position. By moving the pinhole in the beam cross section another average reflectivity curve can be determined. The results are affected by the alignment accuracy of the neutron guide sections because the angle θ is determined with respect to the theoretical position of the mirror surface. The insert in Fig. 3.7 shows the image of the neutron beam extraction guide system. Area d is a direct view of the moderator, in areas aa, ab, bb, and ba the neutrons had one reflection, while in areas ac, ad, bc, and bd two reflections. The outer frame size is 142×142 pixels, the detector pixel size in this experiment is 0.75 mm.

The resulted average reflectivity curve is plotted in Fig. 3.7. In reality, the reflectivity of the individual mirror plates is better than that shown by these results (about 80% at $m = 3$). Dark stripes in the Fig. 3.7 insert are due to the imperfect alignment of the FP12 shutter section, which reduces considerably the average reflectivity.

There is a further possibility to assess the global quality (reflectivity, alignment accuracy) of a neutron guide: to determine its transfer function – the ratio between the incoming and transmitted flux. For this the flux vs. wavelength has to be determined at both the entrance and exit cross section of the guide section investigated. When the direct view of the source is available, the ratio between the brightness of the guide exit and the brightness of the direct

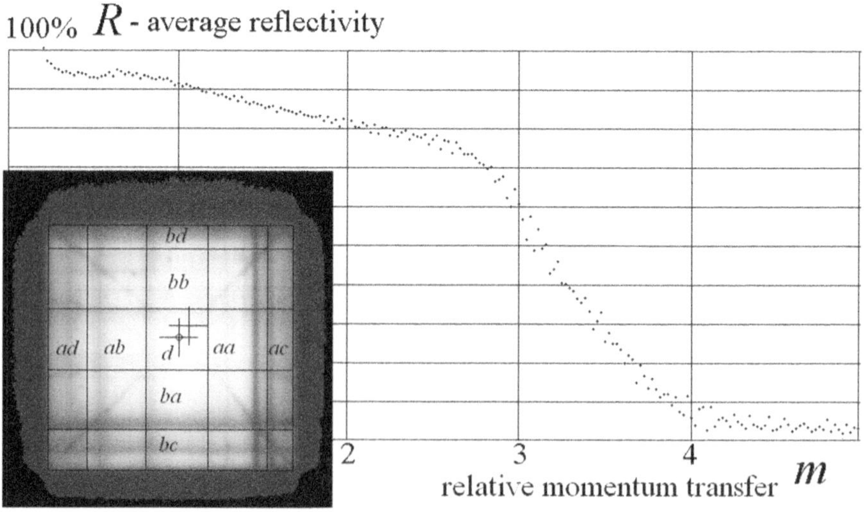

Fig. 3.7. Average reflectivity of the neutron guide supermirror plates at LANSCE. *Inset*: Image of the neutron beam extraction guide system of the FP12 beamline. Wavelength range: 2.74–9.55 Å

view area (Fig. 3.8) gives the guide transfer function for the given position of the pinhole (Fig. 3.9). Repeating the measurement for various positions of the pinhole in the beam cross section the average transfer function can be determined as the mean of the individual results.

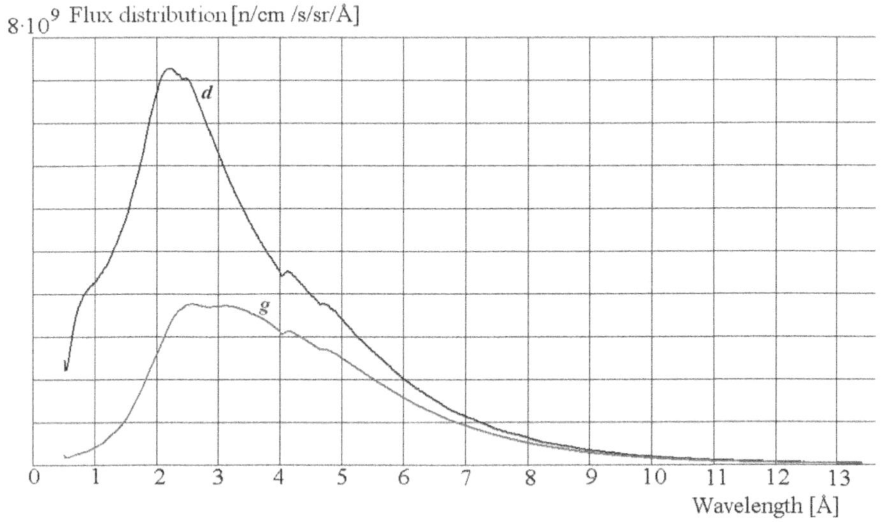

Fig. 3.8. Spectra of the direct view (g) and the entire guide exit (d), showing detected flux, not corrected for gas absorption efficiency, or for detector saturation

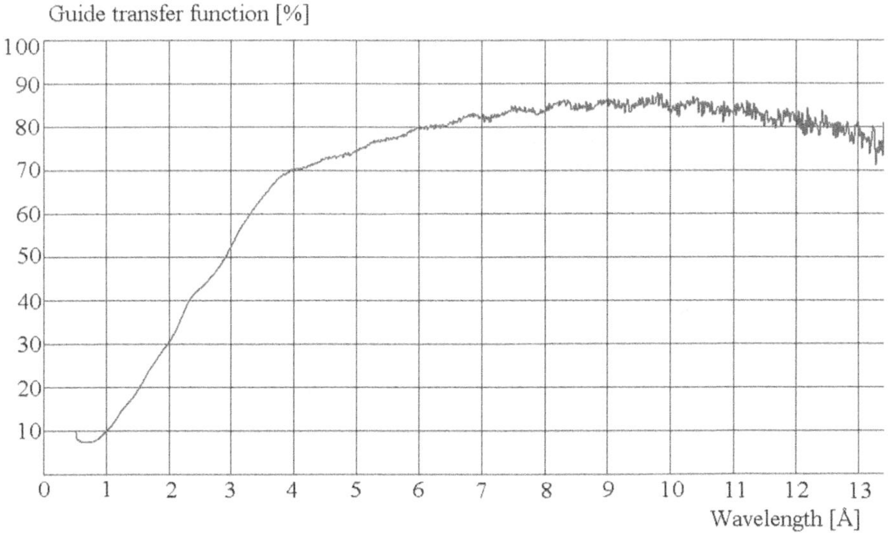

Fig. 3.9. The transfer function of the neutron beam extraction guide system of the FP12 beamline at LANSCE

3.4 Transfer Function of a Velocity Selector

The method is applicable for determination of transfer functions of devices that select a certain range of the incoming beam phase space. As an example a multidisk rotational velocity selector [4] has been investigated. The results obtained for a selector tilted by 2° with respect to the beam axis in azimuthal direction (Figs. 3.10a, b – to be compared to Fig. 3.4 for the incoming beam) show the phase space configuration of the transmitted beam [5].

The beam cross-section image is shown in Fig. 3.10c and the spectra of incoming and transmitted (selected) beam are plotted in Fig. 3.10d.

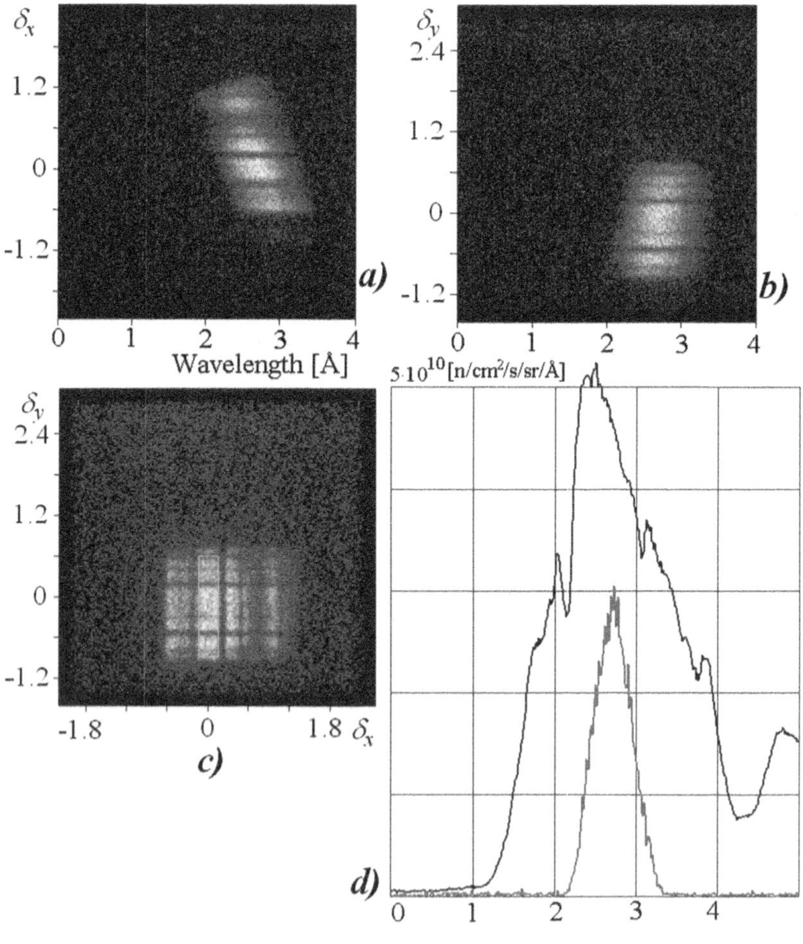

Fig. 3.10. Divergence-wavelength plots (**a** and **b**) of a beam transmitted by a velocity selector, beam cross-section image (**c**) and spectra of incoming and transmitted (selected) beam (**d**)

The method presented offers the means for experimental measurement of the transfer function of velocity selectors (monochromators in general) – the extent to which certain domains of the incoming beam phase space are transmitted. This also allows a quality assessment of the devices to be made. For example, the effect of any false channel due to misalignment of the disks of a selector can be detected.

3.5 Moderator Brightness Evaluation

The reconstruction of a 100 mm high and 25 mm wide cold neutron beamline at BNC offered the possibility to perform a measurement at a position where direct view of the moderator was available.

Figure 3.11a shows the detected images in two wavelength ranges. The epithermal neutrons, shining through both the pinhole mask and the closed chopper disk, outline the footprint of the guide exit. The chopper period has

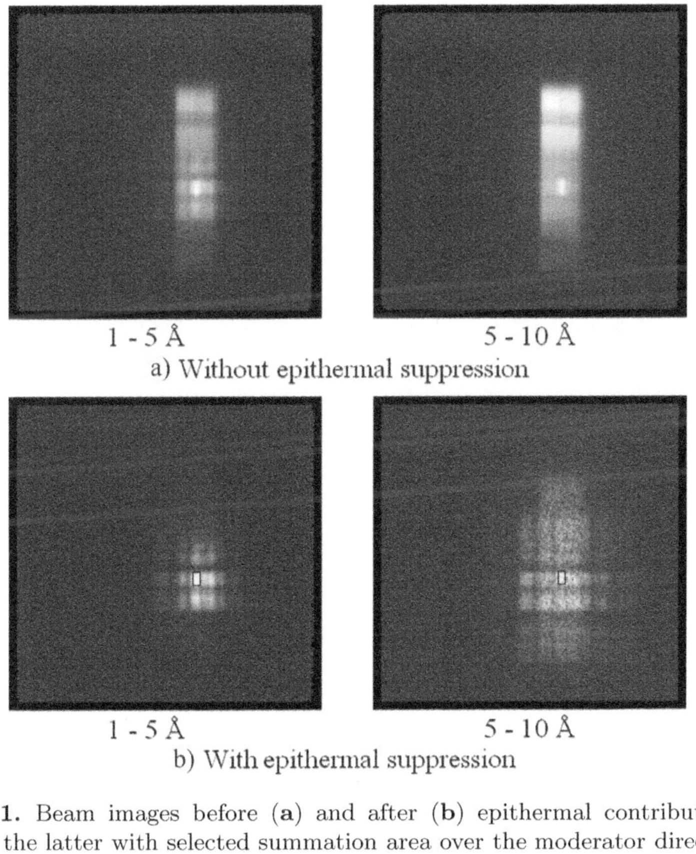

Fig. 3.11. Beam images before (**a**) and after (**b**) epithermal contribution suppression, the latter with selected summation area over the moderator direct view

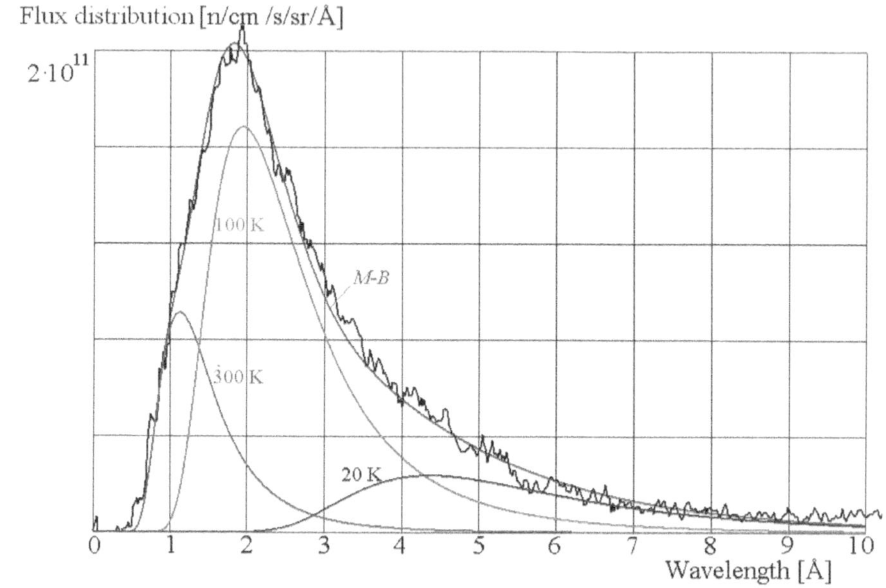

Fig. 3.12. Cold moderator spectrum with Maxwell–Boltzmann distribution fit

been set such that it included a time range where no cold neutrons reached the detector. The events acquired in this time are due to the epithermal neutrons only and their average for every pixel can be subtracted from the raw data at each time bin, allowing the suppression of the epithermal contribution.

The results can be seen in Fig. 3.11b. The horizontal dark lines in Fig. 3.11b are due to an incompletely opened shutter. The brightest region is the direct view of the BNC cold source. The spectrum of the cold source together with the Maxwell–Boltzmann distribution fit is plotted in Fig. 3.12. The Maxwell–Boltzmann distribution of the cold source brightness is

$$\Phi_\lambda = 2\Phi \frac{\lambda_T^4}{\lambda^5} \exp\left(-\frac{\lambda_T^2}{\lambda^2}\right); \quad \lambda_T = \frac{h}{\sqrt{2mkT}} = \frac{30.81}{\sqrt{T}} (\text{Å}), \tag{3.15}$$

with peak coordinates:

$$(\lambda_p, \Phi_{\lambda p}) = \left(\sqrt{\frac{2}{5}}\lambda_T, \frac{25\sqrt{5}}{2\sqrt{2}} \frac{\Phi}{\lambda_T} \exp(-5/2)\right). \tag{3.16}$$

It can be observed that a sum of three distributions is necessary to obtain a proper fit, one at room temperature corresponding to the contribution of thermal neutrons, one at 20 K (the temperature of the cold source) and one at 100 K due to incomplete moderation in the finite cold moderator.

3.6 Conclusions

Energy resolved pinhole imaging offers the possibility to map the neutron beam phase space with good spatial, angular, and wavelength resolution. Direct images of bright sources can be obtained and significant information gained on the status of neutron optical systems hardly accessible by other means of investigation. This is especially the case of moderators and beam extraction systems, both new ones and systems which have been in operation for some time.

Investigation of energy dependent features of neutron optical elements is possible as well as the determination of brightness, uniformity, and spectrum of neutron sources.

References

1. J. Füzi, Physica B **385–386**, 1253 (2006)
2. P.-N. Seo, J.D. Bowman, M. Gericke, G. Greene, J. Long, G.S. Mitchell, S.I. Penttila, W.S. Wilburn, NIM A **517**, 285 (2004)
3. J. Füzi, E. Dávid, T. Kozlowski, P. Lewis, G. Messing, F. Mezei, S. Penttila, L. Rosta, M. Russina, Gy. Török, Physica B **385–386**, 1315 (2006)
4. L. Rosta, Physica B **174**, 562 (1991)
5. L. Rosta, J. Füzi, L. Hományi, Physica B **385–386**, 1283 (2006)

4

Raytrace of Neutron Optical Systems with RESTRAX

J. Šaroun and J. Kulda

Abstract. RESTRAX is an advanced software package for raytrace simulations of neutron optical systems. This chapter reviews the techniques used by its efficient raytrace code, namely sampling strategy, numerical optimizations and details of the random-walk technique used to simulate transport through neutron optical components. In addition, raytrace simulations of resolution functions and model spectra for three-axis neutron spectrometers are briefly described.

4.1 Introduction

Raytrace simulation permits realistic modeling of neutron optics components as well as of complete neutron scattering instruments. While development of neutron raytrace software goes back to the 1970s (MCLIB library, [1]), it has been significantly accelerated by the advent of powerful desktop computers in 1990s and has lead to the creation of several software packages capable of realistic modeling of most of modern neutron optics devices. They include the programs NISP [2] based on the MCLIB library, IDEAS [3], McStas [4], VITESS [5] or RESTRAX [6]. Of course, Monte Carlo (MC) simulation always involves a trade-off between the level of physical reality implemented in the description of neutron transport and computing speed. Consequently, these programs differ in both the physical models underlying simulation of particular components and the structure of their code, depending on the different purposes for which they have been written. While some put emphasis on modularity (McStas, VITESS, NISP), which permits one to easily incorporate new components and to test new ideas of experimental techniques, others, like RESTRAX, trade part of their flexibility for a highly efficient sampling strategy permitting to gain several orders of magnitude in computing speed.

In the first part of this chapter, we review some principles and methods, upon which the efficient raytrace code used by the RESTRAX package[1] is based, namely sampling strategy, numerical optimizations, and details of the

[1] The RESTRAX package is available at http://omega.ujf.cas.cz/restrax

random-walk technique used to simulate neutron transport through neutron optical components. This part is followed by a brief description of the ray-tracing method used to simulate resolution functions of three-axis neutron spectrometers.

4.2 About the RESTRAX Code

The RESTRAX package actually consists of two independent modules. One is designed specifically to simulate three-axis spectrometers (TASs) and includes tools for simulation of resolution functions, convolutions, data fitting, and a set of modules implementing different scattering functions. The other one, called SIMRES, is equipped with a raytrace code which simulates instrument components on a more detailed level and provides more flexibility in varying instrument configuration. In addition, functions for parameter space mapping and numerical optimizations are provided, which makes this module a useful tool for designing novel neutron optical components and instrument configurations.

4.2.1 Instrument Model

The instrument model has a closed structure of a three-axis neutron spectrometer with a fixed number and succession order of the optical elements (Fig. 4.1). Each of the elements is, however, described in considerable detail

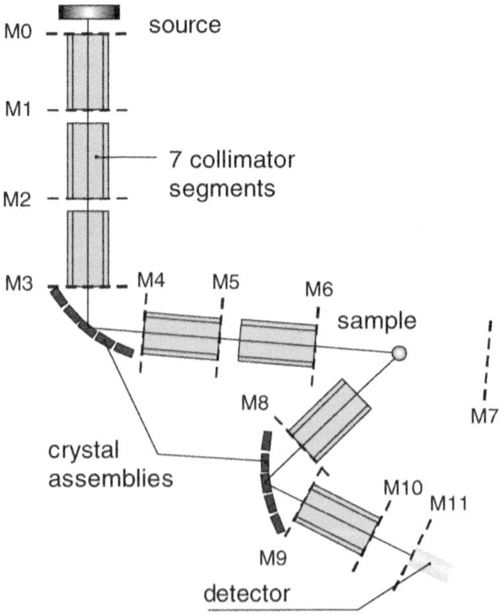

Fig. 4.1. Instrument layout in SIMRES with positions of beam monitors

and some offer a high degree of flexibility, e.g., a collimator with suitably adjusted parameters can be used to describe a neutron guide, Soller collimator, polarizing bender, or a multichannel supermirror focusing guide. As a result, with corresponding simplifications RESTRAX can simulate many types of conventional instruments at steady-state or pulsed neutron sources, except for time-of-flight aspects. One can thus estimate performance of newly developed neutron optical devices realistically as a part of the whole instrument where they are intended to be used. The program relies on lookup tables for the description of properties which are too difficult or time consuming to calculate in real time. These include tables for source flux distribution and reflectivity of supermirror coatings. In this way, simulations can be made very realistically without significantly decreasing computing speed provided, of course, that we have reliable experimental data or calculations for generating the lookup tables.

4.2.2 Sampling Strategy

An efficient sampling strategy is a crucial part of any raytrace code, which can increase computing speed by orders of magnitude without any cost in terms of accuracy and physical relevancy of the results obtained. In RESTRAX, optimization is done at several levels.

First, raytrace can be started at any point and followed either downstream or upstream, yielding equivalent results due to the symmetry of underlying transport equations [7, 8]. In most experiments, the sample dimensions are much smaller than those of any other spectrometer component, so that the sample represents a real bottleneck for the neutron trajectories. Starting the simulation at the sample permits, hence, to significantly reduce the part of rejected events.

Second, the initial estimation of sampling intervals and correlations between phase-space variables is made from the known instrument configuration using analytical formulas. They involve, in particular, the Bragg condition at the (possibly curved) monochromator and analyzer crystals and an assumption of a point-like sample, thereby imposing correlations between spatial and momentum coordinates of neutrons, which are likely to pass through the instrument.

Third, an adaptive covariance estimation technique [9] is employed to refine the initial analytical estimation of the sampling intervals. The phase-space coordinates and other random quantities (e.g., scattering angles or penetration depth in a crystal) determining the neutron trajectory are taken from a common array $\mathbf{Y} \equiv (\mathbf{k}_\mathrm{i},\ \mathbf{r},\ \mathbf{k}_\mathrm{f}, \ldots)$, generated by linear transformation of an array of random numbers $\mathbf{X}$ uniformly distributed over the range $(-0.5; 0.5)$, $\mathbf{Y} = \mathbf{M}{\cdot}\mathbf{X}{+}\mathbf{Y}_0$. The optimum choice of the transformation matrix $\mathbf{M}$ has to lead to a sampling volume $\det(\mathbf{M})$, which is as small as possible in order to minimize the number of wasted events. On the other hand, it must be large enough to intercept all points, which may lead to a successful event

Table 4.1. Comparison of simulation speed for different levels of sampling optimization [10]

Sampling optimization[a]	Intensity (rel. units)	Peak width (°)	CPU time (ms per event)
$1+2+3$	652 ± 3	0.287	2.0
$1+3$	649 ± 3	0.277	2.6
$1+2$	655 ± 4	0.277	3.5
1	650 ± 4	0.284	18.3
$2+3$	655 ± 3	0.283	237.1

[a] Sampling optimization methods: (1) upstream tracing from the sample to the source, (2) initial setting of known phase-space correlations, (3) adaptive covariance estimation

(completed neutron history from the source to the end monitor). Initially, **M** is estimated from the known instrument configuration and correlations, e.g., due to the Bragg conditions for the crystals. After a certain number (~500) of completed histories, **M** is recalculated using the covariance matrix $\langle Y_i Y_j \rangle$. To check that the sampling volume is not too small, a "safety pool" near the limits of sampling intervals is defined. If any successful event is generated from a point inside the safety pool, the simulation is automatically restarted with extended limits. Though this procedure usually calls for several automatic restarts of the simulation before the optimum sampling volume is found, it substantially decreases the total accumulation time.

The importance of sampling strategy is illustrated in Table 4.1 comparing the CPU time per successful event needed for the simulation of a diffraction line on the powder diffractometer D1A, ILL Grenoble. The statistical accuracy of the results is equivalent in all cases, but the simulation speed differs substantially for different combinations of the three sampling strategies. Obviously, the effect of upstream tracing is enormous due to the small gauge volume (~10 mm^3) in this particular case.

4.2.3 Optimization of Instrument Parameters

An efficient sampling strategy permits one to simulate even complex instruments with considerable speed, often exceeding 10^3 of complete neutron histories per second. Under such circumstances, it is feasible to use the ray-tracing code as a generator of the cost function in numerical optimizations. The choice of a suitable optimization algorithm depends on the number of free parameters and character of the cost function. RESTRAX implements the Levenberg–Marquardt method, which is well suited to problems with few free parameters and a well resolved single extremum of the cost function. Despite of these restrictions, this method proves to be useful in practice since only mutually correlated parameters need to be optimized simultaneously, and a reasonably good first guess of the optimum position can often be made on

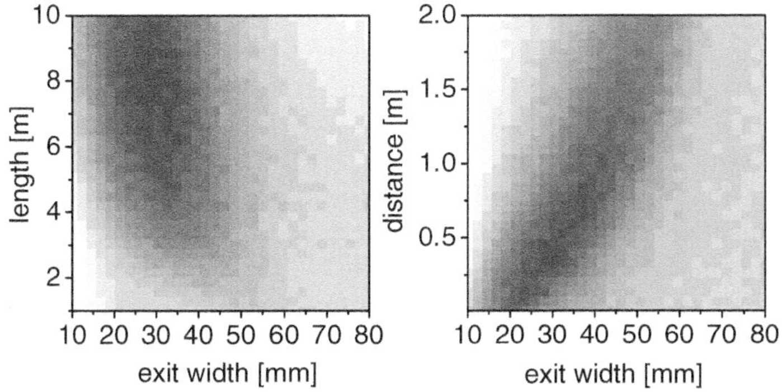

Fig. 4.2. Maps of simulated neutron intensity at the sample after a 100 m long curved neutron guide followed by an elliptically shaped tapered section. The maps are taken in two different sections through the space of the elliptic guide parameters

the basis of physical considerations. As an example, Fig. 4.2 shows two maps of simulated neutron intensity for a system consisting of a 100 m long supermirror guide with a tapered elliptically shaped guide at the end for focusing the neutrons at a sample. The maps are taken at two different section planes in the space of instrument parameters including the elliptic guide width, length, and distance from the sample. They give clear idea of the extent and the position of the optimum in parameter space, although accumulation of such maps by neutron raytrace method is very costly in terms of CPU time. Therefore they cannot be used to find a global optimum for more than two parameters. On the other hand, the iterative optimization procedure can be applied to higher number of free parameters (in practice less than about ten). For our example (Fig. 4.2) with the third free parameter (guide length) added, the optimum was found in less than 20 min on an ordinary desktop computer.

With a growing number of free parameters of the component to be optimized, gradient methods typically fail due to (a) increasing computing time and (b) a tendency to be trapped at local extrema. Recently, methods employing genetic algorithms and particle swarm optimization techniques [11] have been successfully used in a search for the optimum design of neutron optics components, particularly neutron guides, with several tens of free parameters involved. These methods are much less affected by the problem of local extrema and are less bogged down as the number of free parameters increases than are gradient algorithms.

4.3 Simulation of Neutron Optics Components

4.3.1 Neutron Source

The raytrace simulation of a neutron spectrometer does not need to involve propagation of neutrons in a moderator, which is a task for other specialized programs like MCNP [12] or TRIPOLI [13]. Instead, the neutron source is represented by a rectangular, circular, or elliptical area at the interface between moderator and neutron channel with an associated neutron flux distribution. In RESTRAX, the neutron flux is described either analytically as a Maxwellian distribution, or more accurately by a lookup table. In the latter case, a one-dimensional table with wavelength distribution is combined with two-dimensional tables describing correlations between angular and spatial coordinates. Such a table can be easily created by postprocessing of moderator simulation data, which results in a much more realistic model compared to the analytical description and allows for simulations of neutron fluxes on absolute scale.

4.3.2 Diffractive Optics

Simulation of neutron transport through crystals in RESTRAX is based on a random-walk algorithm, which solves intensity-transfer Darwin equations [8] numerically, in principle for any shape of the crystal block. Details of the algorithm are described in [14]. It is based on the assumption of dominant effect of the mosaic structure on the rocking curve width, where mosaic blocks are treated as perfect crystal domains. However, the random walk is not followed through individual mosaic blocks, which would be an extremely slow process in some cases. Instead, the crystal is characterized by the scattering cross section per unit volume, $\sigma(\varepsilon)$, which depends on the misorientation angle, ε of a mosaic block as

$$\sigma(\varepsilon) = Q\eta^{-1}g(\varepsilon/\eta), \tag{4.1}$$

where η is the width of the misorientation probability distribution, $g(x)$ and Q stands for the kinematical reflectivity. The diffraction vector depends on the misorientation angle and, in the case of gradient crystals, also on the position in the crystal, which can be expressed as

$$\mathbf{G}(\mathbf{r}) = \mathbf{G}_0 + \nabla\mathbf{G} \cdot \mathbf{r} + G(\boldsymbol{\varepsilon} + \boldsymbol{\gamma}), \tag{4.2}$$

where the second term describes a uniform deformation gradient and the third one the angular misorientation of a mosaic block parallel ($\boldsymbol{\varepsilon}$) and perpendicular ($\boldsymbol{\gamma}$) to the scattering plane defined by $\mathbf{G}_0$ and incident beam directions. For a neutron with given phase-space coordinates, $\mathbf{r}$, $\mathbf{k}$, we can write the Bragg condition in vector form as

$$[\mathbf{k} + \mathbf{G}_0 + \nabla\mathbf{G} \cdot (\mathbf{r} + \mathbf{k}\tau) + G(\boldsymbol{\varepsilon} + \boldsymbol{\gamma})]^2 - k^2 = 0, \tag{4.3}$$

where $\mathbf{k}\boldsymbol{\tau}$ is the neutron flight-path from a starting point at $\mathbf{r}$. For the random-walk simulation, we need to find an appropriate generator of the random time-of-flight, $\boldsymbol{\tau}$. By neglecting second-order terms in (4.3), we obtain a linear relation between ε and the time-of-flight parameter, $\boldsymbol{\tau}$,

$$\varepsilon = \varepsilon_0 + \boldsymbol{\beta} k \boldsymbol{\tau}, \tag{4.4}$$

where $\varepsilon_0 = \dfrac{(\mathbf{k} + \mathbf{G}_0 + \nabla\mathbf{G} \cdot \mathbf{r} + G\gamma)^2 - k^2}{2Gk\cos\theta_\mathrm{B}}$ and $\beta = \dfrac{(\mathbf{k} + \mathbf{G}_0) \cdot \nabla\mathbf{G} \cdot \mathbf{k}}{Gk^2 \cos\theta_\mathrm{B}}$. (4.5)

Substitution for ε in (4.1) then leads to a position-dependent scattering cross-section, which, by integration along the flight path, yields the probability, $P(\tau)$ that a neutron will be reflected somewhere on its flight path $k\tau$. With the symbol $\Phi(\varepsilon/\eta)$ denoting the cumulative probability function corresponding to the mosaic distribution $g(\varepsilon/\eta)$, we can express this probability as [15]

$$P(\tau) = 1 - \exp\left[-\frac{Q}{\beta}\left(\Phi\left(\frac{\varepsilon_0 + \beta k\tau}{\eta}\right) - \Phi\left(\frac{\varepsilon_0}{\eta}\right)\right)\right]. \tag{4.6}$$

Provided that we know the inverse function to $\Phi(x)$, we can generate τ by transformation from uniformly distributed random numbers, ξ. Let τ_o be the time-of-flight to the crystal exit. Then the next node (scattering point) of the random walk would be

$$\tau = \frac{\eta}{k\beta}\Phi^{-1}\left[\Phi\left(\frac{\varepsilon_0}{\eta}\right) - \frac{\beta}{Q}\ln\left(1 - \xi P(\tau_0)\right)\right] - \frac{\varepsilon_0}{k\beta}, \tag{4.7}$$

while the neutron history has to be weighted by the probability $P(\boldsymbol{\tau}_0)$. In subsequent steps, the random walk continues in the directions $\mathbf{k} + \mathbf{G}(\tau)$ and $\mathbf{k}$ until the neutron escapes from the crystal (or an array of crystals) or the weight of the history decreases below a threshold value. Absorption is taken into account by multiplying the event weight by the appropriate transmission coefficient calculated for a given neutron wavelength and material [16]. In Fig. 4.3, such a random walk is illustrated by showing points of second

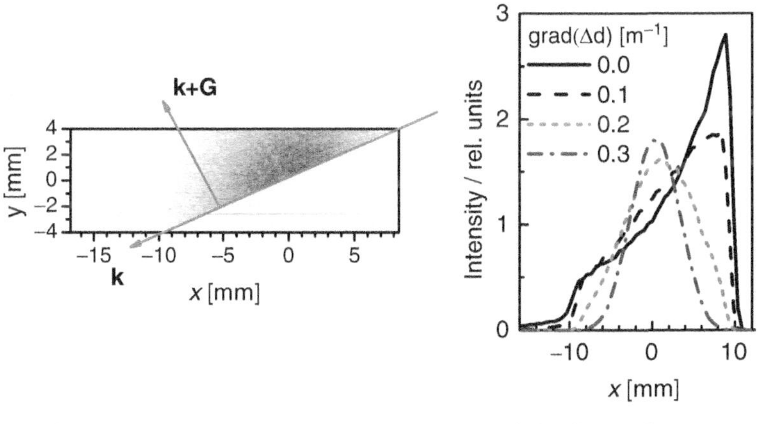

Fig. 4.3. A map of simulated points of second and further reflections inside a Ge crystal, reflection 511, mosaicity $\eta = 6'$, and deformation corresponding to a temperature gradient along y-axis, $|\nabla\mathbf{G}|/G = 0.1\,\mathrm{m}^{-1}$. On the right hand, simulated spatial profiles of reflected neutron beam are plotted for different magnitudes of the deformation gradient

and further reflections in a deformed mosaic Ge crystal and the resulting topography of the reflected beam.

There are two important aspects of this procedure. First is the efficiency, because for usual mosaic crystals, only few steps are made in each history resulting in a very fast procedure. Second, both mosaic and bent perfect crystals can be simulated by the same algorithm. Indeed, in the limit $\eta \to 0$, we obtain $\tau = -\varepsilon_0(k\beta)^{-1}$ and the neutron transport is deterministic, as expected for elastically bent crystals in the quasiclassical approximation [17]. In addition, the weight factor in this case, $P(\infty) = 1 - \exp(-Q|\beta|^{-1})$, is identical to the quantum-mechanical solution for the peak reflectivity of bent perfect crystals [18]. On the other hand, this model fails in the limit of perfect crystals (very small mosaicity and deformation), which would require another approach using dynamical diffraction theory.

The crystal component is flexible enough for modeling most of the contemporary neutron monochromators and analyzers as far as they can be described as a regular array of crystal segments with a linear positional dependence of tilt angles. More sophisticated multianalyzers (e.g., the RITA spectrometer [19]) featuring independent movements of individual segments can only be simulated in a step-by-step manner with the final result being obtained by a superposition of the partials.

4.3.3 Reflective Optics

Raytrace of neutrons through various types of reflecting optics elements is a straightforward task, provided that we can treat the problem in the framework of the geometrical optics approximation and that we know the reflectivity function of the reflecting surfaces. With neutrons, the geometrical approximation is fully adequate for the simulation of transport through elements such as neutron guides or benders and the reflectivity of real Ni and supermirror coatings can be determined experimentally. Mirror reflectivity can be thus stored in lookup tables and the problem is reduced to a geometrical description of the device, apart from computing issues related to numerical precision and convergence problems. Using this approach, RESTRAX can simulate various neutron optics elements, such as curved neutron guides, benders, elliptic or parabolic multichannel guides and most recently also supermirror transmission polarizers.

As an example, we present the simulation of multichannel supermirror guides aimed to focus neutrons onto small samples after passing through a doubly focusing monochromator [20]. Although RESTRAX can simulate two-dimensional grids of reflecting lamellae, for practical reasons we have considered a multichannel device as a sequence of one-dimensional horizontally and vertically focusing sections (Fig. 4.4). Equidistant 0.5 mm thick blades were assumed to be curved either elliptically or parabolically, having reflecting surfaces on the concave sides with the reflectivity of an $m = 3$ supermirror.

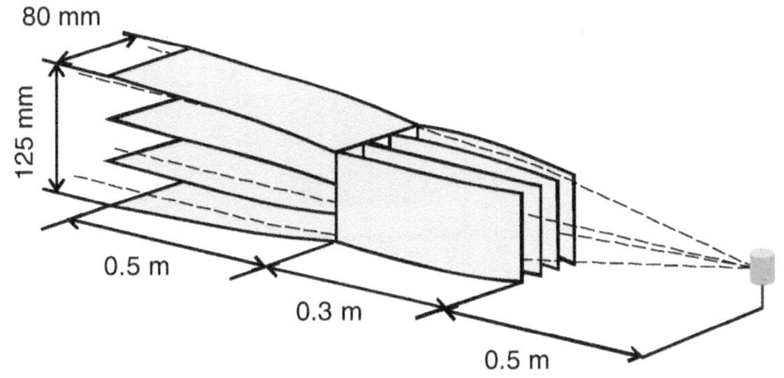

Fig. 4.4. The multichannel supermirror device with the dimensions indicated

For elliptic guides, the number of blades was 20 and 30 for horizontal and vertical focusing, respectively. For the parabolic guide, the respective numbers were 14 and 22. Gaps between the blades and focal distances were defined by entrance and exit widths (or heights) of the guides. We have assumed that the entrance dimensions are equal to the ellipse minor axis in the case of elliptic profile. The simulations involved the entire beam path including a cold source with a tabulated flux distribution, straight ^{58}Ni neutron guide with cross section $6 \times 12\,\mathrm{cm}^2$ and a doubly focusing PG002 monochromator with 7×9 segments at the nominal wavelength 0.405 nm. A lookup table with the measured reflectivity of a real $m = 3$ supermirror was used to achieve a realistic description of the guide properties. Except for the multichannel guide and the horizontally focusing monochromator, the instrument layout corresponded to the IN14 spectrometer at the Institut Laue-Langevin in Grenoble.

It is quite difficult to optimize the parameters of such a device analytically, because it is not obvious how the focusing by the monochromator and the multichannel guide would link to each other and also what the penalty in terms of neutron transmission through the guide and what the effect of the relaxed instrument resolution would be. Some of the relevant parameters (crystal curvatures, guide focal lengths, and spacing between the lamellae) were optimized using the raytrace code and Levenberg–Marquardt techniques implemented in RESTRAX [20]. The results for an optimized parabolically shaped multichannel guide are shown in Fig. 4.5. In contrast to an experiment, Monte Carlo simulation permits one to investigate the beam structure in different phase-space projections quite readily. For example, a projection in the plane of divergence angle and wave-vector magnitude can clearly resolve the directly transmitted and reflected neutrons due to their different dispersion relation, resulting from prior reflection on the monochromator. This effect is entirely hidden in other projections, as illustrated in Fig. 4.5.

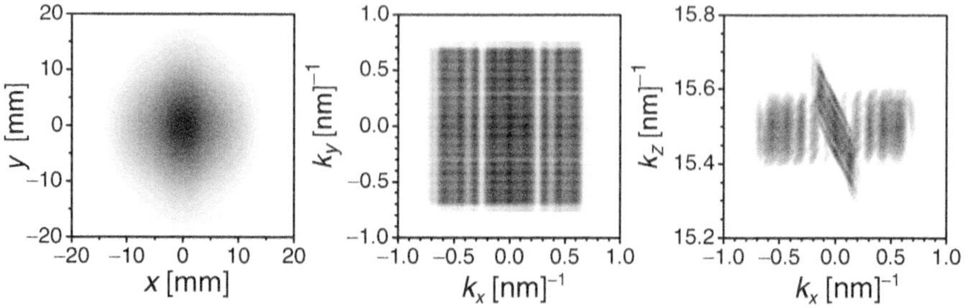

Fig. 4.5. Simulated beam profiles at the sample in different real and momentum space projections for the optimized parabolic guide. The right-hand image permits one to easily distinguish directly transmitted neutrons in the central part from the reflected ones, due to their inverted dispersion relation

4.4 Simulations of Entire Instruments

Ultimately, the matter of concern is in simulations of the entire neutron scattering instrument, which provide data relevant for instrument design and data analysis, such as neutron flux, beam structure in phase-space or resolution functions. Examples of RESTRAX applications in instrument development can be found in the literature [21–26]. In the following section, we give a brief summary of the raytrace method used to simulate TAS resolution functions.

4.4.1 Resolution Functions

The intensity of a neutron beam scattered by the sample with a probability $W(\mathbf{k}_\mathrm{i},\ \mathbf{k}_\mathrm{f})$ and registered by the detector in a TAS configuration with the nominal settings of initial and final wave-vectors, $\mathbf{k}_{\mathrm{i}0}$, $\mathbf{k}_{\mathrm{f}0}$, is given by

$$I(\mathbf{k}_{\mathrm{i}0}, \mathbf{k}_{\mathrm{f}0}) = \int W(\mathbf{k}_\mathrm{i}, \mathbf{k}_\mathrm{f}) \Phi_\mathrm{I}(\mathbf{r}, \mathbf{k}_\mathrm{i}) P_\mathrm{F}(\mathbf{r}, \mathbf{k}_\mathrm{f}) \mathrm{d}\mathbf{r} \mathrm{d}\mathbf{k}_\mathrm{i} \mathrm{d}\mathbf{k}_\mathrm{f}. \tag{4.8}$$

The function $\Phi_\mathrm{I}(\mathbf{r}, \mathbf{k}_\mathrm{i})$ represents the flux distribution of incident neutrons at a point $\mathbf{r}$ inside the sample while $P_\mathrm{F}(\mathbf{r}, \mathbf{k}_\mathrm{f})$ is the distribution of probability that the neutron with phase-space coordinates $(\mathbf{r}, \mathbf{k}_\mathrm{f})$ is detected by the analyzer part of the instrument. Evaluation of this integral by the MC method is advantageous for two reasons: the high dimensionality of the integral and the fact that the latter two distributions in the integrand can be sampled directly by the raytrace technique. For this purpose, we set the scattering probability of the sample $W(\mathbf{k}_\mathrm{i}, \mathbf{k}_\mathrm{f}) = 1$. The instrument response function is then obtained as an ensemble of $(\mathbf{k}_{\mathrm{i,e}}, \mathbf{k}_{\mathrm{f,e}})$ vectors and their weights, p_e, which describe all possible scattering events detected by the instrument. They have the distribution given by the integral

$$R(\mathbf{k}_\mathrm{i}, \mathbf{k}_\mathrm{f}) = \int \Phi_\mathrm{I}(\mathbf{r}, \mathbf{k}_\mathrm{i}) P_\mathrm{F}(\mathbf{r}, \mathbf{k}_\mathrm{f}) \mathrm{d}\mathbf{r}. \tag{4.9}$$

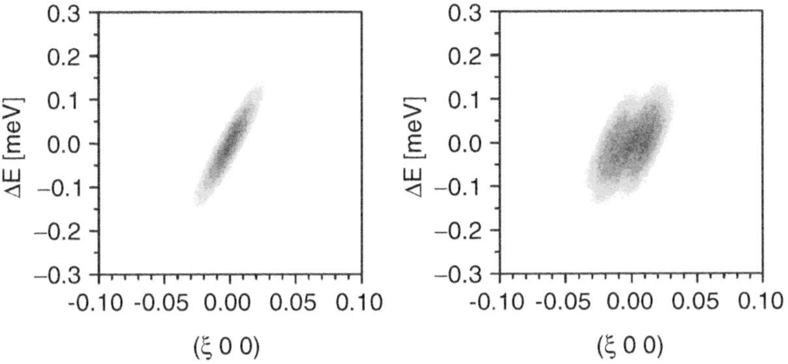

Fig. 4.6. Resolution functions of the whole TAS instrument without (*left*) and with (*right*) the multichannel guide. The center of the resolution function corresponds to elastic scattering at $Q = (0, 0, 10)\ \mathrm{nm}^{-1}$

Convolution with a scattering function, $S(\mathbf{Q}, \omega)$, is carried out in analogy to the integral in (4.8) as a sum of the scattering function values over all events,

$$I(\mathbf{Q}_0, \omega_0) = \sum_e \frac{k_{\mathrm{f,e}}}{k_{\mathrm{i,e}}} S(\mathbf{Q}_\mathrm{e}, \omega_\mathrm{e}) p_\mathrm{e}, \tag{4.10}$$

where $\boldsymbol{Q} = \boldsymbol{k}_\mathrm{f} - \boldsymbol{k}_\mathrm{i}$ and $\omega = \frac{\hbar}{2m}\left({k_\mathrm{i}}^2 - {k_\mathrm{f}}^2\right)$.

Since the events can bear memory of initial and final spin states, this method makes it possible to distinguish resolution functions for the four combinations of initial and final neutron spin states.

In Fig. 4.6, we show the resolution functions simulated for the TAS IN14 at the ILL, Grenoble, equipped with the multichannel guide described in the previous section. Inflation of the resolution volume as a result of beam compression by the multichannel guide is proportional to the gain in neutron flux at the sample. However, the resolution in energy transfer is not affected because the guide can be tuned to the monochromator curvature so that monochromatic focusing condition is fulfilled.

References

1. M.W. Johnson, C. Stephanou, *MCLIB: a library of Monte Carlo subroutines for neutron scattering problems*, RAL Technical Reports, RL-78-090 (1978)
2. P.A. Seeger, L.L. Daemen, Proc. SPIE **5536**, 109 (2004)
3. W.T. Lee, X.L. Wang, J.L. Robertson, F. Klose, C. Rehm, Appl. Phys. A **74**(Suppl.), s1502 (2002)
4. P. Willendrup, E. Farhi, K. Lefmann, Physica B **350**, e735 (2004)
5. G. Zsigmond, K. Lieutenant, S. Manoshin, H.N. Bordallo, J.D.M. Champion, J. Peters, J.M. Carpenter, F. Mezei, Nucl. Instr. Meth. A **529**, 218 (2004)
6. J. Šaroun, J. Kulda, Physica B **234–236**, 1102 (1997)

7. V. Sears, *Neutron Optics* (Oxford University Press, New York, Oxford, 1989) p. 259
8. V. Sears, Acta Cyst. A **56**, 35 (1997)
9. J. Šaroun, J. Kulda, J. Neutron Res. **6**, 125 (1997)
10. J. Šaroun, J. Kulda, Proc. SPIE **5536**, 124 (2004)
11. P.M. Bentley, C. Pappas, K. Habicht, E. Lelievre-Berna, Physica B **385–386**, 1349 (2006)
12. J.F. Breismeister, *MCNPF: A general Monte Carlo n-particle transport code*, Report LA-12625-M (LANL, Los Alamos, NM, 1997)
13. J.C. Nimal, T. Vergnaud, in *Advanced Monte Carlo for Radiation Physics, Particle Transport Simulation and Applications*, ed. by A. Kling, F. Barão, M. Nakagawa, L. Távora, P. Vaz (Springer, Berlin Heidelberg New York, 2001), p. 651
14. J. Šaroun, Nucl. Instrum. Methods A **529**, 162 (2004)
15. H.C. Hu, J. Appl. Cryst. **26**, 251 (1993)
16. A. Freund, Nucl. Instrum. Methods. **213**, 495 (1983)
17. A.D. Stoica, M. Popovici, J. Appl. Cryst. **22**, 448 (1989)
18. J. Kulda, Acta Cryst. A **40**, 120 (1984)
19. K. Lefmann, D.F. McMorrow, H.M. Rønnov, K. Nielsen, K.N. Clausen, B. Lake, G. Aeppli, Physica B **283**, 343 (2000)
20. J. Šaroun, J. Kulda, Physica B **385–386**, 1250 (2006)
21. A. Hiess, R. Currat, J. Šaroun, F.J. Bermejo, Physica B **276–278**, 91 (2000)
22. J. Šaroun, J. Kulda, A. Wildes, A. Hiess, Physica B **276–278**, 148 (2000)
23. R. Gilles, B. Krimmer, J. Šaroun, H. Boysen, H. Fuess, Mater. Sci. Forum **378–381**, 282 (2001)
24. J. Kulda, P. Courtois, J. Šaroun, M. Thomas, M. Enderle, P. Flores P, SPIE **4509**, 13 (2001)
25. J. Kulda, J. Šaroun, P. Courtois, M. Enderle, M. Thomas, P. Flores, Appl. Phys. A **74**(Suppl.), s246 (2002)
26. J. Šaroun, T. Pirling, R.B. Rogge, Appl. Phys. A **74**(Suppl.), s1489 (2002)

5

Wavefront Propagation

M. Bowler, J. Bahrdt, and O. Chubar

Abstract. The modelling of photon optical systems for third generation synchrotrons and free electron lasers, where the radiation has a high degree of coherence, requires the complex electric field of the radiation to be computed accurately, taking into account the detailed properties of the source, and then propagated across the optical elements – so called wavefront propagation. This chapter gives overviews of two different numerical approaches, used in the wavefront propagation codes SRW and PHASE. Comparisons of the results from these codes for some simple test cases are presented, along with details of the numerical parameters used in the tests.

5.1 Introduction

In recent years, there has been an upsurge in the provision of new powerful sources of transversely coherent radiation based on electron accelerators. Free electron lasers (FELs) are providing coherent radiation from THz wavelengths to the ultraviolet, and there are projects in place to build FELs providing X-rays with the XFEL at HASYLAB in Hamburg, the Linear Coherent Light Source LCLS at Stanford and the Spring8 Compact SASE Source SCSS in Japan. Coherent synchrotron radiation (CSR) at wavelengths similar to or longer than the electron bunch is also produced by accelerating electrons. For CSR, the intensity is proportional to the square of the number of electrons in the bunch, hence very intense THz radiation is produced at bending magnets when the bunch length is of the order of a hundred microns, such as is required for FEL operation. Finally, the radiation from undulators, which provide the main sources of radiation in the new storage ring synchrotron radiation (SR) sources from UV to hard X-rays, has a high degree of coherence.

Traditionally, ray tracing, based on geometric optics, has been used to model the beamlines that transport the SR radiation from the source to the experiment. This has provided a sufficiently accurate model for most situations, although at the longer wavelength end of the spectrum some allowances for increased divergence of radiation due to diffraction at slits must be made.

For the coherent sources, interference effects are important as well as diffraction, and one needs to know the phase of the radiation field as well as the amplitude. Hence wavefront propagation, which models the evolution of the electric field through the optical system, is required.

The full solution of the Fresnel Kirchoff equation for propagating the field is possible, but it is computationally intensive and approximate solutions are sought. One approximation applicable to paraxial systems is to use the method of Fourier Optics. The code SRW (synchrotron radiation workshop) generates the source radiation field and also allows for its propagation across "thin" optics. This code is described in Sect. 5.2. Beamlines at UV and shorter wavelengths require highly grazing incidence optics, and in this case the thin optic assumption may not be appropriate. The Stationary Phase method is applicable in this regime and is used to approximate the propagation in the code PHASE, described in Sect. 5.3.

To cross-check both approximations, a Gaussian beam has been propagated across toroidal mirrors of different grazing angles and demagnifications, using both codes, and the size of the focal spots compared. These results are presented in Sect. 5.4 along with a study of the ability of both codes to handle astigmatic focusing.

SRW and PHASE have both been used to model the beamline for transporting THz radiation from the Energy Recovery Linac Prototype (ERLP) at Daresbury Laboratory. This is described in Sect. 5.5. Finally Sect. 5.6 summarizes the results and looks at future needs for wavefront propagation simulations.

The contribution of the COST P7 action has been in making two of these codes, PHASE and SRW, more widely known to the optics community, in running the test cases and in providing documentation to aid the new user. Two of the authors of these codes have joined with the COST P7 participants to write this chapter.

5.2 Overview of SRW

The SRW software project was started at the European Synchrotron Radiation Facility in 1997 [1]. The purpose of this project was to provide users with a collection of computational tools for various simulations involving the processes of emission and propagation of synchrotron radiation. The SRW code is composed of two main parts, SRWE and SRWP, enabling the following:

- Computation of various types of synchrotron radiation *emitted* by an electron beam in magnetic fields of arbitrary configuration, being considered in the near-field region (SRWE)
- CPU-efficient simulation of wavefront *propagation* through optical elements and drift spaces, using the principles of wave optics (SRWP).

Thanks to the accurate and general computation method implemented in SRWE, a large variety of types of spontaneous synchrotron emission by relativistic electrons can be simulated, e.g., radiation from central parts and edges of bending magnets, short magnets, chicanes, various planar and elliptical undulators and wigglers. Either computed or measured magnetic fields can be used in these simulations. Simple Gaussian beams can also be easily simulated. The extension of this part of the code to self-amplified spontaneous emission (SASE) and high-gain harmonic generation (HGHG) is currently in progress. An SRWE calculation typically provides an initial radiation wavefront, i.e., a distribution of the frequency-domain electric field of radiation in a transverse plane at a given finite distance from the source (e.g., at the position of the first optical element of a beamline), in a form appropriate for further manipulation.

After the initial wavefront has been computed in SRWE, it can be used by SRWP, without leaving the same application front-end. SRWP applies mainly the methods of Fourier optics, with the propagation of a (fully-coherent) wavefront in free space being described by the Fresnel integral, and the "thin" approximation being used to simulate individual optical elements – apertures, obstacles (opaque, semi-transparent or phase-shifting), zone plates, refractive lenses.

If necessary, the calculation of the initial electric field and its further propagation can be programed to be repeated many times (with necessary pre- and post-processing), using the scripting facility of the hosting front-end application.

5.2.1 Accurate Computation of the Frequency-Domain Electric Field of Spontaneous Emission by Relativistic Electrons

The electric field emitted by a relativistic electron moving in free space is known to be described by the retarded scalar and vector potentials, which represent the exact solution of the Maxwell equations for this case [2]:

$$\vec{A} = e\int_{-\infty}^{+\infty} \frac{\vec{\beta}}{R}\delta(\tau - t + R/c)\mathrm{d}\tau,\ \varphi = e\int_{-\infty}^{+\infty} \frac{1}{R}\delta(\tau - t + R/c)\ \mathrm{d}\tau, \quad (5.1)$$

where e is the charge of electron, c is the speed of light, $\vec{\beta} = \vec{\beta}(\tau)$ is the electron relative velocity, R is the distance between the observation point $\vec{r}$ and the instantaneous electron position $\vec{r}_{\mathrm{e}}(\tau)$, $R = |\vec{R}(\tau)|$, $\vec{R}(\tau) = \vec{r} - \vec{r}_{\mathrm{e}}(\tau)$, t is the time in laboratory frame, τ is the integration variable having the dimension of time, and $\delta(x)$ is the delta-function. The Gaussian system of units is used in (5.1) and subsequently.

One can represent the delta-function in (5.1) as a Fourier integral, and then differentiate the potentials (assuming the convergence of all integrals) to obtain the radiation field

$$\vec{E} = -\frac{1}{c}\frac{\partial \vec{A}}{\partial \tau} - \nabla\varphi = \frac{1}{2\pi}\int_{-\infty}^{+\infty} \vec{E}_\omega \exp(-\mathrm{i}\omega t)\,\mathrm{d}\omega,$$

$$\vec{E}_\omega = \frac{\mathrm{i}e\omega}{c}\int_{-\infty}^{+\infty}\left[\vec{\beta} - \left(1 + \frac{\mathrm{i}c}{\omega R}\right)\vec{n}\right]\frac{1}{R}\exp[\mathrm{i}\omega(\tau + R/c)]\mathrm{d}\tau, \tag{5.2}$$

where $\vec{E}_\omega$ is the electric field in frequency domain; $\vec{n} = \vec{R}/R$ is a unit vector directed from the instantaneous electron position to the observation point. We note that (5.2) has the same level of generality as (5.1), since no particular assumptions about the electron trajectory or the observation point have been made so far. One can show equivalence of (5.2) to the expression for the electric field containing the acceleration and velocity terms [3]. The exponent phase in (5.2) can be expanded into a series, taking into account the relativistic motion of the electron, and assuming small transverse components of the electron trajectory and small observation angles:

$$\tau + R/c \approx \frac{z - z_{\mathrm{e}0}}{c} + \frac{1}{2}\left[\tau\gamma_{\mathrm{e}}^{-2} + \int_0^\tau (x_{\mathrm{e}}'^2 + y_{\mathrm{e}}'^2)\mathrm{d}\tilde{\tau} + \frac{(x - x_{\mathrm{e}})^2 + (y - y_{\mathrm{e}})^2}{c(z - c\tau)}\right], \tag{5.3}$$

where γ_{e} is the reduced energy of electron ($\gamma_{\mathrm{e}} \gg 1$); x, y, z are respectively the horizontal, vertical, and longitudinal Cartesian coordinates of the observation point $\vec{r}$; x_{e}, y_{e} are the transverse (horizontal and vertical) coordinates of the electron trajectory; x_{e}', y_{e}' are the trajectory angles (or the transverse components of the relative velocity vector $\vec{\beta}$); and $z_{\mathrm{e}0}$ is the initial longitudinal position of the electron. The transverse components of the vector $\vec{n}$ in (5.2) can be approximated as

$$n_x \approx (x - x_{\mathrm{e}})/(z - c\tau),\ n_y \approx (y - y_{\mathrm{e}})/(z - c\tau). \tag{5.4}$$

The dependence of the transverse coordinates and angles of the electron trajectory on τ can be obtained by solving the equation of motion under the action of the Lorentz force in an external magnetic field. In the linear approximation this gives

$$(x_{\mathrm{e}},\ x_{\mathrm{e}}',\ y_{\mathrm{e}},\ y_{\mathrm{e}}')^{\mathrm{T}} \approx \mathbf{A}\cdot(x_{\mathrm{e}0},\ x_{\mathrm{e}0}',\ y_{\mathrm{e}0},\ y_{\mathrm{e}0}')^{\mathrm{T}} + \mathbf{B}, \tag{5.5}$$

where $(x_{\mathrm{e}0},\ x_{\mathrm{e}0}',\ y_{\mathrm{e}0},\ y_{\mathrm{e}0}')^{\mathrm{T}}$ is the four-vector of initial and instantaneous transverse coordinates and angles of the electron trajectory, $\mathbf{A} = \mathbf{A}(\tau)$ is a 4×4 matrix, and $\mathbf{B} = \mathbf{B}(\tau)$ is a four-vector with the components being scalar functions of τ.

Since the approximations used by (5.3) and (5.4) take into account the variation of the distance between the instantaneous electron position and the observation point during the electron motion, the expression (5.2) with these approximations is valid for observations in the near field region.

Consider a bunch of N_e electrons circulating in a storage ring, giving an average current I. The number of photons per unit time per unit area per unit relative spectral interval emitted by such an electron bunch is

$$\frac{\mathrm{d}N_{\mathrm{ph}}}{\mathrm{d}t\mathrm{d}S(\mathrm{d}\omega/\omega)} = \frac{c^2\alpha I}{4\pi^2 e^3 N_e}\left|\vec{E}_{\omega\mathrm{bunch}}\right|^2, \quad (5.6)$$

where $\vec{E}_{\omega\mathrm{bunch}}$ is the electric field emitted by the bunch in one pass in a storage ring, α is the fine structure constant. $\vec{E}_{\omega\mathrm{bunch}}$ can be represented as a sum of two terms describing, respectively, the incoherent and coherent synchrotron radiation [4]:

$$\left|\vec{E}_{\omega\mathrm{bunch}}\right|^2 \approx N_e \int \left|\vec{E}_\omega(\vec{r};\mathrm{X}_{e0},\gamma_{e0})\right|^2 \tilde{f}(\mathrm{X}_{e0},\gamma_{e0})\mathrm{d}\mathrm{X}_{e0}\mathrm{d}\gamma_{e0} +$$

$$N_e(N_e-1)\left|\int \vec{E}_\omega(\vec{r};\mathrm{X}_{e0},z_{e0},\gamma_{e0}) f(\mathrm{X}_{e0},z_{e0},\gamma_{e0})\mathrm{d}\mathrm{X}_{e0}\mathrm{d}z_{e0}\mathrm{d}\gamma_{e0}\right|^2, \quad (5.7)$$

where $\vec{E}_\omega$ is the electric field emitted by one electron with the initial phase space coordinates x_{e0}, y_{e0}, z_{e0}, x'_{e0}, y'_{e0}, γ_{e0}, abbreviated to $\mathrm{X}_{e0}z_{e0}$, γ_{e0} in (5.7); f is the initial electron distribution in 6D phase space, normalized to unity; $\tilde{f} = \int f \mathrm{d}z_{e0}$. The Stokes components of the spontaneous emission can be calculated by replacing the squared amplitude of the electric field in (5.7) with the corresponding products of the transverse field components or their complex conjugates.

5.2.2 Propagation of Synchrotron Radiation Wavefronts: From Scalar Diffraction Theory to Fourier Optics

Let us consider the propagation of the electric field of synchrotron radiation in free space after an aperture with opaque nonconductive edges. Using the approach of scalar diffraction theory, one can find the electric field of the radiation within a closed volume from the values of the field on a surface enclosing this volume by means of the Kirchhoff integral theorem [5]. After applying the Kirchhoff boundary conditions to the transverse components of the frequency-domain electric field emitted by one relativistic electron (see (5.2)), one obtains

$$\vec{E}_{\omega 2\perp}(\vec{r}_2) \approx \frac{\omega^2 e}{4\pi c^2}\int_{-\infty}^{+\infty}\mathrm{d}\tau \int_\Sigma\!\!\int \frac{\vec{\beta}_{e\perp}-\vec{n}_{R\perp}}{RS}\exp[\mathrm{i}\omega[\tau+(R+S)/c]]$$
$$(\vec{\ell}\cdot\vec{n}_R + \vec{\ell}\cdot\vec{n}_S)d\Sigma, \quad (5.8)$$

where $\vec{R} = \vec{r}_1 - \vec{r}_e$, $\vec{S} = \vec{r}_2 - \vec{r}_1$, with $\vec{r}_e$ being the position of the electron, $\vec{r}_1$ a point at the surface Σ within the aperture, and $\vec{r}_2$ the observation point

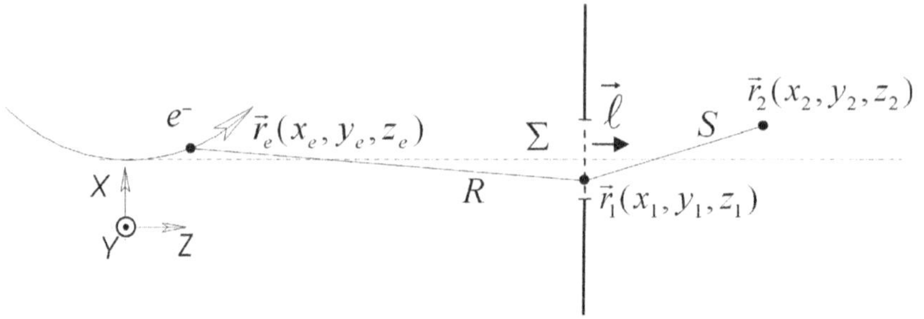

Fig. 5.1. Illustration of the Kirchhoff integral theorem applied to synchrotron radiation

(see Fig. 5.1). $S = |\vec{S}|$, $R = |\vec{R}|$, $\vec{n}_\mathrm{R} = \vec{R}/R$, $\vec{n}_\mathrm{S} = \vec{S}/S$, $\vec{\ell}$ is a unit vector normal to the surface Σ. The expression (5.8) is valid for $R \gg \lambda$, $S \gg \lambda$, where $\lambda = 2\pi c/\omega$ is the radiation wavelength. One can interpret (5.8) as a coherent superposition of diffracted waves from virtual point sources located continuously on the electron trajectory, with the amplitudes and phases of these sources dependent on their positions. This approach allows the calculation of complicated cases of SR diffraction, not necessarily limited by small observation angles.

In the approximation of small angles, the propagation of the SR electric field in free space can be described by the well-known Huygens–Fresnel principle [6], which becomes a convolution-type relation for the case of the propagation between parallel planes:

$$\vec{E}_{\omega 2\perp}(x_2, y_2) \approx \frac{\omega}{2\pi \mathrm{i} cL} \iint \vec{E}_{\omega 1\perp}(x_1, y_1)$$
$$\exp\left[\frac{\mathrm{i}\omega}{c}[L^2 + (x_2 - x_1)^2 + (y_2 - y_1)^2]^{1/2}\right] \mathrm{d}x_1 \mathrm{d}y_1, \tag{5.9}$$

where $\vec{E}_{\omega 1\perp}$ and $\vec{E}_{\omega 2\perp}$ are the fields before and after the propagation and L is the distance between the planes. For efficient computation of (5.9), the methods of Fourier optics can be used.

The propagation through a "thin" optical element can be simulated by multiplication of the electric field by a complex transfer function T_{12}, which takes into account the phase shift and attenuation introduced by the optical element:

$$\vec{E}_{\omega 2\perp}(x, y) \approx \vec{E}_{\omega 1\perp}(x, y) T_{12}(x, y). \tag{5.10}$$

As a rule, the "thin" optical element approximation is sufficiently accurate for (nearly) normal incidence optics, e.g., for slits, Fresnel zone plates, refractive lenses, mirrors at large incidence angles, when the optical path of the radiation in the optical element itself is considerably smaller than distances between the elements or the distance from the last element to the observation plane.

For cases when the longitudinal extent of an optical element (along the optical axis) cannot be neglected, e.g., for grazing incidence mirrors, in particular when *the observation distance is comparable to the longitudinal extent of the optical element*, the "thin" approximation defined by (5.10) may need to be replaced by a more accurate method, which would propagate the electric field from a transverse plane just before the optical element to a plane immediately after it. Such propagators may be based on (semi-) analytical solutions of the Fourier integral(s) by means of asymptotic expansions. The general approach is still valid; the free-space propagator defined by (5.9) can be used immediately after the optic, followed by propagators through subsequent optical element(s), if any.

To take into account the contribution to the propagated radiation from the entire electron bunch, one must integrate over the phase space volume occupied by the bunch, treating the incoherent and coherent terms. For the simulation of incoherent emission (first term in (5.7)), one can sum up the intensities resulting from propagation of electric fields emitted by different "macro-particles" to a final observation plane. In many cases, like imaging by a thin lens, diffraction by a single slit, etc., the intensity in the observation plane is linked to the transverse electron distribution function via a convolution-type relation. In such cases the simulation can be accelerated dramatically. An alternative method for the propagation of the incoherent (partially-coherent) emission consists in manipulation with a mutual intensity or a Wigner distribution [6].

If the synchrotron radiation source is diffraction limited, the wavefront described by the first term in (5.7) is transversely coherent, and therefore it is sufficient to treat the propagation of the electric field emitted by only one "average" electron. Similarly, to simulate propagation of the coherent SR described by the second term in (5.7), it is also sufficient to manipulate with only one electric field wavefront, obtained after integration of the single-electron field over the phase space volume of the electron bunch.

5.2.3 Implementation

The emission part of the code (SRWE) contains several different methods for performing fast computation of various "special" types of synchrotron radiation. However, the core of the code is the CPU-efficient computation of the frequency-domain radiation electric field given by (5.2) with the radiation phase approximated by (5.3) in an arbitrary transversely uniform magnetic field.

The wavefront propagation in SRWP is based on a prime-factor 2D FFT. The propagation simulations are fine-tuned by a special "driver" utility, which estimates the required transverse ranges and sampling rates of the electric field, and re-sizes or re-samples it automatically before and after propagation through each individual optical element or drift space, as necessary for a given overall accuracy level of the calculation. In practice this means that running

SRWP is not more complicated than the use of conventional geometrical ray-tracing.

The SRW code is written in C++, compiled as a shared library, and interfaced to the "IGOR Pro" scientific graphing and data analysis package (from WaveMetrics). Windows and Mac OS versions of the SRW are freely available from the ESRF and SOLEIL web sites.

5.3 Overview of PHASE

In this section, we describe the principle of wavefront propagation within the frame of the stationary phase approximation. The method is complementary to the Fourier Optics technique with the following advantages:

- There is no ray tracing required across the optical elements as is needed in Fourier Optics. Hence, the method is valid also for thick lenses or long mirrors under grazing incidence angles.
- There are no restrictions concerning the grid spacing, the number of grid points, or the grid point distribution in the source and the image planes. One-dimensional cuts as well as images with small dimensions in one direction (e.g., monochromator slits with arbitrary shape) can be evaluated.
- Under certain conditions the propagation across several elements can be done in a single step.
- No aliasing is observed even for strongly demagnifying grazing incidence optics.
- The memory requirements are low.

The disadvantages are the following:

- The speed of simulation is significantly slower for the same number of grid points since the CPU time scales with N^4 rather than $N^2 \ln N$, with N being the number of grid points. On the other hand, the array dimension N needed to propagate $\pm 3\sigma$ of a Gaussian mode with a comparable resolution is generally much smaller as compared to Fourier Optics, which makes the CPU times of both methods comparable.
- The locations of the source plane and intermediate planes can not be chosen arbitrarily (see further).
- The description of the optical surface is restricted to low order polynomials, i.e., randomly distributed slope errors cannot be modelled.

The algorithm has been implemented into the code PHASE [7, 8]. The optical elements are described by fifth order polynomials. All expressions have been expanded up to fourth order in the image coordinates and angles. The FORTRAN code has been generated automatically using the algebraic code REDUCE [9].

Currently, PHASE is being rewritten to be used as a library within a script language. Several existing codes for pre- and post processing the electric fields will be included. This version will provide more flexibility to the user than the existing monolithic program.

5.3.1 Single Optical Element

In the following we assume a small divergence of the photon beam, which allows us to neglect the longitudinal field component. First, we will derive the transformation of the transverse field components from the source plane across a single optical element to the image plane (see Fig. 5.2 for the definition of the variables).

According to the Huygens–Fresnel principle the electric fields transform as

$$\vec{E}(\vec{a'}) = \int h(\vec{a'}, \vec{a})\vec{E}(\vec{a})\mathrm{d}\vec{a}, \tag{5.11}$$

where

$$h(\vec{a'}, \vec{a}) \propto \frac{1}{\lambda^2} \int_{\text{Surface}} \frac{\exp(\mathrm{i}k(r + r'))}{rr'} b(w, l)\mathrm{d}w\mathrm{d}l, \tag{5.12}$$

$k = 2\pi/\lambda$ is the wave vector and $b(w, l)$ is the transmittance function of the optical element, which is not included in further equations. The propagator h includes the integration over the element surface. Principally, the propagator for two elements can be composed of the propagators of two individual elements in the following manner:

$$h(\vec{a''}, \vec{a}) = \int h_2(\vec{a''}, \vec{a'})h_1(\vec{a'}, \vec{a})\mathrm{d}\vec{a'}. \tag{5.13}$$

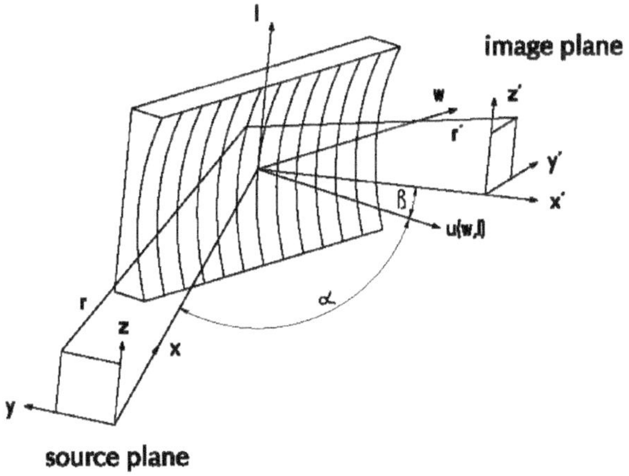

Fig. 5.2. Variables in the source and image plane and at the optical element

The total number of integration dimensions has increased to six, two dimensions for the surface integration over each element and two dimensions for the intermediate plane. The combined propagator can be rewritten in a way that the integration across the intermediate plane is skipped. This is justified since the beam properties are not modified at the intermediate plane. Therefore, the two element geometry requires two additional integrations as compared to the one element case. Each further optical element enhances the integration dimensions by two. Even for the one element geometry a simplification of the propagator is required to carry out the integration within a reasonable CPU time.

The integration over an optical element surface can be confined to a rather narrow region where the optical path length is nearly constant. If the path length changes rapidly, the integrand oscillates very fast and does not contribute to the integral.

We expand the propagator h around a principle ray where the optical path length PL has zero first derivatives with respect to the optical element coordinates and, hence, the phase variations are small.

$$\begin{aligned} h(\vec{a},\vec{a'}) \propto \; & \frac{1}{r_{w_0,l_0} r'_{w_0,l_0}} \exp[\mathrm{i}k(r_{w_0,l_0} + r'_{w_0,l_0})] \\ & \int \exp[\mathrm{i}k(\frac{\partial^2 \mathrm{PL}}{\partial \delta w^2}\frac{\delta w^2}{2} + \frac{\partial^2 \mathrm{PL}}{\partial \delta l^2}\frac{\delta l^2}{2} + \frac{\partial^2 \mathrm{PL}}{\partial \delta w \partial \delta l}\delta w \delta l)] \mathrm{d}\delta w \mathrm{d}\delta l. \end{aligned} \tag{5.14}$$

The quadratic form in δw and δl can be transformed to a normal form where the cross products vanish (principle axis theorem). Then, the double integral can be broken up into two integrals, which can be integrated analytically to infinity:

$$\int\int \cdots \mathrm{d}\delta w \mathrm{d}\delta l = \frac{2\pi \mathrm{i} \cdot \mathrm{sign}}{k\sqrt{|a|\cdot|b|}} = \frac{2\pi \mathrm{i} \cdot \mathrm{sign}}{k\sqrt{|D|}}, \tag{5.15}$$

where a and b are the principle axes. The determinant D is invariant for orthogonal transformations:

$$a \cdot b = D = \begin{vmatrix} \frac{\partial^2 \mathrm{PL}}{\partial \delta w^2} & \frac{\partial^2 \mathrm{PL}}{\partial \delta w \partial \delta l} \\ \frac{\partial^2 \mathrm{PL}}{\partial \delta l \partial \delta w} & \frac{\partial^2 \mathrm{PL}}{\partial \delta l^2} \end{vmatrix} \tag{5.16}$$

Thus, the integral can be solved analytically using the coefficients of the untransformed quadratic form. The second invariant of the quadratic form is the trace T and using D and T the expression *sign* can be evaluated:

sign $= 1$ if $T, D > 0$
sign $= -1$ otherwise

The surface integral of the propagator has been removed and the propagation now scales with the fourth power of the grid size N. This procedure is called the stationary phase approximation [10] and it is justified as long as (1) the optical element does not scrape the beam and (2) the principle rays with identical source and image coordinates are well separated, which means that the quadratic approximation of the path length variation is valid. The latter constraint requires a careful choice of the location of the source plane in order to exclude a zero second derivative of the path length. A proper choice is indicated by a weak dependence of the results on small variations of the position of the source plane. The input data that are given for a certain longitudinal position can easily be propagated in free space to the source plane of the following PHASE propagation. This first step is done by Fourier optics.

For a sequence of optical elements it is useful to describe all expressions in terms of the coordinates and angles of the image plane, which we call initial coordinates in this context (the coordinates of the source plane are named final coordinates). The integration over the source plane is replaced by an integration over the angles of the image plane:

$$\vec{E}(\vec{a'}) = \int h(\vec{a'}, \vec{a}) \vec{E}(\vec{a}) \left| \frac{\partial(y,z)}{\partial(\mathrm{d}y', \mathrm{d}z')} \right| \mathrm{d}(\mathrm{d}y')\mathrm{d}(\mathrm{d}z'). \tag{5.17}$$

The functional determinant containing the derivatives of the old with respect to the new coordinates is expanded in the initial coordinates $(y', z', \mathrm{d}y', \mathrm{d}z')$. Similarly, the expression $1/\sqrt{|D|}$ can be expanded in the same variables. These equations are the basis for the extension of the propagation method to several optical elements in the next section.

For narrow beams the path length derivatives can be replaced by

$$\left| \frac{\partial^2 \mathrm{PL}}{\partial \delta w^2} \frac{\partial^2 \mathrm{PL}}{\partial \delta l^2} - \left(\frac{\partial^2 \mathrm{PL}}{\partial \delta w \partial \delta l} \right)^2 \right| = \frac{\cos(\alpha)\cos(\beta)}{r^2 r'^2} \left| \frac{\partial(y,z)}{\partial(\mathrm{d}y', \mathrm{d}z')} \right|. \tag{5.18}$$

The expression on the right hand side can be generalized to a combination of several optical elements. There is, however, no obvious way to improve the accuracy of this substitution for wider photon beams.

Prior to the wavefront propagation, analytic power series expansions with respect to the initial coordinates $(y', z', \mathrm{d}y', \mathrm{d}z')$ of the five items listed in Table 5.1 are evaluated [7,8]. Items 2 and 3 describe the phase advance $\Delta\Phi$ across the element:

$$\Delta\Phi = \left((\mathrm{PL}(w_0, l_0) - \mathrm{PL}(0,0))/\lambda + \mathrm{mod}(w, 1/n) \right) 2\pi, \tag{5.19}$$

where $1/n$ is the groove separation if the element is a grating. For mirrors the second term in the bracket is skipped. Items 4 and 5 are described by a common set of expansion coefficients and are multiplied together.

5.3.2 Combination of Several Optical Elements

Generally, the transformation of the coordinates and angles across an optical element is nonlinear. On the other hand, the transformation of all cross

Table 5.1. Quantities to be expanded with respect to the initial coordinates and angles

1	The final coordinates (y, z) in the source plane have to be known for the interpolation of the electric fields.
2	The path length differences which determine the phase variations of the principle rays.
3	The intersection points $(w_0,\ l_0)$ of the principle rays with the optical element are needed if the element is a diffraction grating.
4	The functional determinant relating source coordinates and image angles.
5	The expression $1/\sqrt{\|D\|}$ accounting for the surface integration.

products of the coordinates and angles is linear and can be described in a matrix formalism:

$$\overline{Y}_f = \overline{\overline{M}} \cdot \overline{Y i}$$
$$\overline{Y}_{f/i} = (y_{f/i}, z_{f/i}, \mathrm{d}y_{f/i}, \mathrm{d}z_{f/i}, {y_{f/i}}^2, y_{f/i}z_{f/i} \dots). \tag{5.20}$$

Expanding the products to fourth order the corresponding matrix has the dimensions of 70×70. First, the quantities 1–4 of Table 5.1 are derived for each optical element k. Then, the coordinates and angles of the intermediate planes are expressed by the coordinates and angles of the image plane of the complete beamline:

$$\overline{Y_m} = \left(\prod_{k=m+1}^{N} \overline{\overline{M_k}} \right) \cdot \overline{Y}_N = \overline{\overline{\aleph}} \cdot \overline{Y}_N. \tag{5.21}$$

Using these transformation maps the expressions 1–4 of Table 5.1 can be expanded with respect to the variables in the image plane of the beamline, e.g.,

$$\Delta\Phi_m = \overline{\Delta\Phi_{Cm}} \cdot \overline{Y}_m = \overline{\Delta\Phi_{Cm}} \cdot \overline{\overline{\aleph}} \cdot \overline{Y}_N$$
$$w_{0m} = \overline{w_{C0m}} \cdot \overline{Y}_m = \overline{w_{C0m}} \cdot \overline{\overline{\aleph}} \cdot \overline{Y}_N. \tag{5.22}$$

The derivation of the fifth expression in Table 5.1 is more complicated. To account for all possible optical paths a $2N$-dimensional integral A has to be evaluated. Again, all cross terms are removed via a principle axis transformation from the coordinates $(\delta w_1,\ \delta l_1, \cdots \delta l_N)$ to the coordinates $(y_1, \cdots y_{2N})$ and the integral is solved analytically using the stationary phase approximation. The integral is related to the product of the eigenvalues $(\lambda_1,\ \lambda_2, \cdots \lambda_{2N})$ of the matrix $\overline{\overline{G}}$ (see (5.24)) via

$$A(w_{10}, l_{10}, \dots w_{N0}, l_{N0}) = \frac{1}{\prod_{i=1}^{N+1} r_i} \frac{\partial(\delta w_1, \delta l_1 \cdots \delta l_N)}{\partial(y_1 \cdots y_{2N})}.$$
$$\left(\frac{2\pi}{k}\right)^N e^{im\pi/4} - e^{-i(2N-m)\pi/4} \frac{1}{\sqrt{|\lambda_1 \cdot \lambda_2 \cdots \lambda_{2N}|}}. \tag{5.23}$$

The expression $\partial(w_1, l_1 \cdots l_N)/\partial(y_1 \cdots y_{2N})$ equals one if the principle axis transformation is a pure rotation. m is the number of positive eigenvalues.

Using again the invariance of the determinant we get

$$\lambda_1 \cdot \lambda_2 \cdots \lambda_{2N} = \begin{vmatrix} g_{w_1 w_1} & g_{w_1 l_1} & g_{w_1 w_2} & g_{w_2 l_2} & \cdots & 0 \\ g_{l_1 w_1} & g_{l_1 l_1} & g_{l_1 w_2} & g_{l_1 l_2} & \cdots & 0 \\ g_{w_2 w_1} & g_{w_2 l_1} & g_{w_2 w_2} & g_{w_2 l_2} & \cdots & 0 \\ g_{l_2 w_1} & g_{l_2 l_1} & g_{l_2 w_2} & g_{l_2 l_2} & \cdots & 0 \\ \cdots & \cdots & \cdots & \cdots & \cdots & \cdots \\ 0 & 0 & 0 & 0 & \cdots & g_{l_N l_N} \end{vmatrix} . \tag{5.24}$$

The expansion coefficients $g_{p_i q_j}$ represent the second derivatives of the path length with respect to the optical element coordinates p_i and q_j of the elements i and j. They are zero if $|p - q| > 1$. Each coefficient is a fourth order power series of the initial coordinates and angles and using these expansions the square root of the inverse of the determinant can be expanded with respect to the same variables. In principle m can be determined within an explicit derivation of all eigenvalues. We evaluate, however, only the determinant, the product of the eigenvalues, and we can only conclude from the sign of the determinant whether m is even or odd. A sign ambiguity of the integral A remains. This is acceptable since we are finally interested in the intensities rather than the amplitudes. For more details we refer to [11].

5.3.3 Time Dependent Simulations

So far we have discussed the propagation of monochromatic waves that are infinitely long. In reality, finite pulses with a certain degree of longitudinal coherence have to be propagated. The complete radiation field of an FEL can be generated with time dependent FEL codes like GENESIS [12]. Generally, these radiation pulses are described by hundreds or thousands of time slices where each slice describes the transverse electric field distribution at a certain time.

Prior to the propagation these fields have to be decomposed into their monochromatic components. For each grid point in the transverse plane the time dependence of the electric field is converted to a frequency distribution via an FFT. All relevant frequency slices (those for which the intensity is large and the frequency is not blocked by the monochromator) are then propagated as already described. The resulting frequency slices in the image plane are again Fourier transformed providing the time structure of the electric field in the image plane.

A detailed description of the longitudinal and transverse coherence of the FEL radiation is essential if the generation and the propagation of the radiation fields are combined. The spectral content and the time structure of the

FEL output can be modified by a monochromator as demonstrated in the following two cases:

- The signal-to-noise ratio of a cascaded HGHG FEL decreases quadratically with the harmonic number of the complete system. The output power of a four stage HGHG FEL can be significantly improved if the spectrum of the first stage is spectrally cleaned with a monochromator before it passes the following three stages. [13–15].
- The spectral power and purity of a SASE FEL can be enhanced by more than an order of magnitude using a self-seeding technique as proposed for the FLASH facility [16]. The SASE radiation of the first undulators that are operating far below saturation is monochromatized and used in the following undulators as a seed. Combined GENESIS and PHASE simulations have been performed for this geometry [17].

5.4 Test Cases for Wavefront Propagation

To compare the numerical methods used in SRW and PHASE, simple test cases looking at the focusing of Gaussian beams by a toroidal mirror have been carried out for a range of incident angles and demagnifications. For astigmatic optics, where the extent of the field is very different in different directions, it can be difficult to adequately represent the electric field, hence the tests were repeated using astigmatically focusing mirrors. The numerical parameters used in running the codes are given to assist new users.

5.4.1 Gaussian Tests: Stigmatic Focus

The optical system consists simply of a single toroidal mirror placed 10 m beyond the waist of a Gaussian radiation source. Focus distances of 10, 2.5, and 1 m were studied, giving demagnifications of 1, 4, and 10, respectively.

SRW was used to create the radiation field of a Gaussian beam of energy 2 eV, (wavelength 620 nm) with a waist of RMS size 212 μm at 1 m from the waist. Note that for Gaussian sources, the waist size is defined as $\sqrt{2}$ times the RMS size, that is, the input waist value was 300 μm. A small interface code was written to convert the format of the SRW field files to the input required for PHASE. Both codes could then be used to propagate the field along a 9 m drift space, across a toroidal mirror and then along a further drift space to the focus.

Numerical Set-Up

For propagation by SRW, the input field was generated using the automatic radiation sampling option, with an oversampling of 4, giving an input mesh for the propagation of 32×32 points over a 2 mm square. As a starting point for the propagation, automatic radiation sampling was used with the accuracy

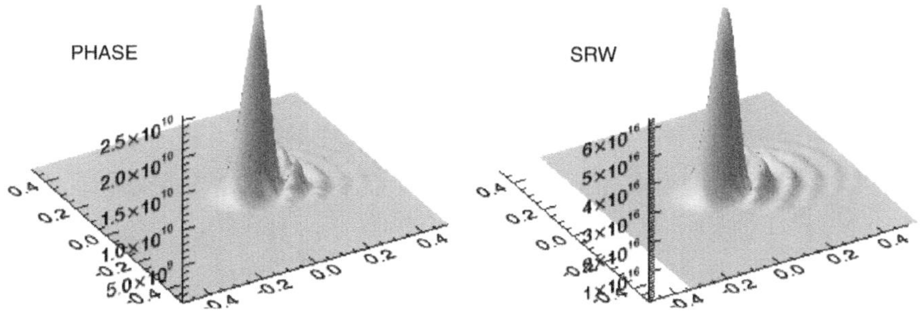

Fig. 5.3. Focal spot for the 87.5° incidence mirror, 10:1 demagnification. Intensities are in arbitrary units

parameter set to 4, and 100 × 100 points were used to define the mirror surfaces. However, for the optics with incidence angles of 85° and 87.5°, more points were required to define the mirror surface otherwise ghosting occurred in the image, and for the 10:1 demagnification the large number of points required to represent the field led to the code running out of memory when attempting to propagate the 1 m to the focal plane in 1 step. The focus shown in Fig. 5.3 for the 87.5° case with 10:1 demagnification was obtained using a grid of 2,000 by 2,000 points on the mirror, and propagating to the focus in three stages, resizing the beam at each stage.

For input to PHASE, a fixed grid of 51 × 51 points was used over the 2 mm square. PHASE requires the range and number of angles used to calculate the image to be set up (see Sect. 5.3.1). For the cases run above, the angular range used varied from ±5 mrads for unit magnification to ±25 mrads in the 10:1 demagnification case. One hundred and one angles were more than sufficient to give an accurate description of the field in the image plane.

Results

Horizontal and vertical cuts through the peak of the beam at the foci have been fitted using Gaussian functions. Table 5.2 compares the RMS sizes of the fitting functions for the PHASE and SRW calculations.

It can be seen that the focal spot sizes are in good agreement (<10%) except for the cases of 10:1 demagnification and the 4:1 demagnification for the most grazing system. It was found that this discrepancy is due to the principle rays not being well separated for highly demagnifying systems, thereby violating the conditions for the stationary phase approximation. The position of the source plane for PHASE can be altered so that the principle rays do not interfere; in this region the details of the image should be independent of the position of the source plane – see Sect. 5.3.1 and [11] for further details. Phase was rerun with the input source field calculated by SRW 1 m before the mirror and then propagated by PHASE. The new results are marked by superscript 1 in Table 5.2 and are in good agreement with those from SRW. The shape of the foci are similar, as can been seen from Fig. 5.3.

Table 5.2. Focal spot sizes from SRW and PHASE for a range of toroidal mirrors

Incidence angle (°)	Demagnification	Focal spot size SRW ($h \times v$) (μm)	Focal spot size PHASE ($h \times v$) (μm)
45	1	210 × 210	213 × 213
45	4	52.7 × 52.7	52.9 × 52.9
45	10	22.3 × 21.8	28.4 × 21.4
75	1	220 × 211	213 × 213
75	4	55.7 × 54.0	54.4 × 53.8
75	10	27.8 × 22.8	28.0 × 23.8
85	1	216 × 212	206 × 206
85	4	63.1 × 59.8	68.9 × 57.3
85	10	36.9 × 24.3	28.6 × 25.4
	10		35.8 × 27.2[1]
87.5	1	220 × 217	205 × 206
87.5	4	72.1 × 68.4	66.6 × 60.2
87.5	10	45.8 × 30.1	31.9 × 25.8
87.5	10		43.4 × 29.1[1]

[1] Result obtained by propagating field in PHASE from 1 m before mirror – see text

5.4.2 Gaussian Tests: Astigmatic Focus

The set-up for these tests is very similar to that described in the previous section, with a single toroidal mirror again placed 10 m from the Gaussian waist point. An incidence angle of 75° was chosen for these tests, with one focus either at 5.0 m or at 2.5 m from the mirrors, and the farther focus at 10 m.

Numerical Set-Up

For SRW, an accuracy parameter of 2 was used with the automatic sampling option for the propagation.

For PHASE, in the case of the 2:1 demagnification for the nearer line focus, a range of ±5 mrads was sufficient in order to reconstruct the wavefront at both line foci. For the farther focus, 301 angles were required in the nonfocused direction, whereas 201 were sufficient in the focused direction. In Fig. 5.4, the effect of using an insufficient number of angles can be seen in the distortion at the edge of the image.

Both programs in principle are able to handle astigmatic focusing. However, in PHASE, when the image plane is out of focus, the angles required to reconstruct the image and transverse position are strongly correlated, leading to an increase in the angle range needed over the whole image. This problem increases with the astigmatism. For the case of 4:1 demagnification at the nearer focus, a range of ±10 mrads was needed at the farther focus to reconstruct the image. However, this was required over a fine mesh of angles for the nonfocused direction, and the maximum number of angles allowed in the code was not sufficient to reconstruct the wavefunction. The inner focus at

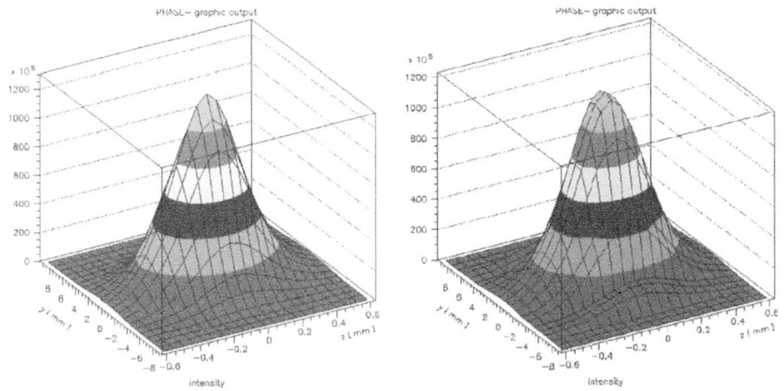

Fig. 5.4. Tangential line focus at 10 m from the toroidal mirror, which has the sagittal focus at 5 m. Left hand image used 301 angles in the nonfocusing direction, where the image on the right used 201 angles, and distortion at the image edge can be seen. Note the order of magnitude difference in the scales on the y and z axes

Table 5.3. Beam widths at the positions of the line foci for the astigmatically focusing toroidal mirrors

Distance to tangential focus (m)	Distance to sagittal focus (m)	Beam size at tangential ($h \times v$) (μm)		Beam size at sagittal ($h \times v$) (μm)	
		SRW	PHASE	SRW	PHASE
10	5	212 × 2,349	205 × 2,269	1,118 × 105	1,123 × 102
10	2.5	212 × 6,975		1,697 × 52.6	1,635 × 51.9
2.5	10	53.7 × 1,713	54.9 × 1,697	6,992 × 212	

2.5 m could be modeled using 501 points in the nonfocused direction over a ±10 mrads range.

This problem in PHASE can be solved by setting the distance, d_foc, from the image to the focal plane. The range of angles used is then centered around y/d_foc, where y is the transverse distance in the image plane, rather than centered round zero. In the version of the code used, this has been done for stigmatic focussing only, but can easily be extended to astigmatic systems.

Results

As before, horizontal and vertical cuts through the beam were fitted with Gaussian functions and the RMS widths are presented in Table 5.3. It can be seen that the agreement between PHASE and SRW is within 5% for the beam sizes at the foci where comparisons can be made. For SRW, the results with 2:1 and 4:1 demagnification in one direction are consistent with each other.

5.5 Beamline Modeling

PHASE has mainly been used to model the highly grazing incidence systems required for XUV and X-ray beamlines. The effect of diffraction gratings can also be included in PHASE as a phase shift across the wavefront. SRWP was designed for "normal incidence" geometries, and has been used for IR and THz wavelength beamlines. For these systems, the monochromator is not part of the beam transport system and SRW does not include any gratings. SRW has also been used for imaging X-rays using X-ray lenses and zone plates.

To compare the codes for modeling a real beamline, the THz beamline on the Energy Recovery Linac Prototype (ERLP) at Daresbury Laboratory [18] has been chosen as it had already been designed using SRW. This example extends the use of PHASE into a regime for which it was not originally designed.

5.5.1 Modeling the THz Beamline on ERLP

The Beamline

A 70 mrad fan of THz radiation will be extracted from ERLP where the electron bunch is shortest, i.e., at the last bend of the dipole chicane bunch compressor, just before the wiggler of the mid-infrared FEL. The THz radiation will be piped to a diagnostic room for analysis before being transported further to an experimental facility. The layout of the accelerator and beamlines is shown in Fig. 5.5; the beamlines have been constrained by the geography of the area, in particular the need to exit the machine room through a labyrinth resulting in the dog-leg in the beamline.

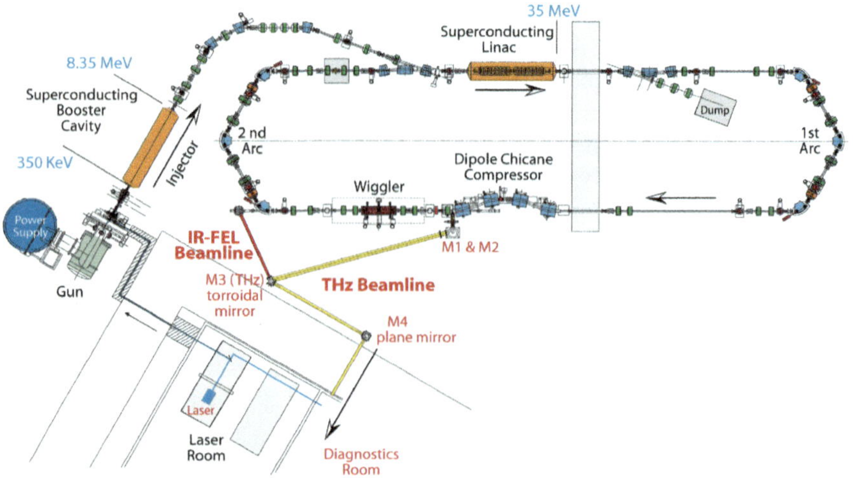

Fig. 5.5. Layout of ERLP, showing the paths of THz and IR-FEL beamlines

The main problem in designing this beamline is that, for longer wavelengths, diffraction becomes significant at the extraction aperture in the dipole magnet vacuum vessel and hence there is no single effective source point to image down the beamline. A second difficulty is that the radiation is naturally very divergent and, for the longer wavelengths, the mirrors will clip the radiation, giving rise to further diffraction.

To start from a known situation, the beamline has been designed to correctly focus the shorter wavelengths, which are not significantly diffracted by the extraction aperture. Two toroidal focusing mirrors, the first mirror (M1) after the aperture and the third mirror (M3) at the start of the labyrinth, are used in the transport path to the diagnostic room. There the radiation is collimated by a third toroidal mirror for input into a purpose-built large-aperture step-scan Fourier Transform Spectrometer. The mirrors have a diameter of 150 mm. The path length from the source to the collimating mirror is 17.3 m.

SRW was used to calculate the radiation emitted by the 35 MeV electrons at the 0.1 T bending magnet in the bunch compressor. The radiation field was generated just before the extraction aperture, at 0.659 m from the source point. Note that the intensity calculations are based on an average current, and so there is no account taken of the finite electron bunch length and hence there is as yet no allowance for the coherent enhancement of the radiation in SRW. Hence the intensities are in arbitrary units. Initial modelling was carried out at 62 μm (20 meV), for which diffraction is not important, to check that the mirror foci occurred at their nominal positions and the radiation could be well collimated. Propagation was then carried out at longer wavelengths, including 620 μm (2 meV) shown below. Of particular interest is the degree of collimation that can be obtained when there is no unique source point as described earlier.

Numerical Considerations

For propagation by SRW of the 620 μm radiation, the radiation field just before the extraction aperture plane was generated using automatic radiation sampling with an oversampling factor of 4. This generated 40×40 points over the initial grid of 45 mm by 45 mm. Propagation through the 37 mm diameter aperture and along the beamline to beyond the collimating mirror was done using the automatic field sampling option with the accuracy parameter set to 4.

For input into PHASE, a fixed grid of 51 by 51 points was used in SRW over the same 45 mm by 45 mm area and the output converted into PHASE field format. The aperture is included in PHASE by setting the radius of a pinhole in the input source plane, defined in the input parameter file (fg34.par). An angular mesh of 201 points over ± 50 mrads was used for the propagation.

In PHASE, if the mirrors do not aperture the beam, the transformation across the whole beamline can be carried out in one step. This was done and the results compared with those from SRW using large optics that do not clip the beam.

To allow for the finite size of the optics in PHASE, the beam should be propagated in stages to each aperturing optic, the radiation field output at the optic, and then used as the source for the next step in the propagation down the beamline, setting the source aperture appropriately. As the phase of the radiation field can vary rapidly, the field should be represented by its phase and amplitude, rather than by the real and imaginary parts. This has been done, e.g., by Bahrdt et al. [19] but not all the codes needed are generally available and hence the effect of the finite size of the mirrors was included using SRW alone.

Results

As the beam profiles are approximately Gaussian, cuts through the beam at three different locations were fitted with Gaussian functions. The RMS widths are presented in Table 5.4. It can be seen that the horizontal widths are in good agreement, but the vertical widths at the collimating mirror are about 15% larger from SRW than those from PHASE. This discrepancy is bigger than the variation of about 5% obtained when different sampling of the field is used. The origin of the difference has not yet been found.

The width of the 620 μm beam at the nominal focus of the first mirror, given in Table 5.4, is about an order of magnitude wider than the (correctly focused) 62 μm radiation. This results from the lack of a unique source to image down the beamline for the longer wavelengths. The effect on the collimation of the radiation can be seen by comparing the fitted beam widths at the collimating mirror and 2 m beyond; the 620 μm radiation after the "collimating mirror" is in fact converging. The percentage decrease in the beam size over the 2 m beyond the collimating mirror is very similar from both codes, and should be slow enough for FTIR measurements to be made. These simulations demonstrate the need to use wavefront propagation for systems where diffraction is important.

The effect of the finite mirror size was then investigated using SRW by setting the mirrors to $150 \times 150\,mm^2$. The intensity distributions at and beyond

Table 5.4. Widths (RMS values) of Gaussian fits of horizontal and vertical cuts through the middle of the 620 μm beam at different locations in the beamline

Position in beamline	Horizontal width (mm)		Vertical width (mm)	
	SRW	PHASE	SRW	PHASE
At nominal "focus" of first mirror	10.9	11.1	12.0	11.3
At position of collimating mirror	38.1	37.0	43.9	36.9
2 m beyond collimating mirror	30.9	30.05	36.3	31.1

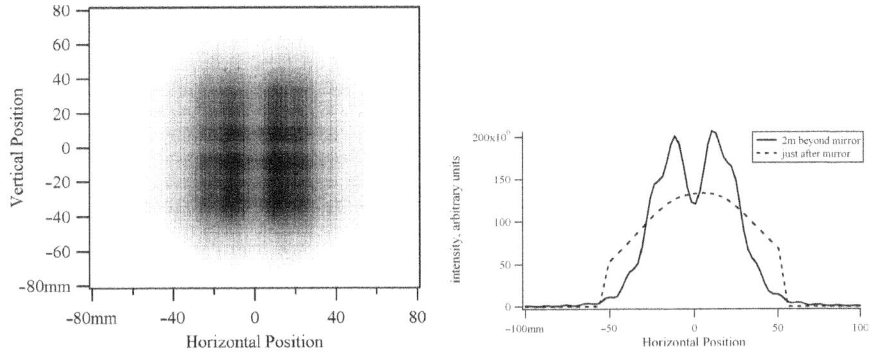

Fig. 5.6. (*LHS*) Image 2 m after the collimating mirror; (*RHS*) Horizontal cuts through the beam at and after the mirror, 620 μm radiation

the collimating mirror are shown in Fig. 5.6, where the diffraction due to significant clipping of the radiation by the collimating mirror is clearly seen. Since this work was carried out, the size of the collimating mirror has been increased to 200 mm.

5.6 Summary

Both methods for wavefront propagation, namely Fourier Optics and the Stationary Phase Approximation, are applicable to a wide range of situations. In general, Fourier techniques are fast, but can become more difficult to implement numerically for highly demagnifying systems. Good agreement was obtained between the results of SRW, which employs Fourier Optics, and PHASE even for highly grazing systems and 10:1 demagnification. Both codes are capable of handling astigmatic systems.

SRW and PHASE have between them been used very successfully to design a wide range of beamlines, from the transport of THz radiation to seeding schemes for XUV FELs. As coherent sources move to ever shorter wavelengths, further developments are required. More types of optical elements need to be included in the codes, in particular crystal optics. Work on wavefront propagation through crystals has for example been carried out by Mocella and co-workers [20] and Oreshko et al. [21]. The effect of imperfect surfaces, which introduce speckle and degrade micro and nanofocusing systems, also needs to be included.

References

1. O. Chubar, P. Elleaume, in *Proceedings of the 6th European Particle Accelerator Conference EPAC-98*, 1998, pp. 1177–1179
2. L.D. Landau, E.M. Lifshifts, *The Classical Theory of Fields*, (Pergamon Press, Oxford, 1975)

3. J.D. Jackson, *Classical Electrodynamics*, 3rd edn. (Wiley, New York, 1998)
4. S. Nodvick, D.S. Saxon, Phys. Rev. **96**, 180 (1954)
5. M. Born, E. Wolf, *Principles of Optics*, 4th edn. (Pergamon Press, New York, 1970)
6. K.-J. Kim, Nucl. Instrum. Methods A **246**, 71 (1986)
7. J. Bahrdt, Appl. Optics **34**, 114 (1995)
8. J. Bahrdt, Appl. Optics **36**, 4367 (1997)
9. A.C. Hearn, *REDUCE 3.5, A General Purpose Algebra System* (RAND, Santa Monica, CA, 1993)
10. L. Mandel, E. Wolf, *Optical Coherence and Quantum Optics*, (Cambridge University Press, Cambridge, 1995)
11. J. Bahrdt, Phys. Rev. Special Topics, Accel. Beams **10**, 060701 (2007)
12. S. Reiche, Nucl. Instrum. Methods A **429**, 243 (1999)
13. M. Abo-Bakr, R. Follath, A. Meseck, in *Proceedings of the 27th International FEL Conference*, Stanford, CA, 2005, pp. 19–22
14. K. Goldammer, A. Meseck, in *Proceedings of the 27th International FEL Conference*, Stanford, CA, 2005, pp. 23–26
15. J. Bahrdt, in *Proceedings of the 27th International FEL Conference*, Stanford, CA, 2005, pp. 694–710
16. J. Feldhaus, E.L. Saldin, J.R. Schneider, E.A. Schneidmiller, M.V. Yurkov, Opt. Comm. **140**, 341 (1997)
17. J. Bahrdt, B. Faatz, R. Treusch, V. Miltckev, R. Reininger, in *Proceedings of the 28th International FEL Conference*, Berlin, Germany, 2006, pp. 150–153
18. M.W. Poole, E.A. Seddon, in *Proceedings of the 9th European Particle Accelerator Conference EPAC-04*, Lucerne, 2004, pp. 455–457
19. J. Bahrdt, U. Flechsig, Proc. SPIE **3150**, 158 (1997)
20. V. Mocella, C. Ferrero, A.K. Freund, J. Hoszowska, L. Zhang, Y. Epelboin, Nucl. Instrum. Methods A **467**, 414 (2001)
21. A.P. Oreshko, V.A. Bushuev, D.V. Novikov, *Hasylab Annual Report* (2000)

6

Theoretical Analysis of X-Ray Waveguides

S. Lagomarsino, I. Bukreeva, A. Cedola, D. Pelliccia, and W. Jark

Abstract. X-ray waveguides are efficient in the generation of nanometre sized beams in the hard X-ray spectral region. This chapter addresses the theoretical behaviour of waveguides in two coupling geometries: resonant beam coupling and front coupling. A quantitative estimation of the transmission efficiency, not elsewhere reported, is presented. For front coupling a detailed study of the internal field structure is reported for a range of structured incident wave-fields.

6.1 Introduction

X-ray waveguides (WG) were first introduced in 1974 by Spiller and Segmüller [1] who demonstrated the propagation of the X-rays in a waveguide composed of a BN film sandwiched between two layers of Al_2O_3. In 1992, exploiting previous studies made by Bedzyk [2] on the X-ray standing waves created above a reflecting mirror surface, Wang [3] studied the resonance-enhanced effect that takes place inside a two layer system. In 1993 Feng [4] re-proposed the three layer structure of Spiller with different materials (Si/polymide/SiO_2) and he showed the higher enhancement effect created inside the middle layer with respect to the two layer structure of Wang. Only in 1995 did Feng et al. [5] and Lagomarsino et al. [6, 7] independently propose the use of this waveguide configuration for X-ray microbeam production. Since then, different configurations and different materials have been used to obtain smaller and smaller beams using both planar and two-dimensional WGs [7–10]. The interest in WGs results from their ability to provide coherent beams as small as 10 nm [11, 12], accepting the full coherent phase space volume of the incoming radiation [13, 14]. Since WGs do not act as a lens, but rather as a spatial and coherence filter, the final beam dimensions do not depend on the source size and on the focal distance, as for the other optical elements generally used. Therefore, nanometer beams can, in principle, be obtained even with table-top sources [15]. Moreover, WGs are essentially achromatic, and their potential use extends from EUV and soft X-rays up to hard X-rays. These characteristics make it possible to use WGs in several applications, from the determination of local strain with 100 nm spatial resolution in materials for microelectronics [16] to the determination of fiber structure [17], to studies of structural properties of engineered bone tissue [18]. Besides the applications that use the waveguide as an optical element, capable of providing nanosize beams, other studies rely directly on the interaction of the guided field with a specimen placed inside the WG. This is the case of studies on confined fluids [19] and on thin macromolecular films [20]. Some recent reviews [21–23] report both the basic principle and some interesting scientific results obtained using X-ray WGs. In this chapter we will present a theoretical analysis of X-ray WGs, analyzing their efficiency, the formation of the resonance modes inside the guiding layer, and the main properties of the exiting beam. We will consider essentially planar WGs. In the case of two-dimensional waveguides with a rectangular section, the two orthogonal directions can be considered as independent.

The incident radiation can be coupled into the guiding layer of X-ray WGs in two different ways: resonant beam coupling (RBC) and front coupling (FC). RBC [1, 5, 6] takes place in a three layer WG, with the incident beam at a grazing angle transmitted by the very thin upper layer and trapped by the intermediate guiding layer (see Fig. 6.1a). With this scheme the incoming beam of several tens of microns can be compressed down to nanometer dimensions and a significant effective gain (>100) can be achieved [8]. In the FC scheme [10] the incoming radiation is directly side-coupled with the WG aperture, and the spatial acceptance is therefore equal to the WG gap (see Fig. 6.1b). Moreover, several modes can be excited simultaneously in the WG. A variant of the front coupling scheme can be adopted with a reflection of the incoming beam just in front of the WG entrance at one of the resonance angles (see Fig. 6.1c). In this way only one particular mode can be selected.

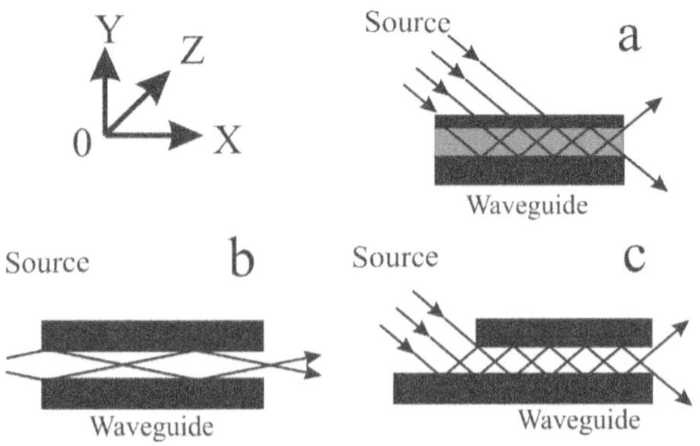

Fig. 6.1. The different coupling modes: (**a**) resonant beam coupling through the cover layer. (**b**) Front coupling directly into the guiding layer. (**c**) Front coupling with prereflection and generation of a standing wave pattern

In Sect. 6.2 we will consider RBC coupling, analyzing in some detail the efficiency of WGs as a function of their geometrical and physical parameters (guiding layer thickness, guiding materials, etc.). In contrast, in Sects. 6.3 and 6.4, the FC schemes with an air (or vacuum) gap will be taken into account. We will make extensive use of a computer simulation code based on solutions of wave-field equations in the parabolic approximation. We will also present an analytical asymptotic solution to describe the wave field formation in an air-gap front-coupling geometry with relatively low absorbing cladding layers.

6.2 Resonance Beam Coupling

This chapter follows the historical development in the field of use of X-ray waveguides, which first saw the use of thin film (slab) waveguides with resonant beam couplers [1, 5, 6]. It will thus only treat the feeding by use of a resonant beam coupler [1] and is based on the more detailed studies in [13, 14].

As far as the output flux is concerned we can easily estimate an upper limit, which is still applicable for all types of waveguide. When the exiting beams from X-ray waveguide terminals were first observed by Feng et al. [5] and by Lagomarsino et al. [6], they were identified by their appearance as interference patterns. In fact an X-ray waveguide is a resonator and as such it acts as a coherence filter, because only coherent radiation can participate in the accumulation of resonantly enhanced intensity in the optical path, especially in a slab waveguide. In any case, the exiting beams were always found to be coherent [5, 7]. If we then want to use these coherent beams in order to obtain high spatial resolution in micro-diffraction experiments, we have to be aware of the effects for this application, as discussed by De Caro et al. [24]. For the output flux we now know that it cannot exceed the spatially coherent part of the incident flux (Note: we do not consider here the temporal, i.e., longitudinal, coherence. As long as one uses a crystal monochromator in a synchrotron radiation beamline, it is always sufficient for the present considerations). As far as X-rays are concerned we do not yet have coherent emitters and thus, at a given distance from a source we will have spatially coherent radiation only if we limit the beam in size and in angular divergence. The latter is usually produced by the finite apparent source size. According to Attwood [25] the phase space volume for spatially coherent radiation is given by

$$A_{\mathrm{FWHM}} \Delta\Phi_{\mathrm{FWHM}} \approx 0.44\,\lambda \tag{6.1}$$

In the following we will express A and Φ as a function of the geometrical and physical characteristics of the WG. Electromagnetic modes can be excited if the refractive indices of a layer system, as presented in Fig. 6.1a, fulfill the condition [26] $n_2 > n_1 \geq n_3$, where the index 2 refers to the guiding (i.e., the median) layer, and 1 and 3, respectively, the upper and lower cladding layers. In the X-ray range the refractive index is usually written as $n = 1 - \delta - \mathrm{i}\beta$, where δ and β are small compared to unity. We can easily obtain them from the internet [27,28], in which they are calculated in the appropriate way from the atomic scattering factors as tabulated by Henke et al. [29] and Chantler et al. [30]. For X-rays δ is a positive number, which varies with the electron density. Generally lighter materials have smaller δ values and thus they have the larger refractive index n compared to heavier materials. Lighter materials usually also absorb less. Hence, X-ray mode excitation is possible in a lighter low absorbing layer sandwiched between heavier materials with higher absorption. The coupling of the beam into the core is very efficient as the air or vacuum environment has the largest refractive index n with $\delta_0 \approx 0$. As shown in Fig. 6.1a, the incident intensity is coupled through a thin enough cover layer directly to permitted modes in the core layer [1]. This gives rise to the name of resonant beam coupling. In fact, the constructive interference, in the entire illuminated beam footprint, between the wavefield coupled into the waveguide and the two internally reflected fields produces standing waves in the direction of the surface normal; as a consequence, the intensity can by far exceed the intensity of the incident wavefield [31]. For the resonance condition to be met, the angle of grazing incidence onto the waveguide surface, Φ_0, and the grazing angle onto the internal interfaces, Φ_2, must be properly related to the thickness of the guiding layer and to the wavelength λ.

The discussion will here be restricted to the fundamental guided mode and to symmetric waveguides with $n_1 = n_3$. The thickness of the absorbing cover layer will always be assumed to be adjusted for maximum internal field enhancement. Obviously, the fundamental guided mode with an intensity maximum in the guiding layer has one node in the vicinity of the top surface and another in the vicinity of the interface with the absorbing underlayer. We assume here that the node separation is given by the effective resonator thickness, defined as $D_{\text{eff}} = d_1 + d_2$, where d_2 is the waveguide aperture and d_1 is the extension of the electromagnetic field into the cladding layers. Then for an incident X-ray beam of wavelength λ and an angle of grazing incidence Φ_2 onto the internal interfaces the mode excitation condition, i.e., the standing wave condition, is given by

$$2D_{\text{eff}} \sin(\Phi_2) = \lambda. \tag{6.2}$$

The optimum cover layer thickness cannot be explained in the present simple model. It needs to be derived from the rigorous calculation of the field progression through the layer system, which can be done with the recursive technique presented by Parratt [32]. For the purpose of the present discussion it is sufficient to know that we always find $d_1 < d_2$ and $d_1 < 10\,\text{nm}$. In general, thicker cores need thinner cover layers, which, however, need to be increased towards higher photon energies, i.e., shorter wavelengths.

We are using here the approximations $\sin(\Phi) \approx \Phi \approx \tan(\Phi)$ and $\cos(\Phi) \approx 1 - 0.5\Phi^2$ as all angles are small. The critical angle for total reflection in air at a mirror made of material "i" is then given as

$$\theta_{\mathrm{c,i}} = \sqrt{2\delta_{\mathrm{i}}}. \tag{6.3}$$

The guiding condition from above can then also be written as $\theta_{\mathrm{c},2} < \theta_{\mathrm{c},1} = \theta_{\mathrm{c},3}$ and the refraction at the interfaces will lead to

$$\Phi_0 = \sqrt{\Phi_2^2 + \theta_{\mathrm{c},2}^2}. \tag{6.4}$$

Finally, we find $\theta_{\mathrm{c},1} > \Phi_0 > \theta_{\mathrm{c},2}$. In other words, for RBC the incident wave field is tunneling through the cover layer.

The limiting case for efficient RBC is when Φ_0 is only slightly smaller than the critical angle of the cladding layer material $\Phi_0 < \theta_{\mathrm{c},1}$. Then, only one guided mode will be excited in the waveguide. Its thickness, i.e., the minimum thickness for efficient beam coupling D_{min}, can be obtained by use of (6.2.) and (6.4):

$$D_{\mathrm{min}} > \frac{\lambda}{2} \frac{1}{\sqrt{\theta_{\mathrm{c},1}^2 - \theta_{\mathrm{c},2}^2}} = \frac{1}{2} \frac{1}{\sqrt{\frac{2\delta_1}{\lambda^2} - \frac{2\delta_2}{\lambda^2}}}. \tag{6.5}$$

Here we will always put lighter cores between heavier cladding layers for which we can assume $\delta_1 \gg \delta_2$. Then (6.5) reduces to

$$D_{\mathrm{min}} > \frac{\lambda}{2\sqrt{2\delta_1}}. \tag{6.6}$$

In the X-ray range ($E > 4\,\mathrm{keV}$) away from the absorption edges δ varies approximately with λ^2. Thus D_{min} is essentially independent of the wavelength and depends only on the cover material chosen. The latter will be a heavier material or metal. In this latter group D_{min} varies very little and is about $D_{\mathrm{min}} = 10\,\mathrm{nm}$. This number and (6.6) are also derived by Bergemann et al. [11] in the rigorous treatment of the minimum spot size obtainable in a tapered double plate X-ray waveguide [33]. The number $D_{\mathrm{min}} = 10\,\mathrm{nm}$ is in agreement with the experimental results of Pfeiffer [34], who could still identify a weak exiting beam in a single-mode Ni/C waveguide with $d_1 + d_2 = (4.1\,\mathrm{nm}) + (10.1\,\mathrm{nm}) = 14.2\,\mathrm{nm}$. Because of the RBC this waveguide provided still six times more output flux [12, 34] than a hypothetical aperture with $d = 10.1\,\mathrm{nm}$. Bergemann et al. [11] argued that this beam size of 10 nm is the ultimate limit to which an X-ray beam can be focused by diffraction. However, other authors later showed that it is not the ultimate limit for X-ray microspots [35–37].

Lengeler et al. [38] introduced the effective aperture as a means of objectively comparing focusing components. It is defined as the component transmission function, $t(y)$, integrated over its geometrical aperture $\{y_1, y_2\}$:

$$A_{\text{eff}} = \int_{y_1}^{y_2} t(y)\,\mathrm{d}y \tag{6.7}$$

The result is the size of an aperture that will provide the same output photon flux [38] as a given optical component behind its geometric aperture $A_{\text{geo}} = y_2 - y_1$. Obviously $t(y) = 1$ leads to $A_{\text{eff}} = A_{\text{geo}}$, while absorbing systems have $A_{\text{eff}} < A_{\text{geo}}$.

For the present geometry it is more convenient to discuss the effective footprint size, l_{eff}, in the reference frame of the waveguide. A ray that impinges onto the waveguide at a distance l upstream of the waveguide exit has a vertical distance from the waveguide exit of $y = \Phi_0 l$. Then we can write

$$A_{\text{eff}} = \Phi_0 l_{\text{eff}} = \Phi_0 \int_{l_2=-\infty}^{l_1=0} t(l)\mathrm{d}l. \tag{6.8}$$

Along the footprint the guided wave will be attenuated because of absorption and because of intensity losses in the reflection processes at the two interfaces (1–2) and (2–3). In a symmetric waveguide the interface reflectivity is $R_{1,2} = R_{2,3} = R$. R is calculated from $R = \tilde{R}_{1,2}\tilde{R}^*_{1,2}$, where the superscript "*" denotes the complex conjugate and the coefficients $\tilde{R}_{1,2}$ are given by [39] $\tilde{R}_{1,2} = \frac{\sqrt{\Phi_0^2-2\delta_1+\mathrm{i}2\beta_1}-\sqrt{\Phi_0^2-2\delta_2+\mathrm{i}2\beta_2}}{\sqrt{\Phi_0^2-2\delta_1+\mathrm{i}2\beta_1}+\sqrt{\Phi_0^2-2\delta_2+\mathrm{i}2\beta_2}}$. For N reflection processes this yields $t(l) = \mathrm{e}^{-\left|\frac{l}{L}\right|}R^N$. Here L is the attenuation length of the core material, which is tabulated [29, 30]. We will substitute the integer N with a continuously varying number $N(l) = \left|\frac{l}{\Delta l}\right|$, where $\Delta l \approx \frac{D_{\text{eff}}}{\Phi_2}$ is the inter-bounce distance in the direction of the beam footprint. In efficient waveguides the reflectivity will be only slightly smaller than 1, so that we can introduce the reflectivity decrement, ΔR, and write $R = 1 - \Delta R$. This allows us to rewrite $R^{N(l)} = \mathrm{e}^{-\left|\frac{l}{\Delta l}\right|\Delta R}$, leading to the exponentially decaying transmission function $t(l)$

$$t(l) = \mathrm{e}^{-|l|\left(\frac{1}{L}+\frac{\Delta R}{\Delta l}\right)}. \tag{6.9}$$

The effective aperture is then

$$A_{\text{eff}} = \Phi_0 \int_{l_2=-\infty}^{l_1=0} t(l)\,\mathrm{d}l = \frac{\Phi_0}{\frac{1}{L} + \frac{\Delta R}{\Delta l}}. \tag{6.10}$$

It is straightforward to extend this discussion to two-dimensional waveguides, i.e., to rectangular channels, the shape that Pfeiffer et al. [9] used in their prototype for the beam confinement in two dimensions. In this case the guided modes also bounce between the horizontal interfaces and are thus subject to additional reflectivity losses. Therefore, the transmission function from (6.9) needs to receive an additional reflectivity term

$$t(l) = \mathrm{e}^{-|l|\left(\frac{1}{L} + \frac{\Delta R_{\mathrm{v}}}{\Delta l_{\mathrm{v}}} + \frac{\Delta R_{\mathrm{h}}}{\Delta l_{\mathrm{h}}}\right)} \tag{6.11}$$

and we have to distinguish between vertical (v) and horizontal (h) bounces. The indices v and h need now to be added also to Φ_0, Φ_2 and D_{eff}.

Even though the transmission function $t(l)$ for the waveguide in (6.9) is not a symmetric function, we will assign a FWHM to it. For a Gaussian transmission function the FWHM, at whose borders the transmission is reduced to half of its maximum value, contains 76% of the ultimately transmittable flux. The same 76% will be used to define the FWHM aperture A_{FWHM} of the waveguide. Then, we have

$$A_{\mathrm{FWHM}} = 1.43\, A_{\mathrm{eff}}. \tag{6.12}$$

With the result from (6.12) we can calculate the angular acceptance of the waveguide by use of (6.1) and we can check its matching to the given experimental conditions.

We will now do this for the successful combination of materials Mo/Be and Cr/C [7]. The first combination is actually found to be the optimum choice for a photon energy of 13 keV by use of a genetic algorithm [40], which, however, does not take into account any matching between the angular acceptances of waveguides and the beam divergence in the experimental setup. Our model uses approximations valid only for light layers sandwiched between heavier layers and can thus not be used for the calculation of waveguides in light materials like C/Be/C, which was prepared and tested by Kovalenko and Chernov [41], or for layer systems produced in carbon of different density [42].

Figure 6.2 presents the FWHM spatial acceptances as calculated by using (6.12) depending on the effective resonator thickness for several photon energies. The apertures level off at larger resonator thicknesses and do not exceed values on the order of 10–30 µm for all photon energies. On the other hand, towards smaller resonator thickness the apertures decrease rapidly. Obviously these small apertures can always be filled by any radiation beam.

The calculations with exchanged core layers in the comparison at 13 keV photon energy indicates that the spatial acceptance is dominated at smaller effective thicknesses by reflection losses, which depend more on the cover than on the core material, while at larger D_{eff} it is dominated by the absorption in the guiding materials. In fact, in the latter case the aperture is simply given by $A_{\mathrm{FWHM,max}} = 1.43\,\theta_{\mathrm{c},2} L$, which is essentially constant, i.e., is independent of the core layer thickness.

At this point it is still too early to discuss the preferable material combinations. In fact, we first need to inspect the angular acceptances of the same waveguide combinations. This is done by use of (6.1), and the results are presented in Fig. 6.3.

For the further discussion we first have to recognize that at state-of-the-art high brightness synchrotron radiation sources the angular spread of unfocused

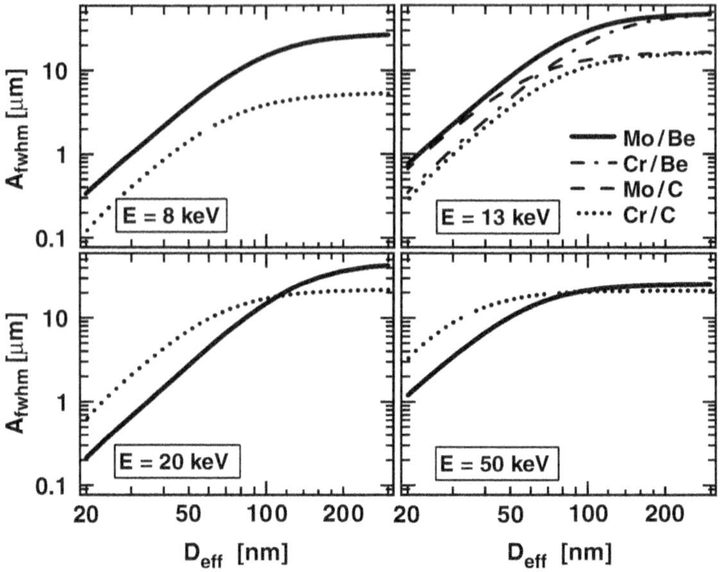

Fig. 6.2. FWHM spatial acceptances depending on effective guiding layer thickness for different photon energies and for X-ray waveguides in Cr/C (*dotted line*) and in Mo/Be (*solid line*). At 13 keV photon energy (*top right*) the calculations for interchanged materials, i.e., for Cr/Be (*dash-dotted line*) and for Mo/C (*dashed line*), are presented (reprinted with permission from [14])

radiation is mostly smaller than $\gamma = 2\,\mu\text{rad}$. Compared to this number the angular acceptances of waveguides with smaller guiding layer thicknesses, D_{eff}, are inconveniently large with numbers that even exceed 100 μrad or about 50 times the available beam divergence. In this case, we cannot efficiently fill these angular acceptances of waveguides, which provide the more interesting smaller beam sizes for microbeam experiments. As a consequence the output photon flux is then smaller than the spatially coherent part of the incident radiation. The divergence matching is not much of a problem for larger resonator thickness and for higher photon energies. Here we mostly overfill the angular acceptance, and thus we expect the output flux to be identical to the spatially coherent content of the incident beam. From (6.1) we see that this spatially coherent contribution decreases towards shorter wavelengths, i.e., towards higher photon energies. Nevertheless, easy-to-produce X-ray waveguides could become the object of choice in this range, where submicron focii are not so easily obtained by use of other microscopy objectives. For the choice of the best resonator thickness we will now also look at the required resonator length, i.e., to the footprint size. This is approximately given by

$$l_{\text{FWHM,max}} = 1.43\,L. \tag{6.13}$$

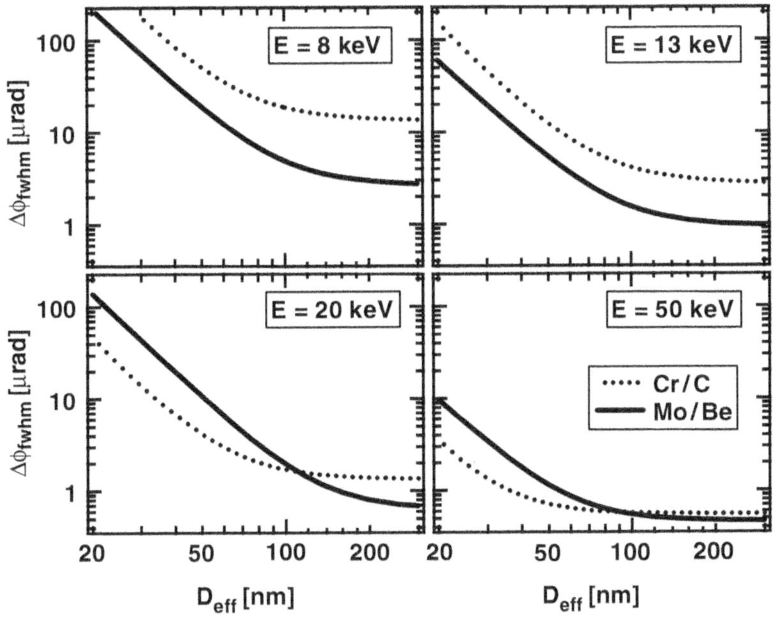

Fig. 6.3. Angular acceptances of the X-ray waveguides in Cr/C (*dotted line*) and in Mo/Be (*solid line*) depending on the effective guiding layer thickness for different photon energies (reprinted with permission from [14])

The related data for some lighter materials for the core layer, i.e., for the materials discussed here, beryllium (Be) and carbon (C) and for PMMA (polymethylmethacrylate, which is $C_5H_8O_2$ with density $1.19\,\mathrm{g\,cm^{-3}}$) are shown in Fig. 6.4.

We see that the mirror lengths remain below about 60 mm. Nevertheless, such objects could be expensive as the tolerance for their flatness and an eventually remaining surface waviness need to be smaller than the angular acceptance. Variations in the tangents to the surface along the footprint on the order of or smaller than 1 μrad push the capabilities of quality control instrumentation based on lasers. With this in mind, carbon core layers offer some advantages if we are interested in higher photon energies. In that case, the divergence matching condition is achieved at relatively small resonator thickness below 100 nm (for $E > 20\,\mathrm{keV}$). Under these conditions, the surface finish required for the substrate is slightly less demanding. More important is the fact that the substrate can be short with a length of the order of 25 mm. To make good use of the relatively higher coherent content in the incident radiation towards smaller photon energies and in thinner waveguides, it is necessary to adjust the phase space volume for spatially coherent radiation by increasing the beam divergence with other prefocusing optical elements. For two-dimensional waveguides we have to be more careful in the

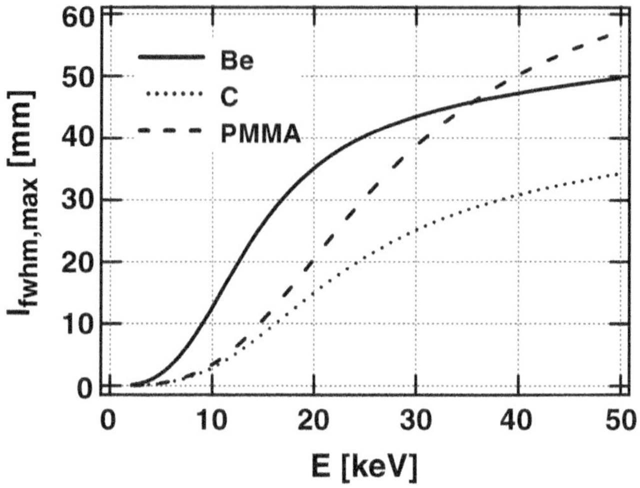

Fig. 6.4. Maximum FWHM footprint size for waveguides with different low Z cores (reprinted with permission from [14])

optimization and the matching to the experimental conditions. Although the mode excitation conditions for both directions are independent, the calculation of the effective travel length needs to take into account all reflection processes. Then, if we want to obtain smaller resonator thicknesses, the smaller of the two dimensions may actually limit the traveling length and thus the correlated effective aperture and angular acceptance.

Up to now we have ignored the spectral bandwidth of the incident radiation. This was possible because the waveguide angular acceptance is found to lead to a spectral acceptance, which is much larger than the bandwidth of crystal monochromators, if operated at synchrotron radiation sources. All reported feasibility studies for the practical application of X-ray waveguides have been made in this combination. However, we have to recognize that the photon flux could be increased if we could afford to worsen the bandwidth of the probing radiation for the experiment.

6.3 Front Coupling Waveguide with Preliminary Reflection

In this section we consider the theoretical investigation of the mode transmission for air-gap FC X-ray waveguides with preliminary reflection (Fig. 6.1c) using both an analytical and a numerical solution approach. The computer simulations were carried out using a computer code based on finite difference scheme solutions of parabolic wave equations [43]. For the incident radiation both synchrotron radiation sources and table-top laboratory sources have been simulated.

6.3.1 Plane Wave Incoming Radiation

We based our analysis of the mode structure inside and outside the planar waveguide on the numerical solution of the parabolic wave equations

$$\begin{aligned} &2\mathrm{i}k\,\partial U/\partial x + \partial^2 U/\partial y^2 = 0, \quad \text{in vacuum} \\ &2\mathrm{i}k\,\partial U/\partial x + \partial^2 U/\partial y^2 + k^2(\varepsilon - 1)\,U = 0, \quad \text{inside WG} \end{aligned} \tag{6.14}$$

$$E(x, y) = U(x, y)\exp(\mathrm{i}kx), \tag{6.15}$$

where $U(x,\ y)$ is the complex amplitude of the electric field $E(x,\ y)$ which changes slowly along the optical axis 0X. ε is the dielectric constant of the materials, and as usual the refractive index $n = \varepsilon^{1/2} = (1 - \delta - \mathrm{i}\beta)$ and $k = 2\pi/\lambda$ is the wavevector. As before, x is parallel to interfaces in the plane of incidence, y is perpendicular to guiding layer, and z is parallel to it. It is assumed that the dimensions along z are much higher than along y; therefore, the field can be considered constant along z (planar waveguide).

Because of the high brilliance and of the large distance between source and WG, synchrotron radiation is well approximated, in general, by a plane wave. In the pre-reflection geometry the plane wave with grazing incidence angle θ_m is totally reflected just in front of the WG entrance, giving rise to a standing wave above the reflecting surface with the periodicity corresponding to a resonance mode. θ_m corresponds to the resonance mode condition

$$\tan\left[\frac{k\sin\theta_m\, d - (m-1)\pi}{2}\right] = \frac{\sqrt{\cos^2\theta_m - \cos^2\theta_\mathrm{c}}}{\sin\theta_m}, \tag{6.16}$$

where d is the width of the vacuum gap between the two cladding layers, m is the mode number; θ_c is the critical angle for total reflection.

Equation (6.16) is equivalent to (6.2), but takes into account in a quantitative way the penetration of the electromagnetic waves into the cladding layers.

Simulations presented in Fig. 6.5a refer to first resonance from a WG with Pt cladding layers and gap of 100 nm at $\lambda \approx 0.154$ nm. The spatial intensity distribution in front, inside and after the WG is shown. Figure 6.5b shows in more detail the comparison for the same WG between the simulation result of the intensity distribution perpendicular to the WG length (bold lines) and the analytical formulation (empty circles) of the solution U_m of (6.15) for the mode m, given by [26]

$$U_m(y) = \begin{cases} \exp(\delta_m y) & y < 0 \\ \cos(\kappa_m y) + (\delta_m/\kappa_m)\sin(\kappa_m y) & 0 \le y \le d, \\ [\cos(\kappa_m d) + (\delta_m/\kappa_m)\sin(\kappa_m d)]\exp[-\delta_m(y-d)] & y > d \end{cases} \tag{6.17}$$

where $\kappa_m = k\sin\theta_m$ and $\delta_m = \left[(1-\varepsilon)k^2 - \kappa_m^2\right]^{1/2}$.

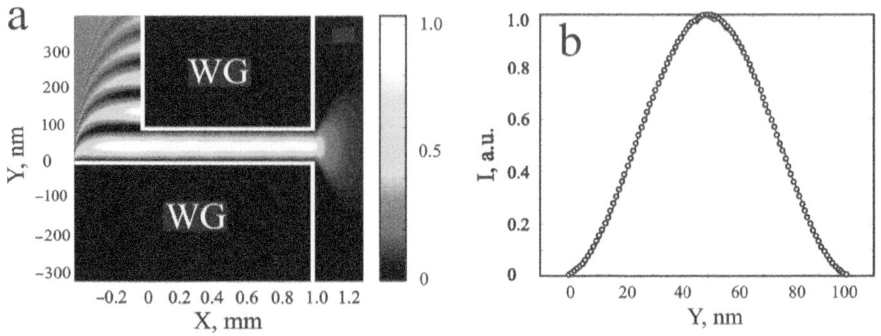

Fig. 6.5. (**a**) Intensity distribution in monomodal regime for the first mode in front of and along a front coupling WG, (**b**) Intensity profile along 0Y of the same waveguide calculated with computer code (*full line*) and with (6.17) (*open circles*)

The distribution of intensity for the first mode in free space close to the WG exit coincides with the diffraction pattern of a confined wave with amplitude given by (6.17) by a slit with size d corresponding to the WG gap. In the near field zone, i.e., $\Delta x < x_{\text{dif}} \sim d^2/4\lambda = 1.6 \times 10^4$ nm, where Δx is the distance from the WG end, the beam cross section is practically constant ($\sim d$). At $\Delta x = x_{\text{dif}}$ it is possible to note a focusing effect (see Fig. 6.7a) with a clear narrowing of the cross section. In the far field zone $x > x_{\text{dif}}$ the beam diverges with the divergence $\approx \lambda/d$.

6.3.2 Radiation from an Incoherent Source at Short Distance

As mentioned before, the WG accepts and transmits the coherent part of radiation. WGs are used mostly with synchrotron radiation sources, but it has been demonstrated that they can provide useful intensity also with table-top laboratory sources [15,44]. For the latter we have to investigate the effect of the illumination of the waveguide entrance with spatially incoherent radiation on the properties of the exiting beam. This may enable us to derive the geometrical condition for optimized coupling. For excitation of independent modes in the front coupling waveguide the next condition, based on the definition of the transverse coherence length, L_c, should be fulfilled:

$$2d \leq L_c = \lambda X/s, \qquad (6.18)$$

where X is the distance between source and waveguide, d is the thickness of the guiding layer, $2d$ is the spatial acceptance of the waveguide, s is the size of the source, and λ is the wavelength of the incoming radiation. If the equality sign holds in (6.18), the distance X can be considered as $X_{\min}$, the minimum distance for coherent illumination. If $d = 100$ nm, $s = 15\,\mu$m, and $\lambda = 0.154$ nm then $X_{\min} = 20$ mm. An incoherent source of radiation in computer code is considered as a set of N radiators with random initial phases distributed

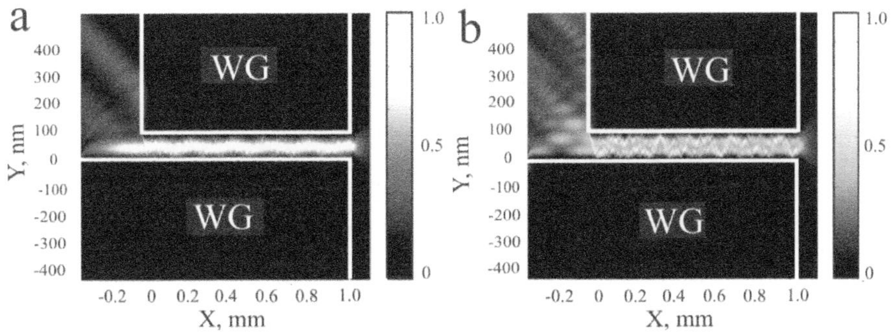

Fig. 6.6. Intensity distribution in front of and inside a WG illuminated by an extended source. Case (**a**) is for a distance $X_{\min} = 20\,\text{mm}$, which corresponds to the lower limit for spatially coherent illumination for a gap $d = 100\,\text{nm}$, a source size $s = 15\,\mu\text{m}$, and $\lambda = 0.154\,\text{nm}$ (Pt cladding). In case (**b**) at $X = 10\,\text{mm}$ the illumination is spatially incoherent under the same conditions

within the interval $[-\pi, \pi]$. The elementary radiator gives rise to a spherical wave with the origin positioned randomly within the source area s.

Figure 6.6a, b illustrates the intensity distribution calculated following the code described above, in front and inside a planar WG with *Pt* cladding layer, for a distance $X = X_{\min}$ and $X_{\min}/2$, respectively. The gap was $d = 100\,\text{nm}$, the source size $s = 15\,\mu\text{m}$, and $\lambda = 0.154\,\text{nm}$. Figure 6.7a, b shows in more detail the distribution at the exit of the WG, and Fig. 6.7c, d the cross-section profiles just at the WG exit at a distance $x = x_{\text{dif}}$ for $X = X_{\min}$ and $X = X_{\min}/2$, respectively. As can be seen, for $X = X_{\min}$ the WG can provide a single mode coherent beam, whilst for $X = X_{\min}/2$ a clear mode mixing takes place.

6.3.3 Material and Absorption Considerations

In the front coupling mode with an air or vacuum gap, absorption losses are due only to the penetration of the tails of the intensity distribution into the cladding material. Rigorous calculations should involve physical optical considerations, but simple ray-tracing can give very good approximate values for the estimation of absorption losses.

In the case of independent mode propagation, the variation of power along the WG length X ($\theta << \theta_c$) is given by [45]

$$W(X) = W_0 {R_{\text{Fr}}}^N(\theta_m) \approx W_0 \exp\left(-\frac{\beta}{\delta^{3/2}} \frac{X}{d\sqrt{2}} \theta_m^2\right), \tag{6.19}$$

W_0 is the energy of radiation coupled by the waveguide, R_{Fr} is the Fresnel coefficient of reflection for the vacuum–cladding layer boundary, $N = X\theta_m/d$

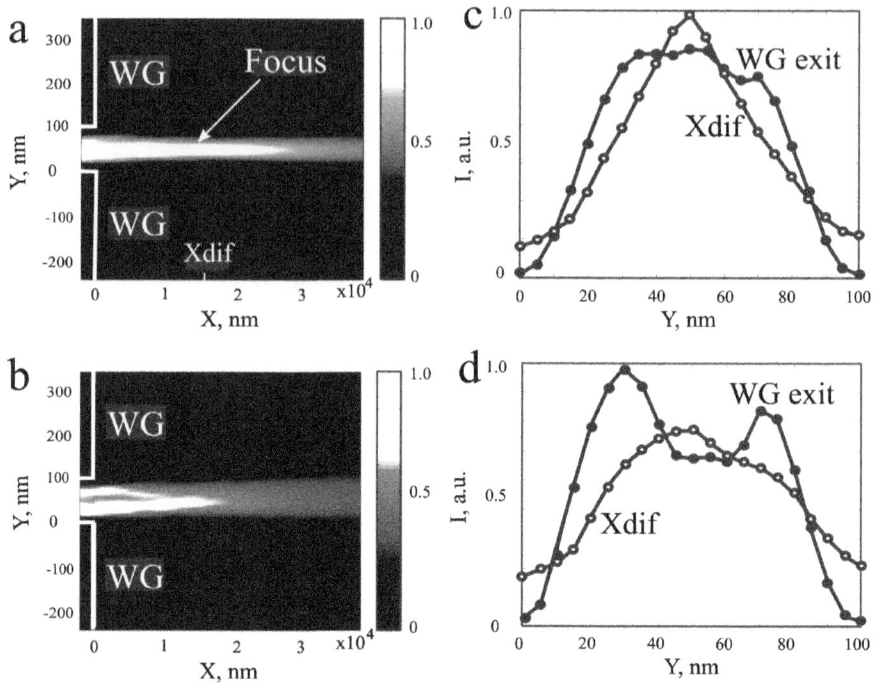

Fig. 6.7. The intensity distribution in the vicinity of the exit of the waveguide illuminated by an incoherent source located at the distance $X_{\min} = 20\,\text{mm}$ (**a**) and $X = 10\,\text{mm}$ (**b**). The vertical intensity distribution just at the exit of the WG and at a distance from the waveguide end corresponding to x_{dif} for $X = 20\,\text{mm}$ (**c**) and $X = 10\,\text{mm}$ (**d**) (see text)

is the number of reflections that the ray undergoes. The same function $W(X)$ can be calculated using the computer code. A very good correspondence is found between the calculated X_{abs} according to (6.19) and the result of a computer simulation based on the solution of parabolic wave equation.

6.4 Direct Front Coupling

When the incoming beam is directly coupled to the waveguide (see Fig. 6.1b), the interaction of the beam with the cladding layers must be considered in detail, especially if the cladding material for the photon energy considered is weakly absorbing. This analysis, which reveals several interesting diffraction and refraction phenomena, substantially modify the wave field in the waveguide. For a more complete treatment see [46].

6.4.1 Diffraction from a Dielectric Corner

Let us have an S-polarized plane wave of wavelength $\lambda = 0.1\,\text{nm}$ incident at right angles to the side of a planar hollow X-ray waveguide. The gap d is limited by two cladding walls with refractive index $n = \varepsilon^{1/2} = (1 - \delta - \mathrm{i}\beta)$. In the following, silicon is considered as the material constituting the walls. At the given photon energies in this paper $\beta << \delta$. We start by considering a single dielectric corner (half of the waveguide in our case). In this case Kopylov and Popov [47] have shown that the diffracted field $U(x,\ y)$ can be expressed, in the paraxial approximation, as

$$U(x,y) = \begin{cases} F(\eta) + M(\eta,\nu), & \text{vacuum} \\ (F(\nu) + M(\eta,\mathrm{i}\nu))\exp(-\nu^2), & \text{material} \end{cases} \tag{6.20}$$

$$M(\eta,\nu) = (\pi\mathrm{i})^{-1}\int_{-\infty-\mathrm{i}a}^{\infty-\mathrm{i}a} \exp(-t^2 - 2t\eta\sqrt{\mathrm{i}})/\sqrt{t^2-\nu^2}\,\mathrm{d}t$$

$$\approx M_1(x,y) + M_2(x,y)$$

$$\eta = |y|\sqrt{k/(2x)}, \quad \nu = \sqrt{kx(\beta + \mathrm{i}\delta)}, \quad k = 2\pi/\lambda,$$

where $F(x,\ y)$ is the Fresnel integral and $M(x,\ y)$ is a new special function whose influence is more significant for weakly absorbing materials. In the approximation of relatively large distances, x, from the WG entrance ($x >> 1/(k\theta_{\mathrm{c}}{}^2)$ with $\theta_{\mathrm{c}} = (2\delta)^{1/2}$ the critical angle for total reflection), the function $M(x,\ y)$ can be expressed asymptotically as a sum of two terms

$$\begin{aligned} M_1(x,y) &\approx \sqrt{\frac{2x}{\pi k}}\exp\left(\mathrm{i}\frac{ky^2}{2x} - \mathrm{i}\frac{\pi}{4}\right)\Big/\sqrt{\theta_{\mathrm{c}}^2x^2 - y^2}, \\ M_2(x,y) &\approx \sqrt{\frac{2x}{\pi k}}\exp\left(\mathrm{i}k(\theta_{\mathrm{c}}y - \frac{\theta_{\mathrm{c}}^2}{2}x) + \mathrm{i}\frac{\pi}{4}\right)\Big/\sqrt{\theta_{\mathrm{c}}x(\theta_{\mathrm{c}}x - y)}. \end{aligned} \tag{6.21}$$

The first one, $M_1(x,\ y)$, is a correction to the Fresnel edge diffraction term due to the material of the wall, and the second one, $M_2(x,\ y)$, represents a lateral plane wave propagating in the wall material along the material–vacuum interface 0X, with the enhanced phase velocity $V_{\mathrm{p}} = c/\varepsilon^{1/2}$ and entering into the vacuum at the critical angle θ_{c} [47]. In (6.20) and (6.21), the origin of the y coordinate is at the vacuum–wall interface.

The superposition of the direct and diffracted beams with the lateral wave gives rise to an interference pattern of successive maxima and minima (see Fig. 6.2 in [46]). Coordinates of maxima in the spatial intensity distribution can be found from the condition for constructive interference of wave fields described in (6.20) and (6.21). The comparison between the spatial distribution of intensity calculated analytically using the above equations, and the result of a computer simulation based on the parabolic wave equation (PWE) numerical solution [43], reveals a very good agreement between the two [46].

6.4.2 Diffraction in a Dielectric FC Waveguide

We extended the same formalism to the analysis of the field at the entrance aperture of the waveguide. In the following, the origin of the y coordinate is in the middle of the gap d, and the cladding walls are at $\pm d/2$. An approximate solution in the far field zone ($x > (d/2)^2/\lambda$) is the superposition of the field

$$\Phi(x,y) \approx \tilde{\Phi}(k\theta)\exp(-\mathrm{i}\pi/4 + \mathrm{i}k\theta^2 x/2)/\sqrt{\lambda x},$$
$$\text{where} \quad \tilde{\Phi}(k\theta) \approx d\left[\frac{\sin(k\theta d/2)}{k\theta d/2} + \frac{\cos(k\theta d/2)}{(kd/2)\sqrt{{\theta_c}^2 - \theta^2}}\right], \tag{6.22}$$

with two lateral plane waves, $M_2(x,\ y)$ (see (6.21)), entering into the vacuum gap from the opposite boundaries $y = \pm d/2$ of waveguide. In (6.22) $\theta = y/x$. The spatial spectral amplitude $\tilde{\Phi}(k\theta)$ in (6.22) includes the sin function of $(k\theta d/2)$, corresponding to the Fraunhofer diffraction of a plane wave from a thin slit and a correction term due to the material of the walls. The correction term shifts the positions of the angular spectrum maxima towards smaller angles. It is easy to show that the spectral amplitude $\tilde{\Phi}(k\theta)$ in (6.22) is equal to

$$\tilde{\Phi}(k\theta) \approx \int_{-\infty}^{+\infty} \varphi(y)\mathrm{d}y, \quad \text{where} \tag{6.23}$$
$$\varphi(y) = \begin{cases} \cos(k\theta y), & |y| < d/2 \\ \cos(k\theta d/2)\exp[-k\mu\,(|y| - d/2)]\,, & \text{else,} \end{cases}$$

where $\mu = ({\theta_c}^2 - \theta^2)^{1/2}$. For θ values equal to the waveguide resonance angles θ_m, the function $\varphi(\theta_m) = \varphi_m$ corresponds exactly to the expression of guided modes. Taking the orthogonal modes $\{\varphi_m\}$ of the waveguide as a basis, the projection of the field $\Phi(x,y)$ on the guided modes at distances $x > x_{\min} = (d/2)^2/\lambda$ is given by [26]

$$\Phi(x,y) = \sum\nolimits_{m=0}^{m=m_{\max}} c_m(\theta_m)\varphi_m(y), \tag{6.24}$$

where the coefficients c_m are given by

$$c_m(\theta_m) = \|\varphi_m\|^{-1}\int_{-\infty}^{+\infty} \varphi_m(y)\mathrm{d}y. \tag{6.25}$$

θ_m are the resonance angles, $\mu_m \approx ({\theta_c}^2 - {\theta_m}^2)^{1/2}$, and $m_{\max}$ is the maximum number of allowed resonance modes. Taking into account the propagation factor $\exp(-\mathrm{i}\chi_m x)$ for each mode, where in the parabolic approximation $(\theta << \theta_c)\chi_m \approx \theta_m^2[k/2 - \mathrm{i}(\beta/\delta^{3/2})/(2^{1/2}d)]$ [45], the wave field $\Phi(x,y)$ at any point of the waveguide is given by

$$\Phi(x,y) = \sum\nolimits_{m=0}^{m=m_{\max}} c_m(\theta_m)\varphi_m(y)\exp(-\mathrm{i}\chi_m x). \tag{6.26}$$

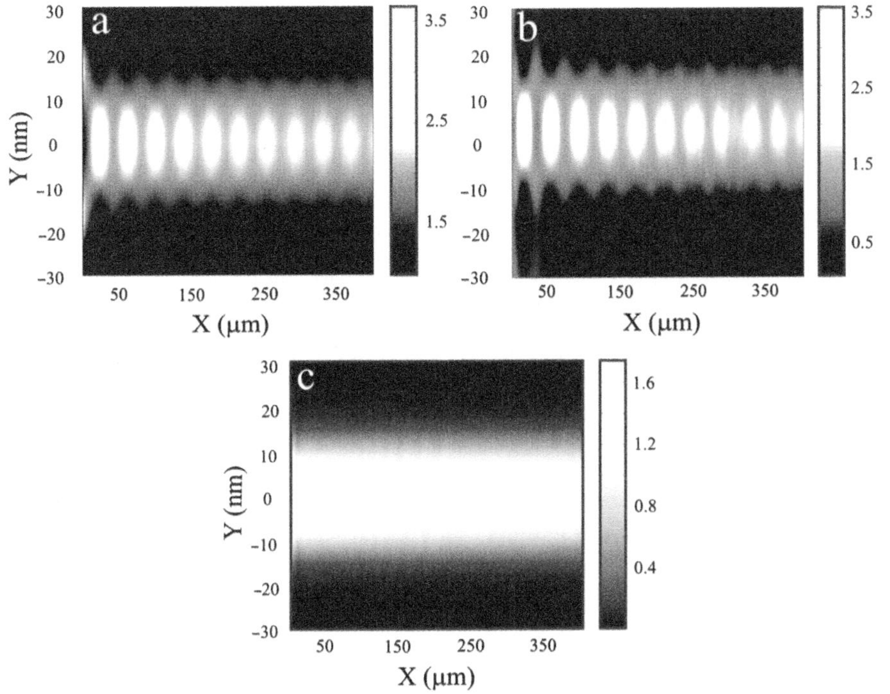

Fig. 6.8. Total field in a waveguide with Si walls and a 30 nm gap (wavelength = 0.1 nm) with a plane wave at the entrance: (**a**) analytical solution; (**b**) computer simulation; (**c**) computer simulation, with a step function field ($U = 1$ in the gap, 0 elsewhere) at the entrance (from Bukreeva et al. [46] with permission of the publisher)

The total field $U(x,\ y)$ is therefore given by the superposition of propagating modes (6.26) and function, which represents a sum of two lateral waves $M_2(x,\ y)$ (see (6.21)) entering into the vacuum gap from the opposite boundaries, $y = \pm d/2$, of the waveguide

$$\Psi\left(x, y\right) \approx \frac{d}{\sqrt{\lambda x}} \exp\left(\mathrm{i}\pi/4 - \mathrm{i}k\theta_{\mathrm{c}}^{2} x/2\right) \frac{\cos\left(k\theta_{\mathrm{c}} y\right)}{\mathrm{k}\theta_{\mathrm{c}} d/2} \exp\left(\mathrm{i}k\theta_{\mathrm{c}} d/2\right). \qquad (6.27)$$

In Fig. 6.8 the global intensity distribution in the vacuum guiding layer for a 30-nm gap waveguide with Si walls and photon wavelength $\lambda = 0.1$ nm is shown. The waveguide supports only one mode. Figure 6.8a depicts the analytical solution given by (6.26) and (6.27), and Fig. 6.8b represents the result of the computer simulation based on the numerical solution of the parabolic wave equation (6.14). The agreement is very good. Figure 6.8c shows the intensity distribution when the field at the waveguide entrance is a step function ($U(0,\ y) = 1$ for $y \in [-d/2,\ d/2]$ and $U(0,\ y) = 0$ elsewhere), and therefore penetration through the cladding walls is excluded. Figures 6.8a–c shows that the interference of a guided mode with lateral waves introduces a strong spatial modulation of the signal.

A more quantitative comparison, not shown here for reasons of space, gives a very good agreement between the asymptotic solution and the computer simulation for both the distribution of the intensity and of the phase of the resulting field along the optical axis 0X (see Fig. 6.1).

The contribution of the field diffracted and refracted by the cladding walls is not only related to the spatial modulation of the signal. Both analytical and computer calculations show that the field penetrating into the waveguide from the weakly absorbing cladding walls significantly increases (approximately 1.5 times) the electromagnetic power in the waveguide compared to the case when the field at the waveguide entrance is a simple step function, thereby eliminating penetration through the cladding walls. In Fig. 6.9 we report the normalized integrated power within the WG gap as a function of propagation distance x.

We have shown here that the calculation of the field in a hollow weakly absorbing X-ray waveguide must take into account the interaction of the incoming beam with the cladding material at the entrance of the waveguide. The total field in the guiding vacuum layer can then be expressed as the superposition of guided modes with nonuniform plane waves penetrating into the guiding gap from the cladding walls at the critical angle of reflection, θ_c (lateral waves). An analytical expression of the total field is given and compared

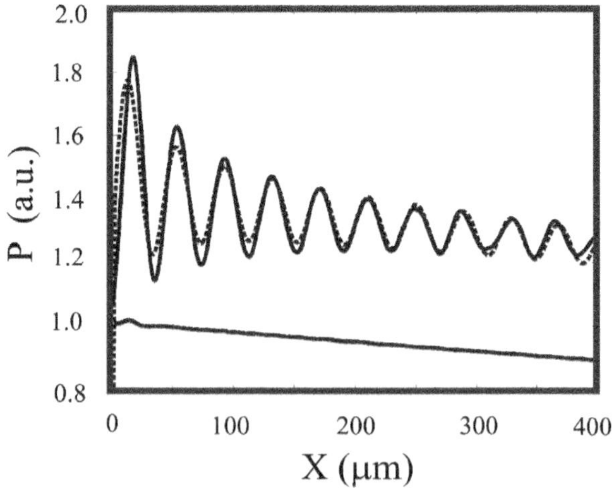

Fig. 6.9. The normalized value of the electromagnetic power integrated within the vacuum gap vs. coordinate X calculated for a step-like entrance function (*bottom lines*) and for the total field calculated following (6.25) and (6.26) (*top lines*). The solid lines are the result of a computer simulation while the dashed lines are the result of an analytical calculation (from Bukreeva et al. [46] with permission of the publisher)

with the results of a computer simulation based on the numerical solution of the parabolic wave equation. The two independent approaches to the diffraction problem, the asymptotic analytical solution and a computer simulation, demonstrate very good qualitative and quantitative agreement.

6.5 Conclusions

The analysis of the behavior of WGs in three different coupling geometries has led us to some conclusions about their optimal use as a function of the source characteristics. We have shown in Sect. 6.2 that the angular acceptance of RBC WGs is by far larger than the beam divergence of most synchrotron radiation beam lines, at least at photon energies below 50 keV and gap values below 100 nm. On the other hand, the spatial acceptance is much smaller than the full beam size at the distances typical of synchrotron radiation facilities. Therefore, to maximize the total flux, a prefocusing optics with quite long focal distance, providing an input beam for WG matched to its spatial and angular acceptances, should be used. Similar considerations are valid for front coupling WGs, but in this case the spatial acceptance is on the order of the gap value, and the angular acceptance is larger than in the RBC case. Therefore, focusing optics with a much shorter focal distance are best matched, as demonstrated also from the experimental point of view in 2D WGs [9]. In any case, irrespective of the coupling mode, the total available flux is limited by the coherent flux, as expressed by (6.1). It is interesting to note that this is a limit applicable to any kind of optics if the requirement of a coherent beam holds. This is an interesting point to consider especially when designing optics for coherent sources such as free electron lasers. The emerging field of coherent diffractive imaging [48] is attracting great interest for its impressive potential in the structural determination of nonperiodic objects with nanometer resolution. WGs, because of their characteristics, are ideally suited for this kind of application. A first demonstration of the possibility to make coherent diffractive imaging experiments using the beam from a WG has been recently carried out [49]. Another aspect to consider in regard to the best coupling mode is that RBC cannot be used in the soft X-ray range, because of the strong absorption from the cover layer. Therefore, in this spectral region only FC (with or without prereflection) must be used. In this case, it has been shown here that careful considerations of all the refractive and diffractive effects resulting from interaction of the incoming beam with the cladding must be taken into proper account to predict the field distribution and the intensity inside the WG.

Acknowledgements

The authors gratefully acknowledge partial financial support from Italian projects SPARC and SPARX.

References

1. E. Spiller, A. Segmüller, Appl. Phys. Lett. **24**, 60 (1974)
2. M.J. Bedzyk, G.M. Bommarito, J.S. Schildkraut, Phys. Rev. Lett. **62**, 1376 (1989)
3. J. Wang, M.J. Bedzyk, M. Caffrey, Science **258**, 775 (1992)
4. P. Feng, S.K. Sinha, H.W. Deckman, J.B. Hastings, D.P. Siddons, Phys. Rev. Lett. **71**, 537 (1993)
5. Y.P. Feng, S.K. Sinha, E.E. Fullerton, G. Grübel, D. Abernathy, D.P. Siddons, J.B. Hastings, Appl. Phys. Lett. **67**, 24 (1995)
6. S. Lagomarsino, W. Jark, S. Di Fonzo, A. Cedola, B.R. Müller, C. Riekel, P. Engstrom, J. Appl. Phys. **79**, 4471 (1996)
7. W. Jark, S. Di Fonzo, S. Lagomarsino, A. Cedola, E. di Fabrizio, A. Bram, C. Riekel, J. Appl. Phys. **80**, 4831 (1996)
8. W. Jark, A. Cedola, S. Di Fonzo, M. Fiordelisi, S. Lagomarsino, N.V. Kovalenko, V.A. Chernov, Appl. Phys. Lett. **78**, 1192 (2001)
9. F. Pfeiffer, C. David, M. Burghammer, C. Riekel, T. Salditt, Science **297**, 230 (2002) and http://www.sciencemag.org/cgi/data/297/5579/230/DC1/1
10. A. Jarre, C. Fuhse, C. Ollinger, J. Seeger, R. Tucoulou, T. Salditt, Phys. Rev. Lett. **94**, 074801 (2005)
11. C. Bergemann, H. Keymeulen, J.F. van der Veen, Phys. Rev. Lett. **91**, 204801 (2003)
12. F. Pfeiffer, T. Salditt, P. Høghøj, I. Anderson, N. Schell, Phys. Rev. B **62**, 16939 (2000)
13. W. Jark, S. Di Fonzo, J. Synchrotron Radiat. **11**, 386 (2004)
14. W. Jark, S. Di Fonzo, in *SPIE – Proceedings: Design and Microfabrication of Novel X-ray Optics II.* vol. 5539, (2004) p. 138
15. D. Pelliccia, A. Cedola, F. Scarinci, S. Lagomarsino, J. Phys. D: Appl. Phys. **38**, 1 (2005)
16. S. Di Fonzo, W. Jark, S. Lagomarsino, C. Giannini, L. De Caro, A. Cedola, M. Müller, Nature **403**, 638 (2000)
17. M. Müller, M. Burghammer, D. Flot, C. Riekel, C. Morawe, B. Murphy, A. Cedola, J. Appl. Cryst. **33**, 1231 (2000)
18. A. Cedola, M. Mastrogiacomo, M. Burghammer, V. Komlev, P. Giannoni, R. Cancedda, F. Rustichelli, A. Favia, S. Lagomarsino, Phys. Med. Biol. **51**, N109, 6 (2006)
19. M.J. Zwanenburg, J.H.H. Bongaerts, J.F. Peters, D.O. Riese, J.F. van der Veen, Phys. Rev. Lett. **85**, 5154 (2000)
20. T. Salditt, F. Pfeiffer, H. Perzl, A. Vix, U. Mennicke, A. Jarre, A. Mazuelas, T.H. Metzger, Phys. B **336**, 181 (2003)
21. S. Lagomarsino, A. Cedola, in *Encyclopedia of Nanoscience and Nanotechnology*, vol. 10, ed. by H.S. Nalwa (American Scientific Publisher, CA, U.S.A., 2004), p. 681
22. A. Cedola, S. Lagomarsino, Synchrotron Radiat. News (special issue) **17**, 30 (2004)
23. A. Cedola, D. Pelliccia, I. Bukreeva, A. Surpi, F. Scarinci, S. Lagomarsino, *X-ray Micro and Nano-probes: Fundamentals and Applications*, Transworld Research Network, Recent Developments in Applied Physics, **9**, ISBN: 81-7895-213-0 (2006)

24. L. De Caro, C. Giannini, S. Di Fonzo, W. Jark, A. Cedola, S. Lagomarsino, Opt. Commun. **217**, 31 (2003)
25. D. Attwood, *Soft X-Rays and Extreme Ultraviolet Radiation: Principles and Applications*, Chap. 8 (Cambridge University Press, Cambridge, 1999)
26. D. Marcuse, *Theory of Dielectric Waveguides* (Academic Press, San Diego, 1991)
27. http://www-cxro.lbl.gov/optical_constants/
28. http://physics.nist.gov/ffast
29. B.L. Henke, E.M. Gullickson, and J.C. Davis, At. Data Nucl. Data Tables **54**, 181 (1993)
30. C.T. Chantler, K. Olsen, R.A. Dragoset, A.R. Inshore, S.A. Kotochigova, D.S. Zucker, *X-Ray Form Factor, Attenuation and Scattering Tables* (version 2.0), http://physics.nist.gov/ffast (National Institute of Standards and Technology, Gaithersburg, MD, 2003)
31. S. Lagomarsino, A. Cedola, S. Di Fonzo, W. Jark, V. Mocella, J.B. Pelka, and C. Riekel, Cryst. Res. Technol. **37**, 758 (2002)
32. L.G. Parratt, Phys. Rev. **95**, 395 (1954)
33. M.J. Zwanenburg, J.H.H. Bongaerts, J.F. Peters, D. Riese, and J.F. van der Veen, Phys. B **283**, 285 (2000)
34. F. Pfeiffer, Diploma thesis, Ludwig-Maximilians-Universität München, Sektion Physik, Munich, Germany, 1999
35. C.G. Schroer, B. Lengeler, B. Phys. Rev. Lett. **94**, 054802 (2005)
36. A.G. Michette, S.J. Pfauntsch, A. Erko, A. Firsov, A. Svintsov, Opt. Commun. **245**, 249 (2005)
37. H.C. Kang, J. Maser, G.B. Stephenson, C. Liu, R. Conley, A.T. Macrander, S. Vogt, Phys. Rev. Lett. **96**, 127401 (2006)
38. B. Lengeler, C.G. Schroer, J. Tümmler, B. Benner, M. Richwin, A. Snigirev, I. Snigireva, M. Drakopoulos, J. Synchrotron Radiat. **6**, 1167 (1999)
39. M. Born, E. Wolf, *Principle of Optics*, 6th edn (Pergamon Press, New York, 1980)
40. P. Karimov, E.Z. Kurmaev, Phys. Lett. A **320**, 234 (2003)
41. N.V. Kovalenko, V.A. Chernov, Budker Institute of Nuclear Physics, Novosibirsk (Russia), private communication (2000)
42. P. Karimov, J. Kawal, E.Z. Kurmaev, Anal. Sci. **21**, 811 (2005)
43. Y.V. Kopylov, A.V. Popov, A.V. Vinogradov, Opt. Commun. **118**, 619 (1995)
44. V.K. Egorov, E.V. Egorov, X-Ray Spectrometry, **33**, 360 (2004)
45. A.V. Vinogradov, I.A. Brytov, A. Ya. Grudsky, M.T. Kogan, I.V. Kozhevnikov, V.A. Slemzin, *Mirror X-Ray Optics* (in Russian) (Mashinostroenie, Leningrad, 1989)
46. I. Bukreeva, A. Popov, D. Pelliccia, A. Cedola, S. Dabagov, S. Lagomarsino, Phys. Rev. Lett. **97**, 184801 (2006)
47. Y.V. Kopylov, A.V. Popov, Radio Sci. **31**, 1815, (1996)
48. J. Miao, P. Charalambous, J. Kirz, D. Sayre, Nature **400**, 342 (1999)
49. L. De Caro, C. Giannini, A. Cedola, D. Pelliccia, S. Lagomarsino, W. Jark, Appl. Phys. Lett. **90**, 041105 (2007)

7

Focusing Optics for Neutrons

F. Ott

Abstract. Neutrons beams are difficult to handle since the neutron is a neutral particle with a very weak interaction with matter. In addition, neutron sources are broad and isotropic, which makes it very challenging to provide high neutron fluxes at sample positions in order to perform scattering experiments. Despite these problems techniques have been developed that allow high neutron fluxes to be brought to the sample position, focused over small areas. The whole range of optical solutions has been considered, namely refraction, reflection and diffraction. Progress in the field of neutron optics has accelerated in the last decade and as a consequence has provided major improvements for all types of neutron spectrometers and their applications.

7.1 Introduction

In the past decade, the request for higher neutron flux at the sample position has driven research for advanced focusing neutron optics. Three main principles have guided the techniques of focusing neutrons:

- Refraction (refractive lenses or magnetic lenses)
- Reflection (focusing guides, reflective lenses, capillary optics)
- Diffraction (Fresnel zone plates)

We discuss and compare the performances of different focusing devices based on these principles. Presently, on neutron scattering spectrometers, the main request is for an illumination spot of the order of a few mm up to 10 mm. Only in exceptional cases is submillimeter focusing requested. Neutron spectrometers are usually rather large and focal lengths of several meters can generally be accommodated. Focal lengths below 1 m are also requested for some applications.

7.2 Characteristics of Neutron Beams

Neutrons are created in a uranium core and then moderated in a tank of heavy water at 50°C. Their energy is then of the order of 25 meV. In a number of experiments, neutrons with lower energies are required and they must be cooled down further to energies below 5 meV. This takes place in so-called "cold sources." This cold source is a metal canister containing about 20 l of liquid hydrogen or deuterium (at 25 K). This source has an isotropic emission, and the neutrons are conducted to spectrometers using neutron guides consisting of neutron reflecting walls. The cross section of the neutron tubes has reached considerable dimensions, up to 100 cm^2. The wavelength distribution of the neutron spectrum is Maxwellian centered on the source temperature. This corresponds to neutron wavelengths ranging from 0.2–2 nm. The beam divergence in the guides is given by the angle of total reflection of the guide coatings. For nickel coatings it is equal to $0.1°\ \text{Å}^{-1}$. Typically for $m = 2$ supermirror guides, the divergence is of the order of $0.2°\ \text{Å}^{-1}$.

There are fundamental differences in the phase space distribution between (a) synchrotron X-ray beams, which are intrinsically small, and well collimated, (b) X-ray tubes in which the source is almost a point, but the divergence is high and (c) neutron sources which are both large in size and have a large divergence (Table 7.1).

If one puts in perspective what is expected from X-ray and neutron sources, a comparison of typical values for the beam size and beam divergence shows that, more or less, the neutron phase space is approximately 1,000 times less dense than X-ray phase space. The ultimate aim in focal spots size is thus also within this 1,000 scaling factor.

The wavelength spectrum of neutrons is quite limited compared to synchrotron radiation, ranging typically from 0.5 up to 30 Å. The optical index of neutrons is of the same order of magnitude as the optical index of X-rays with the same wavelength. One advantage of neutrons is that their absorption by matter is very small. This enables one to build devices which would not be feasible for X-rays. Since the neutron flux is much lower than the photon flux, one must avoid as much as possible neutron losses, and devices with a very high efficiency are required (above 50%).

In order to maximize the number of available neutrons at the sample position, one is usually ready to compromise on the phase-space distribution.

Table 7.1. (**a**) Point source with low divergence (synchrotron); (**b**) divergent point source (X-ray tube); (**c**) divergent and extended source (neutrons)

	Synchrotron	X-ray tube	Neutrons
Beam size	100 μm	100 μm	100 mm
Divergence	10 μrad	1 rad	10 mrad (in guides)
Aimed spot	1 μm	100 μm	1 mm

A number of experiments require high resolution only in the scattering plane. In the direction perpendicular to the scattering, a downgraded resolution can often be accepted. In such a case, a spatially high beam is often vertically focused on the sample position. This results in a divergent beam in the vertical direction at the sample position. Going from a beam size of 100 mm down to 10 mm typically increases the flux by a factor 10 and increases the vertical divergence by a factor of 10.

Two possibilities can be considered to shape the phase space: reduce δx at the expense of $\delta\theta$ (classical focusing on the sample); increase δx to reduce $\delta\theta$, this last solution is never considered on scattering experiments since it would lead to nonpractical sample sizes, but it is being considered for new geometries of neutron guides in order to increase the guide transmission over long distances.

In the case of neutron focusing, a number of parameters affect the choice of a specific focusing solution:

- What is the wavelength spectrum? Short or long wavelengths? Broad or narrow wavelength spectrum?
- 1D or 2D focusing?
- Required focal length? It can range from 20 mm up to 10 m depending on the application (e.g., neutron activation or small angle scattering)
- What is the size of the focal spot?
- Polarized/non polarized?

Depending on these characteristics, different solutions should be considered. In the following, we will present the different devices which have been developed over the years to focus neutron beams. This field is presently evolving more rapidly than ever and a number of solutions are proposed which have not been implemented yet on real spectrometers.

7.3 Passive Focusing: Collimating Focusing

On all neutron spectrometers, the neutron beams need to be shaped to fulfill the requirements of the specific neutron scattering experiments. The beam divergence needs to be tuned with respect to the required resolution. The beam size needs to be tuned with respect to the sample size, in order to minimize the background noise. Thus, one usually "passively" focuses and collimates neutrons by taking advantage of the fact that the neutron beams are large and divergent. Instead of changing the phase space distribution, one simple reduces it. This is usually achieved by using so-called Soller collimators (see Fig. 7.1a, b). These devices work in one dimension only and are among the most frequently used neutron optical components. The collimating channels are defined by absorbing gadolinium oxide on PETP foils (thickness 50 μm). The device transmission is of the order of 90%.

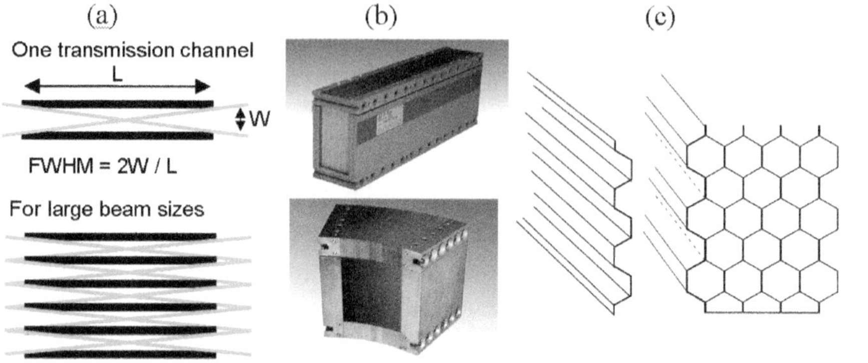

Fig. 7.1. (**a**) Principle of a Soller collimator; (**b**) Real 1D and radial collimators with a beam cross section of several cm^2; (**c**) Embossed aluminum foil (*left*); Honeycomb collimator (*right*) [1]

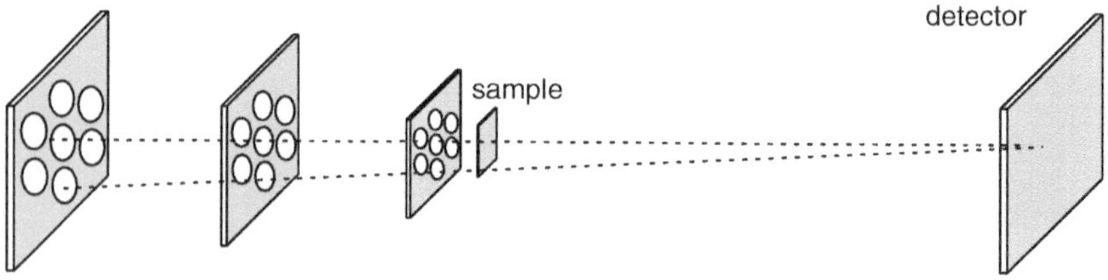

Fig. 7.2. Converging multibeam collimation

More advanced geometries have been developed. In order to perform a collimation in two dimensions, focusing honeycomb collimators have been developed [1]. They are made of embossed aluminum foils (0.1 mm thick) in a hexagonal shape which are coated with Gd oxide. They are then stacked together in order to form a honeycomb structure. The entire structure is 4 m long with channels of 7 mm at the entrance and 6 mm at the exit.

Another efficient geometry is the geometry of focusing small angle neutron scattering (SANS) spectrometers in which several well collimated neutron beams are focused on the detector. Again, this setup takes advantage of the fact that neutron beams are large and divergent. The principle was proposed and implemented very early (Fig. 7.2) [2]. Such a setup is most efficient for scattering at very small Q values which require very small collimations (pinholes $\varnothing \sim$ 1–2 mm). A simple pinhole geometry would provide a very small flux. This geometry has been revived recently for two reasons: (a) the new neutron guides are much bigger than earlier ones and this permits one to significantly multiply the number of incoming channels, even for big pinhole collimations ~10–20 mm; (b) very high resolution SANS spectrometers are required for new types of experiments. S. Désert at the Lab. Léon Brillouin is developing a converging multibeam very small angle neutron scattering

(VSANS) spectrometer which will use about 120 beams, multiplying the luminosity of the spectrometer by an equivalent factor.

7.4 Crystal Focusing

7.4.1 Focusing Monochromator

The earliest active focusing technique was implemented by combining the monochromatization process with the focusing. In typical neutron scattering experiments such as powder diffraction or triple axis spectroscopy (TAS), the sample size is of the order of $1\,\text{cm}^3$. Usually, no resolution is required in the direction perpendicular to the scattering plane. Thus, since the incoming beam is very high (>100 mm), the crystal monochromator can be curved to diffract most of the intensity onto the sample (Fig. 7.3a). Usually, the monochromator is not curved but divided into a number of subelements which have a size of the order of 10–20 mm. In some situations where the measurement is not disturbed by a significant divergence in the scattering plane, such as diffuse scattering or TAS measurements, double focusing monochromators can be used.

Figure 7.3b shows an example of such a monochromator. In this case, the curvature can be tuned with motorized movements. In the case of these focusing monochromators, the focal spot is defined by the monochromator crystal size (of the order of $1\,\text{cm}^2$) and its mosaic (of the order of 0.5° $\sim$10 mrad) which gives an extra broadening of the beam at the sample position ($1{,}000\,\text{mm} \times 10\,\text{mrad} = 10\,\text{mm}$). This technology is thus limited to focal spots of the order of $20 \times 20\,\text{mm}^2$. The system is quite expensive but has proven to be very efficient and reliable over the years. The limitation of the focal spot size is determined by the finite mosaic and the size of the individual monochromator crystals. This can be overcome with a more advanced technology using bent perfect crystals.

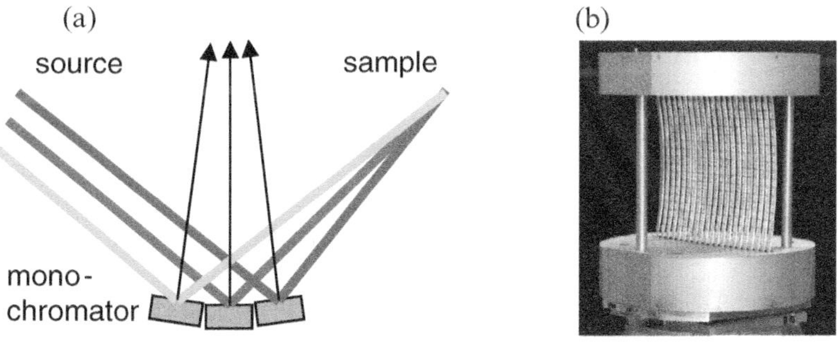

Fig. 7.3. (**a**) Focusing using a composite monochromator. Typically, the monochromator elements are 10–20 mm in size. The total height – width of the monochromator can be larger than $100 \times 100\,\text{mm}^2$. (**b**) Double focusing monochromator consisting of 357 graphite crystals (total surface $1{,}428\,\text{cm}^2$) (MACS, NIST) [3]

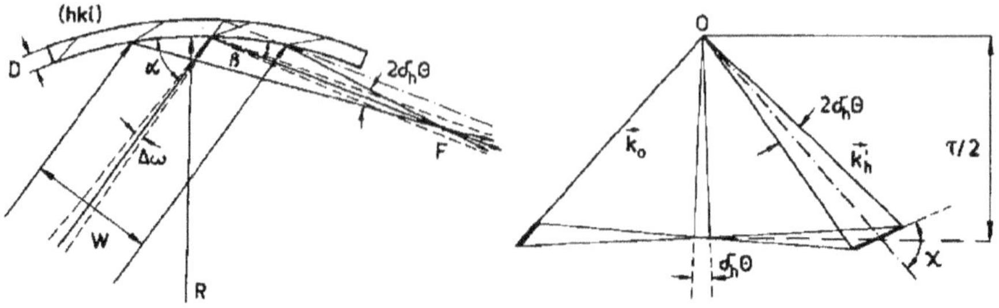

Fig. 7.4. (**a**) Horizontal focusing by a curved crystal in real space; (**b**) momentum-space representation of the diffraction by a bent crystal in the case $2\delta_{\mathrm{h}}\theta \gg \Delta\omega$ [4]

7.4.2 Bent Perfect Crystal Monochromators

Another possibility to focus neutrons consists in using elastically bent perfect crystals. The principle is described in Fig. 7.4 [4]. Bent perfect crystals have no mosaic structure. Their angular reflection profile is rectangular with a width given by simple geometrical factors: bending radius, thickness, and length of the irradiated part of the crystal. One of the crucial advantages of the silicon crystals is the absence of second-order contamination for the hkl-odd reflections, which avoids the use of pyrolitic graphite or beryllium filters. The use of such bent crystals is discussed in [5]. Even though perfect crystals are used, which suggests that their use should be restricted to high-resolution experiments, it is suggested that it can be a more efficient variant to combinations of mosaic crystals with Soller collimators. Such bent crystals have been implemented on strain scanners and on TAS instruments [6, 7].

7.5 Refractive Optics

7.5.1 Solid-State Lenses

A straightforward way of focusing a beam consists in using refractive optics.

Neutron Optical Index

The optical index for neutrons is given by:

$$n = 1 - \frac{\lambda^2}{2\pi}\rho b - \frac{\lambda}{\pi}\rho\sigma_{\mathrm{a}} = 1 - \delta - i\beta$$

where b is the coherent scattering length, ρ is the atomic density and σ_{a} is the absorption cross section. δ is the real part of the optical index, β is the imaginary part of the optical index corresponding to the absorption in the material. Neutron optical indices are extremely small for thermal neutron

Table 7.2. Comparison of the possible elements for refractive optics

Element	δ ($\times 10^{-6}$)	β ($\times 10^{-12}$)	δ/β (a.u.)	Comment
Be	150.0	7.31	**20.5**	Handling difficult
C	186.2	4.9	**38**	Diamond
Mg	36.8	21.5	1.7	
Al	33.0	110	0.3	
Si	33.0	68	0.5	
Ni	149.8	3200	0.05	High absorption
Zr	48.9	63.1	0.8	
Pb	49.4	44.9	1.1	
Al_2O_3	90.4	85	1.1	
MgF_2	80.7	15.20	**5.3**	
MgO	95.2	26.8	3.6	Not stable in air

wavelengths, $\delta \sim 10^{-6}$. Absorption is even smaller $\beta \sim 10^{-11}$. So even if the neutron refraction is very small at an interface, it is possible to multiply the number of lenses while keeping a good transmission. The quality factor of a material can be defined by the ratio δ/β, the ratio of the optical index to the absorption. Table 7.2 summarizes the optical index and absorption of some materials. Among single crystal oxide materials, Al_2O_3 (sapphire) and MgF_2 have rather large optical indices. The absorption of MgF_2 is very low and thus this material appears much better than sapphire. Moreover, MgF_2 is used in UV optics which makes MgF_2 lenses readily available as off-the-shelf components. This is the reason why most of the devices developed until now have used this material. Other materials present a very high potential for refractive optics such as Be and diamond. However beryllium is difficult to handle and machine, and thus Be refractive optics has been developed only for the field of X-rays. Until recently, the use of diamond was limited for cost and availability reasons. However, since 2005, commercial diamond lenses have been made available by Diamond Materials Ltd [8]. Until now they have been evaluated only in the field of X-rays.

Since the optical index is smaller than 1, converging neutron lenses are concave. This is an advantage since the geometry of concave lenses reduces the absorption in the middle of the lens.

Implementations

The first report of the use of a refractive lens on a neutron spectrometer was by Gahler et al. [9] for the search for the neutron electric dipole moment. But the first test of refractive lenses on a neutron scattering spectrometer was performed 20 years later by Eskildsen et al. [10, 11] following the proposal from Snigirev et al. [12] to use Compound Refractive Lenses (CRL) on hard X-ray spectrometers. A CRL made of 30 MgF_2 biconcave lenses was tested for wavelengths ranging from 0.9 to 2 nm. The focal length of the device ranged

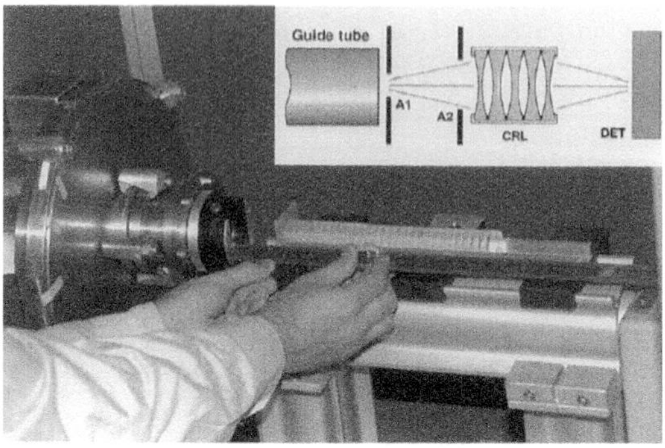

Fig. 7.5. A set of 20 MgF2 lenses implemented on a SANS machine [10]

from 1 to 6 m. Such a lens was implemented in a focusing SANS geometry (see Fig. 7.5). The experimental results showed that gains in intensity above 15 could be achieved.

Besides focusing neutrons in SANS, it has also been proposed that refractive optics could be used to implement a neutron microscope [13].

Advantages-Drawbacks

The potential advantages of such systems are quite clear: the systems are inexpensive (MgF_2 lenses are available as standard optical components) and very flexible (one simply needs to remove lenses to change the focal length as a function of the wavelength).

However, a number of drawbacks have hampered the generalization of the use of refractive lenses on neutron spectrometers.

The low-refractive optical index reduces the use of such optics to long wavelengths ($\lambda > 1$ nm) and long focal lengths (several meters).

Neutron absorption in the optical elements is a second drawback. As soon as one needs to handle large beams (>10 mm) which is the general situation on neutron spectrometers, the transmission of the system decreases very quickly, down to 10% for a beam diameter of 25 mm. This is due to two reasons. On the one hand, the thickness of the lens increases when one goes away of the central optical axis. Attempts have been made to build Fresnel lenses [14, 15]. These attempts have not been successful since spurious refraction effects appear and since the machining of the lenses strongly increases the diffuse scattering. The second problem arises from the fact that experimentally, the absorption is not defined by the intrinsic absorption of the material but more by inelastic scattering in the material. This could be solved by cooling the CRL to low temperatures, typically liquid nitrogen temperature. This would of course cancel the ease of use of room temperature CRL setups.

Diffuse scattering is a major problem. The polish finish of MgF_2 lenses is well suited for visible light or UV light but is not yet good enough for very short wavelengths. Diffuse scattering at the interfaces is important and strongly reduces the signal to noise ratio. This hampers the implementation of such devices on SANS spectrometers. The use of single crystal materials is also compulsory to avoid that the sample SANS signal is overwhelmed by the signal from the optics system.

Another limitation is that these lenses are very chromatic, with a λ^2 dependence. In the case of monochromatic spectrometers this is not a significant issue, even though the wavelength resolution is typically 10%, since one does not want to achieve ultimate spatial resolution on neutron spectrometers except in the case of applications such as neutron microscopes. Chromaticity is however a major problem for all the time-of-flight instrumentation which cannot benefit from such optical systems.

Significant research is still required to achieve viable refractive optics systems, on materials (use of diamond?), on polishing techniques and on the development of suitable cooling systems.

7.5.2 Magnetic Lenses

Neutrons carry a spin which leads to a large neutron-magnetic induction interaction. Thus, neutrons are deflected by magnetic field gradients (Stern–Gerlach effect). The deflection force applied to a neutron is given by $F_x = \pm\mu_\text{n}\cdot$ grad B_x, depending on the neutron polarization parallel and antiparallel with respect to the direction of the magnetic field. In a sextupole magnetic field, the equation of motion can be approximated by $\mathrm{d}^2r/\mathrm{d}t^2 = \pm\omega^2 r$, with $\omega^2 = \alpha G$, $\alpha = |\mu/m| = 5.77\,\mathrm{m}^2\,\mathrm{s}^{-2}\,\mathrm{T}^{-1}$ and the sextupole field $|\mathbf{B}| = (G/2)r^2$ [16]. Such a magnetic configuration acts as a focusing lens for one of the neutron spin states (see Fig. 7.6).

The magnitude of the effect is not very large: it is of the order of the refraction through a solid lens. However, since it is proportional to λ^2, it becomes significant for long wavelengths. The key advantage of magnetic focusing is that no material is put into the beam path. Hence, no diffuse scattering occurs. Extra degrees of freedom are provided by the fact that the magnetic field can

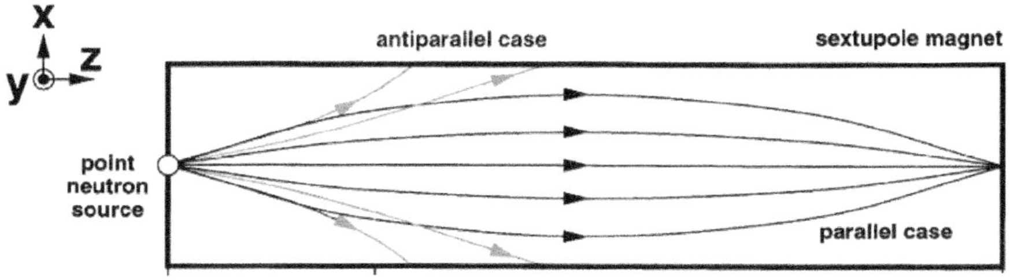

Fig. 7.6. Hexapole magnetic lens. One of the spin states is focused; the other spin state is defocused [17]

be easily tuned or modulated in time. Of course, such a device requires polarized neutron beams since one of the spin-states is focused while the other spin state is defocused. This reduces the efficiency of such a device to a maximum of 50% (unless one is interested in polarized neutrons).

A first sextupole magnet device was built by Shimizu et al. [17] using permanent magnets to demonstrate the feasibility of such a lens (with $G = 32{,}000\,\mathrm{T\ m^{-2}}$). The intensity gain was 30 for $\lambda = 14.4$ Å. One of the limitations was that the lens was very long (2 m) and the inner diameter was limited (9 mm), thus only a small fraction of the beam could be used.

Subsequently, a 2 m long superconducting sextupole magnet (SSM) with an aperture diameter of 46.8 mm was developed [16]. The SSM consists of Nb/Ti superconducting coils and achieves a field gradient of $G = 1.28 \times 10^4\,\mathrm{T\ m^{-2}}$. The focal length is of the order of 1.5 m for $\lambda = 13.5$ Å neutrons. Demonstration experiments have been performed [18].

The latest developments in magnetic lenses consist in modulating the magnetic field in time so as to be able to focus white neutron beams and to be able to implement such lenses on ToF spectrometers and spallation sources [19].

7.5.3 Reflective Optics

A number of different reflective optics systems have been proposed in the past (Lobster eye, stacked layers...). In the last decade, the progress in thin film deposition has made it possible to design neutron mirrors consisting of thousands of layers which artificially increases the optical index by a factor up to 6 compared to nickel. This opens up new possibilities in the field of reflective optics since one is not limited anymore by the very small angle of total reflection. Optical systems employing elliptical or parabolic shapes are being designed, potentially providing tremendous increases in the neutron flux at the sample position.

The field of reflective optics is the field in which the largest variety of solutions have been designed and evaluated. This is mostly due to the fact that the needs are very different from one spectrometer to the other and different solutions must be designed to solve specific issues.

7.5.4 Base Elements

Neutron supermirrors are the base elements for the fabrication of most reflective optics. Multilayer coatings allow one to increase the angle of total reflection to value up to six times the critical angle of natural Ni coatings. The critical angle of reflection can thus go up to $0.6°\ \mathrm{Å}^{-1}$. The reflectivity of neutron supermirrors is usually very high (up to $R = 0.8$) at the critical angle of reflection. This is higher than what can be achieved for X-rays because the absorption of neutrons is very low and makes possible to design supermirrors with several thousand layers. Figure 7.7 illustrates the reflectivity of neutron

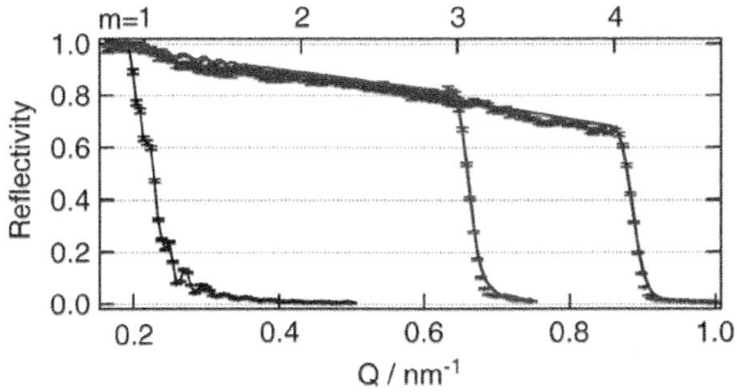

Fig. 7.7. Neutron supermirrors available from Mirrotron Ltd.

Fig. 7.8. Flat tapered guide (2 m long)

supermirrors available from Mirrotron Ltd. The reflectivity is close to 80% at $m = 4$.

7.5.5 Focusing Guides (Tapered: Elliptic: Parabolic)

The easiest way to collect more neutrons from a large source consists in using a flat tapered guide. Such devices are very simple to build since they use standard flat supermirror substrates. Figure 7.8 shows a square tapered guide. The size of the exit is reduced by a factor of two with respect to the entrance. Thus, the flux density at the guide exit is multiplied by a factor of four at the expense of an increased beam divergence. One drawback of such a device is that the beam quickly expands after the exit of the guide, the more so as the beam divergence is increased, and the sample must thus be positioned very close to the guide exit to benefit from the flux increase. Such guides are routinely implemented on neutron spectrometers.

More recently, guide profiles with more complex shapes, parabolic, or elliptical, have been proposed. Such guides can be built at relatively low costs since one is not aiming at sub-μm resolution. The mechanical tolerances are large, and thus deviation from a perfect profile is not an issue.

It is possible to compare the different types of focusing guides in the most usual situation where a neutron beam is available from a large straight guide. Consider a beam which is 80 mm high and which exits from a straight $2\theta_c$

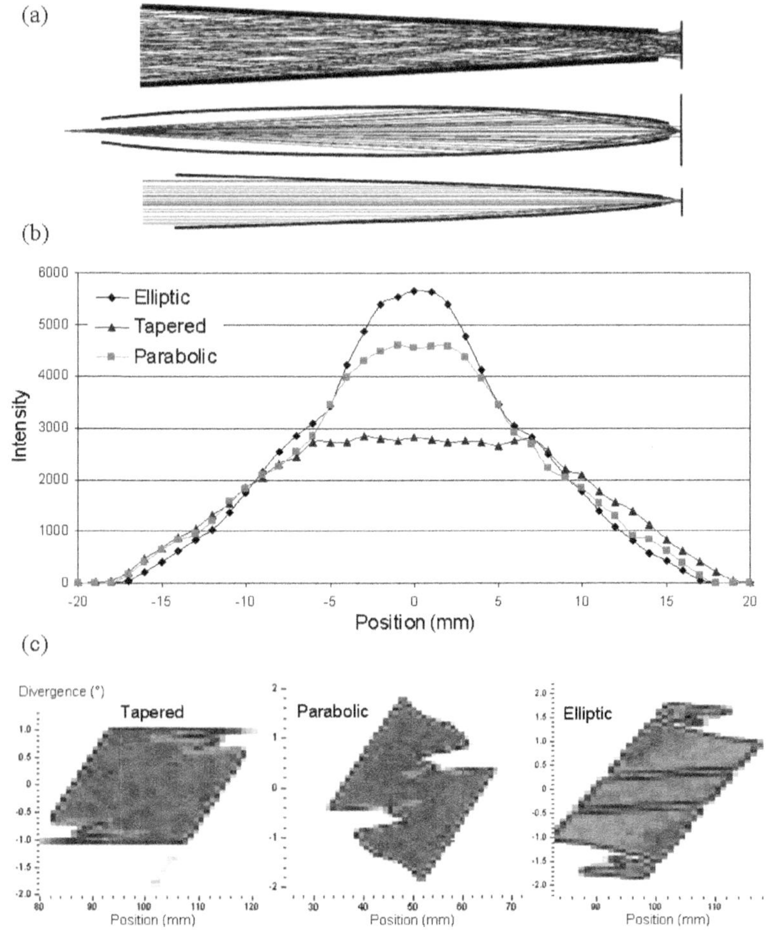

Fig. 7.9. Comparison of a tapered, elliptic, and parabolic guide to focus the exit of a straight neutron guide. (**a**) The different focusing geometries, tapered, elliptic, parabolic; (**b**) flux distribution at the sample position; (**c**) position/divergence plot

guide. The aim is to maximize the flux on a 1 cm^2 sample. It corresponds to the geometry of a focusing reflectometer.

The comparison between different solutions (Fig. 7.9) gives the following results:

- Tapered guide: Total transmission = 68% (30% of the flux over 10 mm)
- Parabolic guide: Total transmission = 81% (43%)
- Elliptical guide: Total transmission = 86% (52%)

The elliptical guide provides the most complex divergence distribution. One must note that the improvement of the performances is only of the order of 70% as compared to the tapered guide. The parabolic guide would be most efficient with a point source. The elliptical guide offers the highest flexibility. The performance could be slightly improved if a closer position of the sample was possible: 51% at 400 mm, 71% at 100 mm. This gain is negligible compared to the accompanying constraints.

The previous situation applies to the case of a facility where the main guides are already in place and where one wants to maximize the flux at the sample position. However, in the case of the design of new spectrometers, the entire beam delivery system should be optimized, including the primary guide.

7.5.6 Ballistic Guides: Neutron Beam Delivery over Large Distances

In general, one wants to transfer as much neutron intensity from the cold (or thermal) source to the sample position. Until recently, this was achieved by using neutron guides (with a square section of the order of $100 \times 50\ \text{mm}^2$) which were bringing the neutrons outside the reactor into an experimental hall where the spectrometers are located. In the previous decade, the spectrometers were optimized to make the best use of the flux in these guides. The trend in the design of new spectrometers is to optimize the setup from the source to the sample and specifically to optimize the guide geometry for the specific need of the spectrometer. This assumes that each spectrometer has its own dedicated guide. The space between the cold source and the spectrometer position is very large (from 10 m to several tens of meters). This makes possible the implementation of complex optical systems.

The first optimization consists in increasing the amount of neutrons effectively brought to the spectrometer. With the implementation of neutron supermirror guides (with $m = 2$ or $m = 3$), the flux at short wavelengths has been significantly increased (Fig. 7.10). However, the problem is that, contrary to nickel coatings, the reflectivity of supermirrors is of the order of 0.8. Thus

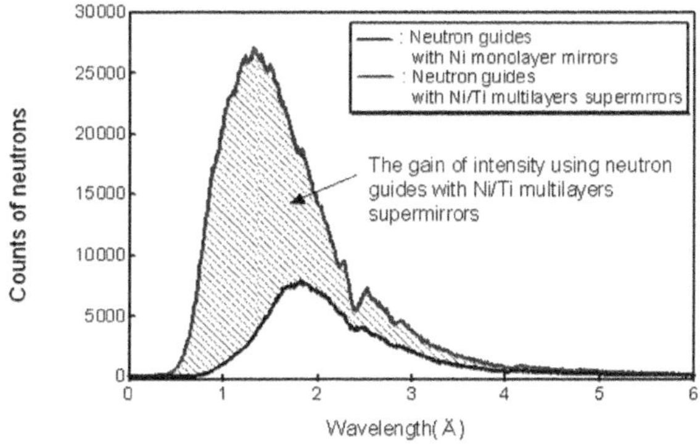

Fig. 7.10. Flux in a nickel guide and in a supermirror guide

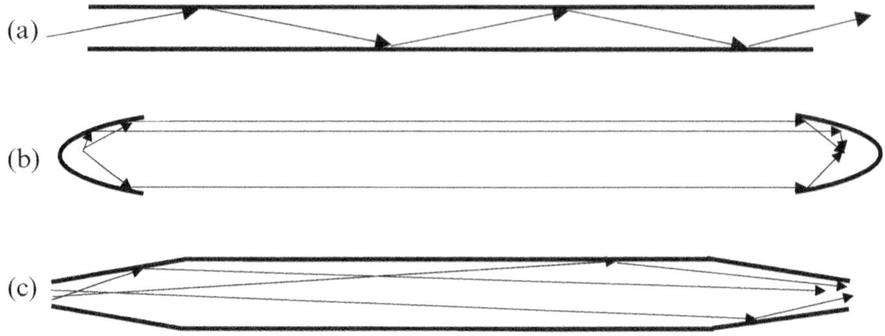

Fig. 7.11. Principle of a ballistic guide: (**a**) straight guide, (**b**) perfect ballistic guide, (**c**) approximation of the ballistic guide

after five reflections, the transmission is reduced to 30%. This means that long supermirror guides are inefficient, besides being expensive. For guides longer than 30 m, the losses outweigh the gains.

Recently, it has been proposed to implement so-called "ballistic guides" [20]. The principle consists in first expanding the neutron beam so as to reduce its divergence, let it travel a long distance, and eventually refocus it onto the sample at the end of the free flight travel. This makes it possible to reduce the number of reflections by getting closer to a parallel beam. Figure 7.11 illustrates this principle. If the source is small, the beam could be made parallel with a parabolic beam expander. In real life, the source is not so small. The first implementation of such a ballistic guide was approximated by using first a "tapered" section which mimics the parabolic expansion of the beam. A straight nickel guide avoids losing the neutrons which are too divergent. This section is quite efficient since the divergence of the beam is small. The beam is then refocused onto the sample with a second tapered section. The cost of such a guide is moderate since a simple nickel coating can be used over most of the guide length.

The first implementation of such a system was done on the guide H113 at the ILL [21]. The initial section is 6 cm and is increased to 9 cm. The gain in flux at the sample position was a factor of 4 for a guide length of 72 m. A detailed description of the characteristics of this guide has been published in [22].

More recently, more complex shapes have been considered for neutron guides. A detailed comparison of the different geometries (tapered, elliptic, and parabolic) has been made [23]. It appears that an ellipse is by far the best geometry which provides the highest neutron flux as well as the best flux homogeneity at the exit. A thorough comparison with other systems has been published [24].

Small scale parabolic guides consisting of four bent supermirrors have recently been made and tested [25]. The entrance was $16 \times 16\,\mathrm{mm}^2$, the exit was $3.2 \times 3.2\,\mathrm{mm}^2$. The focal spot achieved was $1\,\mathrm{mm}^2$. Better result, in terms of the beam homogeneity and focal length, have since then been obtained by using elliptical shapes.

7.5.7 Reflective Lenses

A large number of solutions have been devised to fabricate lenses based on reflection.

Focusing Bender

Silicon is a very transparent material for neutrons. Thus by coating silicon wafers with neutron supermirrors and stacking the substrates, it is possible to build neutron channels with a large angular acceptance. By properly bending the stack, it is possible to focus the neutrons at a given position (see Fig. 7.12). The principle is very similar to polycapillary systems except that (a) it is possible to coat the inner part of the channel walls with supermirrors (possibly polarizing) and thus to have a large angular acceptance and (b) the transmission is close to 90% since Si is transparent. The drawbacks are (a) the focalization is only one dimensional and (b) the focal spot increases quickly in size because of the intrinsic divergence of the beam at the exit of the channel.

Lobster Eye Optics

A number of solutions have been proposed which are based on the principle of the so-called Lobster eye optics (Fig. 7.13) [26]. This principle seems very appealing in theory. In practice, a number of difficulties have been

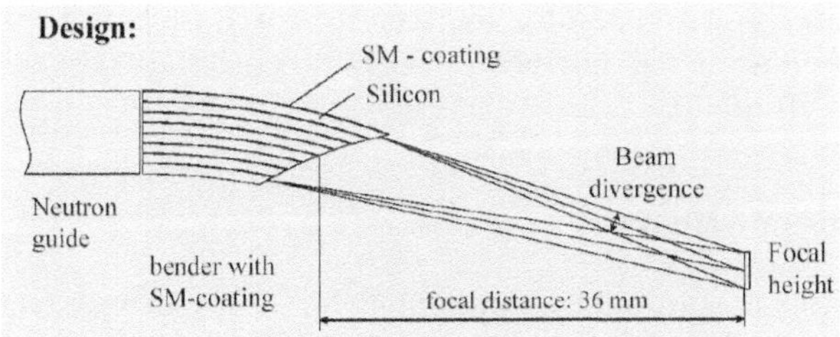

Fig. 7.12. Focusing bender (T. Krist, HMI)

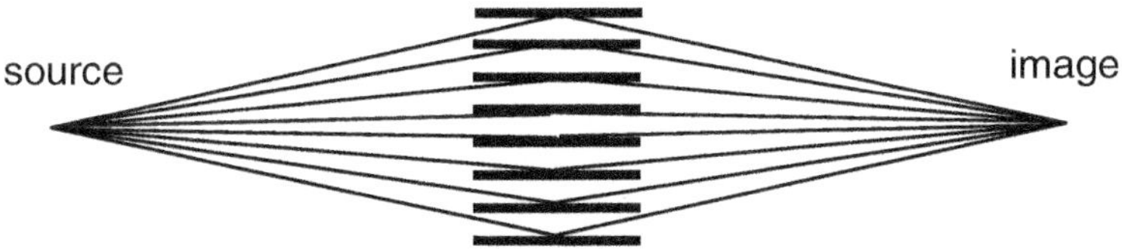

Fig. 7.13. Lobster eye optics [26]

encountered. In the case of 2D lobster eye optics it is not possible at the present time to coat the inside of the reflecting walls. On top of that, artifacts appear because it is not possible to have an optimal geometry over the entire lens surface. The size of the holes is fixed whereas in an optimal geometry it should be variable. In the case of a one-dimensional lens, one gains an extra degree of freedom in the design since it is possible to stack silicon wafers which have high reflective coatings. The prototype built by Daymond and Johnson [27] proved to be efficient but still suffered from artifacts, probably due to defects in the stacking quality. A recent theoretical study [28] has proposed an optimized geometry which could eventually be quite efficient. Such one-dimensional lenses would be fairly compact, 20 mm thick. Thus is would be possible to use two of them in a Kirkpatrick–Baez arrangement: two lenses in series turned at 90° to each other to perform a 2D focalization.

Toroidal Mirrors

The focusing SANS technique has been known for 50 years but until recently it was limited by the quality of the mirrors which were too rough and led to unacceptable diffuse scattering. Only recently, high quality mirrors have been made available as a result of the needs of X-ray technology. The first tests of such a mirror were performed at the ILL by Alefeld et al. [29]. The mirror was 4 m long. The test was satisfactory since the height of the parasitic halo (at σ from the focal spot center) was as low as 6×10^{-5}.

Later on, a USANS spectrometer (KWS3) was built at the FZ Jülich [30, 31]. The improvement over the previous design was that the length of the mirror was much shorter, 1.2 m instead of 4 m, to reduce the spherical aberrations. The mirror was set in a horizontal position to minimize gravity sag. The focal spot was reduced to 1 mm. The implementation of replica mirrors to compensate for the use of a shorter mirror has, however, not been successful yet because it has not been possible to produce replica mirrors with a sufficiently high quality.

7.5.8 Capillary Optics

Capillary optics is a standard component for X-ray beams. Such components are also available for neutrons [32, 33]. A specific section is dedicated to this topic in "Capillary optics for X-rays and neutrons" by A. Bjeoumikhov et al.

One limitation at present is the low angle of reflection in the capillaries since it is not possible to coat them with reflecting materials. The overall transmission is limited but the focal spots can be very small. Such systems are very efficient if the divergence of the beam and the background signal are not a problem. Such devices are typically implemented for PGGA applications. It has also been proposed to use such devices for crystallography [34].

7.6 Diffractive Optics

Within this category one can find Fresnel lenses and gradient super-mirrors. Recently, the progress in microfabrication has made it possible to build efficient diffractive optics for thermal neutrons. The disadvantages of such optics are that they are strongly chromatic and that they generate harmonics.

7.6.1 Fresnel Zone Plates

The principle of Fresnel zone plate is discussed in Sect. 7.9 "Volume effects in zone plates" by G. Schenider et al. The use of Fresnel zone plates for neutrons was demonstrated a long time ago by Kearney et al. [35] for focusing 2 nm neutrons. In the past decade, microfabrication techniques have made tremendous progress and it is now possible to build FZPs which are suitable for neutrons with wavelengths of the order of 0.5 nm [36].

To maximize the efficiency, one should design phase zone plates (Fig. 7.14). In this case, the phase shifter must have a thickness of the order of a few µm depending on the wavelength $t = \lambda/2\delta$ where δ is the optical index of the phase shifter. Single crystal silicon wafers are the ideal substrate since they are almost perfectly transparent to neutrons, and they are easy to handle. By chance, nickel which is the material used for X-ray zone plates is also the most suitable for neutrons since it has the highest optical index. The technology transfer from X-rays to neutrons is thus rather easy. The maximum efficiency for a nickel phase zone plates is of the order of 40%. It can be above 30% for a wavelength band of the order of 4–6 Å.

Large aperture FZPs were fabricated by Altissimo et al. [36] with the following characteristics: use of natural nickel, diameter 5 mm, thickness of 4.8 µm, outermost zone width of 0.4 µm. This corresponds to aspect ratios of 12 which are the highest achievable with the present technology. The focal length of this device is given by: $f[\mathrm{m}] = 20.8/\lambda\,[\text{Å}]$. This corresponds to a focal length of 3.15 m for $\lambda = 6.6$ Å. The theoretical efficiency was 0.28 at the test wavelength. The experimental achieved efficiency was of the order of 0.15.

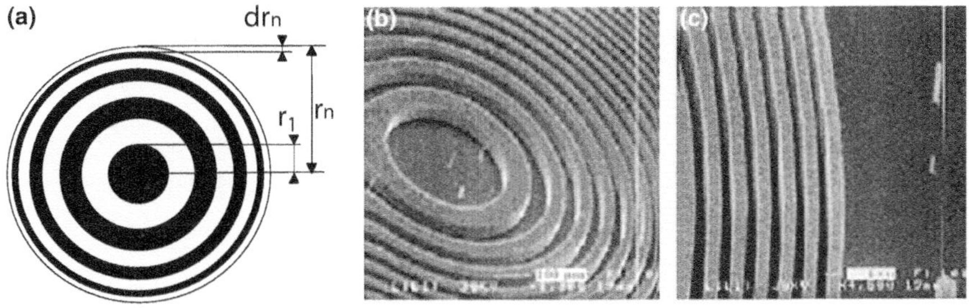

Fig. 7.14. (**a**) Geometry of a phase zone plate. (**b**) SEM image of the central zone. (**c**) SEM picture of the outer most zones

The geometry of a ZP is defined by the following relation:

$$f\lambda = \delta r_n D$$

where f is the focal length, λ the neutron wavelength, δr_N the outermost zone width and D the ZP diameter. The microfabrication techniques impose constraints on the minimum value of δr_N. We have seen that the phase shifter needs to be relatively thick for cold neutrons and that this leads to very high aspect ratios. In this case, δr_N is limited to 0.5 µm.

This technological limitation gives rise to a practical limitation in the FZP diameter. For $\delta r_N = 0.4$ µm, $\lambda = 5$ Å, $f = 4$ m, the diameter is limited to 5 mm. This limitation is very problematic since neutron beams are usually much larger.

Another drawback is that the efficiency is intrinsically limited and thus a large part of the neutron flux is not focused and gives rise to background noise which can be unacceptable in a number of applications, SANS for example. The problem is that the device is chromatic making it unsuitable for ToF applications.

Potential advantages however are that the device is extremely compact and nonabsorbing, very small focal spot sizes (∼µms) can be achieved without effort. With the progress in microfabrication techniques, it is likely that we will eventually see practical devices produced in the next decade.

Since the most severe limitation is the intrinsic size of an individual FZP, arrays of small size FZP ($D \sim 0.3$ mm) have been considered (Fig. 7.15). In this case the fabrication process is easy since small FZPs are much easier to fabricate than large ones. It has been proposed that such arrays could be used to perform ultrahigh-resolution imaging. Presently, neutron imaging is limited by the resolution of the neutron detectors (∼50 µm for CCD cameras). The idea is to scan a highly focused neutron spot ($D \sim 10$ µm) inside an object. Of course, if one would scan the full object with a single spot, this would be too time consuming. The idea is to scan in parallel 1,000 spots in the object while measuring the absorption of these spots within the object using a regular

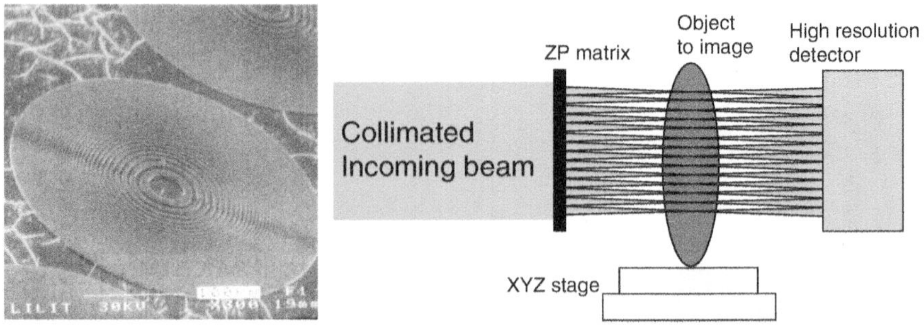

Fig. 7.15. An individual FZP in a 30 × 30 matrix (*left*); high resolution imaging using a matrix of FZP (*right*)

CCD camera (see Fig. 7.15). The interaction with the object is weak but the CCD camera resolution is good enough to discriminate the signals from the different spots. The spots are scanned across the object and the total image is then reconstructed by adding the different images.

7.6.2 Gradient Supermirrors: Goebel Mirrors

Gradient supermirrors have never been built for neutrons for two reasons. First, the neutron source is extended (in contrast to an X-ray tube) and thus the efficiency of a Goebel mirror is limited. Second, since it is possible to produce very efficient neutron supermirrors, one should implement parabolic supermirrors which are far less sensitive to the source extension since there is no angular diffraction condition.

7.7 Modeling Programs

A number of Monte Carlo modeling programs are freely available [37]: McSTAS, Vitess, ResTrax, NISP. All of them offer possibilities for modeling more or less complex optical elements, usually reflective optics. However, none of them handles refractive lenses, magnetic lenses or diffractive optics (FZP).

7.8 Merit of the Different Focusing Techniques

It is impossible to define a general figure of merit for a particular neutron focusing optics. The types of neutron spectrometers are so numerous that each technique requires its own specific optics. The relevant parameters are:

- The sample size, ranging from 10 mm for TAS down to 100 μm for single crystal diffraction.
- The focal length, ranging from 1 m for diffraction to 10 m for SANS.
- The dimensionality: focusing in one direction (diffraction-reflectivity) or 2D (SANS).
- The wavelength band: focusing a white beam (time of flight) or a short wavelength band (monochromatic spectrometer).

Different types of optics should be considered for each specific need. Nevertheless, it is possible to define some of the relevant parameters for a number of focusing techniques.

Another point that should be emphasized is that since neutron beams are very large and divergent, two possibilities exist: either use the large beam size and define a proper divergence for each $\phi(\mathbf{r})$ (multiple focusing collimation technique for SANS) or use the large divergence of a point source and refocus it as a point source. With reflective optics, it is possible to combine both aspects since it can accommodate a large incoming beam.

The focal lengths of these different devices are given by:

- $r/2$ for spherical or elliptical optics where r is the radius of curvature
- $\pi r/\lambda^2\, \rho b$ for refractive optics
- $\delta r_N/\lambda D$ for Fresnel zone plates

Table 7.3 summarizes some of the principal types of focusing optics which exist at the moment. Presently, the most versatile type of optics is reflective focusing since it can be used for large guides down to microguides. A limitation is however the diffuse scattering which can hamper some experiments (typically SANS).

Chromatic optics is not suitable for ToF spectrometers. This is a problem in the case of the new generation of neutron sources based on the spallation principle (SNS in the USA, JPARC in Japan and possibly ESS in Europe) which are coming online. They require to operate in the Time of Flight mode.

Overall, focusing monochromators are the best solution for thermal neutrons and small bandwidth applications. For longer wavelengths and large bandwidth, reflective optics is the optimal and most flexible solution.

It should be pointed out that in the case of neutron scattering, the sample environments can be rather complex: low temperatures (~1 K), high temperature (~1,000° C), high magnetic fields (~10 T), high pressures (~500 Gpa). Thus, a minimal distance between the optics and the sample usually exists which prevents the use of some optics.

7.9 Possible Applications of Neutron Focusing and Conclusion

We quickly summarize the different applications where proper focusing can or has strongly enhanced the performance of neutron spectrometers:

- *Focusing SANS*: the gains can reach up to 100 for VSANS. Focusing also reduces the minimum measurable scattering vector and increases the neutron intensity on the sample. Several solutions are already in operation: toroidal mirrors (KWS3 at FRM2 Münich), multibeam collimation (TPA at LLB Saclay, V16 at HMI Berlin).
- *Focusing monochromators*: gains of 10–50; implemented on most TAS and on strain scanning machines.
- *Reflectivity on small samples*: Focusing of large beams (>100 mm) onto small samples (10 mm) using tapered guides. Gains higher than 10 are routinely achieved without loss of resolution (PRISM at the LLB Saclay, D17 at the ILL Grenoble).
- *Powder diffraction on very small samples (1 mm³)*: Use of focusing guides (MICRO at the LLB Saclay). Elliptic focusing guides have also been successfully tested at FRM2 Münich.

Table 7.3. Comparison chart of different focusing techniques

	Elliptic guide	Focusing bender	Refractive lens	Magnetic lens	Bent crystal	Capillary	Zone plate
Focal spot	<1 mm	1 mm	<1 mm	<1 mm	<1 mm	<50 μm	<50 μm
Wavelength band	2 nm	0.2–1 nm	1 Å	1 Å	0.1 Å	1 Å	5 Å
Efficiency	90%	80%	20–50%	50%	100%	30%	30%
Focal distance	100 mm	40 mm	4 m	4 m	1 m	50 mm	5 m
Input beam size	100 mm	20 mm	5 mm	5 mm	>100 mm	50 mm	5 mm
Angular acceptance	2°	1°	0.3°	0.3°	1°	0.2°	0.3°
Cost	Cheap	Moderate	Moderate	High	Acceptable	Moderate	Acceptable
Comment	Polarization? Scalable Fixed setup		Cooling required Diffuse scattering Flexible Easy to setup	Polarization Time focusing No material in the beam			Fixed wavelength Very compact Difficult to fabricate

- *Single crystal diffraction on ultra small samples*: Use of tapered focusing guides (SXD at ISIS Oxford).

We can see that all the fields of neutron scattering can benefit from focusing techniques. All the gains in flux which have been achieved, often by orders of magnitude, have enabled neutron scattering methods to progress significantly over the recent decades even though the intrinsic luminosity of the sources has only marginally improved. Work in optical and focusing techniques is one of the best investments that can be made to improve neutron spectrometers. In the next decade a further increase of at least one order of magnitude in luminosity can be expected.

References

1. C. Petrillo, E. Guarini, F. Formisano, F. Sacchetti, E. Babucci, C. Campeggi, Nucl. Instrum. Methods A **489**, 304 (2002)
2. C.J. Glinka, J.M. Rowe J.G. Larock, J. Appl. Cryst. **19**, 427 (1986)
3. A. Smee, J.D. Orndorff, G.A. Scharfstein, Y. Qiu, P.C. Bran, L. Broholm, K. Anand, Appl. Phys. A **74**, 255 (2002)
4. P. Mikula et al., J. Appl. Cryst. **19**, 324 (1986)
5. J. Kulda et al., NIM A **379**, 155 (1996)
6. M.-K. Moon et al., Physica B **368**, 70 (2005); Physica B **369**, 1 (2005)
7. M. Kempa et al., Physica B **385–386**, 1080 (2006)
8. http://www.diamond-materials.de/disks_en.htm
9. R. Gahler, J. Kalus, W. Mampe, J. Phys. E **13**, 546 (1980); A.G. Klein et al., Phys. Rev. Lett. **46**, 959 (1981)
10. M.R. Eskildsen et al., Nature **391**, 563 (1998)
11. S.M. Choi, J.G. Barker, C.J. Glinka, Y.T. Cheng, P.L. Gammel, J. Appl. Cryst. **33**, 793 (2000); D.F.R. Mildner, B. Hammouda, S.R. Kline, J. Appl. Cryst. **38**(6), 979 (2005)
12. A. Snigirev, V. Kohn, I. Snigireva, B. Lengeler, Nature **384**, 49 (1996)
13. J.T. Cremer, M.A. Piestrup, C.K. Gary, R.H. Pantell, C.J. Glinka, Appl. Phys. Lett. **85**, 494 (2004)
14. T. Oku, S. Morita, S. Moriyasu, et al., Nucl. Instrum. Methods A **462**, 435 (2001)
15. T. Adachi, K. Ikeda, T. Shinohara, et al., Nucl. Instrum. Methods A **529**, 112 (2004)
16. H.M. Shimizu, T. Adachi, M. Furusaka, et al., Nucl. Instrum. Methods A **529**, 5 (2004)
17. H.M. Shimizu et al., Nucl. Instrum. Methods A **430**, 423 (1999)
18. T. Oku et al., Physica B **356**, 126 (2005)
19. T. Shinohara et al., Proceedings of PNCMI 2006
20. F. Mezei, J. Neutron Res. **6**, 3 (1997)
21. H. Hase et al., NIMA **485**, 453 (2002)
22. H. Abele, D. Dubbers, H. Häse, et al., Nucl. Instrum. Methods A **562**, 407 (2006)
23. C. Schanzer, P. Boni, U. Filges, T. Hils, Nucl. Instrum. Methods A **529**, 63 (2004)

24. T.M. Ito, C.B. Crawford, G.L. Greene. Nucl. Instrum. Methods A **564**, 414 (2006)
25. N. Kardjilov, P. Böni, A. Hilger, M. Strobl, W. Treimer, Nucl. Instrum. Methods A **542**, 248 (2005)
26. B.E. Allman et al., Neutron News **10**, 20 (1999)
27. M.R. Daymond, M.W. Johnson, Nucl. Instrum. Methods A **485**, 606 (2002)
28. A.D. Stoica, X.L. Wang, Rev. Sci. Instrum. **74**, 2463 (2003)
29. B. Alefeld et al., Physica B **234–236**, 1052 (1997)
30. B. Alefeld, L. Dohmen, D. Richter, Th. Brückel, Physica B **283**, 330 (2000)
31. E. Kentzinger et al., Physica B **350**, e779 (2004)
32. M.A. Kumakhov, V.A. Sharov, Nature **357**, 390 (1992)
33. G.I. Borisov, M.A. Kumakhov, Nucl. Instrum. Methods A **529**, 129 (2004); M.A. Kumakhov, Nucl. Instrum. Methods A **529**, 69 (2004)
34. W.M. Gibson, A.J. Schultz, J.W. Richardson, et al., J. Appl. Crystallogr. **37**, 778 (2004)
35. P.D. Kearney et al., Nature **287**, 313 (1980)
36. M. Altissimo, E. Di-Fabrizio, F. Ott, C. Petrillo, F Sacchetti, Nucl. Instrum. Methods A **529**, 148 (2004); M. Altissimo, C. Petrillo, F. Sacchetti, F. Ott, E. Di-Fabrizio, Microelectron. Eng. **73–74**, 644 (2004)
37. McStas: http://neutron.risoe.dk/; VITESS http://www.hmi.de/projects/ess/vitess/; ResTrax http://omega.ujf.cas.cz/restrax/; NISP http://paseeger.com/

8

Volume Effects in Zone Plates

G. Schneider, S. Rehbein, and S. Werner

Abstract. Fresnel zone plates are key optical elements for nano-focusing of soft X-rays and multi-keV radiation. Their diffraction properties are dominated by high aspect ratio phase shifting nanostructures. The three-dimensional shape of high resolution zone structures and their volume diffraction effects are described by electrodynamic theory. In this review, volume diffraction zone plates are analyzed in different orders of diffraction, taking into account different possible imperfections. An efficient way towards sub-10 nm focusing based on stacked volume zone plates is proposed.

8.1 Introduction

All modern high-resolution X-ray microscopes (TXMs) and scanning transmission X-ray microscopes (STXMs) use zone plates as imaging or focusing X-ray optical elements [1, 2]. Therefore, the diffraction properties of zone plates are of particular importance for high-resolution X-ray microscopy. The progress in spatial resolution of the X-ray microscopes depends mainly on the resolving power and the diffraction efficiency – defined as the fraction of the incident intensity diffracted into one selected diffraction order m – of these X-ray optics. Increasing the spatial resolution obtainable with zone plates means increasing their numerical aperture (NA). Because zone plates are made of concentric circular zones with radially increasing line density (see Fig. 8.1, frontal view), their maximum possible diffraction angle depends on the outermost zone width. By comparison, the diffraction angle ϕ realized by a transmission grating depends on the grating period Λ, the wavelength λ, and the diffraction order m:

$$\sin\phi = m\lambda/\Lambda. \tag{8.1}$$

For constant wavelength and diffraction order, the diffraction angle can be increased by employing only smaller grating periods or in the case of zone plates by smaller zones. According to the Rayleigh criterion the point-to-point

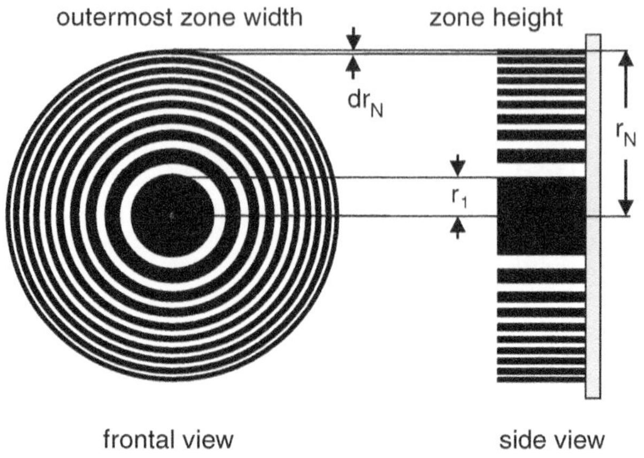

Fig. 8.1. Zone plate geometry in the front and side view showing the increasing line density and aspect-ratio of the zone structures within the radius

resolution, δ, obtainable with an objective is for incoherent imaging conditions given by

$$\delta = 0.61\,\lambda/\mathrm{NA}. \tag{8.2}$$

For illumination with a plane-wave traveling parallel to the optical axis, the numerical aperture of a zone plate is equal to the sine of the grating diffraction angle given by (8.1). Introducing (8.1) into (8.2) and replacing the grating period Λ by the outermost zone period $2\,\mathrm{d}r_N$ of the zone plate leads to

$$\delta = 1.22\,\mathrm{d}r_N \,/\, m \approx \mathrm{d}r_N \,/\, m. \tag{8.3}$$

For this reason the spatial resolution achievable in all X-ray microscopes with zone plates as X-ray objectives is directly correlated with the outermost zone width. The spatial resolution obtainable with zone plates can be improved only by manufacturing smaller zone widths or by using higher orders for X-ray imaging, which will be analyzed by electrodynamic theory in this chapter.

Another important parameter of zone plates is the diffraction efficiency, because the dose and photon density for object illumination scales inversely with the diffraction efficiency. For X-rays the complex refractive indices of matter are slightly different from unity; therefore, absorption and phase shift occur and lead to a certain zone height for an optimal diffraction efficiency. Combining this optimal zone height with an ever higher resolving power means increasing the so-called aspect-ratio – the ratio of the zone height to the zone width – of the zone structures. Earlier, investigations of zone plate diffraction used a geometrical optical approach, which neglects the expansion of the zone structures along the direction of the optical axis [3]. More accurate dynamical calculations of the diffraction of zone plates with high aspect-ratios were performed using coupled-wave theory [4]. In these calculations an ideally smooth rectangular profile of the zones with lines and spaces of equal width was assumed. Furthermore, only the first-order diffraction was analyzed. The aim of this chapter is to extend the theoretical description of zone plates in order to describe their diffraction properties for *arbitrarily* shaped zone profiles with high aspect-ratios and also to describe the diffraction properties of *high orders* for use in X-ray imaging. For this purpose an electrodynamic theory – the coupled-wave theory – is applied, which allows one the quantitative study, under more realistic conditions, of how sensitive the diffraction efficiency of zone plate structures is to parameters such as the aspect-ratio, the deviations of the zone profile from the rectangular shape, the interdiffusion of zone material, and the zone roughness in the direction of the optical axis. Furthermore, it will be concluded from these theoretical investigations which materials are suited for zone plates and which deviations from the ideal rectangular zone profile can be tolerated with respect to optimal diffraction efficiency and high resolving power.

8.2 Transmission Zone Plate Objectives

Zone plates as focusing diffractive optical elements were invented by Lord Rayleigh and independently by Soret about 120 years ago [5]. The geometry for X-ray imaging with zone plates, which is used in this chapter for calculating their diffraction properties, is shown in Fig. 8.2.

Constructive interference is observed when the optical path difference between rays from the points P_1 to P_2 via two subsequent zones with the radii of the boundaries r_n and r_{n-1} is $m\, n\, \lambda/2$,

$$g + b = c + d - m\, n\, \lambda/2, \tag{8.4}$$

or

$$g + b - m\, n\, \lambda/2 = (g^2 + r_n^2)^{1/2} + (b^2 + r_n^2)^{1/2}, \tag{8.5}$$

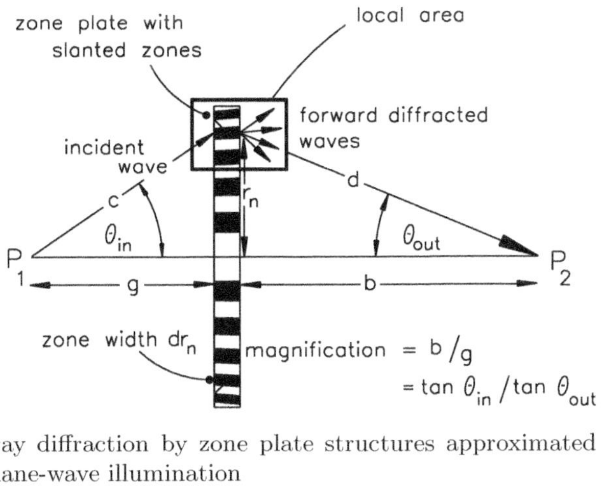

Fig. 8.2. X-ray diffraction by zone plate structures approximated by an infinite grating and plane-wave illumination

where n denotes the zone number and m the diffraction order. The expansion of the square roots in (8.5) gives

$$m\,n\,\lambda/2 = r_n^2\,(g^{-1} + b^{-1}) - \frac{1}{4}\,r_n^4\,(g^{-3} + b^{-3}) \pm .... \tag{8.6}$$

If only aberrations up to the third order are of interest, the approximation $r_n^2 = m\,n\,\lambda\,f_m$ with the focal length $f_m^{-1} = g^{-1} + b^{-1}$ of the mth focusing diffraction order can be used to replace r_n^4 in the second term of (8.6) (see, e.g. [6, 7, 8]). With the magnification $M = b/g$ we obtain for the radii r_n of the nth zone

$$r_n^2 = m\,n\,\lambda f_m + \frac{1}{4}(m\,n\,\lambda)^2 \frac{M^3+1}{(M+1)^3}. \tag{8.7}$$

This equation is only valid for low zone numbers of several hundreds, for example, for zone plate objectives that satisfy the condition $m\,n\,\lambda/f_m \ll 1$. As already known zone plates work in the focusing diffraction orders like thin lenses if the lens law $f^{-1} = g^{-1} + b^{-1}$ is fulfilled. For small diffraction angles we get for the imaging magnification $M = b/g = \tan\theta_{\text{in}}/\tan\theta_{\text{out}} \approx \theta_{\text{in}}/\theta_{\text{out}}$. By differentiating (8.7) with respect to the zone number n, the local zone width $\mathrm{d}r_n$ and the local zone period $\Lambda(r_n)$ are obtained by setting $\mathrm{d}n = 1$:

$$\Lambda(r_n) = 2\,\mathrm{d}r_n = \frac{\lambda\,m\,f_m}{r_n}\sqrt{1 + \left(\frac{r_n}{m\,f_m}\right)^2 \frac{M^3+1}{(M+1)^3}}. \tag{8.8}$$

In the following these relations will be used to calculate the local zone period and the angle of the incident plane-wave θ_{in}, which are parameters in the electrodynamic theory. In addition, the angle of the transmitted plane-wave θ_{out} of the diffraction order used for X-ray imaging has to be evaluated if the zone plate works under Bragg conditions with zone structures slanted by $\psi = (\theta_{\text{in}} - \theta_{\text{out}})/2$ to the optical axis [4]. Together with the refractive indices of the zone materials and the wavelength, these are the relevant parameters for the electrodynamic calculations of the diffracted amplitudes and their corresponding efficiency.

As mentioned earlier, the diffraction efficiency has usually been calculated using the geometric optical approach described by Kirz, which is applicable to thin laminar gratings with arbitrary line-to-space ratio [9]. Let us assume that the zone plate diffraction can be approximated by diffraction of a thin grating, which consists of lines and spaces of the materials A and B. Then the diffraction efficiency $\eta_m(t)$ of the mth diffraction order is obtained from the equation [9]

$$\begin{aligned}\eta_m(t) = {} & \frac{1}{\pi^2 m^2}\sin^2(m\,\pi L/(L+S))\,[\exp(-4\pi\beta_{\mathrm{A}}t/\lambda) \\ & + \exp(-4\pi\beta_{\mathrm{B}}t/\lambda) - 2\cos(2\pi t(\delta_{\mathrm{A}} - \delta_{\mathrm{B}})/\lambda) \\ & \times \exp(-2\pi\,t\,(\beta_{\mathrm{A}} + \beta_{\mathrm{B}})/\lambda)] \\ & \text{for} \quad m = \pm 1, \pm 2, ...,\end{aligned} \tag{8.9}$$

where t is the path length of the X-rays through the zones and L:S denotes the line-to-space ratio of the laminar zone structures. The parameters $\delta(\alpha)$ and $\beta(\alpha)$ represent the phase shift and the absorption properties of matter as a function of the wavelength. This is described by the complex index of refraction $\tilde{n}(\alpha) = 1 - \delta(\alpha) - j\beta(\alpha)$. Equation (8.9) is independent of the zone width. Therefore, the diffraction efficiency $\eta_m(t)$ is always the same for all zone widths. Furthermore, $\eta_m(t)$ is independent of the propagation angle of the waves to be diffracted by the grating. The electrodynamic calculations presented in the forthcoming sections will be compared with the results deduced by the Kirz formula.

Numerical calculations performed with the coupled-wave theory for the first-order diffraction of zone structures with line-to-space ratios of 1:1 have shown different results from those obtained by the geometric optical approach if the aspect-ratios of the zones become large [10]. These results were obtained using a Runge–Kutta algorithm for the numerical integration of the first-order coupled-wave equations by neglecting second-order derivatives of field amplitudes and boundary diffraction.

Here we extend the coupled-wave theory to transmission zone plates with arbitrary line-to-space ratio. Then the limits of validity of the geometric optical approach will directly be seen by comparing the results of (8.9) with the results of coupled-wave theory performed for different line-to-space ratios and high-orders of diffraction. The formulation of the coupled-wave equations can describe gratings with arbitrary line-to-space ratio and includes a "shift parameter" that describes a roughness of the zone interface along the direction of the optical axis. With this extension of the coupled-wave equations realistic zone profiles, which are obtained by reactive ion etching, can be modeled and their diffraction properties can be calculated. Instead of the Runge–Kutta algorithm a matrix formalism will be used for solving the resulting differential equation systems. Although the matrix formalism is not simple to solve, it allows us to also include the second-order derivatives and boundary diffraction, which leads to the rigorous coupled-wave analysis of the zone plate diffraction presented in the last part of this chapter.

8.3 Coupled-Wave Theory for Zone Plates with High Aspect-Ratios

The diffraction of electromagnetic waves by spatially modulated media has been extensively studied by several techniques in recent years. The most common of these methods are the modal theory and the coupled-wave theory, which have been applied mainly to analyse the diffraction of periodic structures for spectroscopy, holography, acousto-optics, and integrated optics. The theoretical description of periodically arranged scattering structures with high aspect-ratios for use in the X-ray domain is a modern application field for these electrodynamical theories. Here the coupled-wave approach is applied to diffractive transmission X-ray optics. The most elegant early use of coupled-wave theory was reported by Kogelnik for volume gratings with low sinusoidal modulation illuminated under Bragg conditions, where only one significant diffraction order exists [11]. Extensions have included higher diffraction orders and higher grating harmonics [12, 13].

The diffraction analysis of zone plate structures requires the inclusion of higher diffraction orders as well as the higher grating harmonics with zone structures slanted to the optical axis. In addition, a parameter describing the line-to-space ratio of the slanted zone structures has to be included. Therefore, at first the coupled-wave equations are developed for slanted gratings with arbitrary line-to-space ratio. In the forthcoming sections this formalism is used to study special problems, e.g., roughness, interdiffusion, and profile simulations.

In the X-ray domain the wavelengths are much smaller than the grating period; therefore, polarization effects can be neglected in good approximation. Here it is assumed that the structures are illuminated by a plane-wave polarized perpendicular to the plane of incidence (H-mode). For this reason it is sufficient to consider only one component $E_y \equiv E$ of the electric vector field $\mathbf{E} = (E_x, E_y, E_z)$. In mathematical terms, the problem involves the solution of a time-independent scalar wave equation in a two-dimensional inhomogeneous medium illuminated by monochromatic X-rays:

$$\nabla^2 E(x,z) + k_0^2\,\varepsilon(x,z)\,E(x,z) = 0 \quad \text{with} \quad k_0 = \frac{2\pi}{\lambda}, \tag{8.10}$$

where $\varepsilon(x,z)$ denotes the permittivity of the modulated region and λ is the X-ray wavelength.

The modulated region has to be periodic for the coupled-wave analysis. Zone plates have an increasing line density as can be seen for example in Fig. 8.1. Introducing the permittivity distribution according to this zone plate law in the wave equation, sophisticated mathematics would be required to find solutions. The problem simplifies if the zone plate pattern is subdivided into local regions, which are approximated by infinite gratings.

The local zone period $\Lambda = 2\,\mathrm{d}r_n$ changes slowly with the zone number n if n is large and the zone plate obeys the law $2\,\mathrm{d}r_n = r_1/\sqrt{n}$. For this reason it is justified to assume that an infinite grating of uniform grating constant Λ has the same diffraction efficiency as a small area of the zone plate with local zone period $\Lambda = 2\,\mathrm{d}r_n$ [10]. Therefore, in this model the X-ray diffraction of the zone plate structures is described by an infinite grating with local periodicity $\Lambda(r_n)$. For such a grating the diffracted amplitudes are evaluated by coupled-wave analysis.

Figure 8.3 illustrates the diffraction occuring in such a local zone plate region. The periodically arranged zone structures in the (x,z)-plane are slanted by an angle ψ against the optical axis. In this case the local zone period has to be replaced by $\Lambda(r_n)/\cos\psi$ [14]. The X-ray diffraction of the local zone plate region is determined with oblique plane-waves illuminating an infinite grating with slanted structures. The forward-diffracted amplitudes

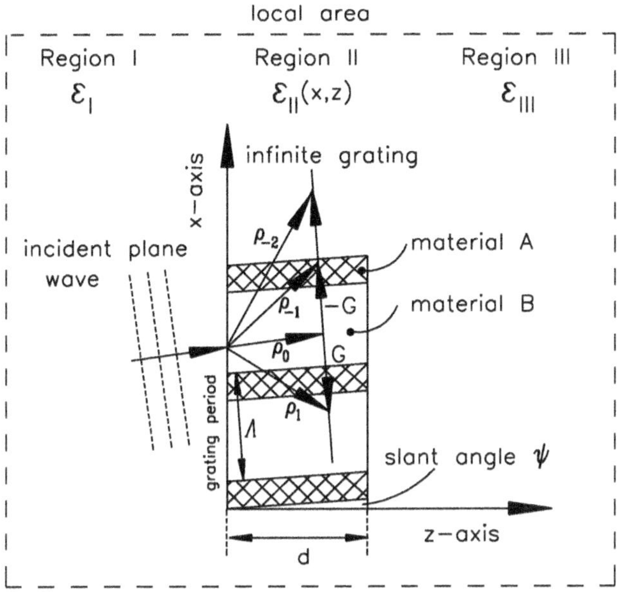

Fig. 8.3. Local zone plate region (see also Fig. 8.2) approximated by an infinite grating with periodicity $\Lambda(r_n)$, grating vector $\mathbf{G}$, slanting angle ψ, and plane-wave illumination with the wave vector $\boldsymbol{\rho}_0$. The wave vectors of the diffracted plane-waves are denoted by $\boldsymbol{\rho}_m$ with $m = \pm 1, 2, \ldots$. The indices for the regions will be used only in the rigorous coupled-wave theory of Sect. 8.8, which requires distinguishing these regions for the boundary conditions

are derived from the solutions of the wave equation describing the modulation of the infinite grating, which refers to the local zone plate region. The permittivities of the regions I and III are only required if boundary diffraction is taken into account in the rigorous coupled-wave theory.

The periodically changing permittivity $\varepsilon(x, z)$ of a grating with arbitrary line-to-space ratio is illustrated in Fig. 8.4. The X-ray optical properties of the grating structures with the permittivities ε_{A} and ε_{B} of material A and B are introduced by

$$\varepsilon(x, z) = \varepsilon_{\mathrm{A}}\, p(x, z) + \varepsilon_{\mathrm{B}}\, q(x, z) \tag{8.11}$$

with

$$p(x, z) = \frac{L}{L+S} + \frac{2\,L}{L+S} \sum_{h=1,2,3\ldots}^{\infty} \operatorname{sinc}\left(h\pi \frac{L}{L+S}\right) \cos\Big(h\,\mathbf{G}\cdot\mathbf{r}\Big) \tag{8.12}$$

and

$$q(x, z) = 1 - p(x, z) \tag{8.13}$$

with

$$|\,G\,| = \frac{2\,\pi}{\Lambda} \quad \text{and} \quad \mathbf{G}\cdot\mathbf{r} = \frac{2\pi}{\Lambda}\,[x\,\cos\psi - z\,\sin\psi] \tag{8.14}$$

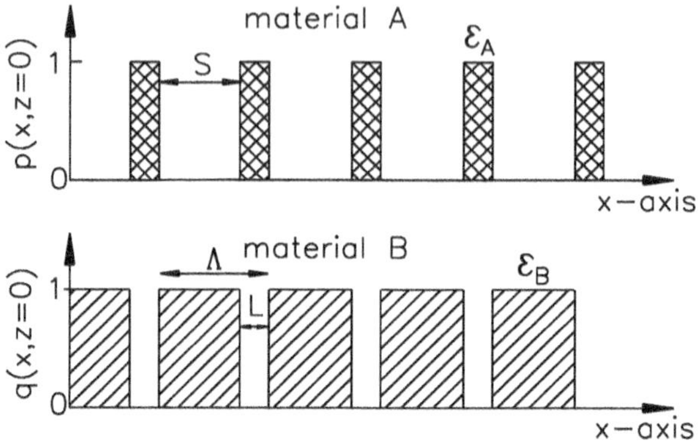

Fig. 8.4. Part of the mathematical functions describing the transmission grating consisting of material A (*upper*) and material B (*lower*) with the permittivities ε_A and ε_B. The local zone plate period Λ is given by $\Lambda = L + S$

where the mathematical functions $p(x, z)$ and $q(x, z)$ denote Fourier series, which are used to describe the spatial distribution of the permittivity of the grating. This is directly related to the material distribution of the grating, which is taken into account in the wave equation by a periodically changing permittivity $\varepsilon(x, z)$, represented by a Fourier expansion with the grating vector $\mathbf{G}$:

$$\varepsilon(x,z) = \bar{\varepsilon} + \Delta\varepsilon \frac{2L}{L+S} \sum_{h=1,2,3...}^{\infty} \operatorname{sinc}\left(h\pi \frac{L}{L+S}\right) \cos\left(h\,\mathbf{G}\cdot\mathbf{r}\right). \tag{8.15}$$

The average permittivity $\bar{\varepsilon}$ of the grating is introduced as

$$\bar{\varepsilon} = \varepsilon_B + (\varepsilon_A - \varepsilon_B)\frac{L}{L+S} = \tilde{n}_B^2 + (\tilde{n}_A^2 - \tilde{n}_B^2)\frac{L}{L+S} \tag{8.16}$$

and the difference $\Delta\varepsilon$ between the permittivities of the materials A and B in terms of the refractive indices is

$$\Delta\varepsilon = \varepsilon_A - \varepsilon_B = \tilde{n}_A^2 - \tilde{n}_B^2, \tag{8.17}$$

which follows from Maxwell's relation $\varepsilon = \tilde{n}^2$. This allows one the calculation of the function $\varepsilon(x, z)$ in terms of the complex indices of refraction usually used in X-ray physics:

$$\tilde{n}_A = 1 - \delta_A - j\beta_A \quad \text{and} \quad \tilde{n}_B = 1 - \delta_B - j\beta_B. \tag{8.18}$$

The X-ray optical parameters δ and β can be determined directly from the atomic scattering factor $(f_1 + jf_2)$. They are tabulated for all relevant elements in the energy range interesting for X-ray microscopy [15, 16, 17]. Introducing the periodically changing permittivity $\varepsilon(x,z)$ in (8.10) leads to the scalar wave equation describing the modulated region

$$\nabla^2 E(x,z) + k_0^2 \left[\bar{\varepsilon} + \Delta\varepsilon \frac{2L}{L+S} \right.$$

$$\left. \times \sum_h^\infty \operatorname{sinc}\left(h\pi \frac{L}{L+S} \right) \cos\left(h\mathbf{G}\cdot\mathbf{r} \right) \right] E(x,z) = 0, \tag{8.19}$$

which is mathematically a linear second-order differential equation with periodic coefficients (Mathieu differential equation). It may be concluded from Floquet's theory that this differential equation has a solution for the electrical field $E(x,z)$ of the form

$$E(x,z) = \sum_{m=-\infty}^{\infty} E_m(x,z) = E_0 \sum_{m=-\infty}^{\infty} A_m(z) \exp\left(-j\, \boldsymbol{\rho}_m \cdot \mathbf{r}\right) \tag{8.20}$$

$$\text{with} \quad \boldsymbol{\rho}_m \cdot \mathbf{r} = \rho_{m,x}\, x + \rho_{m,z}\, z.$$

This solution of the wave equation can be interpreted as an infinite sum of plane-waves with wave vectors $\boldsymbol{\rho}_m$ and spatially varying coefficients $A_m(z)$. Physically, we assume that the electrical field inside the grating can be represented by a sum of diffracted waves traveling in different directions. As a result of the Floquet theorem, the wave vector $\boldsymbol{\rho}_m$ of the mth diffraction order may be represented by using the K-vector closure relationship:

$$\boldsymbol{\rho}_m = \boldsymbol{\rho}_0 + m\,\mathbf{G} \qquad m = 0, \pm 1, ... \tag{8.21}$$

The components of the mth wave vector $\boldsymbol{\rho}_m$ in (x,z)-direction are given by

$$\rho_{m,x} = k \sin\theta_{\text{in}} + m\,G \cos\psi \tag{8.22}$$

$$\rho_{m,z} = k \cos\theta_{\text{in}} - m\,G \sin\psi, \tag{8.23}$$

with $k = 2\pi\bar{\varepsilon}^{1/2}/\lambda$.

An incident plane-wave E_{inc} with wave vector $\boldsymbol{\rho}_0$ is subdivided by X-ray diffraction inside the grating into many different plane-waves, which are propagating in directions given by (8.21) (see Fig. 8.3). Numerical solutions of the modulated wave equation will show, which of the diffracted waves will be damped when propagating into larger depths of the grating and which will be amplified. Such an amplification can be interpreted as an occurrence of constructive interference similar to Laue diffraction in crystals. Physically, equation (8.21) represents the conservation of momentum for the X-ray scattering process inside the grating. This means that constructive interference will occur provided that the change in wave vector is a vector of the reciprocal lattice.

To find the complex amplitudes $A_m(z)$, we need to solve the wave equation in the modulated region. Inserting $E(x,z)$ (8.20) into the scalar wave equation (8.10) and performing the mathematical operations we obtain

$$\sum_{m=-\infty}^{\infty} \exp(-j\,\boldsymbol{\rho_m}\cdot\mathbf{r}) \left\{ \frac{\mathrm{d}^2 A_m(z)}{\mathrm{d}z^2} - 2\,j\,\rho_{m,z}\,\frac{\mathrm{d}A_m(z)}{\mathrm{d}z} \right.$$

$$- (\rho_{m,x}^2 + \rho_{m,z}^2)\,A_m(z) + k_0^2\,\bar{\varepsilon}\,A_m(z)$$

$$\left. + k_0^2\,\Delta\varepsilon\,A_m(z)\,\frac{2\,L}{L+S} \sum_{h=1,2,3\ldots}^{\infty} \mathrm{sinc}\left(h\pi\frac{L}{L+S}\right)\cos(h\,\mathbf{G}\cdot\mathbf{r}) \right\} = 0. \quad (8.24)$$

We also note that the cos-functions can be written by exponential functions

$$\cos(h\,\mathbf{G}\cdot\mathbf{r}) = \frac{\exp(j\,h\,\mathbf{G}\cdot\mathbf{r}) + \exp(-j\,h\,\mathbf{G}\cdot\mathbf{r})}{2}. \quad (8.25)$$

Furthermore, the term $(\rho_{m,x}^2 + \rho_{m,z}^2)$ with the wave vector components is expressed by

$$|\,\boldsymbol{\rho_m}\,|^2 = \rho_{m,x}^2 + \rho_{m,z}^2. \quad (8.26)$$

Thus, we can rewrite (8.24)

$$\sum_{m=-\infty}^{\infty} \exp(-j\,\boldsymbol{\rho_m}\cdot\mathbf{r}) \left\{ \frac{\mathrm{d}^2 A_m(z)}{\mathrm{d}z^2} - 2\,j\,\rho_{m,z}\,\frac{\mathrm{d}A_m(z)}{\mathrm{d}z} \right.$$

$$- (|\,\boldsymbol{\rho_m}\,|^2 - k_0^2\,\bar{\varepsilon})\,A_m(z) + k_0^2\,\Delta\varepsilon\,A_m(z)\,\frac{L}{L+S}$$

$$\left. \times \sum_{h=1,2,3\ldots}^{\infty} \mathrm{sinc}\left(h\pi\frac{L}{L+S}\right) [\exp(j\,h\,\mathbf{G}\cdot\mathbf{r}) + \exp(-j\,h\,\mathbf{G}\cdot\mathbf{r})] \right\} = 0. \quad (8.27)$$

We shall satisfy (8.27) by demanding that the coefficient of $\exp(-j\,\boldsymbol{\rho_m}\cdot\mathbf{r})$ should vanish. Before proceeding further one has to take into account the energy conversion from the mode m to $m \pm h$, which may be seen in view of the relationship:

$$\begin{aligned} \exp(-j\,\boldsymbol{\rho_m}\cdot\mathbf{r})\,\exp(\pm j\,h\,\mathbf{G}\cdot\mathbf{r}) &= \exp(-j\,(\boldsymbol{\rho_m} \mp h\,\mathbf{G})\cdot\mathbf{r}) \\ &= \exp(-j\,\boldsymbol{\rho_{m\mp h}}\cdot\mathbf{r}) \end{aligned} \quad (8.28)$$

or

$$\boldsymbol{\rho_{m\mp h}} = \boldsymbol{\rho_0} + m\,\mathbf{G} \mp h\,\mathbf{G} = \boldsymbol{\rho_0} + (m \mp h)\mathbf{G}. \quad (8.29)$$

Equation (8.27) can be rewritten by introducing the relations between the wave vectors and the grating vector:

$$\sum_{m=-\infty}^{\infty} \exp(-j\,\boldsymbol{\rho_m}\cdot\mathbf{r}) \left\{ \frac{\mathrm{d}^2 A_m(z)}{\mathrm{d}z^2} - 2j\rho_{m,z}\frac{\mathrm{d}A_m(z)}{\mathrm{d}z} \right.$$

$$\left. - (|\,\boldsymbol{\rho_m}\,|^2 - k_0^2\bar{\varepsilon})A_m(z) \right\} + k_0^2\Delta\varepsilon\frac{L}{L+S} \times \sum_{m=-\infty}^{\infty}\sum_{h}^{\infty} \mathrm{sinc}\left(h\pi\frac{L}{L+S}\right)$$

$$\times A_m(z)\,[\exp(-j\boldsymbol{\rho_{m-h}}\cdot\mathbf{r}) + \exp(-j\boldsymbol{\rho_{m+h}}\cdot\mathbf{r})] = 0. \quad (8.30)$$

By applying the formula (in all practical cases $m, h \leq 100$)

$$\sum_{m=-\infty}^{\infty} \sum_{h=1,2,3\ldots}^{\infty} A_m B_{m-h} = \sum_{m=-\infty}^{\infty} \sum_{h=1,2,3\ldots}^{\infty} A_{m+h} B_m, \tag{8.31}$$

we obtain a coupling between the field amplitudes $A_m(z)$ of different modes, which is the reason for the energy exchange between the plane-waves inside the grating in mathematical terms:

$$\sum_{m=-\infty}^{\infty} \exp\left(-j\,\boldsymbol{\rho_m}\cdot\mathbf{r}\right)\left\{\frac{\mathrm{d}^2 A_m(z)}{\mathrm{d}z^2} - 2\,j\,\rho_{m,z}\frac{\mathrm{d}A_m(z)}{\mathrm{d}z}\right.$$

$$\left. - \left(|\,\boldsymbol{\rho_m}\,|^2 - k_0^2\,\bar{\varepsilon}\right) A_m(z)\right\} + k_0^2 \Delta\varepsilon \frac{L}{L+S} \sum_{m=-\infty}^{\infty} \sum_{h}^{\infty} \operatorname{sinc}\left(h\pi\frac{L}{L+S}\right)$$

$$\times \exp(-j\boldsymbol{\rho_m}\cdot\mathbf{r}) \left[A_{m+h}(z) + A_{m-h}(z)\right] = 0. \tag{8.32}$$

By reorganizing the summation over m, we find that the modulated wave equation for the grating is fulfilled if we equate the term in brackets {..........} individually with zero:

$$\sum_{m=-\infty}^{\infty} \exp\left(-j\,\boldsymbol{\rho_m}\cdot\mathbf{r}\right)\left\{\frac{\mathrm{d}^2 A_m(z)}{\mathrm{d}z^2} - 2\,j\,\rho_{m,z}\frac{\mathrm{d}A_m(z)}{\mathrm{d}z}\right.$$

$$- \left(|\,\boldsymbol{\rho_m}\,|^2 - k_0^2\,\bar{\varepsilon}\right) A_m(z)$$

$$\left. + k_0^2 \Delta\varepsilon \frac{L}{L+S} \sum_{h=1,2,3\ldots}^{\infty} \operatorname{sinc}\left(h\pi\frac{L}{L+S}\right) \left[A_{m+h}(z) + A_{m-h}(z)\right]\right\} = 0. \tag{8.33}$$

This gives an infinite set of coupled differential equations, which are the second-order coupled-wave equations we set out to derive

$$\frac{\mathrm{d}^2 A_m(z)}{\mathrm{d}z^2} - 2\,j\,\rho_{m,z}\frac{\mathrm{d}A_m(z)}{\mathrm{d}z} - \left(|\,\boldsymbol{\rho_m}\,|^2 - k_0^2\,\bar{\varepsilon}\right) A_m(z)$$

$$+ k_0^2 \frac{L}{L+S} \Delta\varepsilon \sum_{h=1,2,3\ldots}^{\infty} \operatorname{sinc}\left(h\pi\frac{L}{L+S}\right)\Big[A_{m-h}(z) + A_{m+h}(z)\Big] = 0. \tag{8.34}$$

This system of second-order linear differential equations is mathematically identical to the physical description of the vibrations of masses, which are connected by springs and lead to the well-known *Pendellösung*. For this reason (8.34) can physically be interpreted as describing a set of coupled-waves. The theory derives its name from this interpretation.

8.4 Matrix Solution of the Scalar Wave Equation

Two different coupled-wave approaches are distinguished in the literature: the second-order (rigorous coupled-wave theory) and the conventional first-order coupled-wave approach. In the latter case the first-order differential equations are derived by neglecting the second-order derivatives. An advantage of retaining the second-order derivatives in the rigorous coupled-wave theory is that the boundary conditions can be included for both the electric and the magnetic fields. Therefore, reflected waves can also be evaluated. Transmission X-ray zone plates are usually illuminated with incident angles near normal incidence and the refractive indices of matter are close to unity for X-rays. Therefore, in diffractive transmission X-ray optics forward-diffracted waves are dominant and sufficiently accurate results will be obtained with the first-order approach. However, the limits of validity of the first-order approach can be determined if its results are compared with the calculations that are performed with the rigorous coupled-wave theory.

The solution of the modulated scalar wave equation is now derived by neglecting the second-order derivatives in (8.34). The infinite set of coupled-wave equations can only be solved by restricting the number of grating harmonics to a finite number. In practice, it is sufficient to approximate the grating structures with about $h_{\max} = 30$–50 Fourier components. This means that the matrix consists of $(2h_{\max}+1) \times (2h_{\max}+1)$ complex elements. We obtain a system of first-order differential equations for the forward-diffracted amplitudes $A_m(z)$, which is rewritten in matrix notation:

$$\begin{bmatrix} \vdots \\ \mathrm{d}A_m(z)/\mathrm{d}z \\ \mathrm{d}A_1(z)/\mathrm{d}z \\ \mathrm{d}A_0(z)/\mathrm{d}z \\ \mathrm{d}A_{-1}(z)/\mathrm{d}z \\ \vdots \end{bmatrix} = \begin{bmatrix} \ddots & \ddots & \ddots & 0 & 0 & 0 \\ \ddots & a_m & b_{m,1} & \ddots & 0 & 0 \\ \ddots & b_{1,1} & a_1 & b_{1,1} & \ddots & 0 \\ 0 & \ddots & b_{0,1} & a_0 & b_{0,1} & \ddots \\ 0 & 0 & \ddots & b_{-1,1} & a_{-1} & \ddots \\ 0 & 0 & 0 & \ddots & \ddots & \ddots \end{bmatrix} \begin{bmatrix} \vdots \\ A_m(z) \\ A_1(z) \\ A_0(z) \\ A_{-1}(z) \\ \vdots \end{bmatrix}, \tag{8.35}$$

where according to (8.34) the matrix elements are given by

$$a_m = \frac{j\,(|\,\boldsymbol{\rho}_m\,|^2 - k_0^2\,\bar{\varepsilon})}{2\rho_{m,z}} \tag{8.36}$$

and

$$b_{mh} = \frac{k_0^2\,\Delta\varepsilon\,L}{2\,j\,\rho_{m,z}(L+S)}\,\mathrm{sinc}\left(h\pi\frac{L}{L+S}\right). \tag{8.37}$$

The limited number of spatial harmonics, $h_{\max}$, of the grating is taken into account by the zeros in the truncated matrix, which – in mathematical terms – avoids an energy transfer into matrix elements with indices $h > h_{\max}$.

Equation (8.35) is the expanded form of the matrix equation given by

$$\frac{\mathrm{d}\mathbf{A}(z)}{\mathrm{d}z} = \underline{\underline{\mathbf{M}}}\,\mathbf{A}(z), \tag{8.38}$$

where $\underline{\underline{\mathbf{M}}}$ denotes a complex general matrix, which includes the X-ray optical parameters of the grating as well as the incidence angle of the plane-wave illumination. Linear first-order differential equation systems of this type are solved mathematically by calculating the *eigenvalues* χ_h and the corresponding *eigenvectors* of the matrix $\underline{\underline{\mathbf{M}}}$. The solution can be written as

$$A_m(z) = \sum_h q_{mh}\,[c_h \exp(\chi_h\, z)] \tag{8.39}$$

or in expanded form

$$\begin{bmatrix} \vdots \\ A_m(z) \\ A_1(z) \\ A_0(z) \\ A_{-1}(z) \\ \vdots \end{bmatrix} = \begin{bmatrix} \ddots & \ddots & \ddots & \ddots & \ddots & \ddots \\ \ddots & q_m & q_{m,1} & q_{m,2} & q_{m,3} & \ddots \\ \ddots & q_{1,1} & q_1 & q_{1,1} & q_{1,2} & \ddots \\ \ddots & q_{0,2} & q_{0,1} & q_0 & q_{0,1} & \ddots \\ \ddots & q_{-1,1} & q_{-1,2} & q_{-1,1} & q_{-1} & \ddots \\ \ddots & \ddots & \ddots & \ddots & \ddots & \ddots \end{bmatrix} \begin{bmatrix} \vdots \\ c_h\, e^{(\chi_h z)} \\ c_1\, e^{(\chi_1 z)} \\ c_0\, e^{(\chi_0 z)} \\ c_{-1}\, e^{(\chi_{-1} z)} \\ \vdots \end{bmatrix}, \tag{8.40}$$

where q_{mh} are the elements of the matrix $\underline{\underline{\mathbf{Q}}}$ constructed from the *eigenvectors*. This *ansatz* involves finding the *eigenvalues* and the *eigenvectors* of the complex general matrix $\underline{\underline{\mathbf{M}}}$, which contains up to 101×101 complex elements. The strategy for finding the *eigensystem* is to reduce the balanced matrix to a simpler form, and then to perform an iterative procedure – the Francis QR algorithm – on the simplified matrix. The simpler matrix is a complex upper Hesseberg matrix, which has zeros everywhere below the diagonal except for the first subdiagonal row. Note that the sensitivity of *eigenvalues* to rounding errors can be reduced by the mathematical procedure of *balancing* if the elements of the matrix $\underline{\underline{\mathbf{M}}}$ vary considerably in size. It performs similarity transformations by interchanging rows and corresponding columns, so that the smaller elements appear in the top left hand corner of the matrix (for details, see for example [18]).

In coupled-wave theory the number of differential equations available is always exactly the number of unknowns c_h in (8.39) or (8.40). After computing the matrix $\underline{\underline{\mathbf{Q}}}$ with the eigenvector components q_{mh}, we obtain according to equation (90) a system of linear equations at the zone height $z = 0$:

$$\mathbf{A}(z=0) = \underline{\underline{\mathbf{Q}}}\,\mathbf{C}, \tag{8.41}$$

where $\mathbf{C}$ denotes the vector of the unknown coefficients c_h. The grating is illuminated by a plane-wave of the form

$$E_{\text{inc}} = E_0 \exp\left(-j\,\boldsymbol{\rho}_0 \cdot \mathbf{r}\right), \tag{8.42}$$

which gives the boundary conditions that all amplitudes are equal to zero except the zero-order amplitude

$$A_0(z=0) = 1 \quad \text{and} \quad A_m(z=0) = 0 \quad \text{for} \quad m \neq 0 \quad \text{at} \quad z = 0, \tag{8.43}$$

whose initial value is set equal to one. Thus, the vector $\mathbf{C}$ with the unknown coefficients is derived from the boundary conditions and the inverse matrix $\underline{\underline{\mathbf{Q}}}^{-1}$:

$$\mathbf{C} = \underline{\underline{\mathbf{Q}}}^{-1}\,\mathbf{A}(z=0). \tag{8.44}$$

Using a technique such as Gauss elimination yields the unknown coefficients. Introducing the coefficients c_h into (8.40) allows one the evaluation of the complex field amplitudes $A_m(z)$ of all diffraction orders inside the zone structures as a function of the zone height. The diffraction efficiency, η_m, of the mth diffraction order can be directly calculated from the normalized amplitudes $A_m(z)$ by multiplication with their complex conjugates $A_m^*(z)$:

$$\eta_m(z) = A_m(z)\,A_m^*(z). \tag{8.45}$$

The matrix solution of the coupled-wave differential equation system presented here includes absorption as well as phase shift of X-rays, because both effects were taken into account for the diffraction analysis. As mentioned before, another way to solve the coupled-wave equations is numerical integration by applying the Runge–Kutta algorithm. The results obtained from the matrix solution were compared with the data received by Maser, who performed calculations for ideal rectangular grating profiles with an equal width of lines and spaces with the Runge–Kutta algorithm. Both methods deliver with high accuracy (relative deviation about 10^{-6}) identical results (for details see the thesis [14]). However, as expected, it was found that the numerical integration of the coupled-wave equations is much more time-consuming than the matrix solution, which runs fast on a standard PC. The matrix formalism presented here is therefore superior for studying the diffraction properties of diffractive X-ray optics and for optimizing their parameters, e.g., the zone height and shape of the zone profile. In the following, questions regarding various profiles of zone plate structures with high aspect-ratios will be discussed, and their diffraction efficiency in different orders will be evaluated by applying the theory presented above. In addition, chemical elements that are most suited as materials for high-resolution zone plates for different X-ray energies are determined.

8.4.1 The Influence of the Line-to-Space Ratio

Before we discuss the influence of an *arbitrary* line-to-space ratio on the diffraction efficiency of zone plates, we summarize results for a line-to-space ratio of 1:1, which were published by [14]. We start with a comparison between the first-order diffraction efficiencies of zone structures made from different elements, which are extended parallel to the optical axis. Calculations were performed for nickel, germanium, and silicon zone structures with 20 nm lines and 20 nm spaces. As shown in Fig. 8.5 nickel is more suited than the other elements for high-resolution zone plates, because it combines high efficiency with lower zone height for optimal diffraction efficiency. This means that zone plates manufactured in nickel achieve their optimal diffraction efficiency at lower aspect-ratios of the zones than the zone plates made of the other elements.

It was already shown by coupled-wave calculations that the geometric optical approach is no longer valid for the evaluation of the first-order diffraction efficiency in the 20 nm zone width region [14]. For example, the geometric optical approach delivers optimal first-order diffraction efficiencies of 23.2, 18.8, and 23.5% at 256, 383, and 630 nm height for zone plates with $dr_n = 20$ nm made of Ni, Ge, and Si, respectively. However, the coupled-wave calculations yield for the same L:S-ratio, zone width, wavelength of 2.4 nm and chemical elements optimal efficiency values of 23.2, 16.5, and 14.4% at optimal zone heights of 270, 350, and 470 nm, respectively. Therefore, good agreement of both theories is given only for nickel zone structures with a low aspect-ratio.

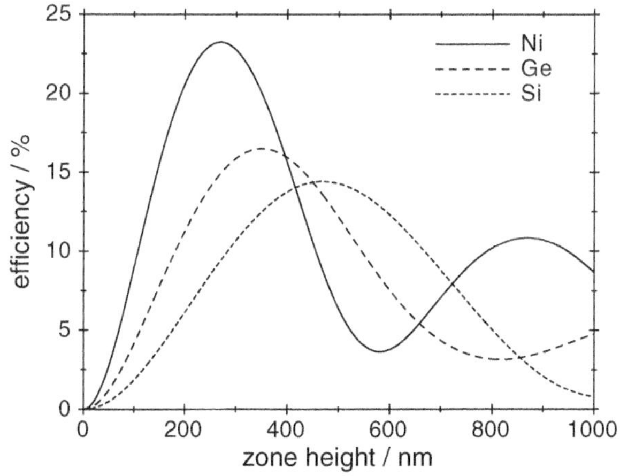

Fig. 8.5. Coupled-wave calculations of the first-order diffraction efficiencies at 2.4 nm wavelength of nickel, germanium, and silicon zone plates with a line-to-space ratio of L:S = 20 nm:20 nm and a rectangular zone profile as a function of the zone height. Parameters: unslanted zone structures and imaging magnification 1,000×. The diffraction efficiency evaluated for Ni is the same as can be calculated with the theory of thin gratings, whereas for Ge and Si significantly smaller values are obtained by coupled-wave theory

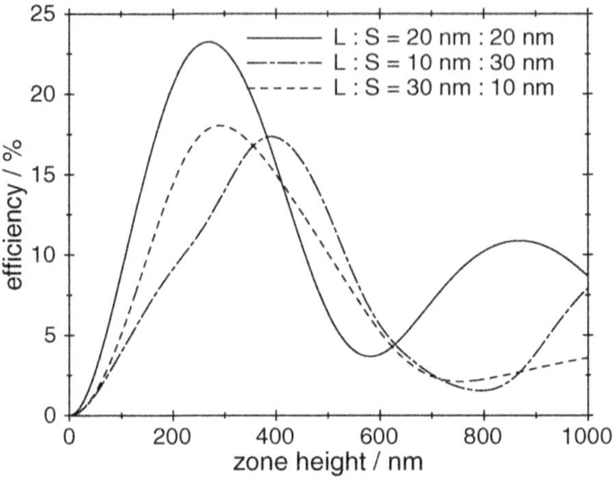

Fig. 8.6. Coupled-wave calculations of the first-order diffraction efficiencies at 2.4 nm wavelength of rectangular nickel zone structures with different line-to-space ratios $L{:}S$ as a function of the zone height. Parameters: unslanted zones, local zone period $\Lambda = 40$ nm and imaging magnification 1,000×. Note that complementary $L{:}S$ ratios yield different diffraction efficiencies

The formalism of the previous sections is now used to study how an *arbitrary* line-to-space ratio $L{:}S$ influences the first-order diffraction efficiency of zones extending parallel to the optical axis. Results of the coupled-wave calculations are plotted in Fig. 8.6. The first-order diffraction efficiency is shown for nickel zone structures with $L{:}S = 20$ nm:20 nm, 30 nm:10 nm, and 10 nm:30 nm as a function of the zone height.

The optimal diffraction efficiencies of 23.2, 17.3, and 18% are achieved at 270, 290, and 390 nm height, respectively. By comparison, according to (8.9) with the theory of thin gratings only 11.6% diffraction efficiency is obtained at 256 nm height for $L{:}S$ = 10 nm:30 nm and $L{:}S$ = 30 nm:10 nm. Note that the theory of thin gratings predicts that the efficiencies are always equal for complementary zone structures (see (8.9)), e.g., for $L{:}S = 10$ nm:30 nm and 30 nm:10 nm we get the same efficiency, which is in contradiction to the results obtained by electrodynamic theory. We can conclude that even for nickel zone plates, which have a comparatively low optimal zone height, the local first-order diffraction efficiency can not be evaluated with sufficient accuracy by applying the theory of thin gratings if the zone period is in the range of $\Lambda = 40$ nm. Therefore, the parameters of such zone plates, which are currently under development, have to be optimized by applying the coupled-wave theory.

The coupled-wave analysis of zone plate diffraction has shown that the first-order diffraction efficiency can be increased if the zone structures are slanted against the optical axis according to the Bragg condition [14]. In the next sections we extend the numerical calculations of slanted zone structures to an *arbitrary* diffraction order with an *arbitrary* line-to-space ratio by using the coupled-wave formalism described in the previous sections. Now we derive the slanting angle, ψ, of the zone structures for arbitrary diffraction orders. If the zones are regarded as the reflecting lattice planes of a crystal, we can write according to the Bragg equation

$$m\,\lambda = 2\,\Lambda\,\sin\alpha = 4\,\mathrm{d}r_n\,\sin\alpha, \tag{8.46}$$

where α denotes the angle between the incident plane-wave and the planes of the periodically arranged X-ray scattering zone structures. If the Bragg condition for an order of diffraction is fulfilled, each zone structure acts as a partly reflecting mirror, which means that the forward-diffracted plane-wave has the same angle between the planes of the zone structures as the incident plane-wave. As can be seen from the Fig. 8.2 and 8.3, we get for the Bragg angle α

$$\alpha = \theta_{\mathrm{in}} - \psi = \theta_{\mathrm{out}} + \psi, \tag{8.47}$$

which leads to the slanting angle, ψ, of the zone structures expressed in terms of the local zone width $\mathrm{d}r_n$ and the imaging magnification M:

$$\psi = \frac{\theta_{\mathrm{in}} - \theta_{\mathrm{out}}}{2} \approx \frac{m\lambda}{4\,\mathrm{d}r_n}\left(\frac{M-1}{M+1}\right), \tag{8.48}$$

with

$$\theta_{\mathrm{in}} = \arctan(r_n/g) = \arctan\left[\frac{m\lambda}{2\,\mathrm{d}r_n}\left(\frac{M}{M+1}\right)\right] \approx \frac{m\lambda}{2\,\mathrm{d}r_n}\left(\frac{M}{M+1}\right) \tag{8.49}$$

and

$$\theta_{\mathrm{out}} = \arctan(r_n/b) = \arctan\left[\frac{m\lambda}{2\,\mathrm{d}r_n}\left(\frac{1}{M+1}\right)\right] \approx \frac{m\lambda}{2\,\mathrm{d}r_n}\left(\frac{1}{M+1}\right). \tag{8.50}$$

Note that the slanting angle, ψ, increases within the radius of the zone plate as can be seen from (8.48). Therefore, each local zone plate area has a different local slanting angle.

The influence of the line-to-space ratio on the first-order diffraction efficiency of zone structures slanted to the optical axis and fulfilling the Bragg condition is shown for nickel zone plates working at 2.4 nm wavelength with an imaging magnification of $M = 1{,}000\times$. It is seen from Fig. 8.7 that the first-order diffraction efficiency can be enhanced drastically by reducing the zone width of the nickel structures and increasing their spaces if the Bragg condition is fulfilled for the first-order radiation. It was found from additional calculations that the line-to-space ratio can be chosen in such a way that very high diffraction efficiencies can be realized and zone plates can become in the first-order nearly as efficient as refractive lenses for visible light. As the Bragg condition can be fulfilled for any diffraction order, the diffraction efficiencies of slanted zones with *arbitrary* $L{:}S$ are now investigated for *arbitrary* diffraction orders.

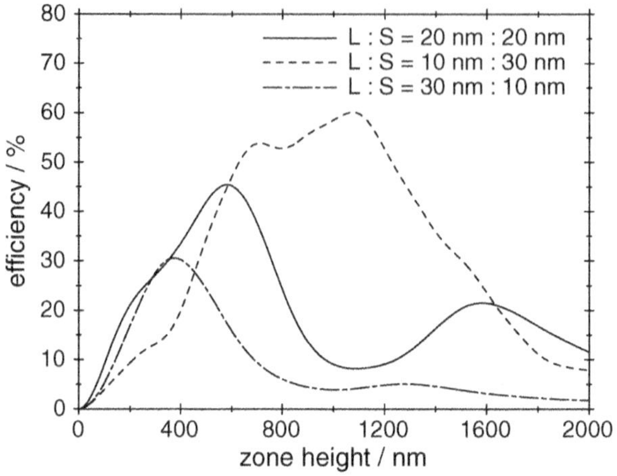

Fig. 8.7. Coupled-wave calculations of the first-order diffraction efficiencies at 2.4 nm wavelength as a function of the zone height of rectangular nickel zone structures slanted against the optical axis fulfilling the Bragg condition. Line-to-space ratios are 20 nm:20 nm, 10 nm:30 nm, and 30 nm:10 nm. Imaging magnification: 1,000×. The best diffraction efficiency is obtained for smallest L:S ratio

8.4.2 Applying High-Orders of Diffraction for X-ray Imaging

The resolving power of zone plates can be increased in two different ways. The conventional way is to use smaller zone periods in the first-order of diffraction. Another way is to use high-order diffraction, because the obtainable resolution scales inversely with the diffraction order m used for X-ray imaging (see (8.3)). However, if optically thin gratings are assumed, it results that an increased resolving power is achieved at the cost of a drastically reduced diffraction efficiency (see (8.9)). One intention of this chapter is to demonstrate that high-orders are nevertheless highly efficient under special conditions, which are determined in this section by coupled-wave theory.

As shown above for the first-order radiation, the X-ray optical parameters of nickel make it more suited in this wavelength range than many other elements. Therefore, we are especially interested in the high-order diffraction at 2.4 nm wavelength of zone plates manufactured in nickel as optical elements – condensers and objectives – in X-ray microscopes. To approximate the permittivity of the zone structures with an arbitrary line-to-space ratio with high accuracy, all the calculations of the high-order diffraction presented in the following were performed up to the 50th spatial harmonic of the Fourier expansion, which leads to a general complex matrix consisting of 101 × 101 elements. The energy distribution of the radiation in different orders is shown in Fig. 8.8 for the case that the Bragg condition is fulfilled for the 6th order. One can see that this order is very efficient at about 1,500 nm zone height,

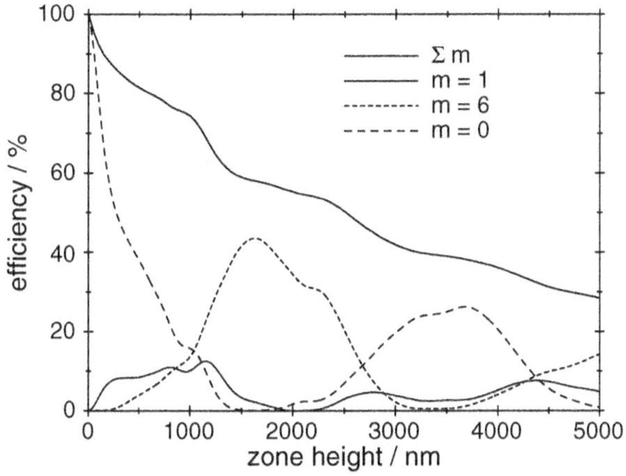

Fig. 8.8. Diffraction efficiencies of the zero-order, the 1st-order, the 6th-order, and the sum over all orders for nickel zone structures with $L{:}S = 24\,\text{nm}{:}96\,\text{nm}$. Parameters: 2.4 nm wavelength, unslanted zone structures and imaging magnification $M = 1\times$. High efficiency is obtained even in the 6th-order of diffraction

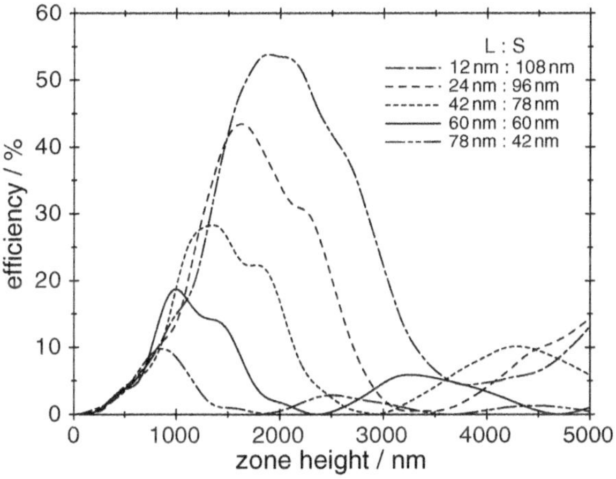

Fig. 8.9. Diffraction efficiencies $\eta_6(z)$ of the 6th order at 2.4 nm wavelength for nickel structures with different $L{:}S$ for $\Lambda = 120\,\text{nm}$ zone period as a function of the zone height. Parameters: rectangular zone structures parallel to the optical axis and imaging magnification $M = 1\times$. It shows that the smallest $L{:}S$ leads to the best η_6

whereas the contribution of all other orders to the total amount of radiation is small at this zone height.

Furthermore, as expected, the sum of the diffraction efficiencies of all orders is continuously attenuated with increasing zone height, which is due to the photoelectric absorption in the zone structures. Figure 8.9 shows the diffraction efficiency, $\eta_6(z)$, of the 6th order as a function of the zone height

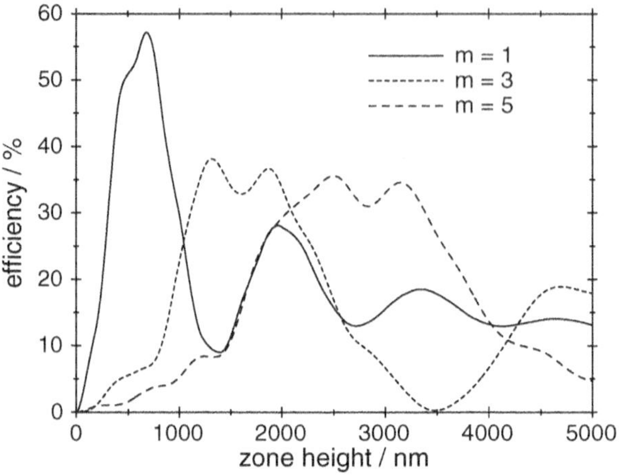

Fig. 8.10. Diffraction efficiencies of the 1st, 3rd, and 5th-order at 2.4 nm wavelength of nickel zone structures with a rectangular zone profile and $L{:}S = 9\,\text{nm}{:}21\,\text{nm}$, $27\,\text{nm}{:}63\,\text{nm}$, and $45\,\text{nm}{:}105\,\text{nm}$, respectively. Parameters: slanted zone structures according to the Bragg condition and imaging magnification $M = 1{,}000\times$. The resolution is $1.22 \times 15\,\text{nm}$ in all cases. Note that the optimal zone height increases significantly if high order focusing is regarded

for different $L{:}S$ of 78 nm:42 nm, 60 nm:60 nm, 42 nm:78 nm, 24 nm:96 nm, and 12 nm:108 nm. In these calculations the imaging magnification is $M = 1$ and therefore the Bragg condition for the incident and the 6th diffracted wave is satisfied for zone structures parallel to the optical axis (see (8.48)). Under these conditions the intensity of the high diffraction order selected can be increased up to 54% of the incident intensity by decreasing $L{:}S$. By comparison, calculations using the cited geometric optical approach result in optimal $\eta_6(z)$ of only 0.06, 0, 0.06, 0.22, and 0.58% for the same $L{:}S$ values as above.

If the magnification is not equal to one, the zones have to be slanted according to the Bragg condition against the optical axis, since they work similarly to small partially reflecting planes. Figure 8.10 shows $\eta_m(z)$ in the 1st, 3rd, and 5th order for slanted Ni structures with $L{:}S = 9\,\text{nm}{:}21\,\text{nm}$, 27 nm:63 nm, and 45 nm:105 nm zone width, respectively.

Note that all three cases can deliver the same spatial resolution. It shows that the high-orders as well as the first-order become highly efficient if the Bragg condition is satisfied. Furthermore, it is shown that the optimal zone height and, therefore, the aspect-ratio increases for high-orders. Additional calculations have shown that in high-orders diffraction efficiencies of 30–50% are also possible for harder X-rays in the sub-1 nm wavelength range, which is also interesting for hard-X-ray microscopy. Note that high-order zone plates for use at shorter wavelengths require very high aspect-ratios far beyond the aspect-ratios manufactured by reactive ion etching techniques nowadays. No limitation in the aspect-ratio is given if the zone plates are generated by the so-called sputtered sliced technique (see Sect. 8.5) [19].

Summing up, the diffraction properties in high-orders change with increasing height of the zone structures. Therefore, the geometric optical approach is no longer valid if η_m of high-orders are calculated for zone structures with high aspect-ratios. In addition, in this case the diffraction efficiency $\eta_m(z)$ also depends critically on satisfying the Bragg condition. The plots of Figs. 8.9 and 8.10 demonstrate that high-order zone plates can be used as X-ray condensers as well as X-ray objectives. For example, a nickel zone plate objective with 120 nm smallest zone period with a $L{:}S$ of 30 nm:90 nm, averaging η_6 over the whole zone plate, yields about 40% diffraction efficiency at 2.4 nm wavelength and can resolve about 10 nm features.

By comparison, zone plate objectives to be used in the first diffraction order with smallest zone structures with $L{:}S = 30$ nm:30 nm, aspect-ratios up to 6:1, and measured total η_1 of 15% (theoretical maximum 23.2% for nickel zone plates) at 2.4 nm wavelength were already processed using electron-beam lithography and reactive ion etching techniques [20].

In the same manner highly efficient condensers with high numerical aperture and monochromatizing properties can be developed to collect photons from a large solid angle from laser generated microplasma sources for object illumination in laboratory X-ray microscopes. Such small plasma X-ray sources are currently under development. The X-ray source can be imaged by the high-order condenser with an imaging magnification of $M = 1$. Therefore, the Bragg condition is fulfilled for zones parallel to the optical axis, which makes it possible to manufacture such zone plates by electron-beam lithography and reactive ion etching.

First measurements of the high-order diffraction of zone plate condensers produced by these techniques [21] indicate that the coupled-wave method is convenient to describe their X-ray diffraction properties. Therefore, it can be expected that the theoretical considerations presented here will lead to the new type of high-order diffractive X-ray optics with a high numerical aperture.

8.5 The Influence of Interdiffusion and Roughness

Up to now X-ray microscopes mainly operate in the soft X-ray wavelength range between the K-absorption edges of oxygen and carbon (2.34–4.37 nm wavelength). However, the third generation of electron storage rings are highly intense X-ray sources, and their insertion devices, e.g., undulators, emit sufficient photon flux at sub-1 nm wavelengths to be well suited as X-ray sources for microscopy. As the X-ray optical constants of matter change with wavelength, zone plates suited for shorter wavelengths have to be developed. At these wavelengths the maximum diffraction efficiency η_m – the fraction of the incident intensity diffracted into one selected diffraction order m – of zone plates is in general achieved at much larger zone heights. Therefore, zone plates with much higher aspect-ratios are required for the sub-1 nm wavelength region and new techniques have to be used to manufacture zone plates for these wavelengths. Zone plates for use at soft X-ray wavelengths are manufactured by electron-beam lithography and reactive ion etching (RIE) techniques. In this technique the zones are supported by a thin foil; no stabilizing spacer material is located between the zones. It is very difficult to use this technique for aspect-ratios of significantly more than 10:1 at very small zone widths of, e.g., $\mathrm{d}r_N = 20\,\mathrm{nm}$. Therefore, zone plates for use at short wavelengths are manufactured with another technology. A thin microwire is used, which is coated alternately with two materials of different X-ray scattering properties. The so-called "sputtered sliced zone plate" is then obtained from the coated microwire by slicing it perpendicular to its axis and thinning the slices down to the required zone plate height [19].

The diffraction efficiency, η_m, of zone plates manufactured by either method is diminished if the zone walls are not ideally smooth. Furthermore, the efficiency also depends on the shape of the profile. In the case of sputtered sliced zone plates, two different zone materials are in contact at their interface. This can lead to interdiffusion and roughness at the interfaces of the two materials, which diminishes the efficiency further. In general, a zone roughness can be regarded as a local positioning error of segments of the zone material. This will reduce η_m and possibly the spatial resolution obtainable with zone plates.

Here it is the intention to perform coupled-wave calculations for zone plates with interdiffusion at the interfaces and to include a zone roughness in the radial direction as a function of the zone height. Such a roughness can be observed in sputtered sliced zone plates, because the deposition process on the microwire causes roughness of the material interface ranging over many zones. Mathematically, this effect is taken into account by subdividing the zone structures in their height (z-axis) into N layers and by randomly shifting the ith layer by Δx_i in the radial direction for roughness simulation (see Fig. 8.11).

In mathematical terms the function $p_i(x,z)$ shown in Fig. 8.11 (without interdiffusion $\Delta x_{\mathrm{diff}} = 0$) and Fig. 8.12 (interdiffusion region $\Delta x_{\mathrm{diff}} \neq 0$) can be expanded in a Fourier series, which is used to describe the spatial distribution of the permittivity in the grating structures of the ith layer consisting of the materials A and B with the permittivities ε_{A} and ε_{B}

$$
\begin{aligned}
p_i(x,z) = {} & \frac{L_i}{L_i+S_i} + \frac{2\,L_i}{L_i+S_i} \\
& \times \sum_{h=1,2,3\ldots}^{\infty} \mathrm{sinc}\left(h\pi\frac{\Delta x_{\mathrm{diff}}}{\Lambda}\right) \mathrm{sinc}\left(h\pi\frac{L_i}{L_i+S_i}\right) \\
& \times \cos\left(h\,G\,[(x-\Delta x_i)\cos\psi - z\sin\psi]\right),
\end{aligned}
\tag{8.51}
$$

where $L_i{:}S_i$ denotes the line-to-space ratio of the ith layer and G is the magnitude of the grating vector $\mathbf{G} = 2\pi/\Lambda\,(\cos\psi, -\sin\psi)$. Interdiffusion is

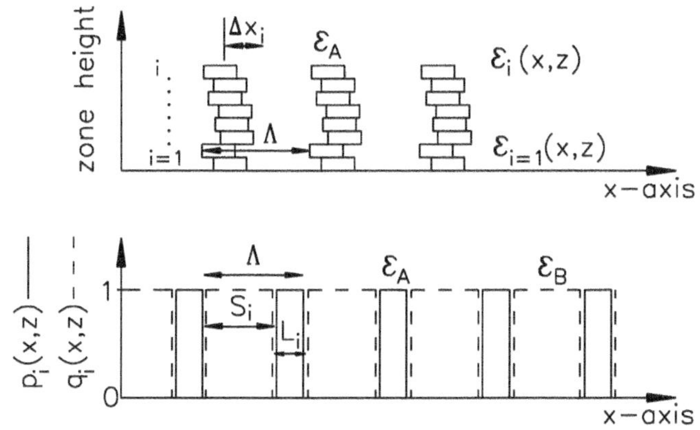

Fig. 8.11. (*Lower*) Fourier expansions $p_i(x, z)$ and $q_i(x, z)$ used to describe the permittivity of the ith layer consisting of two different materials A and B with corresponding permittivities ε_A and ε_B. The local zone plate period Λ is given by $\Lambda = L_i + S_i$. (*Upper*) Roughness of the grating is simulated by shifting the layers in radial direction

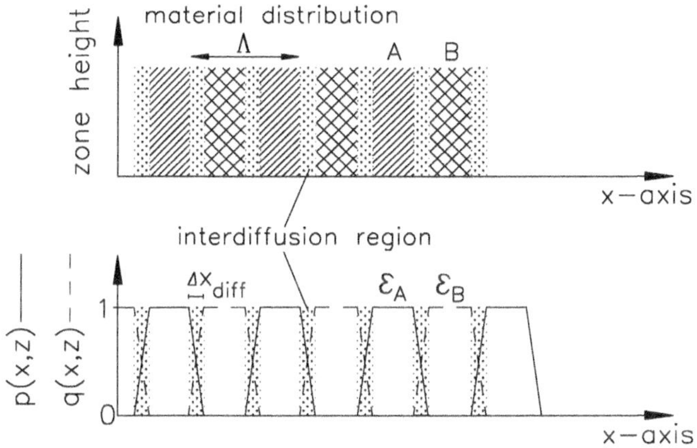

Fig. 8.12. (*Lower*) Fourier expansion describing the interdiffusion region between the grating structures consisting of two different materials A and B. (*Upper*) Simulation of interdiffusion in the interface region between the zones manufactured from two different X-ray scattering materials

described by the mixture of the materials A and B, which is denoted in (8.51) by the width of the interdiffusion region Δx_diff (see Fig. 8.12).

It is assumed that the permittivity changes linearly with the position x in the interdiffusion region. Using the relation $q_i(x, z) = 1 - p_i(x, z)$ (see Figs. 8.11 and 8.12) we obtain for the periodically changing permittivity $\varepsilon_i(x, z) = \varepsilon_\mathrm{A}\, p_i(x, z) + \varepsilon_\mathrm{B}\, q_i(x, z)$ of the ith layer:

$$\varepsilon_i(x,z) = \bar{\varepsilon}_i + \Delta\varepsilon \frac{2\,L_i}{L_i + S_i}$$

$$\times \sum_{h=1,2,3\ldots}^{\infty} \operatorname{sinc}\left(h\pi \frac{\Delta x_{\text{diff}}}{\Lambda}\right) \operatorname{sinc}\left(h\pi \frac{L_i}{L_i+S_i}\right)$$

$$\times \left[\cos(h\,\mathbf{G}\cdot\mathbf{r})\cos(h\,G\,\Delta x_i\,\cos\psi) + \sin(h\,\mathbf{G}\cdot\mathbf{r})\sin(h\,G\,\Delta x_i\,\cos\psi)\right],$$

where $\mathbf{G}\cdot\mathbf{r}$ denotes the scalar product of the vectors $\mathbf{G}$ and $\mathbf{r} = (x, z)$. Again the angle ψ allows one to slant the zones against the optical axis of the zone plate. Here the average permittivity $\bar{\varepsilon}_i$ of the ith grating layer is given by

$$\bar{\varepsilon}_i = \varepsilon_{\text{B}} + (\varepsilon_{\text{A}} - \varepsilon_{\text{B}})\frac{L_i}{L_i+S_i} = \tilde{n}_{\text{B}}^2 + (\tilde{n}_{\text{A}}^2 - \tilde{n}_{\text{B}}^2)\frac{L_i}{L_i+S_i} \tag{8.52}$$

and

$$\Delta\varepsilon = \varepsilon_{\text{A}} - \varepsilon_{\text{B}} = \tilde{n}_{\text{A}}^2 - \tilde{n}_{\text{B}}^2 \tag{8.53}$$

is the difference between the permittivities of the materials A and B in terms of the refractive index.

As described in Sect. 8.3, the transmitted waves are scattered by the periodically arranged structures into many different directions. Thus, the resulting electromagnetic field in each layer is developed in terms of plane-waves with spatially varying coefficients $A_{m,i}(z)$:

$$E_i = E_{0,i} \sum_{m=-\infty}^{\infty} A_{m,i}(z)\,\exp\left(-j\,\boldsymbol{\rho_m}\cdot\mathbf{r}\right), \tag{8.54}$$

with

$$\boldsymbol{\rho_m} = \boldsymbol{\rho_0} + m\,\mathbf{G} \qquad m = 0, \pm 1, ..., \tag{8.55}$$

where $\boldsymbol{\rho_m}$ is the wave vector of the mth diffraction order and $\mathbf{G}$ is the grating vector of the local zone plate period. Introducing (8.52) and (8.54) into the scalar wave equation (8.10) and performing the mathematical operations shown in Sect. 8.3, we obtain a linear second-order differential equation system:

$$\frac{\mathrm{d}^2 A_{m,i}(z)}{\mathrm{d}z^2} - 2\,j\,\rho_{m,z}\frac{\mathrm{d}A_{m,i}(z)}{\mathrm{d}z} - (|\,\boldsymbol{\rho_m}\,|^2 - k_0^2\,\bar{\varepsilon}_i)\,A_{m,i}(z) + k_0^2\frac{L_i}{L_i+S_i}\Delta\varepsilon$$

$$\times \sum_{h=1,2,3\ldots}^{\infty} \operatorname{sinc}\left(h\pi \frac{\Delta x_{\text{diff}}}{\Lambda}\right) \operatorname{sinc}\left(h\pi \frac{L_i}{L_i+S_i}\right)$$

$$\times \left[\cos(h\,G\,\Delta x_i\,\cos\psi)\left(A_{m-h,i}(z) + A_{m+h,i}(z)\right)\right.$$

$$\left. + j\sin(h\,G\,\Delta x_i\,\cos\psi)\left(A_{m+h,i}(z) - A_{m-h,i}(z)\right)\right] = 0. \tag{8.56}$$

As discussed earlier, transmission zone plates as X-ray optics are commonly used near normal incidence and therefore, to a good approximation, no waves are reflected. Thus, the second-order derivatives in (8.56) can be neglected and linear first-order differential equation systems expressed in matrix notation are obtained:

$$\frac{\mathrm{d}\mathbf{A}_i(z)}{\mathrm{d}z} = \mathbf{M}_i\ \mathbf{A}_i(z), \tag{8.57}$$

where $\mathbf{M}_i$ denotes the complex general matrix containing the X-ray optical and illumination parameters of the ith layer. As derived above, the solution of this matrix equation is of the form

$$A_{m,i}(z) = \sum_h Z_{mh,i}\ [c_{h,i}\ \exp(\chi_{h,i}\, z)] \quad \text{for} \quad z_i \leq z \leq z_{i+1}. \tag{8.58}$$

Note that the *eigenvalues* $\chi_{h,i}$ and *eigenvectors* have to be calculated separately for each layer. The unknown coefficients $c_{h,i}$ are determined using the boundary conditions. The initial values for the layer $i = 0$ are given by

$$A_{0,0}(z=0) = 1 \ \text{ and } \ A_{m,0}(z=0) = 0 \ \text{ for } \ m \neq 0. \tag{8.59}$$

Then the coefficients $c_{h,i+1}$ of the $(i+1)$th layer are calculated from

$$\mathbf{c}_{i+1} = \ \boldsymbol{Z}_{i+1}^{-1}\ \ \mathbf{A}_i(z_i) \ \text{ for } \ i = 0, 1, 2, ..., \tag{8.60}$$

where $\mathbf{A}_i(z)$ is determined from (8.56). The normalized amplitudes $A_{m,i}(z)$ are propagated through the layer system and yield the diffraction efficiency

$$\eta_{m,i}(z) = A_{m,i}(z)\, A^*_{m,i}(z) \tag{8.61}$$

of the forward-diffracted orders.

8.6 Numerical Results for Zone Plates with High Aspect-Ratios

The development of zone plates that are suited for the photon energy range of 2–10 keV is a current field of activity in X-ray optics. One aim is to manufacture zone plates with high resolving power and high diffraction efficiency by the sputtered sliced technique (see also Sect. 8.5). Here some criteria for tolerable interdiffusion and roughness values are derived from coupled-wave calculations for the zone structures.

We start our considerations with zone plates made from alternating Ni and SiO_2 zones, which were sputtered with slight modifications on a glass microwire and sliced afterwards [22]. From the theoretical point of view these zone materials are well suited for the energy range mentioned above. Also at soft X-ray wavelengths in the water window a high efficiency is achieved if the

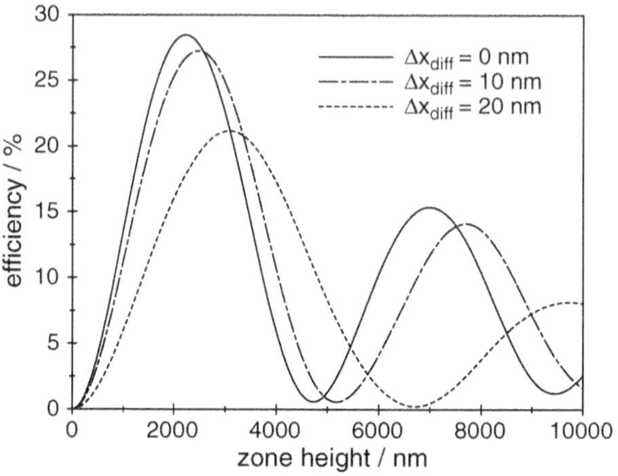

Fig. 8.13. First-order diffraction efficiencies for different interdiffusion widths $\Delta x_{\text{diff}} = 0$, 10, and 20 nm at $\lambda = 0.3$ nm of zone structures made by alternately depositing nickel and silicon dioxide with $L{:}S = 20\,\text{nm}{:}20\,\text{nm}$ as a function of the zone height. Parameters: unslanted zone structures and imaging magnification $M = 1{,}000\times$. The plot shows that the optimal zone height increases with increase in interdiffusion

zone height is adapted to the required optimal values, which are much smaller than that for higher X-ray energies.

Now the effect of interdiffusion and roughness in zone structures made of the materials Ni and SiO_2 is studied for $\Lambda = 40$ nm zone period at 0.3 nm wavelength. Figure 8.13 shows the first-order efficiency, $\eta_1(z)$, for interdiffusions of $\Delta x_{\text{diff}} = 0$, 10, and 20 nm. It is found that the interdiffusion should not exceed approximately $\Delta x_{\text{diff}} \leq \Lambda/3$, otherwise the first-order diffraction efficiency significantly decreases and the maximum efficiency is achieved only at larger zone heights. Roughness, given as a function of the zone height, is included in the calculations of the Ni/SiO_2 system mentioned above by composing the zone structures from layers of 20 nm height. With $N = 500$ layers a maximum zone height of 10 μm is achieved. Each layer is randomly shifted in the x-direction as shown in Fig. 8.11, which can quantitatively be characterized by the root-mean-square (RMS) roughness σ:

$$\sigma = \sqrt{\frac{1}{N}\sum_{i=1}^{N}(\Delta x_i)^2}. \tag{8.62}$$

Figure 8.14 shows the first-order efficiency $\eta_1(z)$ for the Ni/SiO_2 layer system as a function of the zone height for different RMS-values at $\lambda = 0.3$ nm wavelength.

It can be seen how the maximum diffraction efficiency decreases with increasing RMS roughness. Furthermore, if roughness is present the optimal

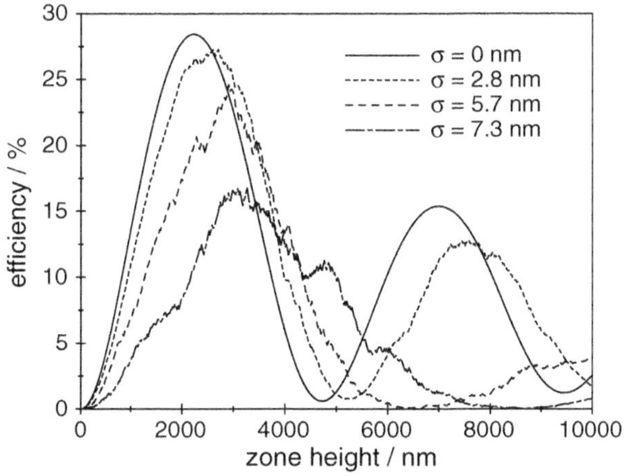

Fig. 8.14. First-order diffraction efficiencies $\eta_1(z)$ for different root-mean-square roughnesses $\sigma = 0$, 2.8, 5.7, and 7.3 nm at $\lambda = 0.3$ nm of zone structures made of nickel and silicon dioxide with $L{:}S = 20\,\text{nm}{:}20\,\text{nm}$ as a function of the zone height. Parameters: unslanted zone structures and imaging magnification $M = 1{,}000\times$. Note that the optimal zone height increases with increasing roughness, whereas η_1 decreases

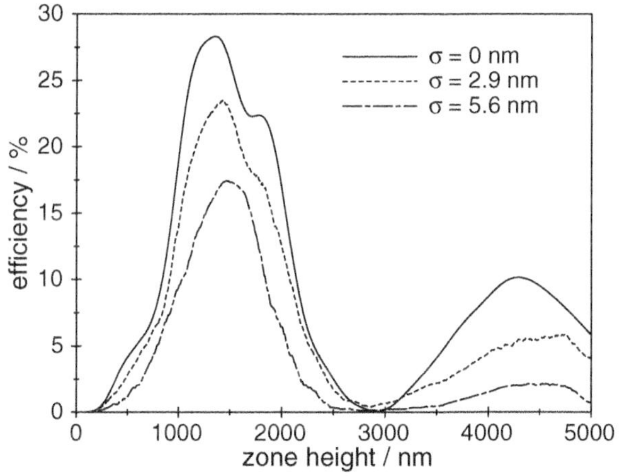

Fig. 8.15. Sixth-order diffraction efficiency $\eta_6(z)$ for different RMS values of $sigma = 0$, 2.9, and 5.9 nm. Parameters: rectangular nickel zone structures with $L{:}S = 42\,\text{nm}{:}78\,\text{nm}$ parallel to the optical axis, 2.4 nm wavelength and imaging magnification $M = 1\times$. Note that the optimal zone height is almost not altered, but η_6 decreases

zone height is achieved at larger zone heights. Therefore, it is proposed that $\sigma \leq \Lambda/7$ should be selected as a criterion. If it is fulfilled, the maximum diffraction efficiency is reduced by less than 20%.

The effect of roughness on the high-order diffraction efficiency of zone structures is shown in Fig. 8.15, where the Bragg condition is fulfilled for the sixth-order. This roughness simulation for nickel zones with $L{:}S$ = 42 nm:78 nm shows a different behavior for the sixth order diffraction efficiency than that obtained for the first-order radiation. The optimal zone height is not significantly shifted by roughness, only the diffraction efficiency decreases with increasing roughness. If we tolerate again an efficiency reduction of 20% due to roughness, we get – from Fig. 8.15 and additional calculations with other parameters – for the high-order diffraction the criterion $\sigma \leq \Lambda/(7\,m)$.

The effect of roughness on the imaging quality of zone plates is not evaluated here. However, it can be argued that the image quality is degraded if the complex field amplitudes $A_m(z)$ vary randomly from local zone plate area to local zone plate area. This means that also the wave-front generated from the local zone plate areas varies randomly, whereas the wave-front of a perfect focusing optic is a monotonic function of the radius of the optic. It is therefore recommended that the effect of roughness on the resolving power of sputtered sliced zone plates is studied in future works. Summing up, it is shown by model calculations using coupled-wave theory that both interdiffusion and roughness can reduce η_m [23]. In practice, the two materials for the zone structures have to be selected very carefully, because low interdiffusion and roughness have to be realized at the interfaces at the same time. Since the maximum of the first-order diffraction efficiency is shifted in the presence of interdiffusion and roughness, an exact knowledge of these parameters yields the zone height to be processed for an optimal diffraction efficiency η_1.

8.7 Nonrectangular Profile Zone Structures

The diffraction efficiency of diffractive optical elements depends on the zone profile realized by nanostructuring processes. The influence of the profile of the structures on the diffraction efficiency will now be studied. For this purpose the theory described in Sect. 8.5 for roughness simulation is applied to describe *arbitrarily shaped zone profiles.* A change of the profile in the direction of the optical axis (z-axis) often can be observed for zone plates processed by electron-beam lithography and reactive ion etching techniques. These profiles cannot be described by a single Fourier expansion. (A single Fourier expansion was used to describe the modulated permittivity in the wave equation for rectangular profiles.) However, *arbitrarily shaped* zone profiles may be subdivided into thin layers. Each layer is again described by a rectangular profile with line-to-space ratio $L{:}S$ and the shift parameter, Δx_i, for the radial position (see Fig. 8.11 and (8.52)). With this method any profile can be synthesized. Similar to the roughness calculations, the field amplitudes are propagated through the layer system using the previous formalism of the coupled-wave theory. Note that the *eigenvalues* and *eigenvectors* of the grating matrix have to be calculated separately for each thin layer. In practice, it is sufficient to approximate zone profiles by about 50 layers.

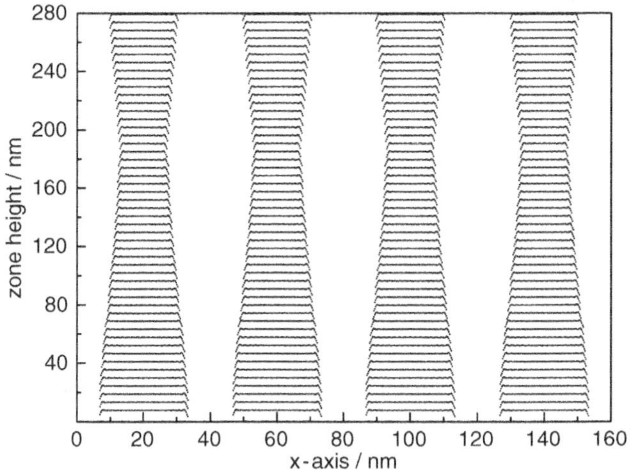

Fig. 8.16. Profile of zone structures with nonrectangular shape, periods of 40 nm, and 280 nm height approximated by 50 grating layers of 5.6 nm thickness

Zone profiles with high apect ratios, which are manufactured by reactive ion etching (RIE) techniques, can show deviations from the ideal rectangular shape, which has been demonstrated for grating structures manufactured in germanium [24]. An example of an etched profile, e.g., for organic polymer structures, which can be filled with nickel by electroplating, is illustrated in Fig. 8.16. After electroplating, the organic polymer structures are removed by RIE. Therefore, the inverse structures of Fig. 8.16 are relevant for the diffraction analysis of nickel zone plates. The profile of Fig. 8.16, its inverse profile, and an ideal rectangular profile were used to calculate the first-order diffraction efficiency of nickel structures with 40 nm period by coupled-wave theory. The results are plotted in Fig. 8.17.

No significant loss of diffraction efficiency is obtained for the first-order. Therefore, one can expect that a nanostructuring process for nickel as the material for high-resolution zone plates working in the first-order should deliver high diffraction efficiency even if a rectangular zone profile cannot be realized perfectly.

The results presented in Figs. 8.5 and 8.17 suggest that nickel is well suited for the diffracting structures of zone plates. In particular nickel requires the lowest aspect-ratio for an optimal diffraction efficiency.

8.8 Rigorous Electrodynamic Theory of Zone Plates

In the previous sections the diffraction efficiency of zone plate structures was analyzed by applying the first-order coupled-wave approach. Because of the absorption of X-rays in matter, the calculated field amplitudes are

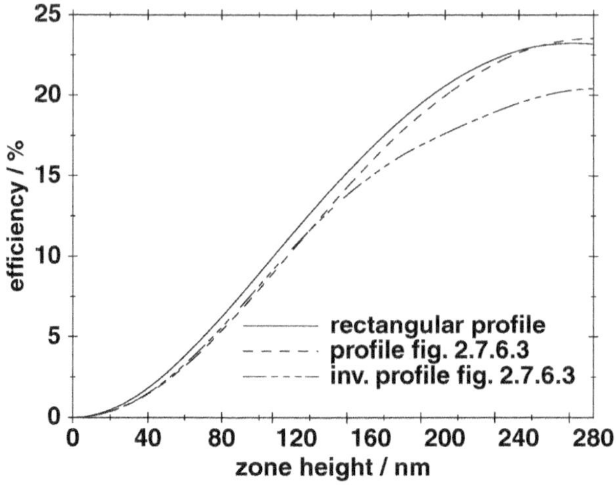

Fig. 8.17. First-order diffraction efficiencies as a function of the zone height for nickel structures with an ideal rectangular profile (L:S = 20 nm:20 nm), the profile shown in Fig. 8.16 and the inverse profile of Fig. 8.16. Parameters: 2.4 nm wavelength, zone structures parallel to the optical axis, and imaging magnification $M = 1{,}000\times$. No significant change in the first-order diffraction efficiency is obtained if the profile slightly alters its shape

attenuated in the zone structures and the sum of the diffraction efficiencies of all orders should continuously decrease with increasing zone height. However, it was found by numerical calculations that the solutions of the first-order coupled-wave approach do not converge if the zone width becomes as small as the wavelength. To overcome this weak point, the first-order coupled-wave approach can no longer be applied. The diffraction of zone plates with very high resolving power and numerical aperture has to be analyzed by the rigorous coupled-wave theory (RCWT).

In the RCWT, the second-order derivatives in the differential equation systems are retained and boundary diffraction is included. This was described at first in [25] for the calculation of *cosine*-modulated phase gratings without loss by absorption. Applying the RCWT in the X-ray domain to describe dielectric gratings requires the inclusion of absorption as well as phase shift, because the imaginary part and the real part of the complex refractive index are almost equal. Another advantage of the RCWT is that the illumination angle of the incident plane-wave is no longer restricted to values that do not lead to significant intensities of reflected waves. Therefore, accurate calculations of reflection gratings can also be performed, because waves, diffracted forward as well as backward, are taken into account in the electrodynamic description. The rigorous coupled-wave formalism is now derived for infinite gratings, in order to describe zone plate diffraction with zone structure widths near the wavelength more accurately.

We proceed further with infinite gratings and study the effect of retaining the second-order derivatives of the field amplitudes in the rigorous coupled-wave equations (see (8.34) and (8.56)) given by

$$\frac{\mathrm{d}^2 A_m(z)}{\mathrm{d}z^2} - 2\,j\,\rho_{m,z}\,\frac{\mathrm{d}A_m(z)}{\mathrm{d}z} - (|\,\boldsymbol{\rho_m}\,|^2 - k_0^2\,\bar{\varepsilon})\,A_m(z) + k_0^2\frac{L}{L+S}\,\Delta\varepsilon$$

$$\times\sum_h^{\infty}\mathrm{sinc}\left(h\pi\frac{\Delta x_{\mathrm{diff}}}{\Lambda}\right)\mathrm{sinc}\left(h\pi\frac{L}{L+S}\right)\Big[A_{m-h}(z) + A_{m+h}(z)\Big] = 0, \quad (8.63)$$

which is a set of second-order linear differential equations with constant coefficients. The problem leads to an eigenvalue problem, which is solved numerically after including the boundary conditions and restricting the number of grating harmonics to a finite number.

An advantage of retaining the second-order derivatives in the RCWT is that the boundary conditions can be included for both the electric and magnetic fields. Therefore, reflected waves can also be evaluated. However, the validity of the first-order approach can be verified by comparing the results with (more time consuming) calculations performed by RCWT. Numerical results from the RCWT calculations are shown in Fig. 8.18. The calculation is

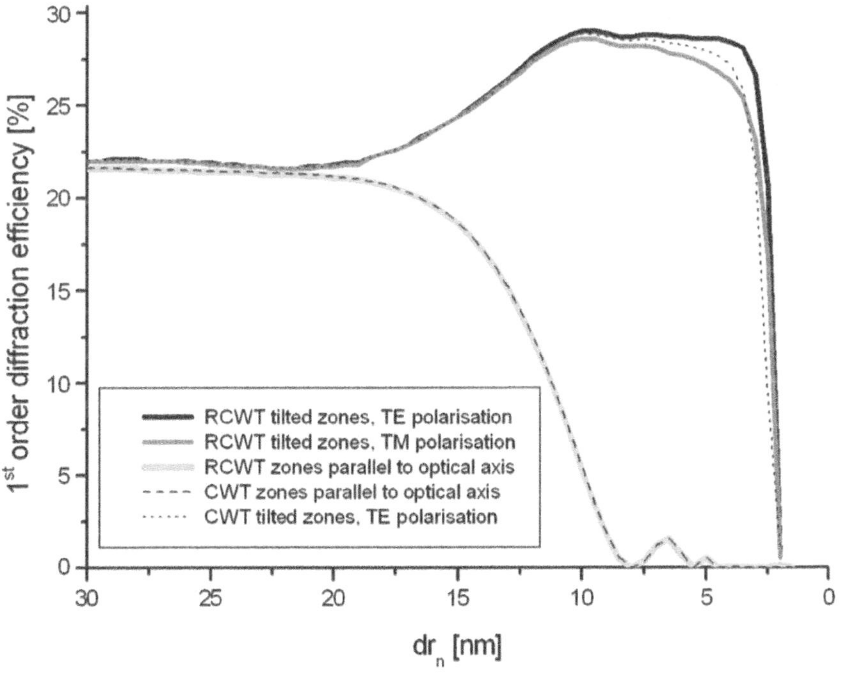

Fig. 8.18. Fisrt-order diffraction efficiency of a nickel zone plate with 200 nm high zones as a function of the zone width calculated by RCWT and CWT. X-ray wavelength is 2.4 nm

performed for the first-order diffraction efficiency of a nickel zone plate with a zone height of 200 nm, an X-ray wavelength of 2.4 nm, and a line-to-space ratio of $L{:}S = 50{:}50$. Additionally, to RCWT calculations results of CWT calculations are plotted for comparison.

Scalar theory predicts a constant efficiency of about 22% independent on the zone width. The RCWT predicts for zones parallel to the optical axis that the diffraction efficiency decreases continuously if the lateral dimensions of the zone width approach the wavelength used for imaging. For an outermost zone width of $\mathrm{d}r_n = 8\,\mathrm{nm}$, the diffraction efficiency is close to 0%. Note that a maximum diffraction efficiency of about 4.6% is obtained for parallel zones with $\mathrm{d}r_n = 8\,\mathrm{nm}$ if the nickel zone height is reduced to 96 nm, i.e., volume effects are reduced.

The RCWT calculations plotted in Fig. 8.18 also show that the difference in diffraction efficiency is small between TE and TM polarization. By comparing the RCWT and the CWT calculations (Fig. 8.18), it can be concluded that the CWT gives a sufficiently accurate prediction of the diffraction efficiency for a zone width larger than 5 nm.

8.9 Proposed Fabrication Process for Volume Zone Plates

The resolving power of zone plates scales with the order of diffraction m. By applying high-order of diffraction, it is possible to increase the resolution without the need to manufacture increasingly smaller outermost zone widths far below 20 nm. High-orders for imaging requires manufacturing tilted zone structures with aspect ratios of about 20:1. This is far beyond the current nanostructuring processes used for zone plate fabrication. To overcome the extremely difficult problem of manufacturing tilted zones with such high aspect-ratios ($\geq$20:1), we propose manufacturing zone plates on top of each other with slightly decreasing zone radii (see Fig. 8.19).

To a good approximation – depending only on the number of layers – the zones can be tilted according to the local Bragg condition and each single layer requires only structures with moderate aspect-ratios. An ideally tilted zone plate pattern will be approximated by a stack of zone plates. The quality of the approximation and, therefore, the expected diffraction efficiency will depend on the number of fabricated layers and on the overlay accuracy for the layers. Therefore, we study the influence of both effects on the diffraction efficiency in the following.

Theoretical results on the dependence of the number of layers are shown in Fig. 8.20. The calculations are performed for the whole stack similar to the theory applied to the roughness calculations performed in Sect. 8.5. The fifth-order diffraction efficiency is plotted as a function of the number of layers for different line-to-space ratios. To obtain 10 nm Rayleigh resolution, the period

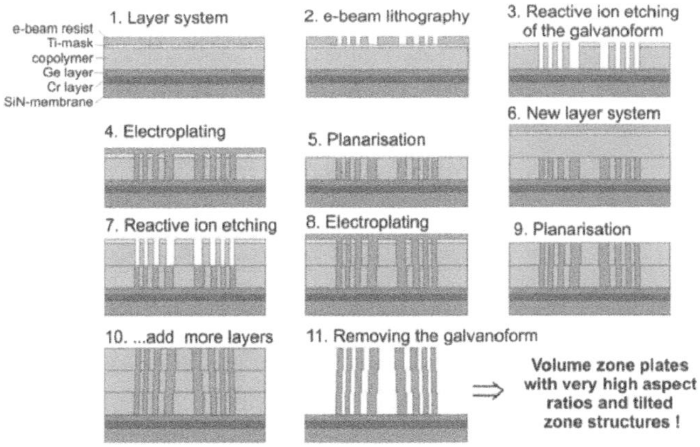

Fig. 8.19. Proposed fabrication process for zone plates with tilted zones and high aspect-ratio

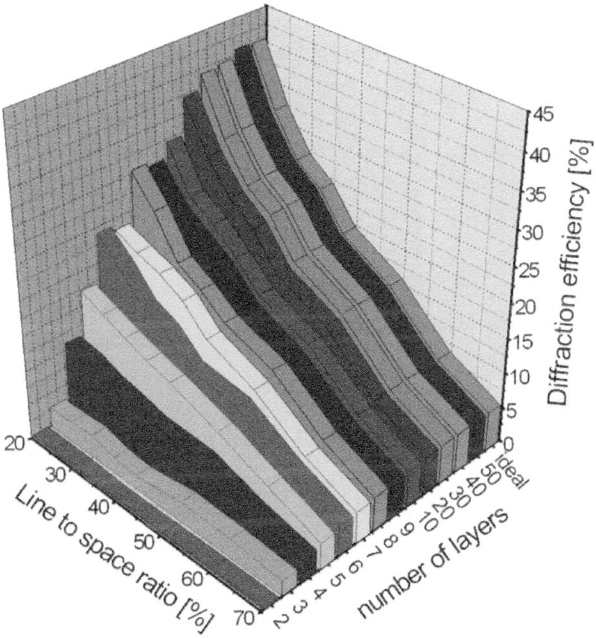

Fig. 8.20. Fifth-order diffraction efficiency plotted as a function of the number of layers and the line-to-space ratio of the minimum structures

that has to be manufactured is 84 nm. For the plotted example with a line-to-space ratio of 30%, 25-nm-wide nickel zones have to be fabricated, which are separated by 59 nm wide spaces. The total height of the nickel zones is 900 nm. This means for five layers, the thickness of a single layer is 180 nm.

Referring to a line-to-space ratio of 30%, e.g., the 59-nm-wide spaces and the proposed fabrication process, the aspect-ratio of the plating mould structures is only 3:1.

The line width that has to be exposed by electron-beam lithography is 25 nm, which can be done routinely. The plot in Fig. 8.20 also shows that the efficiency converges with an increasing number of layers to the optimal diffraction efficiency for ideally tilted zones, which has almost the same value as calculated for 50 layers. About 16% diffraction efficiency is obtained with five layers, which is about half the value of ideally tilted zone structures.

However, the fabrication process proposed in Fig. 8.19 requires a very high overlay accuracy for electron-beam writing. The plot in Fig. 8.21 shows the efficiency of the fifth-order of diffraction for five layers as a function of the maximum positioning error of the layers during electron-beam lithography for a line-to-space ratio of 25%. For the calculation, a linear grating is assumed instead of a circular zone plate pattern. The first layer of the stack is expected to be at its ideal position. Each of the following four layers is shifted randomly, relative to its ideal position by a positioning error of 0 nm, ±1 nm, ±2 nm, ±3 nm, or ±4 nm. For all possible combinations of the 5 layers the corresponding diffraction efficiency is calculated. These combinations are sorted by the respective maximum absolute value of the positioning error, which occurs in one combination and are plotted in Fig. 8.21. For a certain positioning error the possible minimum and maximum efficiency values as well as the average efficiency value obtained from all possible combinations are plotted.

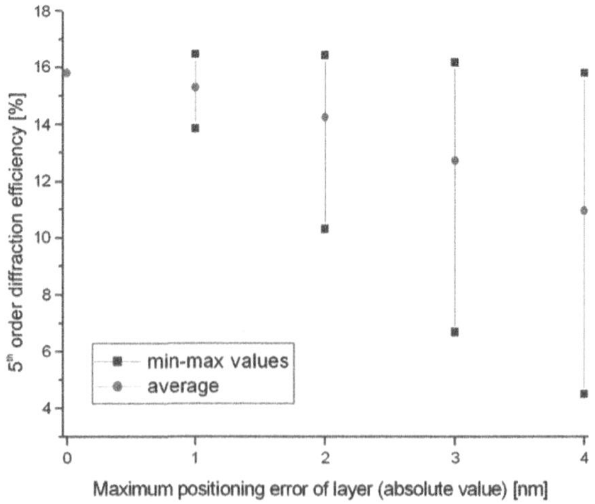

Fig. 8.21. Fifth-order diffraction efficiency (five layer, 30% line-to-space ratio) plotted as a function of the maximum efficiency error of a layer

We conclude that an overlay accuracy of $\leq 2\,\text{nm}$ is required to obtain sufficient efficiency for a volume zone plate with 10 nm Rayleigh resolution. Recently, such an overlay accuracy in the range of the required value for the volume zone plate was reported [26].

References

1. G. Schmahl, D. Rudolph, P. Guttmann, O. Christ, in *X-Ray Microscopy*, ed. by G. Schmahl, D. Rudolph, Springer Series in Optical Sciences, vol. 43 (Springer, Berlin Heidelberg New York, 1984), pp. 63–74
2. J. Kirz, C. Jacobsen, M. Howells, Q. Rev. Biophys. **28**, 33 (1995)
3. J. Kirz, J. Opt. Soc. Am. **64**(3), 301 (1974)
4. J. Maser, in *X-ray Microscopy IV*, ed. by V.V. Aristov, A.I. Erko (Bogorodskii Pechatnik Publishers, Chernogolovka, Moscow Region, 1994), pp. 523–530
5. J. Soret, Annalen der Physik und Chemie **156**, 99 (1875)
6. D. Attwood, *Soft X-Rays and Extreme Ultraviolet Radiation – Principles and Applications*, (Cambridge University Press, New York, 2000)
7. B. Niemann, D. Rudolph, G. Schmahl, Opt. Commun. **12**, 160 (1974)
8. A.G. Michette,*Optical Systems for Soft X-Rays* (Plenum Press, New York, 1986)
9. J. Kirz, J. Opt. Soc. Am. **64**(3), 301 (1974)
10. J. Maser, G. Schmahl, Opt. Commun. **89**, 355 (1992)
11. H. Kogelnik, Bell Syst. Tech. J. **48**, 2909 (1969)
12. R. Magnusson, T.K. Gaylord, J. Opt. Soc. Am. **67**, 1165 (1977)
13. L. Solymar, D.J. Cooke, *Volume Holography and Volume Gratings* (Academic Press, London, 1981)
14. J. Maser, Dissertation, University of Göttingen, 1994
15. B.L. Henke, P. Lee, T.J. Tanaka, R.L. Shimabukuro, B.K. Fujikawa, At. Data Nucl. Data Tables **27**, 1 (1982)
16. B.L. Henke, et al., *A Preliminary Report on X-Ray Photoabsorption Coefficients and Atomic Scattering Faktors for 92 Elements in the 10–10,000 eV Region* (Lawrence Berkeley Laboratory, Berkeley, California, 1988)
17. B.L. Henke, E.M. Gullikson, J.C. Davis, At. Data Nucl. Data Tables **54**, 181 (1993)
18. G.H. Golub, C.F. van Loan, *Matrix Computations*, 2nd edn. (The John Hopkins University Press, Baltimore, MD, 1989)
19. D. Rudolph, G. Schmahl, B. Niemann, SPIE Proc. **316**, 103 (1981)
20. D. Weiss, M. Peuker, G. Schneider, Appl. Phys. Lett. **72**(15), 1805 (1998)
21. D. Hambach, G. Schneider, J. Vac. Sci. Technol. B **17**(6), 3212 (1999)
22. B. Kaulich, Dissertation, University of Göttingen, 1996
23. G. Schneider, Appl. Phys. Lett. **73**(5), 599 (1998)
24. T. Matthies, C. David, J. Thieme, J. Vac. Sci. Technol. B **11**(5), 1873 (1993)
25. M.G. Moharam, T.K. Gaylord, J. Opt. Soc. Am. **17**, 811 (1981)
26. W. Chao, B. Harteneck, J.A. Liddle, E.H. Anderson, D. Attwood, Nature **435**, 1210 (2005)

Part II

Nano-Optics Metrology

9

Slope Error and Surface Roughness

F. Siewert

Abstract. This chapter describes the diffraction theory relationships between figure errors of grazing incidence mirrors and their imaging performance at synchrotron radiation beam lines. A practical illustrative example is the topographic errors of a synchrotron mirror as a function of spatial frequency. The basic idea of figure error inspection by direct slope measurement is described.

Mirrors of high shape accuracy used in grazing incidence are essential to focus or collimate the light in synchrotron radiation (SR) beamlines. Thus, many of the synchrotron laboratories have been established the expertise and equipment to measure the critical characteristics of high performance optical elements. Better knowledge describing the shape of an optical element allows better modeling, optimization, and in the final analysis, performance of optical systems. The quality of reflective optical elements can be described by their deviation from ideal shape at different spatial frequencies. Usually one distinguishes between the figure error, the low spatial error part ranging from aperture length to 1 mm frequencies, and the mid- and high spatial error part from 1 mm to 1 μm and from 1 μm to some 10 nm spatial frequencies, respectively [1, 2]. While the figure error will affect the imaging properties of the system the higher spatial frequency errors will cause light to be deflected or scattered away from the spectral image. The quantitative description of the surface errors of a reflecting optical element can be statistically interpreted in terms of power spectral density function, PSD, and summarized as the slope error and the surface roughness. Based on Kirchhoff's theory of diffraction, Church and Takacs [2, 3] have developed a model to describe this (see also [4, 5]). The natural "system coherence length," W, is given by

$$W = \sqrt{2}\frac{\lambda}{\Theta \cos \theta_\mathrm{i}} \tag{9.1}$$

with Θ as the angular radius of the system image (customer given), θ_i is the angle of incidence relative to the mirror plane, and λ is the operating wavelength. W is the surface spatial wavelength that diffracts to the $1/e$ intensity point in a Gaussian system image [2, 3]. In the case of diffraction limited optical elements W is approximately the length of the illuminated area of the mirror of the length L. In the case of system limited optical components assumed here it is much less than L [2]. In the case of synchrotron optics W is a wavelength between the spatial wavelengths ranging from λ to L. We assume that the influence of the mirror errors on the image quality can be described by the on-axis Strehl factor for the grazing incidence case, (1.3) [2]. The on-axis Strehl factor

$$\frac{I(0)}{I_0(0)} \tag{9.2}$$

is given as the ratio of the on-axis intensity in the presence of real surface errors to its value for zero errors. For the grazing incidence case can be written [2]:

$$\frac{I(0)}{I_0(0)} \approx 1 - \frac{8}{\Theta^2}\delta^2 - \left(\frac{4\pi}{\lambda}\cos\theta_{\mathrm{i}}\right)^2 \sigma^2, \tag{9.3}$$

where δ and σ are the bandwidth limited values of the rms values of the slope error (δ) and the roughness (σ) given by [2]

$$\delta^2 = (2\pi)^2 \int_{1/L}^{1/W} \mathrm{d}f_x S_1(f_x) f_x^2 \tag{9.4}$$

and

$$\sigma^2 = \int_{1/W}^{1/\lambda} \mathrm{d}f_x S_1(f_x). \tag{9.5}$$

$S_1(f_x)$ is the profile spectrum of the surface under test (SUT).

The surface roughness can be measured using an interferometric microscope or atomic force microscope (AFM). To inspect the slope error, a different measuring technique is used: interferometry or various types of surface profilers. The use of an interferometer, to inspect optical components, requires a reference of complementary shape and excellent quality. Due to the fact that many of the necessary optical components are of aspherical shape and are designed individually for each SR-beamline, such reference objects would cause an extraordinary expense for the metrology. In addition, the quality of the references used limits the accuracy of the result of the metrology. Because of the flat (grazing) angles used in SR-beamlines, synchrotron mirrors may have a length of up to 1 m and more and have a large ratio of length to width. Thus, basic conditions have lead to the idea of inspecting the shape of optical elements in the long dimension by slope measurement. A few slope measuring instruments have been developed during the last two decades, foremost among them, the long trace profiler (LTP) [6, 7], and recently the nanometer optical component measuring machine (NOM) [8, 9] and the extended shear angle difference method (ESAD) [10, 11]. These methods are based on the principle of direct measurement of slope deviation and curvature and, in contrast to

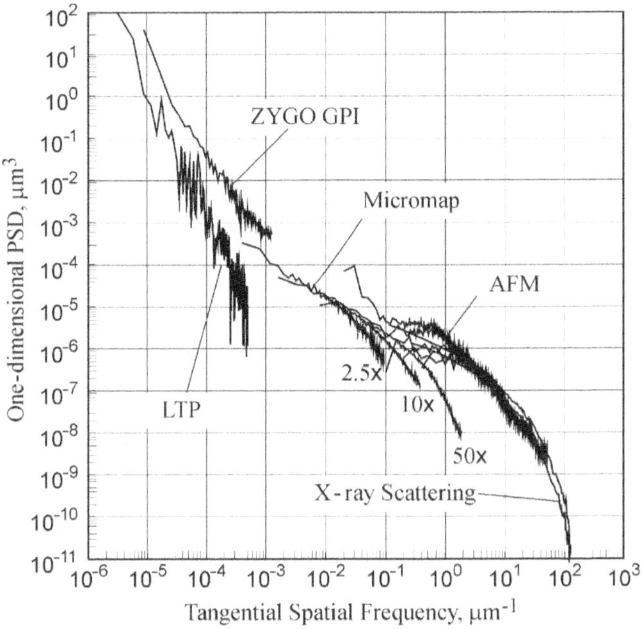

Fig. 9.1. Different measuring instruments cover different ranges of spatial frequency. Tangential 1D PSD spectra obtained with the 6-in ZYGO-GPI^{TM} interferometer, the LTP-II, the Micromap-570TM, the AFM and the CXRO reflectometry and scattering experimental facility at the advanced light source (ALS). The figure is a compilation of data published in [5, 12] (courtesy of Valeriy V. Yashchuk, LBNL/ALS)

other methods, yield results without the need for a reference. The measurement result is directly traced back to SI base units angle and length (SI – is the International System of Units). The measurement is a noncontact scan by using a laser source to create a measurement beam. Depending on the angular acceptance of the instrument, it is possible to measure the geometry of any reflective surface.

Figure 9.1 shows the range of spatial frequencies to be inspected by using different metrological instruments [5]. In the case shown, a stainless steel mirror was investigated using different measuring techniques over a spatial frequency range from $\sim 10^{-6}$–10^{-2} μm [5, 12].

9.1 The Principle of Slope Measurements

To measure the shape of an optical element a test beam from a laser source is reflected from the SUT. The relative position after reflection from the SUT is determined by the local shape and is detected on a sensor. The reflection of a test beam from a mirror along the axis of the instrument will depend on

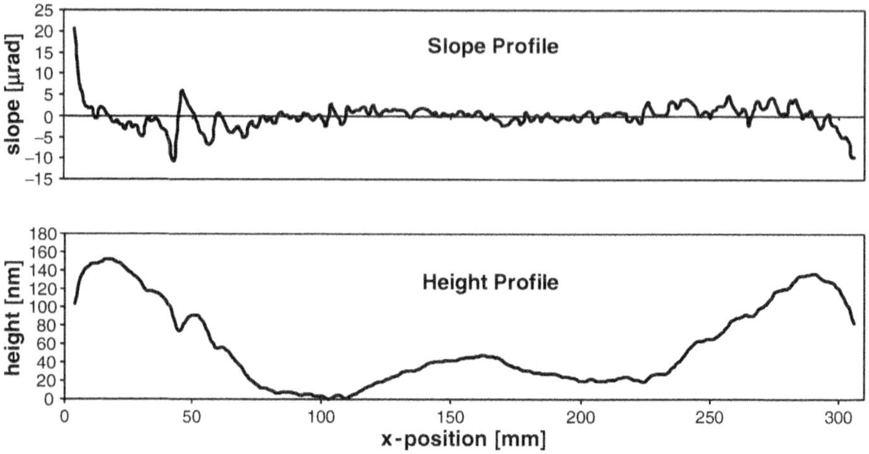

Fig. 9.2. Result of a direct slope measurement (NOM-scan): slope profile of a plane mirror (*top*) and the corresponding profile of height (*below*), obtained by integration of the slope data

the angle θ, of the mirror's normal with respect to the propagation direction of the laser beam [13, 14]. The slope δ is

$$\delta(x) = \tan\theta = \mathrm{d}y/\mathrm{d}x. \tag{9.6}$$

Of interest is the relative slope along the line of inspection. What is detected in the sensor is the change in angle of reflection from one position, x, on the SUT to the next position, $x + \Delta x$. From these data the height profile can be obtained by an integration of the slope function $\delta(x)$ over the abscissa x, see also Fig. 9.2. The height function is given by

$$h(x) = \int_{x=\text{scanstart}}^{x=\text{scanstop}} \delta(x)\mathrm{d}x \tag{9.7}$$

The commonly used criteria for the characterization of the shape quality is the figure slope error or residual slope error, obtained by subtracting the profile for the theoretically perfect geometry from the raw slope data.

Acknowledgments

The author gratefully acknowledges Daniele Cocco (Elettra) and Valeriy V. Yashchuk (ALS) for helpful discussions.

References

1. J.S. Taylor, G.E. Sommargren, D.W. Sweeney, R.M. Hudyama, UCRL-JC-128290 Rev. **1** (1998)
2. E.L. Church, P.Z. Takacs, Opt. Eng. **34**(2), 353 (1995)

3. E.L. Church, P.Z. Takacs, Appl. Optics **32**(19), 3344 (1993)
4. V.V. Yashchuk, M.R. Howells, W.R. McKinney, P.Z. Takacs, in *Proceedings of The 3rd International Workshop on Metrology for X-Ray Optics*, Daegu, 2006
5. V.V. Yashchuk, S.C. Irick, E.M. Gullikson, M.R. Howells, A.A. MacDowell, W.R. McKinney, F. Salmassi, T. Warwick, in *Advances in Metrology for X-Ray and EUV Optics*, ed. by L. Assoufid, P.Z. Takacs, J.S. Taylor, Proceedings of SPIE, vol. 5921, 2005, p. 18
6. P. Takacs, S. Qian, J. Colbert, SPIE **749**, 59 (1987)
7. P.Z. Takacs, S.-N. Qian, United States Patent 4,884,697, 1989
8. F. Siewert, T. Noll, T. Schlegel, T. Zeschke, H. Lammert, in *AIP Conference Proceedings*, vol. 705, Mellvile, New York, 2004, p. 847
9. H. Lammert, T. Noll, T. Schlegel, F. Siewert, T. Zeschke, in *Patentschrift DE10303659 B4 2005.07.28*
10. I. Weingärtner, M. Schulz, C. Elster, in *Proceedings of SPIE*, vol. 3782, 1999, p. 306
11. C. Elster, I. Weingärtner, Appl. Opt. **38**, 5024 (1999)
12. V.V. Yashchuk, S.C. Irick, E.M. Gullikson, M.R. Howells, A.A. MacDowell, W.R. McKinney, F. Salmassi, T. Warwick, *Advances in Metrology for X-Ray and EUV Optics*, ed. by L. Assoufid, P.Z. Takacs, J.S. Taylor, Proceedings of SPIE, vol. 5921, 2005, p. 18
13. S.C. Irick, Rev. Sci. Instrum. **63**(1), (1992)
14. R. Signorato, M. Sanchez del Rio, in *Proceedings of SPIE*, vol. 3152, 1997, p. 136

10

The Long Trace Profilers

A. Rommeveaux, M. Thomasset, and D. Cocco

Abstract. The Long Trace Profiler (LTP) is the most commonly employed instrument for measuring grazing incidence optics used in synchrotron radiation. This is a direct slope measurement device, able to detect root mean square slope variations of the order of 0.1 μrad. It was originally developed at the Brookhaven National Laboratory in Upton, NY, but several custom modified devices are used at laboratories around the world. In this chapter the main principles as well as various modifications are described, in order to give a general overview of what is possible with such instruments.

10.1 Introduction

In this section, we will describe the characteristics of the long trace profiler (LTP), the most commonly used instrument for measuring grazing incidence optics used in synchrotron radiation. Some features are common to all the LTPs available around the world. Many laboratories have modified their instrument according to their own particular needs or to try to improve the performance. However, the underlying principle is more or less the same and will be described here. Some variations for improving or customizing the instrument will be also highlighted.

This section together with the section on the NOM, are intended as a guideline for the choice of a particular trace profiler or its configuration, depending on the requirements and available budget.

10.2 The Long Trace Profiler

The most frequently used instrument for grazing incidence optics is the LTP. This instrument was developed at Brookhaven National Laboratories by Takacs et al. [1–4], and marketed by Continental Optical Corporation (later on by Ocean Optics [5]). It is basically a double pencil, slope-measuring interferometer, able to directly measure the slope of optical surfaces of any shape

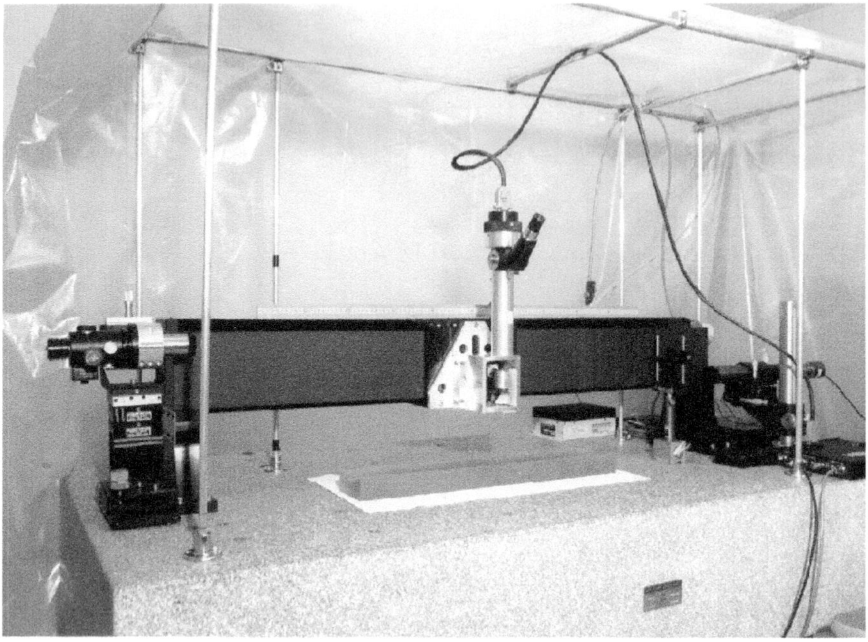

Fig. 10.1. Picture of the first LTP. The optics head moved on a ceramic bar and no reference mirror was used. Picture courtesy of Peter Takacs (Brookhaven)

up to 2 m in length (depending on the setup). Under optimum operating conditions, very precise measurements, with a repeatability better than 2 nm P–V (or 0.1 μrad rms) can be made.

Figure 10.1 shows the first ever LTP to be produced. It was installed and tested at Brookhaven National Laboratory. In this set up, an optics head moved along a ceramic beam, using air–vacuum bearing pads. A linear encoder was used (as today) to determine the position of the measurement.

The most important part of the instrument is, of course, the optics head. The schematic shown in Fig. 10.2 includes some improvements to the original design. Originally light coming from a laser diode is collimated and sent to a beam splitter and corner reflectors so that two parallel, collimated, coherent beams emerge downward toward the surface under test (SUT). The beams reflected by the SUT return into the optical system, and a beam splitter directs the returning beams into the Fourier transform lens (FTL), which produces an interference pattern at the linear detector array. The interference pattern contains a sinusoidal component whose phase depends on the angle of the beam pair with respect to the optic axis, or on the phase difference between the two beams, produced by the local angle of reflection of the SUT. Therefore, the position of the sinusoidal component will directly depend on the slope of the surface at the position where the laser hits the surface.

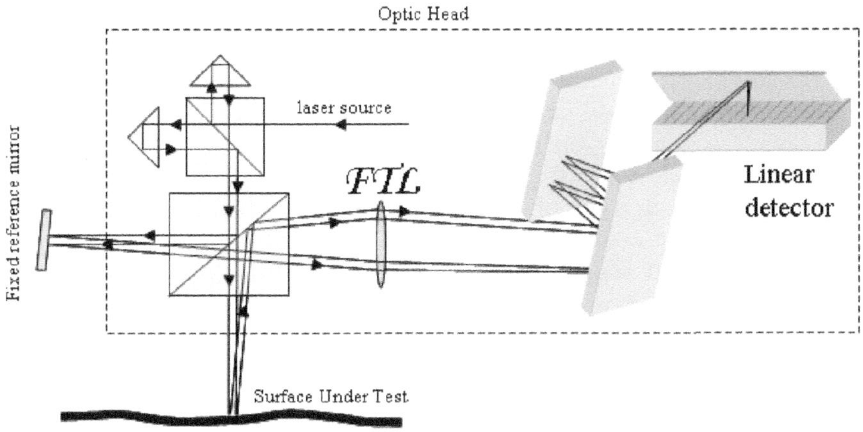

Fig. 10.2. Drawing of the optical setup for the LTP 2. The laser beam is split into two pairs of beams. One is sent to a fixed reference mirror (to compensate for the nonlinear movement of the head) and the other is sent to the SUT

In other words, the FTL converts a tilt induced in the laser beam pairs, by the mirror local slope, into a variation of the position of the interference pattern in the focal plane itself. Using a linear array detector to measure the position of the minimum of the interference pattern, one can directly measure the slope of the optic, point by point.

In principle, the measured pattern will depend only on the slope of the SUT but, in reality, it depends on a number of additional factors, like the beam pointing instability of the laser, the nonlinearity of the motion of the carriage on the ceramic bar, imperfect optics inside the optics head, random errors.

Over the years, most of these problems have been overcome. A major source of error was the imperfect movement of the optics head on the ceramic. As the beam propagation direction at the laser changes, the angle of the beam with respect to the optic axis also changes. However, if one can simultaneously measure the SUT and a fixed reference mirror, one can subtract the reference signal from that of the SUT and, apart from random noise, one can perfectly compensate for this effect. This first modification drastically improved the performance of the LTP, down to the microradian level (this is the main improvement of the LTP 2). To measure the reference mirror, the laser beam pair is divided in two by a further corner cube beam splitter. One of the two beam pairs is sent to a second mirror, rigidly mounted on one of the legs supporting the ceramic bar and the beam reflected by it is measured together with the beam reflected by the SUT.

Other sources of error include refractive index changes in the air where the beams propagate (due to air turbulence induced by sound and thermal waves), thermal instability which act on the laser as well as causing deformation of the mechanical mounting, mechanical relaxation. Several in-house modifications were necessary over the years to improve the performance of the different LTPs to meet the increasingly demanding precision requirements of mirrors in the synchrotron radiation field [6]. In the following, some of the most frequently adopted modifications will be described. In some LTPs, the solid-state laser source was substituted with a He–Ne laser tube, connected to the optics head by means of a polarization-preserving optic fiber. This is a very stable source in terms of wavelength, which is a critical requirement when the groove density variation of diffraction gratings has to be measured [7–9]. Moreover, Qian [10] suggested a different method of scanning by moving only a pentaprism. In this case the optics head is stationary and the laser is sent to the scanning pentaprism. The main advantages of such a configuration were the introduction of an angle-maintaining pentaprism (less sensitive to vibrations and to the tilting errors of the scanning translation stage), a significant weight reduction of the movable part of the interferometer (with an obvious decrease of the mechanical flexure of the scanning slide), and an easy switching between side-mounting and upward facing configurations for the SUT. Nevertheless the use of a pentaprism can reduce the precision of the profilometer since the beam has to cross it twice. In fact, an imperfect surface, in particular in terms of microroughness, introduces a phase shift between the two beams retro reflected by the SUT. This is seen as a false slope of the mirror and therefore introduces systematic errors which are not easily removed. This problem can be overcome by using a pair of superpolished mirrors mounted at 22.5° acting in the same way as the reflecting part of the pentaprism. Unfortunately these are not the only optical components along the path of the laser beam. Therefore, one of the main limitations of the LTP is the presence of systematic errors due to imperfect optics used in the profilometer.

Another important, and fundamental, hardware improvement is related to environmental condition control (mainly temperature stability and air turbulence along the laser beam path). The fundamental need is to shield completely the area where the scan is made from the external laboratory. This avoids air turbulence which introduces random noise in the measurement. In addition thermal stabilization is very important, and several solutions have been adopted with or without active temperature control. If the temperature changes during a scan, an artificial slope is introduced, resulting in an incorrect measurement of the radius of curvature of the optic. Precise thermal monitoring at different points inside this inner room or, even better, by looking at a reference mirror to be sure that no thermal effect are present is therefore mandatory if sub-microradian level of precision is required.

With such environmental control, it is possible to reach a very good level of repeatability. An example, obtained at Elettra, is plotted in Fig. 10.3, where the same mirror was measured on different days, but the measured residual slope errors differ by less than 0.04 μrad.

Another important source of error is the misalignment of the different optics present in the optics head, in particular relative to the position of the linear array detector which has to be "exactly" in the focal position of the

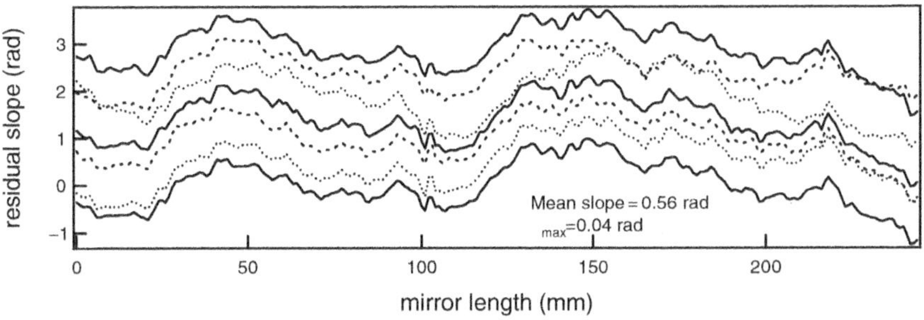

Fig. 10.3. Different measurements of a 250 mm long mirror. The measured residual slope errors differ by less than 0.04 µrad. The curves are shifted to be easily distinguished

FTL [11]. The detector must be in the focal plane of the lens with a precision better than 0.1 mm. If the beam is focused either before or after, it will impinge on the detector in a different position with respect to the focused one. This is again a systematic error that can be eliminated by proper alignment which can be performed in several ways. One of the easiest ways is to measure a mirror at different positions of the detector. Any discrepancy among the measurements means that there is a misalignment of the lens and, by trial and error, one can correct it.

A better way to calibrate an LTP, to estimate and if possible eliminate the systematic errors, is to use a calibrated reference mirror, i.e. a mirror with a well-known profile. Of course, this raises the problem of finding the real profile of the reference mirror. This can be overcome by making cross measurements at several laboratories and with different instruments.

A cross calibration measurement campaign started in 2004, among some of the major European laboratories [12], with the aim of defining a set of reference optics, of calibrating the different instruments, and of finding the limit of their performances (under the European funded action COST P7 [13]). The results and the procedure adopted in such a round robin will be described in detail in chapter 14.

10.3 Major Modifications of the Original Long Trace Profiler Design

There are a number of "homemade" LTPs in various laboratories. One of these, the NOM machine developed at Bessy, is described in detail in the chapter 11. Other designs use the pencil beam concept of Von Bieren [1] but with major modifications with respect to the original LTP solution. An example of this is the completely homemade design of the Soleil Profilometer, described below.

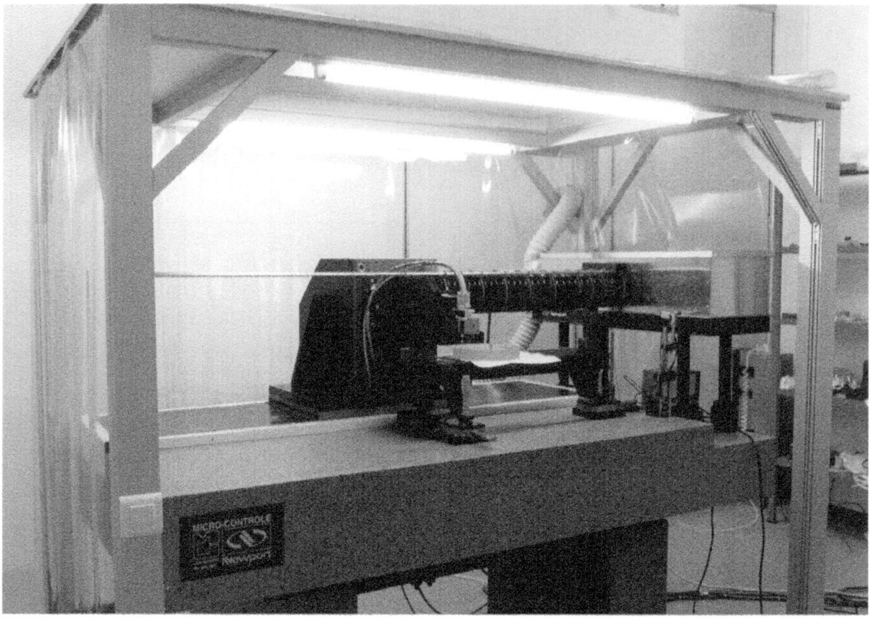

Fig. 10.4. Picture of the Soleil LTP with the environmental control enclosure

The SOLEIL LTP was constructed from a custom 1-m translation stage on air bearings powered by a linear motor (Fig. 10.4). It was designed to make "on the fly measurements." The guide surfaces of the air bearing are directly manufactured in the granite optical table. Large distances between the bearings ensure a high stability of the three angle components. Pitch, roll, and yaw errors have been measured along the whole carriage travel and are lower than 5 μrad rms. The position of the carriage is read by a Heidenhain optical encoder with 1 μm precision. The absolute accuracy has been controlled with an Hewlett-Packard interferometer and is better than 5 μm over the whole travel range. The carriage is able to support an overall weight of 45 kg.

The optics scheme is a variant of the pentaprism LTP described by Qian et al. [10] and is shown in Fig. 10.5. A fixed optical system creates a collimated light beam which is sent to the movable optical head, parallel to the motion direction. The beam is then reflected by the combination of three reflectors (mirrors or total reflection prisms) toward the SUT.

The advantage of the roof configuration of M1 and P1 is to allow a variable distance between these two elements and therefore an easy way of changing the transverse position of the measured track on the SUT without moving the latter. The prisms P1 and P2 are glued together for better stability and ease of manipulation. The stabilized beam is then reflected by the SUT and the local slope affects the return direction which is measured by the lateral position of the image spot on the CCD camera. Between the prisms and the SUT a

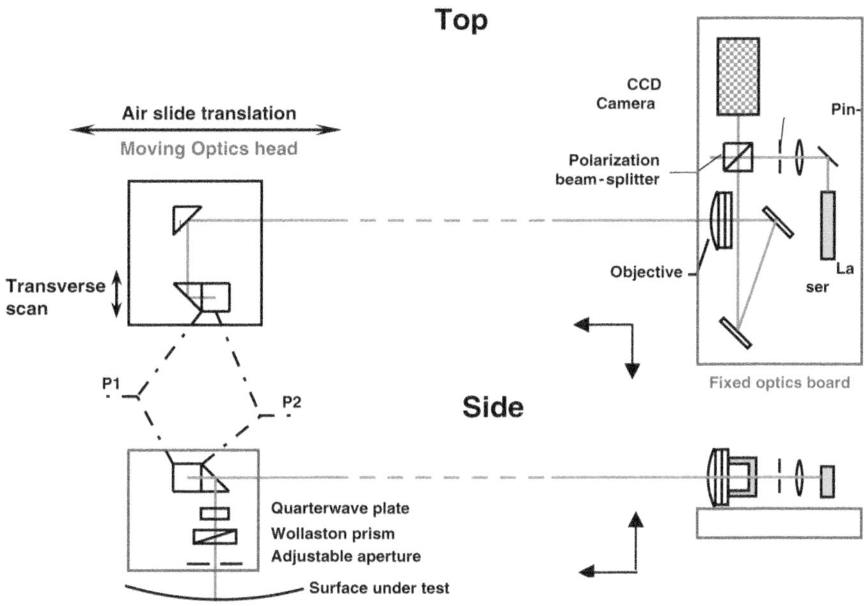

Fig. 10.5. Setup of the LTP of Soleil which was made in-house

compact polarization interferometer is inserted whose main components are a quarter wave plate and a Wollaston prism. The interferometer is completed by the polarization beam splitter which acts as a polarization filter.

The vertically polarized beam from the laser source is sent to the optics head while the horizontal component of the returning beam is sent to the camera.

The cut angle of the Wollaston prism is calculated to produce a sinusoidal fringe pattern with a period of about 2.5 mm when used in double pass (5 mm in single pass) between cross polarizers. The size of the beam is defined by an iris aperture located as close as reasonably possible to the SUT so that the probe beam size is precisely defined. The only precaution is to "clean" the beam by focusing the laser source ($\lambda = 532$ nm) on a small pinhole.

The camera is a C8040-96 Hamamatsu digital CCD camera, with $1{,}280 \times 1{,}024$ pixels at $6.7 \times 6.7\,\mu$m pixel pitch and integrated microlenses. A 2D detector allows the alignment of toroidal mirrors in the longitudinal direction with a precision better than 20 μrad. Objective lenses of different focal lengths can be used in the fixed optics part. A 500 mm focal length is normally used.

The Wollaston interferometer has been chosen for its ease of implementation and very high stability. In order to get a symmetrical image with a sharp central minimum on the CCD camera, the dark fringe of the Wollaston must be properly centered on the aperture. This is done by centering the Wollaston and by fine adjustment of the quarter wave plate orientation. However it can be shown that the spurious signal resulting from a slightly offset sinusoidal fringe is in quadrature with the signal resulting from the centered fringe. The depth of the image minimum is affected but its position does not change. The position of the minimum is interpolated from nine bracketing points.

In some cases, namely when measuring gratings [7–9], the SUT reflectivity will be different for polarization along or perpendicular to the track direction. In order to minimize the loss of fringe contrast in this case we use a specially cut Wollaston prism arrangement where the optical axes of the two prisms are set at 45° to the wedge direction and therefore parallel to the quarterwave plate axes, instead of being parallel and perpendicular to the wedge as it is usually constructed. Due to the symmetry, the reflected components for the two principal directions of polarization are equal and the fringe contrast is preserved. Finally the direction of the probe beam can be chosen by different arrangements of the mirrors and prisms in the moving head. By rotating P2 by 180° around the X-axis before gluing, we obtain an upward pointing stabilized beam. The actual configuration used to measure downward facing surfaces is obtained by inserting between M1 and P1 a periscope composed of two flat and parallel mirrors which brings the beam up without changing its direction. Side illumination is realized using the same principle with M1 and the following prisms in an upward pointing configuration, turned 90° around the incoming beam so that the lateral direction of the equivalent roof reflector is now along Y instead of Z.

A 500 m long instrument of the type described above is able to measure slopes in the range of about ±5 mrad corresponding to a radius of 10 m in a 100 mm long mirror [14, 15]. When this range is not enough, it is still possible to extend the measurement length by stitching a series of successive scans with different inclinations of the surface. A limited number of scans can be stitched without degrading the accuracy as they can be overlapped sufficiently.

Another important issue is to be able to measure very long mirrors, up to 2 m. With this target in mind, the European Synchrotron Radiation Facility (ESRF) constructed its own trace profiler.

The ESRF LTP is a homemade instrument. The first version was built in 1993 with the help of Takacs to measure long mirrors up to 1.5 m [3]. Many modifications have been made to the original design: the source and the detector are now separate from the moving optical head and fixed to the table (Fig. 10.6), the source is a helium–neon stabilized laser fitted to the optics head through a polarization-preserving optics fiber, a mirror assembly equivalent to a pentaprism is carried by the linear motor stage guided by the 2.5 m long ceramic beam.

The error in the linearity of the translation is optically corrected by the pentaprism. A fixed reference mirror corrects for any source instabilities. The detector is a 1,024 pixels photodiode linear array from Hamamatsu which gives a maximum measurable range of 12 mrad. Placed at the focal plane of the lens (800 mm focal lens), the sensor detects a fringe pattern intensity profile resulting from the interference of the two beams coming from the Michelson

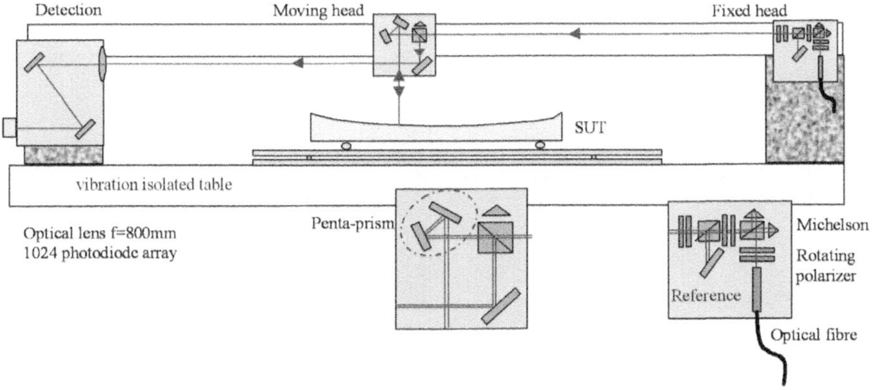

Fig. 10.6. Optical setup of the ESRF long trace profiler

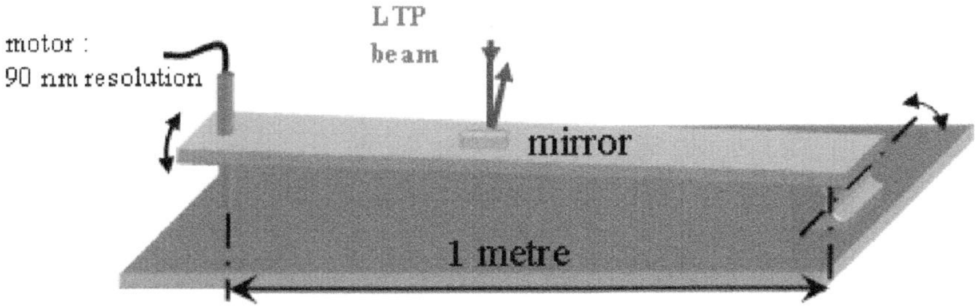

Fig. 10.7. ESRF LTP calibration setup

interferometer. The algorithm used to define its position on the detector is based on a fast Fourier transform calculation. The software has been developed using Labview® as programming language and can be easily adapted for specific needs. In the standard measurement configuration, the sample under test is reflecting upward but an optical bracket can be added to this setup if the SUT is reflecting downward.

Measurements are taken "on the fly"; the data are collected while the optical head is smoothly moving above the mirror at a constant speed of $40\,\mathrm{mm\,s^{-1}}$. The LTP is surrounded by a Plexiglas enclosure which reduces greatly the air turbulence. Measurements can be carried out faster, thus repeatability has been improved and is better than $0.05\,\mu$rad rms, while the slope accuracy on flat mirrors is better than $0.2\,\mu$rad. To ensure a reliable measurement, an important issue is the determination of the calibration factor. At the ESRF a method based on the well-known wedge angle technique is used; Fig. 10.7 shows the setup used for calibration. A motor displacement of $1\,\mu$m induces a $1\,\mu$rad angular deviation. The precision achieved is $0.1\,\mu$rad.

The mirror to be characterized may be integrated on a static or bending holder system. When no mechanical mounting system is provided, the mirror

Fig. 10.8. *Left*: mirror facing down under LTP measurement – *Right*: detail of the split retro reflector

is lying with its surface facing up on three balls or two cylinders separated by a well-known distance. Thus the deformation induced by gravity can be analytically calculated and subtracted from the measurement. Gravity can have a strong influence on the slope error profile.

Nevertheless it is always preferable to measure a mirror as close as possible to its future working conditions on the beamline in terms of mounting and the X-ray beam reflecting direction. For mirrors reflecting downward an additional bracket with a split retro reflector is added to on the LTP moving head (Fig. 10.8) in order to redirect the beam toward the surface through a roof prism and a right angle prism. This combination keeps the number of reflections needed to preserve the pentaprism correction. For further details on the characteristics of this instrument, please see [16].

References

1. K. Von Bieren, Proc. SPIE, **343**, 101 (1982)
2. P.Z. Takacs, S.N. Qian, J. Colbert, Proc. SPIE, **749**, 59 (1987)
3. P.Z. Takacs, S.N. Qian, U.S. Patent 4,884,697, 5 Dec 1989
4. S.C. Irick, W. Mckinney, D.J. Lunt, P.Z. Takacs, Rev. Sci. Instrum. **63**, 1436 (1992)
5. http://www.oceanoptics.com/
6. G. Sostero, A. Bianco, M. Zangrando, D. Cocco, Proc. SPIE, **4501**, 24 (2001)
7. D. Cocco, G. Sostero, M. Zangrando, Technique for measuring the groove density of diffraction gratings using the long trace profiler, Rev. Sci. Instrum. **74–7**, 3544 (2003)
8. S.C. Irick, W.R. McKinney, AIP Conf. Proc. **417**, 118 (1997)
9. J. Lim, S. Rah, Rev. Sci. Ins. **75**(3), 780 (2004)
10. S. Qian, W. Jark, P. Takacs, Rev. Sci. Ins. **66**(3), 2562 (1995)
11. S.N. Qian, G. Sostero, P.Z. Takacs, Opt. Eng. **39–1**, 304 (2000)

12. A. Rommeveaux, D. Cocco, V. Schoenherr, F. Siewert, M. Thomasset, Proc. SPIE, **5921**, (2005)
13. http://costp7.free.fr/
14. M. Thomasset, S. Brochet, F. Polack, Proc. SPIE, **5921–2**, 2005
15. J. Floriot et al., in *European Optical Society Annual Meeting*, Paris, 2006
16. A. Rommeveaux, O. Hignette, C. Morawe, Proc. SPIE, **5921** (2005)

11

The Nanometer Optical Component Measuring Machine

F. Siewert, H. Lammert, and T. Zeschke

Abstract. The Nanometer Optical component measuring Machine (NOM) has been developed at BESSY for inspection of the surface figures of grazing incidence optical components up to 1.2 m in length as in synchrotron radiation beam lines. It is possible to acquire information about slope and height deviations and the radius of curvature of a sample in the form of line scans and in a three dimensional display format. For plane surfaces the estimated root mean square measuring uncertainty of the NOM is in the range of 0.01arcsec. The engineering conception, the design of the NOM and the first measurements are discussed in detail.

11.1 Engineering Conception and Design

The nanometer optical component measuring machine (NOM) (Fig. 11.1) was developed at BESSY for the purpose of measuring the surface figure of optical components up to 1.2 m in length used at grazing incidence in synchrotron radiation beamlines [1–3]. With it, it is possible to determine slope and height deviations from an ideal surface and the radius of curvature of a sample in the form of line scans and in a three-dimensional display format. With the NOM surfaces, up to 600 cm^2 have been measured with an estimated measuring uncertainty in the range of 0.05 μrad rms and with a high reproducibility. This is a five- to tenfold improvement over the previous state of the art of surface measuring techniques such as achieved using the Long Trace Profiler (LTP-II) [3,4]. The NOM is basically a hybrid of two angle measuring sensor units, a Long Trace Profiler (LTP-III) and a modified high resolution autocollimating telescope (ACT). The latter (ACT) has been developed with a very small aperture of about $d = 2$ mm [1] (Fig. 11.2). The measuring principle of both sensors is noncontact deflectometry. In both cases, no reference surface is needed. The LTP III head is a BESSY-specified development by Ocean Optics Ltd. in cooperation with Peter Takacs (BNL) who created the optical design. The autocollimator used is a special development by Möller Wedel Optical GmbH. The two sensors are mounted stationary and opposite to each other on a compact stone base (Fig. 11.2) [1,3]. The two test beams are adjusted in a

Fig. 11.1. The nano optic measuring machine NOM at BESSY. To insure stable environmental conditions the instrument is enclosed in a double walled housing

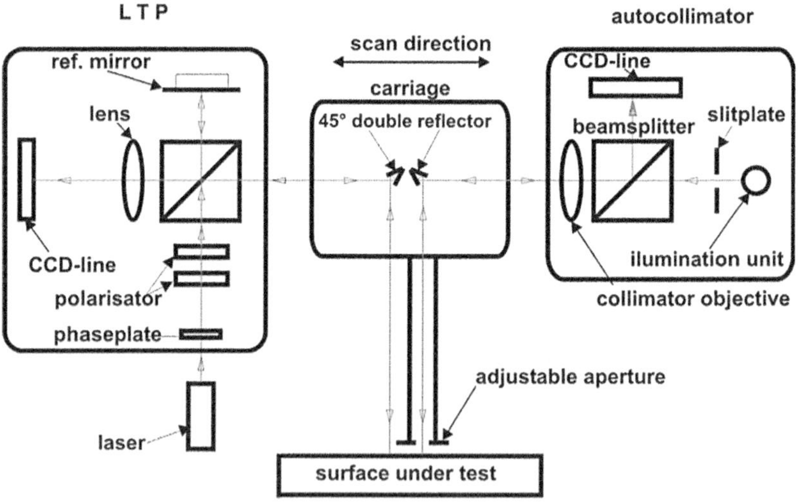

Fig. 11.2. Optical set up of the NOM

straight line to each other and are guided by a pentaprism or double reflectors to and from the specimen. The influence of the pitch tilt on the measurement is compensated for by the 45°-pentaprism design. The reflector unit is mounted on a movable air-bearing carriage system on the upper member of the stone frame. It consists of two parts: (a) one carriage for the motor, which is linked

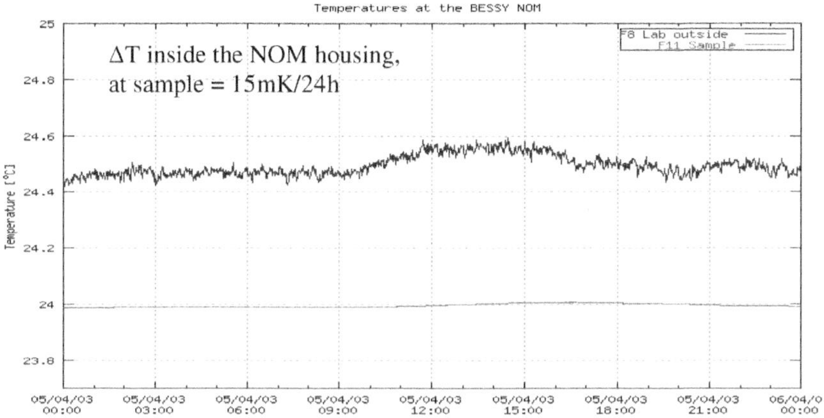

Fig. 11.3. Thermal stability at the BESSY metrology-Laboratory (*blue line*) and inside the NOM housing (*green line*)

by a torque-free coupling to the second, (b) the main carriage with the open pentaprism. A second air-bearing movable Y-table below positions the sample laterally. The drive units are linear motors. Both a step-by-step and an on-the-fly modus are available for data acquisition. To guarantee a maximum of thermal stability, the complete heat load of the NOM is limited to less than 2 W. Furthermore, the NOM is enclosed by a thermally stable, double-walled, and thermal-bridge-free housing in a temperature controlled measuring lab. The housing also limits the influence of air turbulence on the measurement. During measurement a temperature stability of 0.1 mK min^{-1} is maintained. The material of choice for the mechanical part of the NOM is stone (Gabbro) characterized by a sluggish response for thermal change. The use of metallic parts among the mechanical parts is avoided as far as possible. The weight of the compact stone parts of about 4,000 kg is a simple but very useful technique to damp the influence of vibrations on the measurement over a wide range of frequencies. A monitoring system recording the mean environmental data such as temperature, air pressure, humidity, and vibrations, as detected on the measurement table close to the specimen, is part of the established conception of metrology at BESSY. The measured temperature stability inside the test housing of the NOM is as low as 15 mK per 24 h (Fig. 11.3).

11.2 Technical Parameters

The measuring area of the NOM covers 1,200 mm in length and 300 mm laterally. The accuracy of guidance of the scanning carriage system is about ±1 μm for a range of motion of 1.3 m. A correspondingly high accuracy of guidance is also achieved with the y-positioning carriage over 0.3 m. The reproducibility of the scanning-carriage movement is in the range of 0.05 μrad rms. This

Table 11.1. Technical parameters of the NOM sensors

	LTP	Autocollimator
View angle	±6.6 mrad	±5 mrad
Measurable radius of curvature	1 m	10 m
Spatial resolution	about 1 mm	2 mm

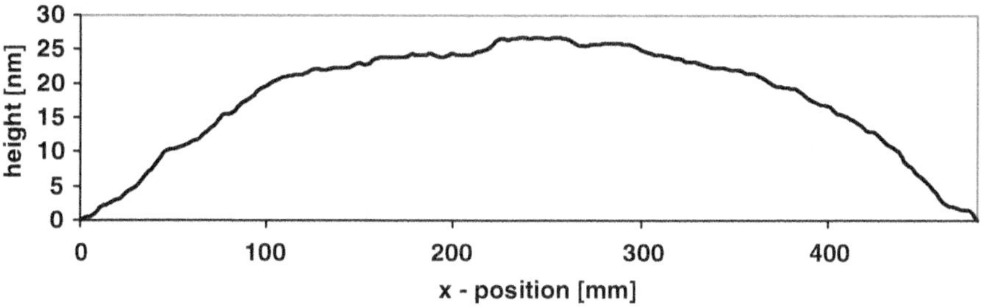

Fig. 11.4. Height profile of the center line of a 510 mm reference mirror (substrate material Zerodur*). Scan length = 480 mm. Peak to valley = 26.5 ± 0.6 nm. Spatial resolution for this measurement: 5 mm

reproducibility, combined with the insensitivity of the 45°-double-reflector for pitch, is an essential condition for the excellent measurement uncertainty achieved. Table 11.1 shows the parameters of the two optical heads. Both offer the possibility to scan plane, spherical, or aspherical surfaces. In the case of a surface curvature of 10 m or less the specimen is scanned by the LTP alone.

11.3 Measurement Accuracy of the NOM

To minimize the measurement uncertainty, possible systematic errors of the measuring device must be determined. Systematic errors can be determined by making a cross check using different methods for the measurement. This approach has been realized here [7, 8]. A plane reference surface of 510 mm in length (substrate material Zerodur*) has been measured using the NOM at BESSY by the PTB (Physikalisch Technische Bundesanstalt) with the extended shear angle difference (ESAD) method [9] and by stitching interferometry at Berliner Glas KG, the manufacturer of the reference. The ESAD method is the national reference for flatness in Germany. Additionally, two different measuring heads, based on different measuring principles, are an integral part of the NOM itself. The influence of random deviations such as mechanical vibration, instabilities caused by thermal effects, electronic noise, changes of the refraction index by thermal change, variation of air pressure, and humidity has been determined by comparing measurement data gained under essentially identical conditions. The reproducibility achieved is better than 0.01 µrad rms or 0.5 nm rms in height over a scan length of 480 mm at the center line of the sample (Fig. 11.4).

Table 11.2. Summary of uncertainty terms for a 480 mm line scan at the NOM on a plane reference surface (substrate material: Zerodur[1])

Error source	
ACT	0.015 μrad rms
Air turbulence	0.015 μrad rms
Beam guiding optics	0.005 μrad rms
Mechanical instability	0.005 μrad rms
Other random noise	0.010 μrad rms
Uncertainty overall u_c	0.025 μrad rms
expanded uncertainty: ($k = 2$)	0.05 μrad rms

[1] Zerodur is a trade mark of Schott Glass Mainz/Germany

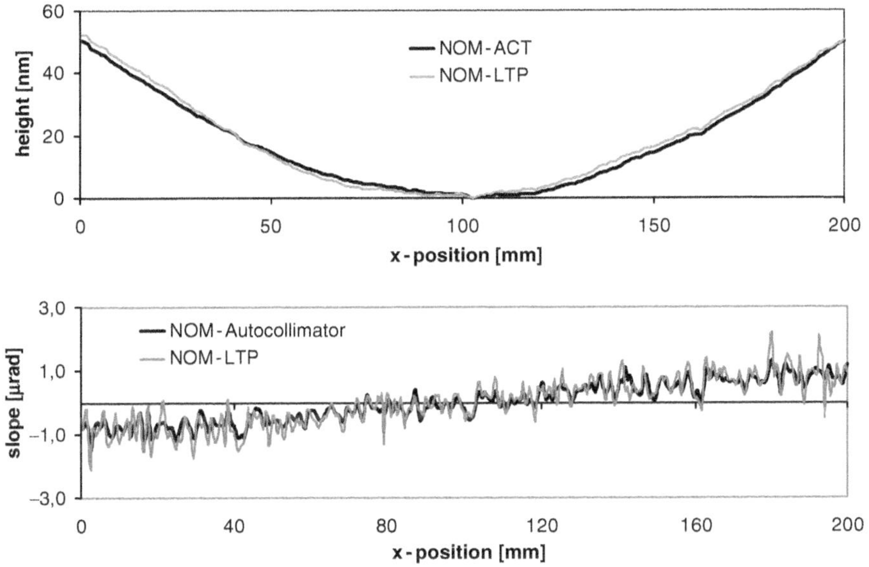

Fig. 11.5. Slope profile (*above*) and height profile (*below*) of NOM-ACT and NOM-LTP line scans, step size 0.5 mm on a 200 mm plane mirror. The LTP-slope profile is the result of 26 averaged line scans. The reproducibility is about 0.12 μrad rms. The ACT measurement consists of 14 averaged line scans with a reproducibility of 0.03 μrad rms. The estimated measurement uncertainty is 0.25 μrad rms for the LTP and 0.05 μrad rms for the ACT result

It is difficult to eliminate all sources of systematic errors. However, comparing fundamentally different methods, NOM, ESAD, and interferometry, is a very reliable test. The measurement uncertainty determined for the NOM measurement is in the range of 0.05 μrad rms. Table 11.2 shows the estimated uncertainty budget for the measurement result. Compared with the measurements of the other partners in the round-robin procedure, a

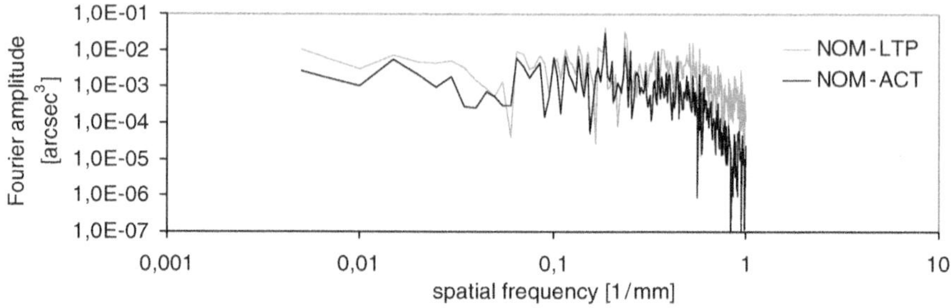

Fig. 11.6. Power surface density (PSD) curve of NOM-ACT and NOM-LTP line scans on a 200 mm plane mirror

conformity in the range of 0.7 nm rms compared to ESAD and of 1.3 nm rms to the result of the stitching interferometry has been achieved [10].

Figures 11.5 and 11.6 show the results of slope measurements on a 200-mm-long plane mirror (substrate material: single crystal silicon) by use of the two optical sensor units of the NOM. For both measurements a measuring point spacing of $dx = 0.5$ mm was chosen. The conformity of both unfitted results is in the range about 0.3 or 1.1 nm rms. The reproducibility of 0.03 μrad rms for the NOM-ACT measurement is about four times better than the reproducibility of 0.12 μrad rms achieved for the NOM-LTP.

11.4 Surface Mapping

Highly accurate topography measurements of an optical surface are required if optical elements are to be characterized in detail or to be reworked to a more perfect shape. Figure 11.7 demonstrates in principle a three-step "union jack" like method to scan the complete surface of a rectangular sample. To generate a 3D-data matrix two sets of surface scans, each consisting of a multitude of equidistant parallel sampled line scans, are traced orthogonally to each other in the meridional and in the sagittal direction successively. Each single surface line scan is taken on the fly. Between two single line scans the sample is moved laterally by the Y-position table. The scan velocity selected determines the measuring point spacing of the traced line. The lateral step size is defined by selecting the lateral shift between the lines scans in the start menu of the scanning software. In a final step the two diagonals have to be measured as two individual line scans. After taking the data of the two surface mapping scans, the root mean squares of the height data, obtained by integration of the slope measurements, are minimized and the points of the topography that lie on each of the measured diagonals are selected. Using the directly measured diagonal as a reference, the rms values of the difference between these two are obtained. In this way, a twisting of the surface, which is recognized and measured in the direct measurement, is superimposed onto the generated diagonal and correspondingly onto the entire array of x- and y-data, yielding the genuine shape of the sample. The agreement of the diagonals

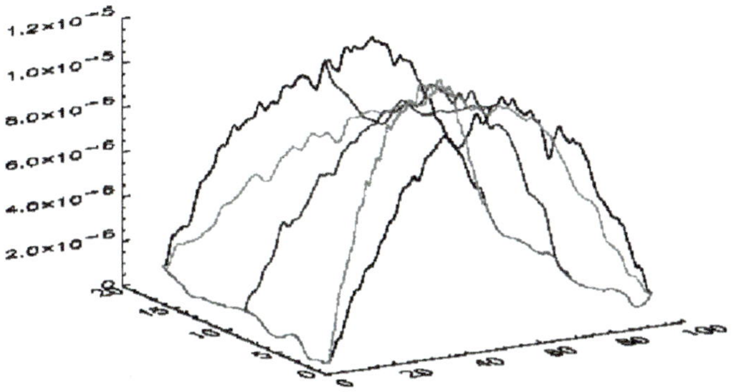

Fig. 11.7. Principle of 3D-mapping (dimensions in millimeter)

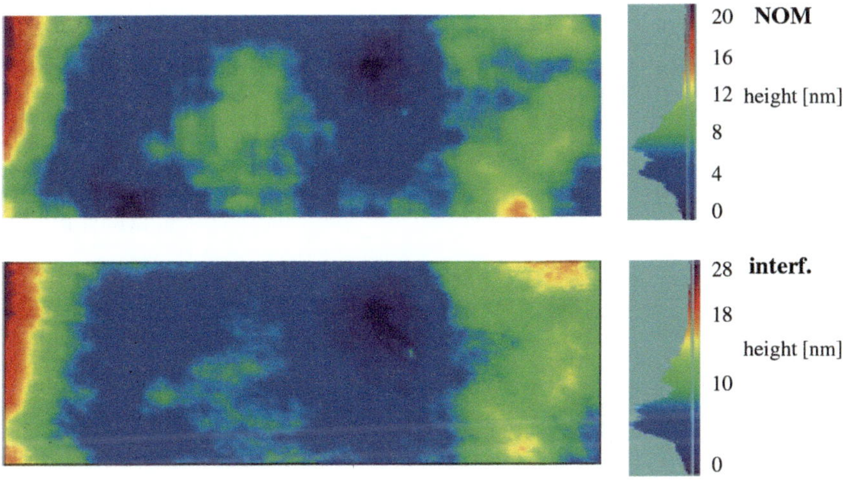

Fig. 11.8. NOM 3D-measurement on a $310 \times 118\,\text{mm}^2$ Zerodur reference compared to a measurement result gained by stitching interferometry. Result of the NOM-measurement: height, 20.8 nm pv per 3.1 nm rms, and interferometry: height, 27.8 nm pv per 4.4 nm rms

gained from the calculated surface map and the directly measured line scans is taken as a criterion of accuracy of the measurement. In the case of plane surfaces an agreement in the sub-nanometer range is achieved. Figure 11.8 shows the result of a comparison of a NOM-measurement with an interferometrical measurement.

Acknowledgments

The authors gratefully acknowledge Tino Noll, Thomas Schlegel (BESSY), Ingolf Weingärtner, Michael Schulz, Ralf Geckeler (Physikalisch Technische Bundesanstalt), and Ingo Rieck, Chris Hellwig (Berliner Glas KG) for scientific cooperation.

References

1. F. Siewert, T. Noll, T. Schlegel, T. Zeschke, H. Lammert, in *AIP Conference Proceedings*, vol. 705, Mellvile, New York, 2004, pp. 847–850
2. H. Lammert, T. Noll, T. Schlegel, F. Siewert, T. Zeschke, Patentschrift DE10303659 B4 2005.07.28
3. F. Siewert, H. Lammert, T. Noll, T. Schlegel, T. Zeschke, T. Hänsel, A. Nickel, A. Schindler, B. Grubert, C. Schlewitt, in *Advances in Metrology for X-Ray and EUV-Optics*, Proc. of SPIE, vol. 5921, 2005, p. 592101
4. P. Takacs, S. Qian, J. Colbert, Proc. SPIE **749**, 59 (1987)
5. P.Z. Takacs, S.-N. Qian, US Patent 4884697, 1989
6. H. Lammert, T. Noll, T. Schlegel, F. Siewert, T. Zeschke, Patentschrift DE10303659 B4 2005.07.28
7. F. Siewert, H. Lammert, in *HLEM on Production metrology for Precision Surfaces*, Braunschweig, 2004
8. R.D. Geckeler, I. Weingärtner, Proc. SPIE **4779**, 1 (2002)
9. I. Weingärtner, M. Schulz, C. Elster, Proc. SPIE **3782**, 306 (1999)
10. R. Geckeler, Proc. SPIE **6293**, 629300 (2006)

12

Shape Optimization of High Performance X-Ray Optics

F. Siewert, H. Lammert, T. Zeschke, T. Hänsel, A. Nickel, and A. Schindler

Abstract. A research project, involving both metrologists and manufacturers has made it possible to manufacture optical components beyond the former limit of 0.5 µrad in the root mean square (rms) slope error. To enable the surface finishing, by polishing and finally by ion beam figuring, of optical components characterized by a rms slope error in the range of 0.2 µrad, it is essential that the optical surface be mapped and the resulting data used as input for the ion beam figuring. In this chapter the results of metrology supported surface optimization by ion beam figuring will be discussed in detail. The improvement of beam line performance by the use of such high quality optical elements is demonstrated by the first results of beam line commissioning.

12.1 Introduction

To benefit from the improved brilliance of third generation synchrotron radiation sources and sources such as energy recovery linacs (ERL) or free electron lasers (FEL), optical elements of excellent precision characterized by slope errors clearly beyond the state of the art limit of 0.5 µrad rms for plane and spherical shapes are needed [1,2]. The challenging specifications for such beam-guiding elements can be fulfilled by deterministic technology of surface finishing, for example, by ion beam finishing (IBF) or computer controlled polishing (CCP) [3,4]. It is essential that the surface finishing be supported by metrology instruments of accuracy 3–5 times superior to that of the desired end product.

12.2 High Accuracy Metrology and Shape Optimization

Here a short description of the optimization of the surface of optical components based on ion beam technology is given. To demonstrate the capability of IBF supported by advanced metrology, three demonstration components have been shape-optimized after classical and chemical–mechanical polishing

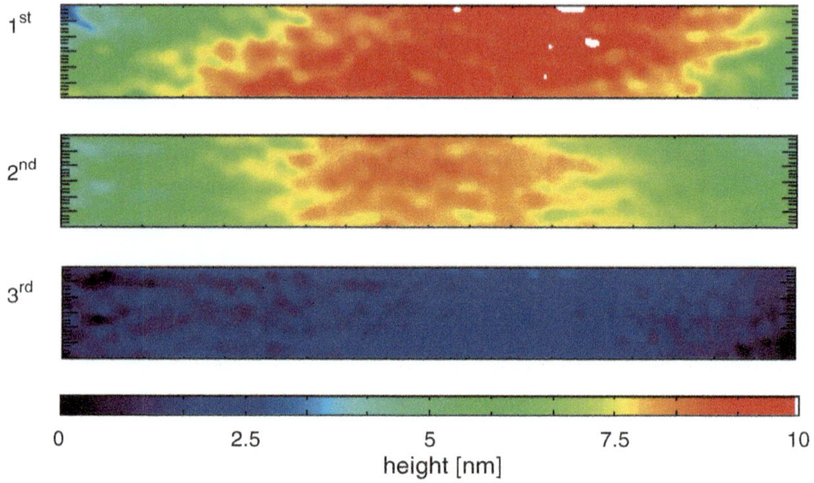

Fig. 12.1. Three iterations of ion beam finishing on a 100 × 20 mm grating blank (substrate material: Si). NOM measurement, spatial resolution: 2 mm

First iteration:	11.8 nm pv
Second iteration:	5.1 nm pv
Final state:	3.3 nm pv
Residual slope error:	0.1 μrad rms
measured at the center line	

by IBF technology. The demonstration components are one plane mirror of 310 mm in length, one grating blank of 100 mm in length, and a refocusing mirror of plane–elliptical shape, 190 mm in length [3]. To obtain an optimal result of the surface finishing, the initial state of the substrate had to have a microroughness essentially of that required at the end: 0.2–0.3 nm rms for the plane elements and <0.8 nm rms for the plane–ellipse. To finish the plane grating blank, the substrate was measured by interferometry and on the BESSY-NOM. To define the macroscopic shape of the surface, the NOM 3D-data were used. In addition, to have an optimized spatial resolution in the range of 80–100 μm, required for the IBF, the interferometric data have been fitted into this matrix. The progress in the shape optimization and the final state of the blank of 0.1 μrad rms for the residual slope error is illustrated in Fig. 12.1. In the case of this grating blank, the residual height deviation of 0.38 nm rms and the microroughness of 0.2 nm rms, which were finally achieved, are of the same order of magnitude. For the 310 mm plane mirror this procedure was in use for the first two iterations of ion beam treatment. The last three steps were done based on interferometer data. In a completing step the final state of about 0.2 μrad rms for the slope error was determined by NOM measurements (Fig. 12.2)

The refocusing mirror was finished based on the data of NOM measurements only (Fig. 12.3). For this purpose a measuring point spacing of

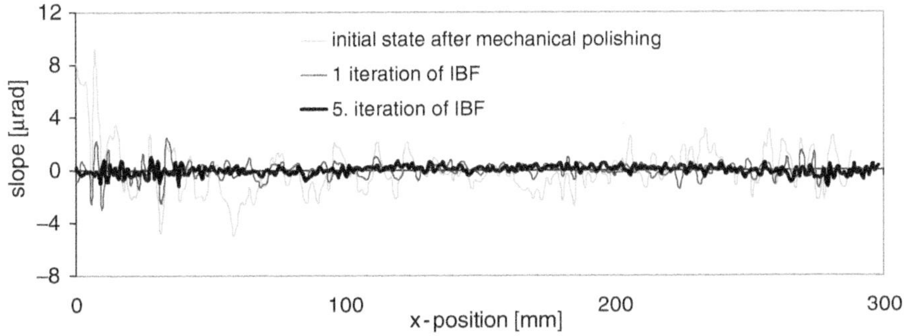

Fig. 12.2. NOM-measurements on a 310 mm plane mirror (spatial resolution: 2 mm, substrate material single crystal silicon, 5 iterations of IBF were used). The residual slope profile of the center line was the following: initial state, 1.69 µrad rms; after 1.IBF, 0.63 µrad rms; final state, 0.2 µrad rms

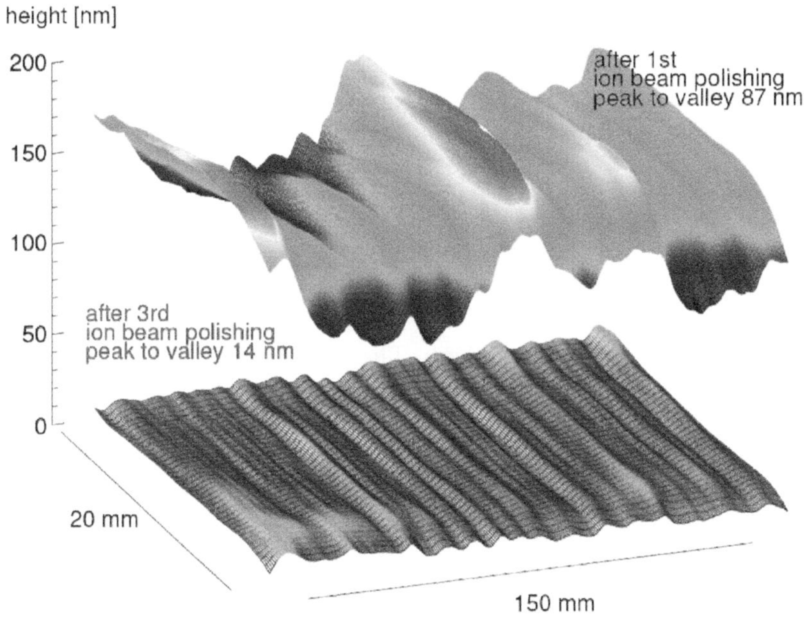

Fig. 12.3. Map of residual height of a plane–elliptical refocusing mirror after 1st iteration of ion beam polishing and final state. The residual slope error after three iterations of IBF is 0.67 µrad rms measured at center line

$0.2 \times 0.2\,\mathrm{mm}^2$ was chosen [6–9]. An interferometric measurement of this substrate would require a number of partial surface measurements to be stitched, a time consuming option of questionable reliability. The figuring process was realized by a computer controlled scanning of a small-sized ion beam with an ion beam of near-Gaussian profile across the surface. The linewidth and the dwell time have been varied in proportion to the amount of material

Table 12.1. Final results of surface finishing by IBF compared to the initial state after chemical–mechanical polishing

Optical element	Initial state residual slope (μrad rms)	Final state after IBF residual slope (μrad rms)
Plan grating blank (Si) $100 \times 20\,\text{mm}^2$	0.6	0.1
Plane mirror (Si) $310 \times 30\,\text{mm}^2$	1.7	0.2
Plane–elliptical mirror (Zerodur) $190 \times 37\,\text{mm}^2$	5.9	0.67 (0.5 is possible)

to be removed [8]. The simulation of the figuring is based on a modification of van Citter deconvolution in the local coordinate space using the Fourier transformation and contains an optimal turn and smoothing of the output topology, a graphic output of the topologies and profiles as well as the generation of the dwell times. A 40 mm Kaufmann-type ion source with a focusing grid system was used [6]. The ion source parameters for the figuring using Ar as the etch gas were ion beam voltage, 800 eV; ion beam current, 20 mA. The positive charged ion beam was neutralized by a hot filament neutralizer. Because of the high requirements for X-ray optics these optical elements have to be finished by tools working at different optically relevant spatial frequency ranges. The size of the rotational symmetric Gaussian beam has been adjusted with the help of circular diaphragms of different hole diameters. The beam profiles and the etch rates have been determined by etching a "footprint" for a certain time into a test blank. The "footprint" was than measured by interferometry. The mirror substrate was figured in three IBF steps with the following ion current density profiles:

- For IBF steps 1 and 2 a beam size of 6 mm FWHM (diaphragm hole diameter: 4 mm) was used
- For the final IBF step a beam size of 2.1 mm (diaphragm hole diameter: 2 mm) was used

In the case of the three demonstration objects the substrates were moved relative to the fixed ion beam position. In Table 12.1 a general view on the capability of surface finishing by ion beam technology is shown.

12.3 High Accuracy Optical Elements and Beamline Performance

The performance of a SR-beamline is ultimately determined by the quality of the optical elements in use to guide the light from the source to the experiment at the focus. The shape-optimized plane–elliptical demonstration

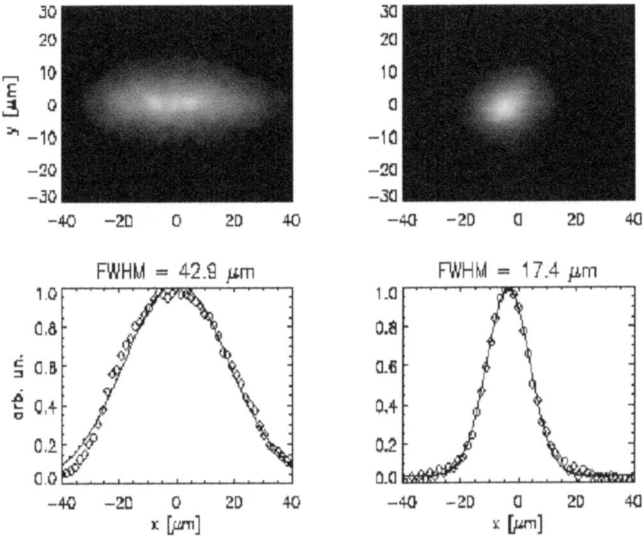

Fig. 12.4. Foci and horizontal energy distribution of two different refocusing mirrors characterised by a slope error of (*left*) 7.22 µrad rms and (*right*) 0.67 µrad rms

mirror described above serves as a refocusing mirror at the UE52-SGM1 beamline at the BESSY-II storage ring. By measurements of the focus size while commissioning the beamline the improvement achieved has been determined [8,9]. Figure 12.4 shows the optimized focus and the horizontal energy distribution FWHM measured for the previous refocusing mirror and for the IBF improved mirror. A focus size of less than $20 \times 20\,\mu\text{m}^2$ for the energy range inspected (350–1,100 eV) at an exit slit width of 3–4 µm has now been achieved. Compared to the previously obtained horizontal focus size of about 43 µm (FWHM) the present value of about 17 µm (±10%) represents a more than twofold improvement. Because of the characteristics of the undulator source at this beamline, the potentially smallest dimension of the focus size has been reached. A further surface optimization of this refocusing element beyond the limit of 0.1 arcsec rms would not provide an improvement of beamline performance.

References

1. F. Siewert, H. Lammert, G. Reichardt, U. Hahn, R. Treusch, R. Reininger, in *AIP Conference Proceedings*, Mellville, New York, 2006
2. L. Assooufid, O. Hignette, M. Howells, S. Irick, H. Lammert, P. Takacs, Nucl. Instrum. Methods Phys. Res. A **467–468**, 399 (2000)
3. A. Schindler, T. Haensel, A. Nickel, H.-J. Thomas, H. Lammert, F. Siewert, *Finishing procedure for high performance synchrotron optics*, in *Proceedings of SPIE*, **5180**, 64 (2003)

4. T. Hänsel, A. Nickel, A. Schindler, H.J. Thomas, in *Frontiers in Optics*, OSA Technical Digest (CD) (Optical Society of America, 2004), paper OMD5
5. H. Lammert, T. Noll, T. Schlegel, F. Senf, F. Siewert, T. Zeschke, *Breakthrough in the Metrology and Manufacture of Optical Components for Synchrotron Radiation*, BESSY Annual Report 2003, www.bessy.de, Berlin, 2004
6. F. Siewert, K. Godehusen, H. Lammert, T. Schlegel, F. Senf, T. Zeschke, T. Hänsel, A. Nickel, A. Schindler, *NOM measurement supported ion beam finishing of a plane-elliptical refocussing mirror for the UE52-SGM1 beamline at BESSY*, BESSY Annual Report 2003, www.bessy.de, Berlin, 2004
7. F. Siewert, H. Lammert, T. Noll, T. Schlegel, T. Zeschke, T. Hänsel, A. Nickel, A. Schindler, B. Grubert, C. Schlewitt, in *Advances in Metrology for X-Ray and EUV-Optics*, Proc. of SPIE, vol. 5921, 2005, p. 592101
8. K. Holldack, T. Zeschke, F. Senf, C. Jung, R. Follath, D. Ponwitz, *A Microfocus Imaging System*, BESSY Annual Report, www.bessy.de, Berlin 2000, pp. 336–338
9. H. Lammert, *NOK-NOM-Schlussbericht, Nanometer-Optikkomponenten für die Synchrotronstrahlung, Messen und Endbearbeitung bis in den Subnanometer-Bereich unter λ/1000*, Berlin, 2004. TIB Hannover: http://edok01.tib.uni-hannover.de/edoks/e01fb05/500757100.pdf

13

Measurement of Groove Density of Diffraction Gratings

D. Cocco and M. Thomasset

Abstract. The use of diffraction gratings with variable groove density is becoming increasingly common. This is because it has become possible to preserve the beam divergence, reduce aberrations and improve the focal characteristics of such gratings. The demands in terms of optical performance are becoming even greater and, to be sure that a grating as manufactured is close to that required, techniques to measure accurately the groove density variation have had to be developed. In this chapter, one such method, arguably the most accurate, is described, although it has some limitations which will also be discussed.

13.1 Introduction

In this chapter, we describe a way to precisely measure the groove density variation of a diffraction grating. Diffraction gratings are widely used to monochromatize and even to focus the soft X-ray radiation produced by the high brilliance third generation synchrotron radiation sources. They consist of a periodic structure on a substrate which can be completely constant along the grating surface or can change according to a particular polynomial law. In this second case, the groove density variation is used to change the focal property of a grating or to reduce the third-order aberration. The instrument employed for this work is the long trace profiler [1–4].

13.2 Groove Density Variation Measurement

A diffraction grating is an artificial periodic structure with a well-defined period, d. The incoming and outgoing radiation directions are related by a simple formula:

$$\frac{n\lambda}{d} = \sin(\alpha) - \sin(\beta), \tag{13.1}$$

where α is the angle of incidence and β the angle of diffraction, both with respect to the normal, n the diffraction order, and λ the wavelength of the selected radiation. An alternative description of the same law is given by the following:

$$nK\lambda = \sin(\alpha) - \sin(\beta), \tag{13.2}$$

where $K = 1/d$ is the groove density.

Diffraction gratings can be mechanically ruled or holographically recorded. It is also possible to replicate them from a master. In all these cases some errors occur during the manufacturing process. These defects can be periodic, quasiperiodic, or completely random. The final effect of these defects can be a reduction of the ability of the grating to select the proper photon energy, a reduction of the photon flux (due to light scattering), or the presence of unwanted diffracted energy in the focus together with the selected energy (ghosts).

Sometimes a variable line spacing (VLS) grating is requested. The groove density $K(w) = K_0 + K_1 w + K_2 W^2 + \ldots$ along the direction of the optical axis, w, perpendicular to the grooves and centered on the pole of the grating can be measured by the long trace profiler.

Since our LTP is able to detect small angle deviations of the reflected laser beam due to a slope variation of the mirror under test, it is equally able to detect angle deviations of a laser beam diffracted (instead of reflected) by a grating. Nevertheless, to properly work with an LTP, the direction of the beam impinging the optics under test and the reflected one must coincide. For this reason, the incoming and diffracted beams must be superimposed on each other.

This condition is the so-called Littrow condition, where, the incoming beam and the diffracted one coincide (Fig. 13.1).

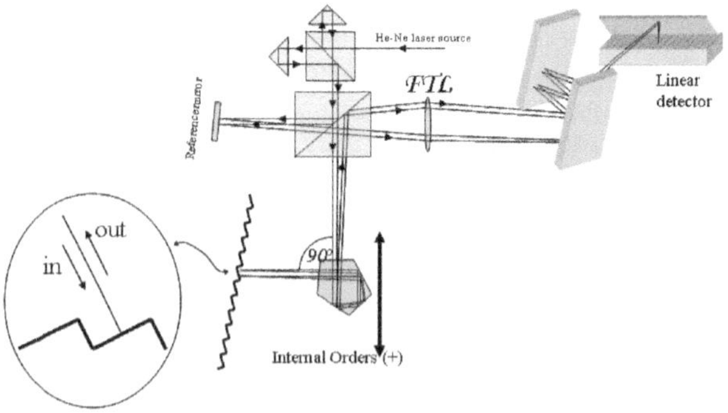

Fig. 13.1. Sketch of the measurement setup. The beam coming from the optics head of the LTP is directed via a pentaprism to the grating surface. The grating is rotated in such a way to superimpose the diffracted beam with the incoming one (in the oval inset an enlarged view of the diffraction configuration). The diffracted beam travels back to the LTP optics head where a Fourier transform lens focuses it on a linear array detector

The incoming and diffracted beam coincide when

$$\beta = -\alpha \rightarrow 2d \sin\alpha = n\lambda \rightarrow 2\sin\alpha = nK\lambda. \tag{13.3}$$

If this equation has a real solution, (with $\lambda = 632.6\,\text{nm}$, i.e., our He–Ne laser source) one is able to measure the groove density, d, of the grating. Practically, one must rotate the grating by a well-defined angle α_0 and after that make a scan with the LTP (Fig. 13.1), exactly as if it were a mirror. Therefore, by measuring α, one directly can measure the groove density of the grating.

The precision of an LTP, when used to measure a mirror, is of the order of $0.5\,\mu\text{rad}$ rms or even better on a 1 m long mirror. Even if this is an underestimation of the accuracy of the instrument, with this kind of error in the slope measurement, the equivalent groove density constancy error ($\delta K/K$) that is measurable is less than 10^{-5}. Alternatively, one can measure the d-spacing variation with a precision of the order of 1 Å rms or better.

To estimate the error induced, for instance, in the parameter K, one has to derive it with respect to the measured value, i.e., the back diffracted beam angle β:

$$\frac{\partial K}{\partial\beta} = \frac{\partial}{\partial\beta}\left(\frac{\sin\alpha_0}{n\lambda}\right) - \frac{\partial}{\partial\beta}\left(\frac{\sin\beta}{n\lambda}\right) = -\frac{\cos\beta}{n\lambda} \rightarrow \delta K = -\frac{\cos\beta}{n\lambda}\delta\beta. \tag{13.4}$$

Alternatively, the precision in the determination of the parameter d can be derived similarly from the previous equation:

$$\delta d = \frac{d^2}{n\lambda}\cos\beta\delta\beta = \frac{1}{n\lambda K^2}\cos\beta\delta\beta. \tag{13.5}$$

In Fig. 13.2, the expected precision of this method is plotted for both the groove density and for the d-spacing.

The two graphs demonstrate that this technique is a powerful method to determine very small deviations from the ideal values of the groove density of a grating. It is important to recognize that there are some limits to these

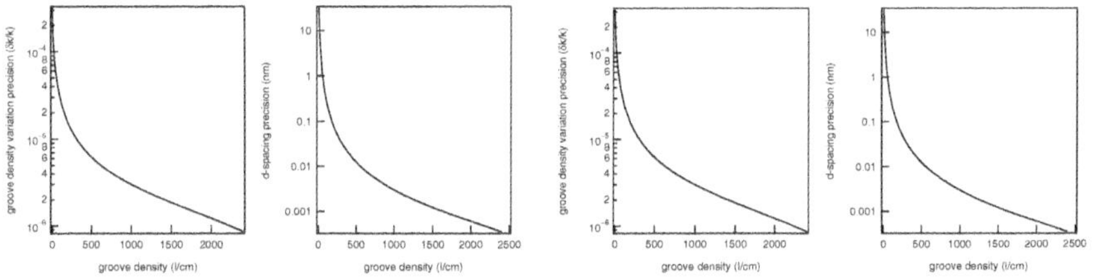

Fig. 13.2. *Left*: Estimation of the error $\delta K/K$ as a function of K in first diffraction order. An overestimated error of $1\,\mu\text{rad}$ in the measurement of the diffracted angle is supposed. *Right*: Estimation of the error in the measurement of the parameter d (groove spacing) as a function of the groove density

measurements. One is the spot size of the laser beam whose typical dimension is 1 mm. Since one has to rotate the grating to satisfy the Littrow condition, the projection of the laser spot on the grating will increase by a factor equal to $1/\cos(\alpha)$. This means that for a high groove density, when α_0 becomes considerable, the projection on the grating surface could be of the order of several mm, and therefore there is a reduction of the measurable spatial frequency. Moreover, it is impossible to measure groove densities K larger than $2/n\lambda$ because (13.3) has no real solution. With a He–Ne laser (632.8 nm), the maximum measurable groove density does not exceed 32,000 l cm^{-1}.

Another problem is the maximum groove density variation measurable in a single scan. If the diffracted direction changes, because of the groove density variation, and is no longer fully captured by the angular acceptance of the lens or of the linear detector, one cannot measure the entire grating in a single scan. In this case, it is necessary to stitch several measurements which introduces a further source of errors.

In the system described earlier, the groove density is measured by rotating the grating in front of the laser beam. However, there is another possibility: the one adopted in the Soleil metrology laboratory.

The measurement is made, also in Littrow condition, as described in (13.3). It is possible to make the measurement without changing anything in the LTP but instead by simply inclining the grating with respect to the optics table in order to obtain the proper incidence angle. However, the Littrow angle can be quite large, e.g., 25° for a 1,600 l mm^{-1} grating. This strong inclination obliges one to increase the distance between the optics head and the surface under test which is an additional source of errors. Moreover the X position along the grating has to be corrected according to the incidence law, and the sampling interval is no longer given by the translation indexing. It was found to be easier to slightly modify the optical setup with the simple attachment described in Fig. 13.3.

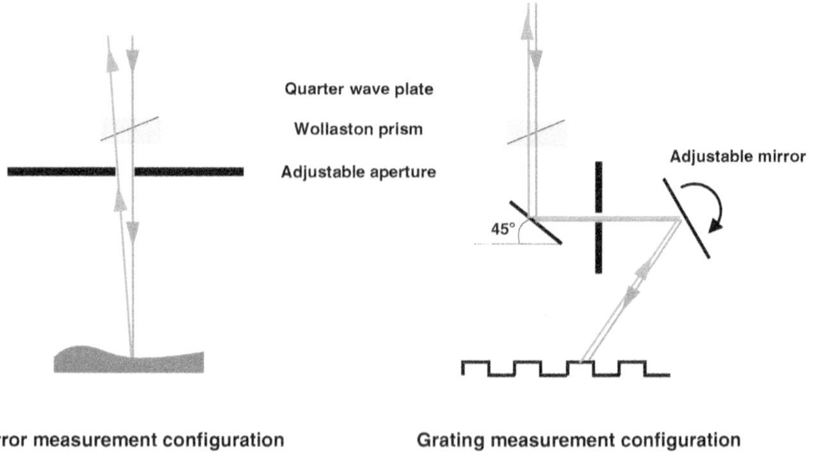

Fig. 13.3. The modified optical path in the grating measurement attachment

Fig. 13.4. View of the optics head with the grating measurement attachment in place

This attachment is composed of two flat mirrors deflecting the beam in the measurement track plane. The first mirror has a fixed 45° incidence angle; the second can be rotated around a horizontal axis to adjust the Littrow angle. The attachment (Fig. 13.4) is set at the place of the normal aperture. A series of interchangeable apertures is provided between the two mirrors in order to keep the field stop as close as possible to the surface under test.

With this device it is easy to work either at normal incidence to determine the grating surface shape, or at Littrow incidence to measure the line density variation law. Both measurements are made with the same sampling interval. The maximum departure of the line density with respect to its central value does not depend significantly of the mean line density and is close to $\pm 10\,\mathrm{l\,mm^{-1}}$. When this variation range is exceeded, the stitching method can be also applied. To be accurate, one should perform the data stitching on the line density values rather than on angle deviations.

References

1. K. Von Bieren, Laser Diagn. Proc. SPIE, **343** (1982)
2. P.Z. Tacaks, S.N. Qian, in *Metrology: Figure and Finish*, ed. By B.E. Truax, Prod. SPIE, vol. 749, 1987
3. P.Z. Takacs, S.N. Qian, U.S. Patent 4,884,697, 1989
4. S.N. Qian, W. Jark, P.Z. Takacs, Rev. Sci. Instrum. **66** (1995)

14

The COST P7 Round Robin for Slope Measuring Profilers

A. Rommeveaux, M. Thomasset, D. Cocco, and F. Siewert

Abstract. As part of the COST P7 Action, the metrology facilities of four European synchrotrons – Bessy, Elettra, ESRF and Soleil – instigated a round-robin programme of instrument inter-comparison. Other synchrotrons will later join this programme. The metrology instruments involved are various direct slope measurement devices, such as the well known Long Trace Profiler (either custom built or modified from commercial devices) and the Bessy Nanometer Optical component measuring Machine (NOM). The round robin was realized by measuring two flat and three spherical mirrors (made of either Zerodur or fused silica) made available by Bessy, Elettra and Soleil. The programme has been a significant aid in the characterization of each of the instruments and could readily be extended to other devices as a calibration tool. The results and advantages are described in this chapter.

14.1 Introduction

Most of the synchrotron radiation (SR) sources have developed their own metrology laboratory to meet the need of optics characterization in terms of microroughness, radius of curvature, slope errors, and shape errors. The instrumentation used consists mainly of commercial instruments: phase shift interferometers for microroughness characterization or Fizeau interferometers for bidimensional topography and optical profilometers for measurements of long optical components like the long trace profiler (LTP) or the nanometer optical component measuring machine (NOM). The LTP was developed at the Brookhaven National Laboratories by Takacs et al. [1], and marketed by Continental Optical Corporation (now Ocean Optics). It is basically a double pencil slope-measuring interferometer, for determining the slope error and radius of curvature and, through integration, the height profile for optical surfaces larger than 1 m in length. Optimally, precise data can be obtained, with reproducibility on the order of 2 nm $P{-}V$ (or 0.1 µrad RMS). What, however, is about the absolute precision of these profilometers? This is directly linked to instrument calibration, and up to now there is no standardization of calibration. In this round-robin endeavor, typical X-ray mirrors provided by the laboratories, plane, spherical, or toroidal are examined by the several laboratories using their own instrumentation in order to better understand the accuracy achievable with them.

The ultimate goal of this Round Robin is to create a database of the measurement results in order to provide these references as calibration tools available for metrology community.

14.2 Round-Robin Mirrors Description and Measurement Setup

Five mirrors have been involved in the present Round-Robin, two plane and three spherical, with varied parameters: reflectivity, material, radius of curvature, dimensions. Their main characteristics are given in Table 14.1.

The mirrors were measured with their optical surface up or on the side according to the standard instrument setup of each laboratory. To limit mechanical stress (sag) due to gravity in case of mirror facing up, the measurement procedure consisted in supporting the mirror with three balls placed at the Bessel points. The trace centered on the optical surface is perfectly defined on each mirror by lateral marks as well as is the scan direction. Each laboratory was free to define the appropriate number of scans to achieve the best accuracy of its instrument. Measurement procedures and parameters are summarized in Table 14.2.

14.3 Measurement Results

For each mirror, the resulting data consist in an array of mirror coordinates and corresponding measured slope. The same calculation method has been applied to process all these data in order to avoid discrepancies due to differences in fitting or integration methods. Slope errors and shape errors correspond to residual slopes and heights after best sphere subtraction. For plane mirrors (Table 14.3) there are important differences on radii values, but it is important to underline that each laboratory obtains a good repeatability of its value. The radius of curvature is obtained from the mirror slope profile. Obviously for plane mirrors with very large radius, the slope linear trend is affected by the intermediate frequencies measured. For this reason the radii results are not in a good agreement.

The graphical results (Fig. 14.1) for mirror P1 show an impressive consistency between residual slopes measured by each laboratory.

For spherical mirrors, the slope variation over the mirror length is obviously greater, implying a stronger influence of the individual characteristics of the different instruments on the measurement results. For LTPs, systematic errors can be corrected by averaging several measurements using different area

Table 14.1. Characteristics of the mirrors tested

Label	P1	P2	S1	S2	S3
Owner	BESSY	Elettra	Elettra	BESSY	Soleil
Dimensions $L \times W \times H$ (mm^3)	$310 \times 30 \times 60$	$400 \times 60 \times 80$	$270 \times 40 \times 30$	$210 \times 40 \times 40$	$120 \times 20 \times 50$
Bulk material	Zerodur	Zerodur	Fused silica	Zerodur	Silicon
Coating	none	Gold	Gold	None	None
Shape	Plane	Plane	Spherical $R = 83\,\text{m}$	Spherical $R = 44\,\text{m}$	Spherical $R = 1{,}280\,\text{m}$
Residual slope error rms	$\approx$1.1 µrad	$\approx$0.7 µrad	$\approx$1 µrad	$\approx$1 µrad	$\approx$0.5 µrad
Scan length	290	390	240	198	110

Table 14.2. Measurement parameters and scanning conditions

	BESSY	ELETTRA	ESRF	SOLEIL
Instrument	Autocollimator	LTP	LTP	LTP
Mirror position	Face up	On the side	Face up	Face up
Number of scans averaged	10	6 with mirror tilt	4	1
Systematic errors correction	Not applied	By mirror rotation	By mirror rotation	Not applied
Scanning method	Point by point	Point by point	On fly	On fly over sampling
Scanning velocity	$1\,\mathrm{mm\,s^{-1}}$	$1\,\mathrm{mm\,s^{-1}}$	$40\,\mathrm{mm\,s^{-1}}$	$0.2\,\mathrm{mm\,s^{-1}}$

Table 14.3. Statistical results obtained for P1 and P2 mirror

	P1 mirror			P2 mirror		
	R (km)	Slope error rms (µrad)	Height error rms (nm)	*R* (km)	Slope error rms (µrad)	Height error rms (nm)
BESSY	427	1.16	24	−1,951	0.88	24
ELETTRA	454	1.12	23	753	0.81	19
ESRF	136	1.12	24	242	0.97	28
SOLEIL	193	0.88	17	−100	1.64	52

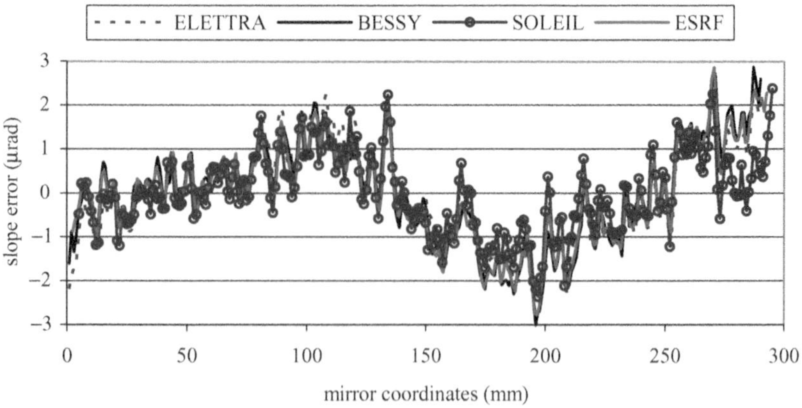

Fig. 14.1. P1 residual slopes after best sphere subtraction

in the internal optics. Table 14.4 shows the statistical results obtained for the three spherical mirrors.

The values for radius of curvature is in agreement by better than 0.3%. The concordance of residual errors is better for S3, which has the shortest length and the longest radius, than for S2 which has the opposite features. The rms agreement of the slope errors varies from 0.13 µrad (1.6 nm) for S3 to 0.26 µrad (4.1 nm) for S2. A gain, the residual slope profiles obtained at each facility are in excellent agreement in particular between BESSY, using the autocollimator sensor of the NOM, and LTP at the ESRF (Figs. 14.2–14.4).

Table 14.4. Statistical results obtained for S1, S2, and S3 spherical mirrors

	S3			S1			S2		
	R (m)	Slope error rms (µrad)	Height error rms (nm)	R (m)	Slope error rms (µrad)	Height error rms (nm)	R (m)	Slope error rms (µrad)	Height error rms (nm)
BESSY	1,280	0.44	3.2	83.01	0.87	11.7	44.52	1.08	17.4
Elettra	1,274	0.53	4.6	83.21	1.05	15.4	44.67	0.86	13.3
ESRF	1,278	0.40	3.0	83.34	0.99	15	44.76	0.82	13.3
Soleil	1,272	0.51	2.8	83.11	0.92	12.7	44.63	1.41	20.3

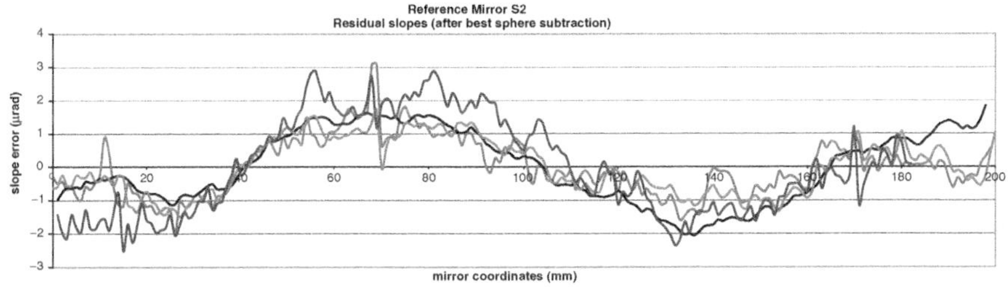

Fig. 14.2. S3 ($R \approx 1.3$ km) residual slopes after best sphere subtraction

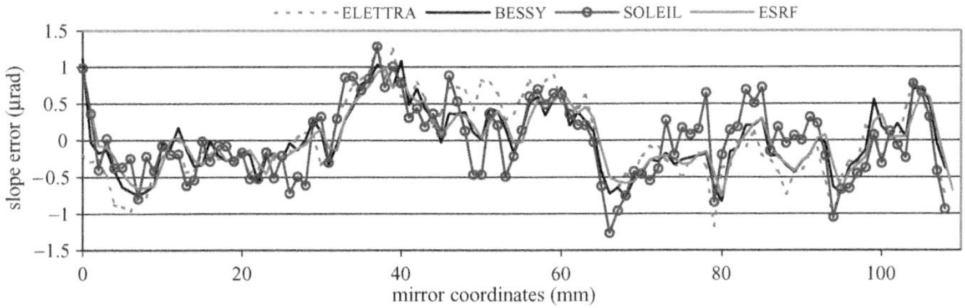

Fig. 14.3. S1 ($R \approx 83$ m) residual slopes after best sphere subtraction

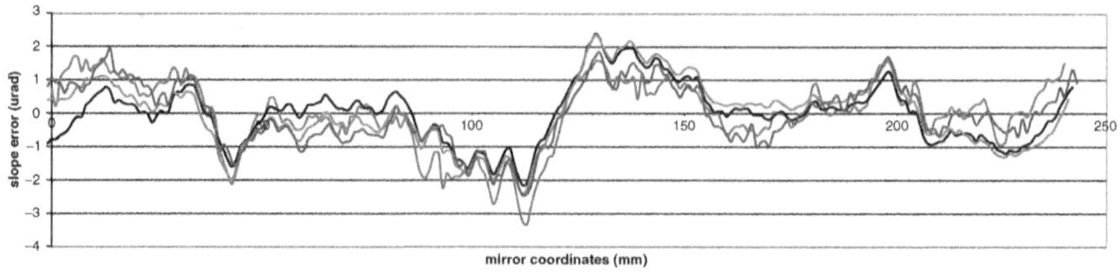

Fig. 14.4. S2 ($R \approx 44\,\mathrm{m}$) residual slopes after best sphere subtraction

14.4 Conclusions

The five mirrors involved in this Round-Robin are good representatives of the kinds of SR optical components to be characterized by slope measuring instruments. These results are in very good agreement with each other, despite the fact that different instruments have been used, in terms of optical setup, hardware, and environmental conditions. Even for the spherical mirrors with a short radius of curvature, which push the measurement accuracy of the instruments to their respective limit, due to the quality of their optical components (mirrors, prisms, lenses), the radii determined agreed better than 0.3%. The curves of the residual slopes after best sphere subtraction are quite superimposable.

These five mirrors cross measured with high consistency can be considered as reference mirrors for instrument calibration. The round-robin is going to be continued, including additional facilities and increasing the number of reference mirrors to be tested [2].

Acknowledgments

The authors wish to thank all their respective colleagues for their help, in particular Sylvain Brochet, Heiner Lammert, Thomas Zeschke, and Giovanni Sostero. They are grateful to the mirror owners for lending them during this round-robin.

References

1. P.Z. Takacs, S.N. Qian, J. Colbert, Proc. SPIE **749**, 59 (1987)
2. F. Siewert et al., in *SRI 2006 AIP Conference Proceedings*, Mellville, New York, **879**, 706 (2007)

15

Hartmann and Shack–Hartmann Wavefront Sensors for Sub-nanometric Metrology

P. Mercère, M. Idir, J. Floriot, and X. Levecq

Abstract. The recent development of third generation synchrotron radiation facilities has led to unprecedented progresse in X-ray applications such as microscopy and photolithography. To optimize performance in such research metrology tools with capabilities in the nanometre and even the sub-nanometre range, in order to characterize the surface figure errors of the optics used to focus or collimate the X-ray beams, to align them on the beam lines and to perform diagnostics of the beam spatial profile. To answer these needs, Synchrotron Soleil and Imagine Optic have developed in partnership a Shack-Hartmann long trace profiler (SH-LTP) which performs the bidimensional surface figure measurement of X-ray mirrors with nanometer precision and an increased dynamic range compared to earlier instruments. The SH-LTP offers a more complete diagnostic for highly curved surfaces compared to standard LTPs. This partnership has also led to the development of X-ray Hartmann wavefront sensors to measure and control the spatial quality of X-ray beams. Compared to interferometric tools, these sensors have better flexibility and can be integrated into closed-loop adaptive optical systems.

15.1 Introduction

Recent evolution of X-ray sources has opened a large field of research and applications in this spectral range. As in the visible range, one can distinguish two main centers of interest: high resolution X-ray imaging on the one hand, and improvement of the spatial quality of collimated X-ray beams on the other hand. Applications are, for example, EUV photolithography, X-ray microscopy, tomography, phase contrast imaging, or material probing.

With the development of such applications, an increasing need of optical components with better surface figures has appeared. Today, optical shape requirements in terms of slope errors are typically in the microradian and sub-microradian ranges. For the characterization of such high-quality components, the long trace profiler (LTP), used in mostly all synchrotron radiation facilities, has become the state-of-the-art off-line metrology tool. However, this instrument presents some drawbacks: 1D measurement, limited dynamic range (small radii of curvature below 1 m cannot be measured), and poor flexibility. To overcome these drawbacks, a stitching Shack–Hartmann long trace profiler (SH-LTP) has been developed. Resulting from a joint collaboration between Imagine Optic and SOLEIL, this new technology fulfils all the new requirements of an enhanced metrology. Bidimensional sub-nanometric characterization of optical surfaces, high precision, high sensitivity, and large dynamic are among the main advantages of the SH-LTP. First section of this chapter will be devoted to a description of this new instrument and the latest results obtained at the Optical Metrology Laboratory of SOLEIL will be presented.

Although off-line characterization of optical components is important (to know exactly if an optic is suitable for the requirements of one's application, or to know exactly what one is putting on the beamline), it cannot replace "at wavelength" metrology. A fact is that "at wavelength" metrology is increasingly required on EUV and X-ray beamlines, not only for in situ characterization of individual optics or optical systems, but also for their precise alignment with respect to the beam. Many kinds of wavefront sensors, mostly based on interferometry, have been developed so far. Up to now, the phase-shifting point diffraction interferometer (PSPDI) and the shearing interferometer (SI) are among the most effective interferometric tools. The PSPDI benefits from a very high precision ($<\lambda_{\text{X-UV}}/350$), a high sensitivity ($<\lambda_{\text{X-UV}}/1{,}000$) and a good spatial resolution, but its low dynamic range is unsuitable for strongly misaligned or very imperfect optics. The SI has the advantage of a good dynamic (about $\lambda_{\text{X-UV}}$), a good sensitivity ($<\lambda_{\text{X-UV}}/500$), and a good spatial resolution, but its poor precision is not sufficient for the measurement of high-quality diffraction limited optics. Both are unsuitable with real time active or adaptive optics, and, as for any interferometer, they are chromatic: the fringe spacing is linked to the wavelength, requiring realignment of the interferometer after any wavelength change. Finally, they are expensive and difficult to setup.

Following these statements and in response to the need of highly capable and flexible wavefront metrology tools, we started, in early 2002, working on the development of X-ray Hartmann wavefront sensors (HWS). First experiments were performed at 13.4 nm wavelength, demonstrating the high metrological performance of the technique: accuracy better than $\lambda_{\text{EUV}}/120$ rms and sensitivity better than $\lambda_{\text{EUV}}/600$ rms. Many improvements have been realized since then, including the extension to shorter and shorter wavelengths. Today, HWS are routinely working between 6.4 eV (193 nm) and 8 keV (0.155 nm), with accuracies as good as 0.04 nm rms. They are used in industry, by research laboratories and by synchrotron and FEL facilities. Simultaneously, the coupling of HWS with grazing incidence deformable mirrors for in situ X-ray active and adaptive optics has become one of our main topics today. First wavefront closed-loop corrections were successfully realized at $E = 3.64$ keV ($\lambda = 0.341$ nm), using a soft X-ray HWS and a 4-actuator Kirkpatrick–Baez (KB) active optical system. The next steps include the closed-loop control of bimorph and mechanical multiactuator deformable mirrors to allow for correction of higher frequency distortions. Together with these developments, we are also currently studying the use of scintillators as detection devices for working energies above 10 keV. Of course, this principle could be extended to lower energies for low-cost HWS, in applications where high precision is not absolutely required. In the second section, we will give a report of our main improvements in this field.

15.2 Generalities and Principle of Hartmann and Shack–Hartmann Wavefront Sensing Techniques

In the Hartmann (Shack–Hartmann) wavefront sensing technique, a beam passes through a hole array (microlens array) and is projected onto a CCD camera that detects the beamlet sampled by each hole (microlens). The positions of the individual spot centroids are then measured and compared with reference positions. This enables the local slopes of the wavefront (i.e., its derivative) to be measured at a large number of points within the beam. Figure 15.1 shows the principle of both techniques. We will use the term "subpupil" to define one hole or one microlens of the sampling array, and the term "pupil" to define all illuminated subpupils which corresponding spot centroids can be calculated. The different subpupils of the sampling array will be identified by two indices, i and j, i being associated with the X dimension and j with the Y dimension.

Let us call $\overrightarrow{k_{ij}^0}$ and $\overrightarrow{k_{ij}}(\neq\overrightarrow{k_{ij}^0})$ the local wave vectors coming from the subpupil (i, j), when sending an incident wavefront without and with phase distortions, respectively. The corresponding spot centroids are detected in (x_{ij}^0, y_{ij}^0) and (x_{ij}, y_{ij}), respectively. For each subpupil, the wavefront's local slopes, noted S_{ij}^x and S_{ij}^y, are given in radians by the following equations:

$$\begin{cases} S_{ij}^x = \tan\,\theta_{ij}^x = \frac{\Delta x_{ij}}{L} \\ S_{ij}^y = \tan\,\theta_{ij}^y = \frac{\Delta y_{ij}}{L} \end{cases}, \tag{15.1}$$

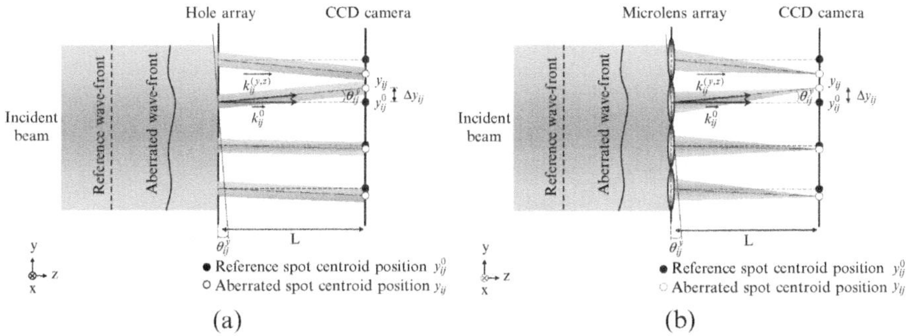

Fig. 15.1. Principle of the Hartmann (**a**) and Shack–Hartmann (**b**) wavefront sensing techniques

- where $\theta_{ij}^x = \left(\overrightarrow{k_{ij}^{(x,z)}}, \overrightarrow{k_{ij}^0}\right)$ and $\theta_{ij}^y = \left(\overrightarrow{k_{ij}^{(y,z)}}, \overrightarrow{k_{ij}^0}\right)$, with $\overrightarrow{k_{ij}^{(x,z)}}$ and $\overrightarrow{k_{ij}^{(y,z)}}$ the respective projections of $\overrightarrow{k_{ij}}$ onto the (X, Z) and (Y, Z) planes.
- where $\Delta x_{ij} = \left(x_{ij} - x_{ij}^0\right)$ and $\Delta y_{ij} = \left(y_{ij} - y_{ij}^0\right)$ represent the displacements of the spot centroids, in the detection plane, with respect to the reference positions.
- where L is the distance between the sampling and detection planes.

The local slopes of the wave front can also be written according to the following expressions:

$$\begin{cases} S_{ij}^x = \tan\,\theta_{ij}^x = \dfrac{\mathrm{d}W\left(x_{ij}^0, y_{ij}^0\right)}{\mathrm{d}x} = \dfrac{\lambda}{2\pi}\dfrac{\mathrm{d}\varphi\left(x_{ij}^0, y_{ij}^0\right)}{\mathrm{d}x} \\ S_{ij}^y = \tan\,\theta_{ij}^y = \dfrac{\mathrm{d}W\left(x_{ij}^0, y_{ij}^0\right)}{\mathrm{d}y} = \dfrac{\lambda}{2\pi}\dfrac{\mathrm{d}\varphi\left(x_{ij}^0, y_{ij}^0\right)}{\mathrm{d}y} \end{cases} \tag{15.2}$$

where $W(x, y)$ represents the optical path difference in meters and $\varphi(x, y)$ the spatial phase in radians. Integration of the measured derivative function enables reconstruction of the incident beam wave front [1].

For absolute wavefront measurements, HWS need to be calibrated with help of a well-known reference wave. Typically this wave is obtained by spatial filtering, wherein an incident beam is focused onto a small pinhole, diffracting a high-quality spherical wave. The reference positions $\left(x_{ij}^0, y_{ij}^0\right)$ are obtained during this preliminary calibration phase of the system.

15.3 Shack–Hartmann Long Trace Profiler: A New Generation of 2D LTP

15.3.1 Principle of the SH-LTP

SH-LTP uses a 26 × 26 sampling points Shack–Hartmann wavefront sensor (HP26/Imagine Optic). This sensor is optimized for working wavelengths around 400 nm, and offers a high performance with an accuracy of better than $\lambda/1{,}000$ rms and sensitivity better than $\lambda/5{,}000$ rms. The light source is a single-mode fiber laser diode (405 nm). Once collimated with an achromatic doublet, the light is sent to the surface under test and reflected on the sensor for analysis (Fig. 15.2). The basic principle is the same as for a conventional LTP [2]. However, the analysis pupil size of the sensor is about $12 \times 12\,\mathrm{mm}^2$, with a spatial resolution of 450 μm (size of the microlenses), and at each point the local slopes are measured in both directions X and Y.

For the characterization of long mirrors and/or improvement of the spatial resolution, a stitching process has to be applied. Its principle consists in the overlapping of adjacent surface measurements by translation of the optical head or of the mirror under test. The overlapping (or coupling ratio) is adjustable. Redundancy of the information is used to subtract all systematic errors including measurement errors induced by the imperfections of the

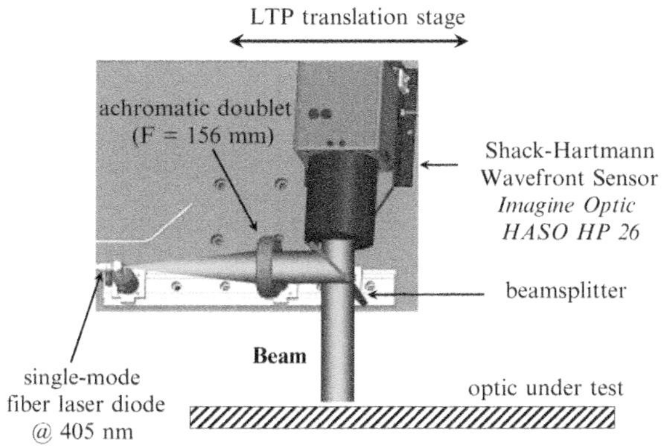

Fig. 15.2. SH-LTP design and principle

translation stage. Thus, the shape of the optical surface can be isolated. Stitching methods have already been extensively used with interferometers [3], but not yet with Shack–Hartmann wavefront sensors.

15.3.2 2D Long Trace Profile of a Plane Reference Mirror

To demonstrate the capabilities of this new technique, we first characterized a plane reference mirror with specified slope errors about 0.6 µrad rms. Stitching measurement was performed over $45 \times 12\,\text{mm}^2$, with a coupling ratio of 90%.

Over the entire surface analyzed, height errors were measured of about 2 nm rms (9.5 nm PV) (Fig. 15.3a); X and Y slope errors were measured at 0.58 µrad rms (10.76 µrad PV) and 0.63 µrad rms (8.21 µrad PV). The averaging of three adjacent traces (each trace is 450 µm large) from the SH-LTP Y slope measurement gave a slope error of 0.5 µrad rms (2.79 µrad PV), while the LTP (1.5 mm size trace) gave 0.6 µrad rms (2.47 µrad PV) (Fig. 15.3b). We can clearly observe a low-frequency pattern with a spatial period of about 50 mm, probably coming from the polishing process. Results of both techniques are in good agreement. Nevertheless, the SH-LTP provides bidimensional cartography of the surface and is much less noisy than the LTP.

Reproducibility of the SH-LTP was measured to be about 0.09 µrad rms (1 µrad PV) over the entire surface, and 0.1 µrad rms (0.44 µrad PV) over three averaged traces. The reproducibility of the LTP was measured to be 0.15 µrad rms (0.78 µrad PV).

15.3.3 2D Long Trace Profile of a Toroidal Mirror

We also characterized a toroidal mirror with radii of curvature specified to be 25.3 m (tangential) and 1.35 m (sagital). Alignment of the mirror axes with respect to the sensor axes was performed by minimizing the 45° astigmatism

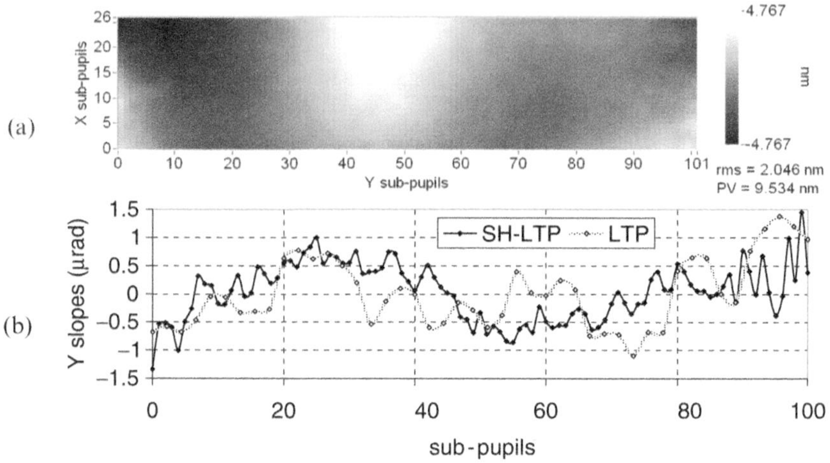

Fig. 15.3. Stitching measurement of a plane reference mirror (**a**) heights and (**b**) comparison (in terms of slopes) with SOLEIL LTP

Zernike coefficient. Stitching measurements were then performed along the tangential direction (i.e., the long radius of curvature), over $45 \times 12\,\mathrm{mm}^2$. Once again, we used a coupling ratio of 90%.

From the measured toroidal shape (Fig. 15.4a), radii of curvature were measured to be 24.98 m (tangential) and 1.41 m (sagittal), and are in good agreement with the specifications given by the manufacturer. By subtracting this perfect toroidal shape, one can recover the surface figure errors (Fig. 15.4b). Deviations from the perfect shape in terms of heights, X and Y slopes were measured to be 19 nm rms (76.1 nm PV), 3.6 µrad rms (32.9 µrad PV), and 4.3 µrad rms (37.4 µrad PV), respectively, over the whole surface analyzed. No residual astigmatism could be observed, showing that the axes of the sensor and those of the mirror were well aligned with each other. We can notice the presence of low- and high-frequency patterns. The averaging of three adjacent traces from the SH-LTP measurement gave a slope error of 5.43 µrad rms (20.35 µrad PV), while the LTP gave 5.1 µrad rms (17.4 µrad PV) (Fig. 15.4c).

Stitching measurements were also performed along the sagittal direction (i.e., the small radius of curvature), leading to similar results: the same radii of curvature and the same slope errors were obtained. With the LTP, measurements of such small radii of curvature (1.41 m) are not possible, the reflected beam from the surface falling outside of the analysis path.

15.3.4 Conclusion

The SH-LTP makes possible precise noncontact bidimensional measurement of the surface figure. At a conventional working wavelength ($\lambda = 405\,\mathrm{nm}$), high accuracy (0.1 µrad), high repeatability (<0.1 µrad) and sensitivity, high

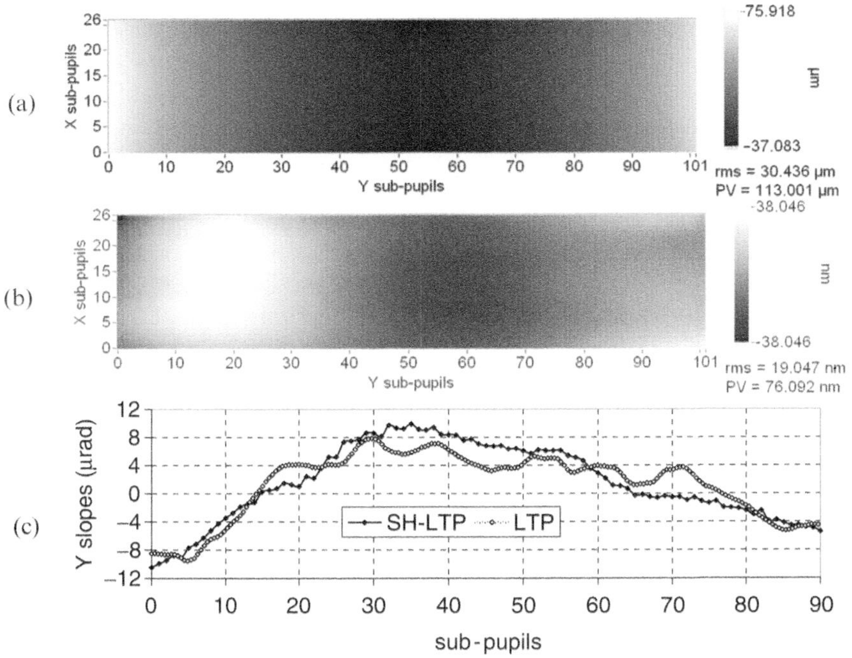

Fig. 15.4. Stitching measurement of a toroidal mirror along its tangential direction. (**a**) Nominal heights; (**b**) heights after best toroidal shape subtraction; (**c**) comparison (in terms of slopes) with SOLEIL LTP

dynamic, high-spatial resolution (450 μm), and insensitivity to vibrations are among the main advantages of this instrument. Radii of curvature down to 0.7 m can be measured with 0.1% accuracy, and large optics can be measured with sub-microradian performances thanks to the stitching approach.

Potential applications of the SH-LTP are various and are not limited to synchrotron topics. This technology fulfils all the new requirements of an enhanced metrology [4], and may offer, in the near future, even better performance in terms of spatial resolution (down to 100 μm), dynamic (cylinder with radius of curvature as small as few cm) and accuracy (down to 0.05 μrad).

15.4 X-Ray Wavefront Measurements and X-Ray Active Optics

The Shack–Hartmann technique is exhaustively used in the IR to UV spectral ranges, where the important diffraction effects limit the performance of the Hartmann technique. However, light properties at shorter wavelengths are reversing this behavior. At X-ray wavelengths, the Shack–Hartmann technique needs the use of Fresnel or Bragg–Fresnel zone plate (FZP) arrays. These diffractive optics are strongly chromatic, making the sensor poorly versatile.

This is the most limiting factor to the use of SHWS at higher energies, because potential users are interested in using a same sensor on different X-ray sources. Moreover, to keep high sensitivity on the local slope measurement of the wave front, one needs a long distance between the sampling and detection planes, i.e., FZP arrays with long focal lengths. At EUV wavelengths, this leads to FZP with large diameters (few hundreds of μm to 1 mm) in order to keep high diffraction efficiency, and results in a poor spatial sampling of the beam. It is only in the hard X-ray regime that spatial resolution vs. efficiency of SHWS could become interesting and competitive. However, the efficiency of FZP in the first diffraction order is low, yielding only a small fraction of the incident flux, all other diffraction orders contributing only in a reduction of the signal-to-noise ratio.

On the other hand, hole arrays for Hartmann tests can be manufactured very easily. Typical sizes of the holes range from 30 to 120 μm, with spacings from 60 to 300 μm, depending on the working wavelength, the geometrical properties of the beam to be analyzed, the pixel size of the CCD and the desired performance of the sensor in terms of accuracy, sensitivity, dynamic, and spatial resolution. HWS are totally achromatic, offer a much better performance, and are robust and easy to align, as compared to SHWS. Following these statements, we choose to develop EUV and X-ray wavefront sensors based on the Hartmann technique.

15.4.1 Hartmann Wavefront Measurement at 13.4 nm with $\lambda_{EUV}/120$ rms Accuracy

In 2002, under a close collaboration with the Center for X-Ray Optics (CXRO), we used ALS beamline 12.0 to perform the first experimental demonstration of wavefront analysis via the Hartmann technique in the EUV spectral range.

ALS beamline 12.0 is an undulator beamline designed for experiments relevant to the development of EUV lithography near 13 nm wavelength [5]. The geometry of ALS beamline 12.0 and the HWS setup are represented schematically in Fig. 15.5. Focusing of the monochromatic beam is performed by Kirkpatrick–Baez (KB) optics. The output-side NA of this focused beam is approximately 0.006, and the dimensions of the focal spot are typically $10 \times 15\,\mu m^2$ FWHM [6, 7].

HWS employed a hole array made in an 80-μm thick nickel plate with 65×65 holes over a $15 \times 15\,mm^2$ area. The holes were 80-μm size squares, spaced by 225 μm and rotated by 25° to minimize the overlap of the diffraction from adjacent holes in the measurement plane [8]. A back-illuminated, thinned, 16-bit EUV CCD camera with $1{,}024 \times 1{,}024$ pixels (24 μm per pixel) was placed at 400 mm behind this hole array.

The reference spherical wave front needed for HWS calibration was generated by placing a small pinhole at the focus of the KB optics, creating nominally spherical wave illumination for the HWS. To guarantee a reference

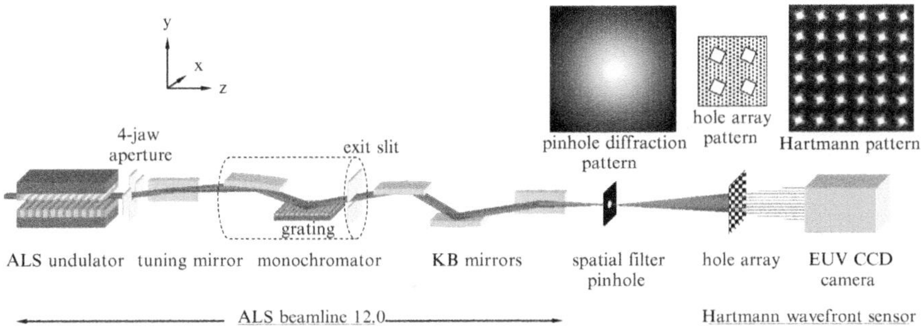

Fig. 15.5. ALS beamline 12.0, with the spatial filtering and the Hartmann wavefront sensor. Images of the beam diffracted by the pinhole, the hole array pattern, and the Hartmann pattern are shown

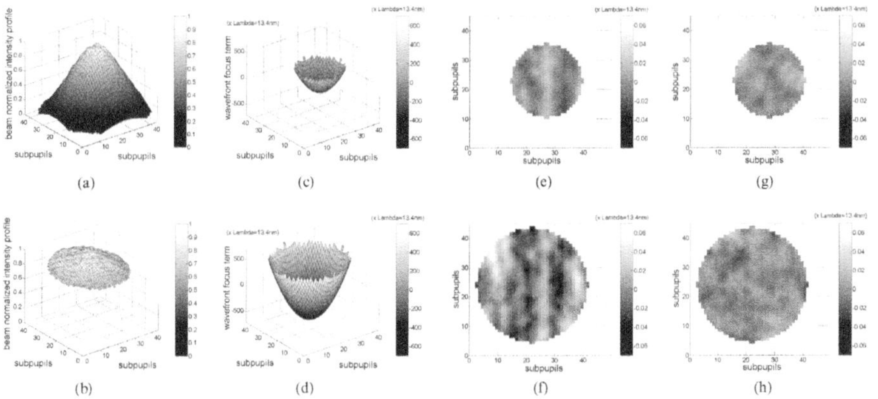

Fig. 15.6. Normalized beam intensity profiles obtained with the 1.7 µm pinhole (**a**) and the 0.6 µm pinhole (**b**). With the 1.7 and 0.6 µm pinholes, respectively, wavefront focus terms (**c**) and (**d**), residual absolute wavefronts (**e**) and (**f**), and repeatability measurements (**g**) and (**h**)

wave front better than $\lambda_{\mathrm{EUV}}/100$ rms, we used only half of the central Airy disk. Two series of experiments were performed with 1.7 and 0.6 µm pinholes. In both configurations, HWS hole array was placed a distance of 610 mm from the KB focus (NA = 0.025), the operational wavelength was 13.4 nm, and the exposure times were about 150 ms typically.

Figures 15.6a, b show the beam's normalized intensity profiles obtained at 13.4 nm with the 1.7 and 0.6 µm pinhole sizes, respectively. With both pinholes, calibration of the sensor was performed over the largest square pupil illuminated (44 × 44 subpupils) centered on the CCD chip. For wavefront analysis, we used the largest round pupil inscribed in the beam FWHM, corresponding to a circular analysis pupil of 26 (38) subpupils in diameter (NA ∼0.0096 and 0.014, respectively).

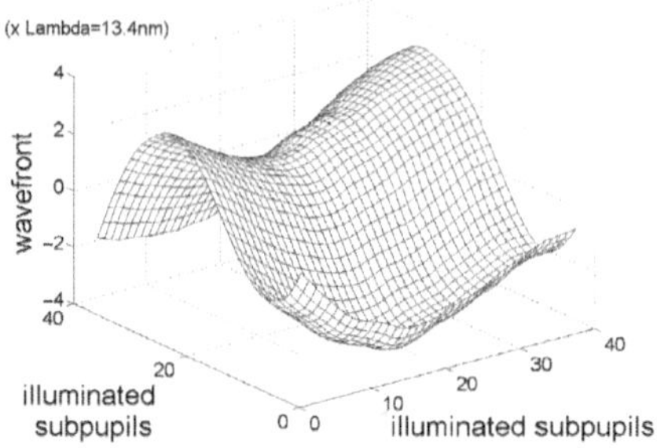

Fig. 15.7. ALS beamline 12.0 wavefront measurement without spatial filtering at 13.4 nm

After calibration of the system we first performed absolute wavefront measurements. Figures 15.6c, d, show the wavefront focus terms obtained with the 1.7 and 0.6 μm pinholes, respectively. In both configurations the diffracted beams are spherical waves with identical radii of curvature, measured by HWS as 610.161 ± 0.009 mm. Figures 15.6e, and f display the respective residual wave fronts obtained after removing the tilt and focus terms. With the pinhole of 1.7 (0.6) μm the residual absolute wavefront root-mean-square (rms) and peak-to-valley (PV) values reached at best $0.012\lambda_{\mathrm{EUV}}$ ($0.021\lambda_{\mathrm{EUV}}$) and $0.055\lambda_{\mathrm{EUV}}$ ($0.113\lambda_{\mathrm{EUV}}$), respectively. As shown in Fig. 15.6g, h, the repeatability of the sensor (obtained by subtracting two successive wavefronts measurements) was measured about $\lambda_{\mathrm{EUV}}/125$ rms, independently of the analysis pupil size. The sensitivity was measured about $\lambda_{\mathrm{EUV}}/600$ rms over 50 successive measurements. No degradation of these performances could be observed between 7 and 25 nm, demonstrating the achromaticity of the sensor [9].

As an application of the Hartmann technique, we removed the spatial-filter pinhole to measure the wave front of the focused beam produced by the KB optics. The wave front displayed in Fig. 15.7, has residual-aberration magnitudes of $1.879\lambda_{\mathrm{EUV}}$ rms and $6.99\lambda_{\mathrm{EUV}}$ PV. Mostly astigmatism can be observed. By convolution of the calculated point-spread function with the geometrical image of the beamline source, we estimated the size of the KB focal spot to be $23 \times 26\,\mu\mathrm{m}^2$ at $1/e^2$, while direct imaging, by use of a YAG:Ce crystal and a (10×) magnification visible imaging system, gave $21 \times 25\,\mu\mathrm{m}^2$ at $1/e^2$. Thus, HWS can also predict the focal spot properties of highly distorted beams.

15.4.2 Wavefront Closed-Loop Correction for X-Ray Microfocusing Active Optics

Most synchrotron beamlines use KB optics for tight focusing of the X-ray beams. So far, these mirrors are manually aligned by the knife-edge technique or by use of well-adapted focal spot imaging systems based on wavelength

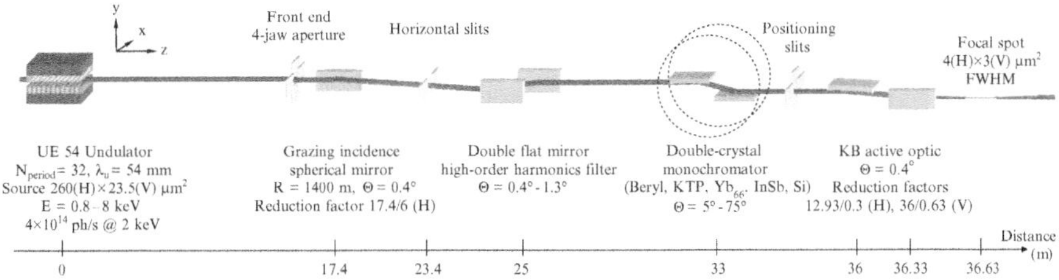

Fig. 15.8. SOLEIL's beamline LUCIA at the SLS

conversion to the visible range. This alignment is complicated and time consuming. For easy microfocusing of the synchrotron beams, we coupled HWS with a KB active focusing optics.

First tests were performed on SOLEIL's beamline LUCIA (Line for Ultimate Characterizations by Imaging and Absorption), which is presently implemented on the Swiss Light Source (SLS). Represented in Fig. 15.8, LUCIA is dedicated to X-ray absorption spectroscopy between 0.8 and 8 keV [10]. The undulator beam passes successively through a grazing incidence spherical mirror, a double plane mirror high-order harmonics filter, and a double crystal monochromator. The monochromatic beam is then focused by a 4-actuator Kirkpatrick–Baez (KB) active optical system. Based on a European Synchrotron Radiation Facility conception [11], this system is composed of two plane silicon mirrors, held at their tips by active jaws. Working at 0.4° incidence, these mirrors (170 × 30 × 8mm) are designed to take their perfect elliptical shapes once bent [12].

For beam diagnosis and to control the KB mirrors, we used a soft X-ray HWS composed of a hole array made in a 30-µm thick nickel plate, with 75 × 75 holes over a 10 × 10 mm^2 area. The holes were 50-µm size squares, spaced by 135 µm and rotated by 25°. A back-illuminated, thinned, 16 bit X-ray CCD camera, with 1,024 × 1,024 pixels (13 µm per pixel), was placed 400 mm behind this hole array. The end station with the KB optic and HWS is given in Fig. 15.9.

First experiments were performed with help of an imaging system of the KB focal spot, and a dedicated genetic algorithm, to achieve best focusing of the beam on a removable YAG:Ce crystal. The corresponding wavefront was then used as reference for calibration of HWS. Closed-loop correction, performed with HWS, enabled convergence in a single iteration to the KB configuration taken as reference [13]. However, knife-edge measurements of the focal spot showed that focusing of the X-ray beam was limited by the "relative" calibration of HWS, i.e., by the resolution of the system imaging the KB focal spot.

To overcome this limitation, an absolute calibration of the sensor was performed by use of the spherical wave diffracted by a 1 µm spatial filter pinhole placed at the focus of the KB. For this calibration phase, the energy

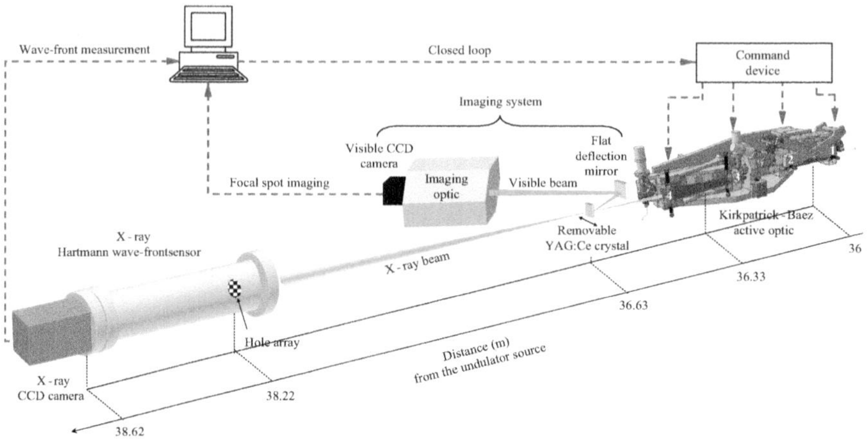

Fig. 15.9. Beamline LUCIA end station with the KB active optical system and the soft X-ray HWS

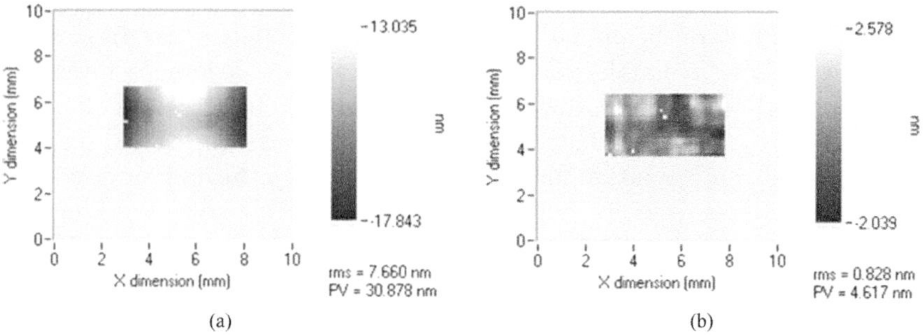

Fig. 15.10. Absolute residual wavefront measurements (single CCD image treatment) (**a**) before and (**b**) after closed-loop correction

of the photon beam was tuned down to 700 eV to ensure a large illumination of the sensor with the central Airy disk.

A closed-loop correction was then performed at $E = 3.64\,\text{keV}$ ($\lambda = 0.34\,\text{nm}$), the spatial filter pinhole having been removed. In a single iteration, we succeeded in correcting the phase distortions from 7.7 nm rms and 30.9 nm PV down to 0.8 nm rms and 4.6 nm PV (Fig. 15.10).

With the KB system correctly aligned, we performed knife-edge scans in both dimensions to characterize the beam. At the focal spot position, the beam sections were measured at $2.4 \times 2.86\,\mu\text{m}^2$ FWHM (Fig. 15.11). These dimensions are close to the theoretical limit given by the source size, the geometry of the beamline, and the slope errors of the KB mirrors (measured about 1.1 μrad).

The performance of HWS at these high energies, in particular, the signal-to-noise ratio and the accuracy of the sensor, is strongly limited by shot noise,

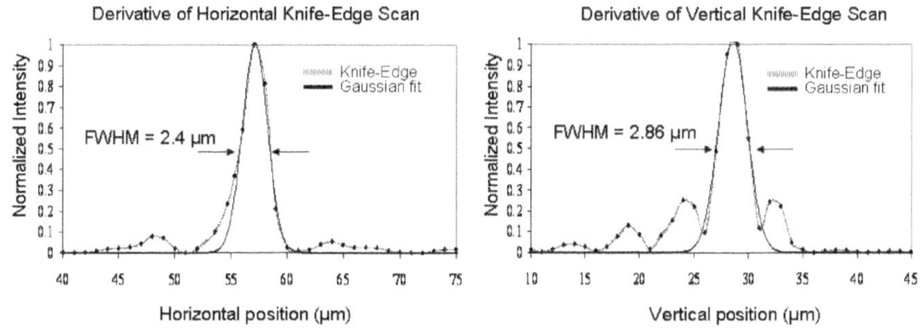

Fig. 15.11. Beam knife-edge measurements at the focal spot position after closed-loop correction with HWS

from the photon-to-electron conversion process in CCDs. The residual wavefront that can be observed after correction in Fig. 15.10b is, for example, only the result of shot noise. To overcome this problem, accumulation of several images is required. The signal-to-noise ratio and the repeatability of wavefront measurements were studied at 2.1 keV as a function of the number of images integrated. To achieve a signal-to-noise ratio of about 100, at least 50 CCD images had to be integrated. By integrating 500 CCD images per wavefront measurement, we improved the repeatability of the sensor to better than 0.04 nm rms. Therefore, high-readout rate CCD cameras may be preferred in the soft to hard X-ray spectral ranges, when an optimal performance of HWS is required.

15.4.3 Conclusion

In the EUV spectral range, wavefront measurements were performed over a wide wavelength range from 7 to 25 nm. The accuracy of the sensor was proved to be better than $\lambda_{\mathrm{EUV}}/120$ rms ($\lambda_{\mathrm{EUV}} = 13.4$ nm), and the sensitivity better than $\lambda_{\mathrm{EUV}}/600$ rms, demonstrating the high metrological performance of this system.

In the soft X-ray range, HWS was successfully used to align a 4-actuator Kirkpatrick–Baez (KB) active optical system. A wavefront closed-loop correction was performed at $E = 3.64$ keV, which led to beam focusing down to $2.4 \times 2.86\,\mu\mathrm{m}^2$ FWHM in a single iteration. Variation of the KB focal length is easily possible by the addition of a curvature term to the closed-loop wavefront target.

Today, HWS are routinely working between 6 eV (193 nm) and 8 keV (0.155 nm), with accuracies as good as 0.04 nm rms. The use of high readout rate CCD cameras for fast accumulation of images and the use of luminescent screens for a visible Hartmann analysis of the beams, especially for energy ranges above 10 keV, are currently under investigation. The coupling of HWS with mechanical and bimorph multiactuator deformable mirrors should also be done in a very near future, to allow easy correction of higher frequency distortions on synchrotron beamlines.

Acknowledgements

The authors are greatly indebted to "REGION ILE DE FRANCE" for funding part of the SH-LTP development project. The authors would like to thank the scientific and technical staffs of ALS beamline 12.0 and SLS beamline LUCIA (including Markus Janousch) for their support. Finally, the authors would like to thank several contributors to these works, including Sylvain Brochet, Samuel Bucourt, Gilles Cauchon, Guillaume Dovillaire, Thierry Moreno, François Polack, Muriel Thomasset, and Philippe Zeitoun.

References

1. W.H. Southwell, J. Opt. Soc. Am. **70**, 998 (1980)
2. M. Thomasset, S. Brochet, F. Polack, in *Advances in Metrology for X-Ray and EUV Optics*, ed. by L. Assoufid, P. Takacs, J. Taylor. Proc. SPIE, vol. 5921 (2005), p. 12
3. M. Otsubo, K. Okada, J. Tsujiuchi, Opt. Eng. **33**, 608 (1994)
4. F. Siewert et al., Third International Workshop on Metrology for X-Ray Optics, Daegu, Korea, 2006
5. D. Attwood, P.P. Naulleau, K.A. Goldberg, E. Tejnil, C. Chang, R. Beguiristain, P. Batson, J. Bokor, E.M. Gullikson, M. Koike, H. Medecki, J.H. Underwood, IEEE Quantum Electron. **35**, 709 (1999)
6. K.A. Goldberg, Ph.D. Dissertation, University of California, Berkeley, 1997
7. D. Attwood, *Soft X-Rays and Extreme Ultraviolet Radiation – Principles and Applications* (Cambridge University Press, Cambridge, England, 1999)
8. Imagine Optic patent, PCT/FR02/02495, July 2002
9. P. Mercère et al., Opt. Lett. **28**(17), 1534 (2003)
10. M. Janousch, R. Abela, Th. Schmidt, J.F. van der Veen, R. Wetter, A.-M. Flank, P. Lagarde, G. Cauchon, *PSI Scientific Report 2002*, Volume VII (2002)
11. O. Hignette, G. Rostaing, P. Cloetens, A. Rommeveaux, W. Ludwig, A.K. Freund, Proc. SPIE, vol. 4499-19 (2001)
12. H.A. Padmore, M.R. Howells, S.C. Irick, T. Renner, R. Sandler, Y.-M. Koo, Proc. SPIE, vol. 2856 (1996), p. 145
13. P. Mercère et al., Opt. Lett. **31**(2), 199 (2006)

16

Extraction of Multilayer Coating Parameters from X-Ray Reflectivity Data

D. Spiga

Abstract. Detailed analysis of X-ray reflectivity (XRR) angular scans of multilayer coated samples has been recognized as a powerful tool to investigate their stack structure. Even though the interpretation of multilayer XRR scans is made complex by the difficulty of managing the large number of parameters that characterize the stack, computer programs can be used to address the problem of the multi-parametric fit of experimental XRR scans of multilayers. This chapter describes a possible strategy to extract the layer thickness values of a multilayer coating from accurate fitting of XRR scans, based on the Python Program for Multilayers coded. The results of a best-fit analysis of XRR with transmission electron microscopy data are also discussed.

16.1 Introduction

The development of multilayer structures intended to enhance the reflection of radiation with wavelengths in the range of 10–0.01 nm, from extreme ultraviolet to X-rays, and of thermal neutrons, is at present being very actively pursued. In particular, the use of wideband multilayer coatings is foreseen in the next generation of soft ($E < 10$ keV) and hard X-ray ($E > 10$ keV) telescopes with imaging capabilities, like SIMBOL-X [1], Constellation-X [2], XEUS [3]. The reflection process in multilayers is a complex one, arising from the interference of the radiation reflected at each interface, beyond the critical angle for total external reflection. The reflection/focusing performance over a wide energy band depends essentially on the thickness precision of all layers and on the smoothness, homogeneity, and sharpness of all interfaces. It is therefore easy to understand how, in order to improve deposition techniques, methods to investigate the internal structure of multilayer stacks are needed, and criteria to evaluate the feasibility of the adopted process in terms of repeatability, uniformity, smoothness, durability must be established.

In this chapter we will compare two techniques that can be used to achieve a detailed characterization of a multilayer coating: the stack section *imaging* with TEM (transmission electron microscope) and the *analysis* of the

XRR (X-ray reflectivity) curves by means of a powerful computer program, PPM (Pythonic Program for Multilayers), developed by *A. Mirone* at ESRF (European Synchrotron Radiation Facility, Grenoble, France). Although the usefulness of the XRR curves are already recognized as important diagnostic tools for multilayers, the exact interpretation is made difficult by their complexity and by the large number of parameters characterizing a multilayer. Therefore, the matching between the experimental and a modeled XRR curve with manually adjusted parameters can be only qualitative in most cases. Consequently, the description of the stack structure is often a poorly detailed approximation of the real one.

On the other hand, the application of PPM to the analysis of XRR data returns *very detailed fits and a realistic description* of the multilayer stack. The advantages of this technique are an *effective, quick, nondestructive, in-depth* probing of the distribution of thicknesses throughout the stack.

In the following sections we review some features of X-ray reflection from multilayers. Then we describe some methods that can be used to extract information from the XRR curves and apply PPM to the reflectivity data of a multilayer. Finally, the PPM results are compared with those of TEM and the difficulties that can arise in such a comparison, due to artifacts in TEM images, are discussed.

16.2 A Review of X-Ray Multilayer Coatings Properties

The usefulness of multilayer coatings resides in their capability of reflecting radiation with wavelength λ in the nanometer/sub-nanometer range when the incidence angle and the energy exceed the conditions for total external reflection. The X-ray amplitude reflectivity, r, of a single interface between two layers with a difference in refractive index, Δn, decays rapidly with increase in incidence angle ϑ_{i} (measured from the surface plane):

$$r(\lambda) \approx \frac{\Delta n(\lambda)}{2\sin^2\vartheta_{\mathrm{i}}}. \tag{16.1}$$

Owing to the very small deviation of the real part of n from unity in X-rays ($\delta = 10^{-4} \div 10^{-5}$, depending on the photon energy and the composition of the reflecting coating), r is usually very small when the incidence angle is larger than the critical one. However, if the spacing of the interfaces of layers in a multilayer is properly conceived, the constructive interference of reflected rays at each interface enhances the reflectivity at definite photon energies.

The reflectance of a multilayer with $2N$ layers with thickness $t_1,\ t_2, \ldots t_{2N}$ and refractive indexes $n_1,\ n_2, \ldots n_{2N}$ can be computed by recursive application of the single-layer reflection formula [4]:

$$R_{m+1} = \frac{r_{m,m+1} + R_m \exp(-\mathrm{i}\Delta\phi_m)}{1 + r_{m,m+1} R_m \exp(-\mathrm{i}\Delta\phi_m)}. \tag{16.2}$$

In the last equation, $r_{m,m+1}$ is the reflectance of the electric field amplitude at the mth/$(m+1)$th layer interface, $\Delta\phi_m = 4\pi n_m t_m \sin\vartheta_m/\lambda$ is the phase shift between reflected rays at the mth and the $(m+1)$th interface, R_m is the amplitude reflectivity of the first m layers. The final X-ray reflectance of the multilayer is $|R_{2N+1}|^2$.

The *d-spacing* $d_j (j = 1 \ldots N)$ is the total thickness of the jth couple of layers (*bilayer*). Multilayers with constant d-spacing, $d_j = d$, are suited to reflect narrow bands of the spectrum, whose locations are *approximately* (neglecting the beam refraction) determined by *Bragg's law*,

$$2d \sin\vartheta_\mathrm{i} \approx k\lambda. \tag{16.3}$$

In (16.3), k is an integer and λ is the wavelength of the radiation in use.

Multilayers able to reflect a continuous energy band are characterized by a variable d-spacing throughout the stack (*graded* multilayers). Radiation with wavelength λ is reflected when it propagates across bilayers whose d-spacing satisfies approximately Bragg's law. A well known possibility is to decrease gradually the d-spacing according to a power-law [5]:

$$d(j) = \frac{a}{(j+b)^c}. \tag{16.4}$$

We denote with $j = 1, 2 \ldots N$ the index of the jth bilayer, ordered from the multilayer outer surface. Wide-band multilayers of the described type, initially developed to reflect neutron beams, are called *supermirrors* and are utilized also for X-ray mirrors, although in this case the absorption is more severe than that for neutrons. The coefficients a, b, c, as well as the number of bilayers, N, and the ratio high-Z material/d-spacing, Γ, have to be optimized in order to obtain the desired reflectivity as a function of the photon energy. For graded multilayers Γ can be constant or slowly variable in order to maximize the reflection efficiency over the energy band to be reflected, i.e., to find the best trade-off between constructive interference and photoelectric absorption.

As an example, we show in Fig. 16.1 a comparison of the reflectivity as a function of the photon energy at 0.2° grazing incidence for a constant d-spacing W/Si multilayer with 200 bilayers, $d = 8.7\,\mathrm{nm}$, $\Gamma = 0.46$ and a supermirror with 200 bilayers, $a = 12\,\mathrm{nm}$, $b = 1.85$, $c = 0.3$, and constant $\Gamma = 0.46$. The supermirror stack was especially designed to provide a reflectivity as uniform as possible in the energy band 1–70 keV. The reflectivity is improved at low energies by adding a capping layer of tungsten and a final layer of carbon [6]. Multilayer stacks of the described type [5, 7] are foreseen for the optics of future hard X-ray imaging telescopes (SIMBOL-X, Constellation-X, XEUS).

Imperfections of the interfaces, such as microroughness and layers interdiffusion, cause a broadening of the interface width. When the two effects can be

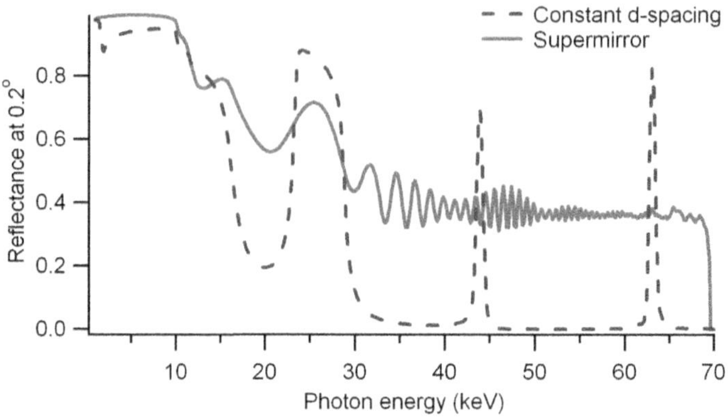

Fig. 16.1. Calculated X-ray reflectivity of a constant d-spacing W/Si multilayer (*dashed line*) and a W/Si supermirror (*solid line*) in the energy range 1–70 keV at the grazing incidence angle 0.2°. The computation supposes zero roughness

considered to be independent of each other, the total interface width, σ, can be computed as the quadratic sum of the roughness, σ_{r}, and the interdiffusion, σ_{d}:

$$\sigma^2 = \sigma_{\mathrm{r}}^2 + \sigma_{\mathrm{d}}^2. \tag{16.5}$$

One of the effects of interface broadening is the *exponential* reduction of the "specular" reflectivity (reflection angle equal to the angle of incidence), following the *Névot–Croce formula* [8]:

$$R_\sigma = R_0 \exp\left(-\frac{16\pi^2\sigma^2 n_{\mathrm{h}} n_{\mathrm{l}} \sin\vartheta_{\mathrm{h}} \sin\vartheta_{\mathrm{l}}}{\lambda^2}\right). \tag{16.6}$$

In this formula, n_{l}, n_{h} are the refractive indexes and ϑ_{l}, ϑ_{h} are the incidence angles in the two components of the multilayer. The two angles are not equal due to beam refraction. The reduction is much more severe for high energies (small λ). The interfacial roughness, σ_{r}, has also another effect, the *X-ray Scattering* in directions around the specular one. This effect has an important role in the degradation of imaging quality of X-ray optics.

High precision in the thickness of the layers and a low roughness are required to ensure a good reflectivity in the energy band of interest. Deviations of the thickness of the layers from the nominal ones can destroy the ordered phase shift distribution that generates the high reflectivity or/and the energy resolution, e.g., for narrow-band multilayers used as monochromators. Therefore, the reflectivity scan of a multilayer is very sensitive to thickness drifts and irregularities, and it is easily understood how the deposition facility has to be carefully calibrated. Furthermore, the deposition rate has to be very steady.

As we shall see in the next section, the sensitivity of X-ray reflectance to small deviations of the multilayer thickness from the nominal one makes X-ray reflectivity scans a powerful tool for the investigation of the internal structure of a multilayer, and consequently, for the evaluation of the improvement of a deposition technique. We shall, moreover, see how a detailed description of the multilayer can be extracted by means of PPM.

16.3 Determination of the Layer Thickness Distribution in a Multilayer Coating

16.3.1 TEM Section Analysis

A possible technique that can be used to visualize the structure of a multilayer coating is the use of a *Transmission Electron Microscope* (TEM). In TEM images, the high-density layers appear dark, whereas the low-density layers are bright. For instance, we show in Fig. 16.2 the TEM sections of a Pt/C multilayer deposited by e-beam evaporation onto a Si wafer ($\sigma \approx 0.3$ nm) substrate at *Media-Lario technologies* (Bosisio Parini, Italy); the layered structure is clearly visible and the thickness of single layers can be directly measured. For example, the TEM image in Fig. 16.2 highlights the presence of a much thicker carbon layer due to an instability of the electron beam evaporator. Indeed, the increase of Pt layers at the right side is an image artifact. It will be explained in Sect. 16.3.3.

The information provided by the TEM analysis is often useful in helping to improve the stability of the deposition system. For example, this sample was a very important test because it constituted the final calibration of the deposition facility for the manufacturing of a hard X-ray optic prototype [9].

In addition to the layers thickness of the multilayer, TEM images also provide useful information concerning the crystallization state of the layers, the interdiffusion between adjacent layers, and sometimes the undulations of

Fig. 16.2. TEM section of a Pt/C multilayer deposited by e-beam evaporation onto a Si wafer. Pt layers are the dark bands. The section thickness, perpendicular to the page, decreases from the right to the left side. The growth direction is from bottom to top (image by *L. Lazzarini and C. Ferrari, IMEM-CNR*, Parma, Italy)

the interfaces due to the *microroughness growth* (see Fig. 16.8). This is a well-known phenomenon, resulting from the combined effect of the replication of topography of the underlying layers and the random fluctuations of the deposition process [10].

The TEM images presented in this work are obtained from a JEOL-2000-FX installed at *IMEM-CNR* (Parma, Italy). The accuracy in layers thickness measurements is ~0.5 nm for multilayers with abrupt interfaces.

16.3.2 X-Ray Reflectivity Analysis

The TEM technique is expensive and the sample preparation is complex and destructive; therefore, it can be utilized only for selected samples.

However, a large amount of information can be extracted from the analysis of the *X-Ray Reflectivity (XRR) scan* of the multilayer. This technique is a commonly performed test of the reflectance efficiency and consists of probing the multilayer by means of a thin X-ray beam incident on the coating and measuring the reflectivity in the specular direction at different incidence angles.

The usefulness of the XRR measurement as a diagnostic tool is also well known: the reflectivity as a function of the grazing incidence angle, resulting from the interference of the radiation reflected at each interface, is usually very sensitive to the details of the multilayer structure, namely all the values of thickness, density, and roughness of the layers. For instance, if the multilayer has a high periodicity and smooth, abrupt surfaces, it will generally exhibit high, sharp, clearly defined interference peaks (16.3). Conversely, irregularities of *d*-spacing will cause the peaks to be "spread" on the angular scale (see Fig. 16.3), whereas rough or diffuse interfaces will reduce the intensity of peaks (16.6).

This technique is not destructive, it is quick, and it does not require any particular preparation of the sample. In addition, the probed surface is usually large (several cm^2) even with very thin beams because the measurement is usually performed in grazing incidence. This reduces selection effects because local fluctuations of *d*-spacing are averaged out.

The requirements for XRR measurements for deriving the multilayer structure are a monochromatic X-ray source with a small divergence (a few 10 arcsec) in order to guarantee a good angular resolution. In addition, the incident X-ray beam has to be very thin (a few tenth/hundredth microns, depending on the sample size) in order to be entirely collected by the sample at very small incidence angles ($\vartheta_i > 500$ arcsec).

Interpretation of X-Ray Reflectivity Data

Although the analysis of XRR curves is a widespread tool, their exact interpretation is a complex problem. Because of the sensitive dependence of XRR measurements on the thickness, density, roughness of *all* layers, the

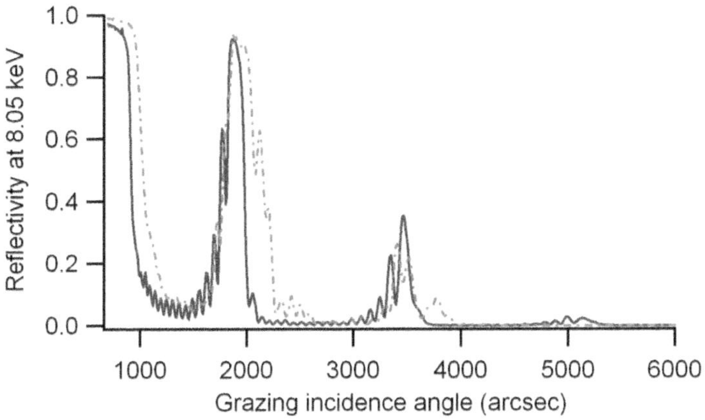

Fig. 16.3. Comparison of the measured X-ray reflectivity scans of two Ni/C multilayers with the same average value (9 nm), but different dispersion of the d-spacing (deposited in 2003 by e-beam evaporation at *Media-Lario technologies*). The smaller dispersion in the case of the solid line curve is made apparent by the narrower and more regular peaks. The approximate Γ factor is 0.2 for the solid line and 0.4 for the dashed line

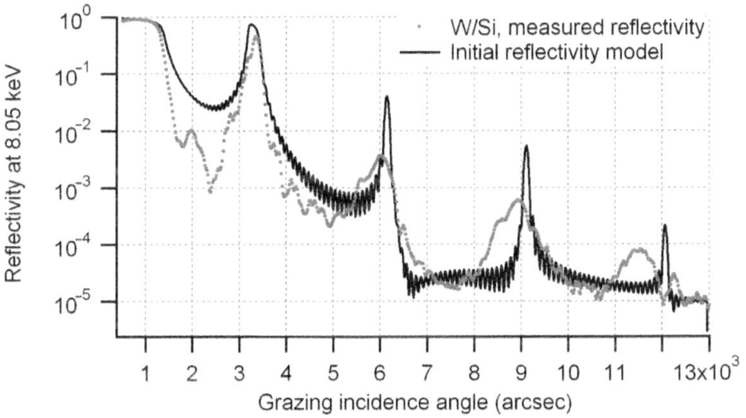

Fig. 16.4. Experimental X-ray reflectivity of the W/Si multilayer with 30 bilayers deposited by e-beam with ion assistance (*grey dots*). The black solid line is the initial reflectivity model, computed with the IMD package [11], assuming a multilayer with constant d-spacing

interpretation of XRR scans is not trivial. The XRR of a multilayer with N bilayers can be computed by applying recursively (16.2) including (16.6) to account for the roughness/interdiffusion. However, to fit the reflectance modeling to the experimental dataset it would be necessary to handle $4N$ parameters, namely *all* the thickness and roughness values of *all* layers, assuming at least constant density values throughout the stack.

We show in Fig. 16.4 an example we adopt in the following pages: the experimental XRR scan at 8.05 keV (measured at *INAF/Osservatorio*

Astronomico di Brera) of a W/Si multilayer (30 bilayers). The sample was deposited in 2004 by e-beam evaporation with Ar+ ion etching onto a Si wafer ($\sigma \approx 0.3$ nm) at *Media-Lario technologies*. The deposition was the first test of the ion-etching facility: this test confirmed that the ion beam is effective in reducing the roughness of a W/Si multilayer, as proven by AFM measurements. The XRR scan was measured with a BEDE-D1 X-ray diffractometer with a Cu-anode X-ray tube as source. The Cu Kα X-ray line is filtered by a Si Channel-Cut crystal and collimated by a system of slits, obtaining a thin (70 μm wide, 25 arcsec divergent) and monochromatic ($\Delta E/E \approx 10^{-4}$) X-ray beam. The reflected beam is collected by a photon counter, a scintillator with high linearity.

The measured XRR curve (dots) exhibits a complex structure with broad peaks: the solid curve in Fig. 16.4 is the best fit that could be reached using a constant d-spacing model with $d = 5.3$ nm, $\Gamma = 0.43$, and $\sigma = 0.5$ nm. The density values were assumed to be 18.1 g cm^{-3} for W and 1.8 g cm^{-3} for Si, lower than the natural ones (19.3 g cm^{-3} for W and 2.3 g cm^{-3} for Si), in agreement with previous single-layer calibrations. However, the disagreement is apparent. Therefore, the multilayer has a variation of d-spacing in the stack: it is likely to be ascribed to a variation of the evaporation rate, combined with fluctuations of the etching rate in the ion-etching facility. However, the exact determination by means of manual fits of the trend of thickness of layers is very difficult.

Fitting Algorithms

The problem of extracting the stack parameters from an XRR scan can be solved by means of numerical codes, which are able to explore a very wide range of parameters in order to find the best solution. The first step is the assumption of an appropriate model for the multilayer stack, as defined by a set of independent parameters. For example, one can assume a continuous drift or an irregular variation of the thickness and roughness of the layers. In the first case the free parameters are the coefficients of the function describing the drift. In the second, each thickness value is a free parameter. Alternatively, the model may consist of a drift superimposed onto a fluctuating term.

The second step is the choice of a *figure of merit* (FOM) to measure the closeness of the experimental curve to the computed one using a standard method with variable parameter values. The problem of searching for the best fit is therefore reduced to the minimization of the FOM, and the best solution is the set of values for parameters corresponding to the *global minimum of the FOM*. A possibility for the FOM could be the χ^2 of the measured-simulated data. However, since in a reflectivity *minimum* the X-ray beam probes a much larger depth of the stack, we can recover more information concerning the thickness of the deepest layers in reflection minima than at reflectivity peaks, even though the reflectivity signal is usually very weak. The FOM should then be calculated from the logarithm of the reflectivity in order to make the algorithm sensitive to the XRR features located near the reflection minima. A possible FOM fulfilling this requirement is

$$\mathrm{FOM} = \sum_i (\log R_\mathrm{m}(i) - \log R_\mathrm{c}(i))^2. \tag{16.7}$$

Here $R_\mathrm{m}(i)$ and $R_\mathrm{c}(i)$ are the measured and calculated reflectivity at the ith angular position of the X-ray mirror. Another possibility is

$$\mathrm{FOM} = \sum_i |\log R_\mathrm{m}(i) - \log R_\mathrm{c}(i)|. \tag{16.8}$$

These and other FOMs have been adopted in the literature (see [12] for a detailed discussion).

Several algorithms have been studied in the past years aimed at FOM minimization working on a large number, F, of parameters:

- *Downhill simplex.* Starting from an initial guess for the parameters values, a set of $F+1$ point (the *simplex*) makes a series of moves selecting the values with the smallest value of the FOM [13]. This method converges to the nearest *local* minimum, where it gets trapped. A more global minimum can be found by iterating this procedure from different initial guessed values and selecting the best result. The method is then called *Iterated Simplex*: this method has been utilized for the optimization of X-ray multilayers for the optics of XEUS [14].
- *Levenberg–Marquardt (LM).* Starting from guessed values, the FOM minimum is searched through a combination of inversions of the Jacobian matrix of R_c with respect to the parameter set [15, 16]. This method works better when initial values lie near the global minimum, e.g., the initial values can be computed analytically from the experimental reflectivity curve [17]. The LM can be utilized to refine the calculated parameter values [18].
- *Genetic Algorithms (GA).* This very powerful class of minimization algorithms has been used by several authors (see e.g. [12, 19, 20]) in facing the problem of fitting the reflectivity of multilayers. This method generates a large number of sets of parameter values, called *individuals*, and simulates the evolution of the population through random mutation and exchange of subsets of parameter values between individuals. The population is also subjected to a "Darwinian" selection in that the poorer performing individuals are suppressed. The survivors generate the next generation of individuals. The evolution of the population should lead to the best fit after a sufficiently large number of generations.
- *Downhill Annealing.* This method combines the local minimization of the Downhill Simplex with the capability of the "Simulated Annealing" to escape from local minima [21–23]. The convergence of the Simplex to the nearest local minimum is compared with the thermalization process at a "temperature" T. Each movement of the simplex is compared with an

energetic transition governed by Boltzmann statistics, with the FOM playing the role of the energy. Owing to the tendency of physical systems to reach the minimum energy, the Simplex will preferably move down the FOM gradient. However, transitions that increase the FOM are also statistically possible. Initially T is high, and so is the rate of transitions that increase the FOM. In this phase the program has a high capability of escaping from local minima. When T is slowly decreased (the "annealing"), the likelihood of occurrence of transitions that increase the FOM becomes smaller and smaller, and the system approaches the global minimum. This algorithm has already been implemented in a program developed at ESRF for multilayer stack analysis [24].

PPM (Pythonic Program for Multilayers)

The program we adopted to perform the XRR fit of the reflectivity curve is PPM (*Pythonic Program for Multilayers*), developed by A. Mirone (*ESRF*). PPM can perform detailed fits of XRR curves at one or more photon energies at the same time. PPM takes as input an XML file that describes the modeling of the stack (thickness drift, free variation of each layer, increasing roughness, etc.) with values initially set by the user. The parameters can vary within fixed limits, also set by the user. The comparison of the calculated XRR curve(s) with the measured one(s) is made quantitative by evaluating the FOM, where (16.8) was adopted. PPM then searches recursively the global minimum of the FOM by means of the *Downhill Annealing* algorithm.

The fitting capabilities of PPM were used to perform fits of U/Fe multilayers in order to measure the optical constants of uranium [25]. Moreover, in a previous SPIE volume [26] we utilized PPM to analyze some XRR curves of X-ray multilayers, which yielded very accurate fits and a detailed description of the coatings. In that work we also compared the results obtained from the PPM analysis with TEM images of sections of the same samples, finding a good agreement within the error of TEM. In this work we will apply PPM to another example.

16.3.3 Stack Structure Investigation by Means of PPM

Application of PPM to the XRR Curve of a Multilayer

The analysis with PPM has been applied to the XRR curve at 8.05 keV in Fig. 16.4 in order to derive the internal structure of the stack. PPM was run in a *LINUX* environment with an *AMD Sempron64 3400* processor (2 GHz). Initially, we assumed a second-order drift of W and Si, and the two trends were considered to be independent. The density values were known from previous X-ray measurements on single layer samples obtained with the same deposition facility: 18.1 g cm^{-3} for W and 1.8 g cm^{-3} for Si (see Sect. 16.3.2). The rms

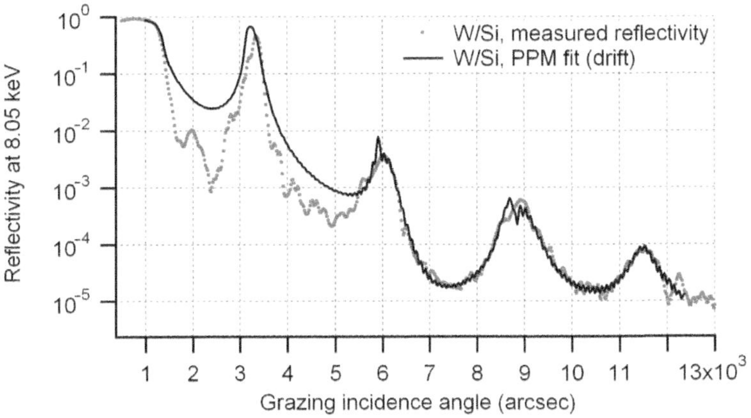

Fig. 16.5. The analysis performed via PPM on the XRR scan of the W/Si multilayer, assuming independent, second-order polynomial drifts of thickness values of W and Si, and a quadratic drift of the roughness

roughness was also assumed to drift throughout the stack with a second-order polynomial trend. No difference between the roughness of the W and Si layers was assumed. PPM was then run on the experimental data after the subtraction of the instrumental noise and the smoothing of apparent noisy features in the reflection minima.

The fit was performed starting from values obtained from the parameters adjusted manually (Fig. 16.4), and the final achieved fit is shown in Fig. 16.5. The shape of the peaks is now fitted better, but not perfectly in particular at the smallest incidence angles. The fitting procedure required just 10 min.

The imperfections in the fit should be ascribed to the assumed continuous drift, which cannot simulate all the irregularities of the d-spacing. Therefore, the fit has been repeated by letting all the layer thicknesses to vary freely within 1 nm, starting again from the values inferred from the manual fit (2.3 nm for W, 3.0 nm for Si). Only the first deposited Si layer was not included in the model because its presence does not noticeably affect the XRR diagram. The roughness is still assumed to have a second degree polynomial drift, and the density values are still the same as used in the previous step. However, because of the huge increase in the number of free parameters, the fitting procedure needed to be restarted several times, with changes in the limits of the allowed values when the fitting value came too close to one of them. The computation required 2 h, but the experimental XRR curve is now fitted accurately (see Fig. 16.6).

The fit results are summarized in Fig. 16.7, where the distribution of thicknesses of W and Si layers is plotted as a function of their bilayer index, numbered from the substrate. The thickness of Si layers oscillates around the 3.0 nm value and that of W around 2.3 nm. Moreover, the rms roughness exhibits an apparent, almost linear *increase* from 0.31 to 0.55 nm, going from

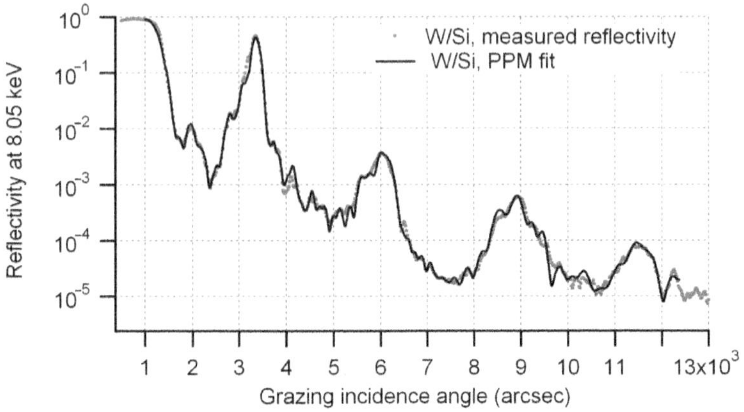

Fig. 16.6. The XRR curve fit with PPM by letting each layer thickness value vary freely

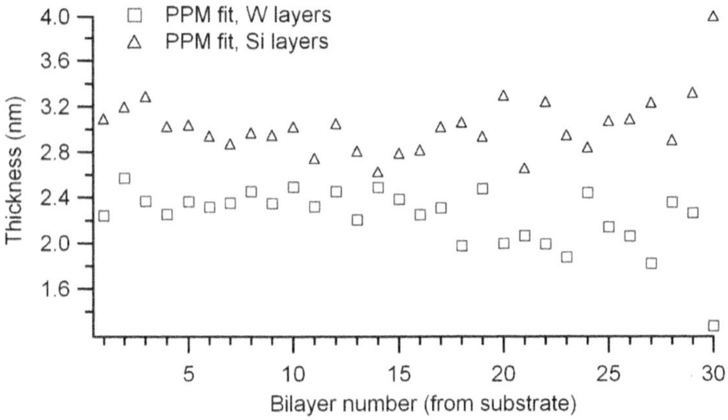

Fig. 16.7. The W/Si multilayer stack structure derived by PPM

the substrate to the multilayer outer surface. This increasing evolution of the roughness is expected classically, even if the use of the ion etching device contributed to restrict the roughness growth.

In the next section we shall compare the analysis results with those of the TEM image (Fig. 16.8).

Interpretation of TEM Images and Comparison with PPM

The results of the fit in Fig. 16.7 can be cross-checked with the results of the sample section taken with TEM (Fig. 16.8). From the image we can see directly the microroughness growth detected by the PPM analysis: undulations with ~25 nm period with increasing amplitude in the growth direction. From a profile of the TEM image the thickness values were extracted, except for the

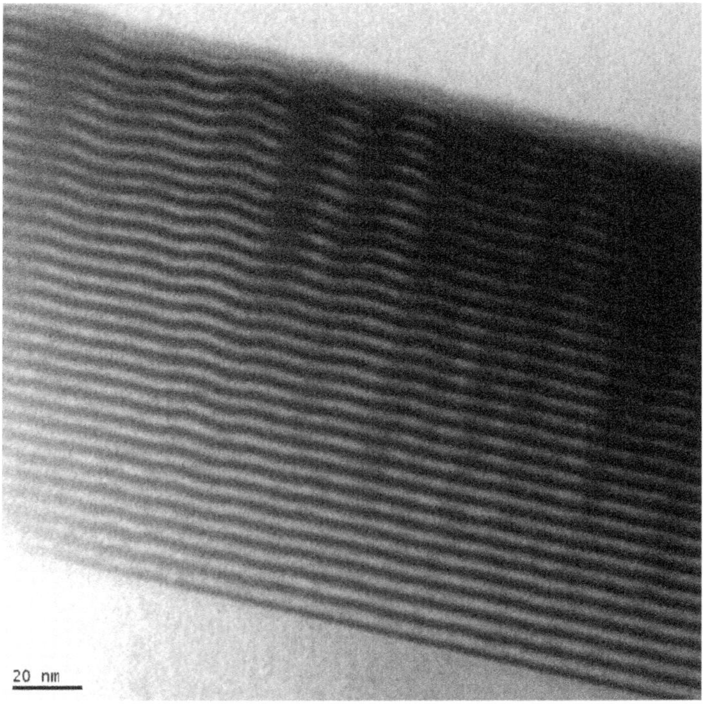

Fig. 16.8. TEM image of the section of the W/Si multilayer sample deposited by e-beam evaporation with ion etching (at *Media-Lario technologies*). The growth direction is from bottom to top (image by *L. Lazzarini and C. Ferrari, IMEM-CNR*)

bilayers 10 and 11, and the comparison of PPM results with those of TEM is shown in Fig. 16.9.

When directly compared, the two distributions of layers as derived using the two methods would be at first glance in disagreement. On the average, the W layer thickness values measured with TEM are larger than the values inferred by PPM by 0.5 nm, whereas the contrary occurs for Si layers.

However, as anticipated in Sect. 16.3.1, the thickness values can be altered by image artifacts: the roughness of the high-Z element tends to obscure the low-Z element layers by superposition of rough profiles on the image plane. Thus, the W layers in Fig. 16.8 appear thicker, and the Si layers appear thinner than they actually are. This is more clearly seen in Fig. 16.2 with the Pt/C multilayer sample, where the obscuring of the C layers increases rapidly with the thickness of the TEM section up to the point of disappearing completely on the right side of the image.

In fact, this artifact depends on the TEM section thickness. When the TEM sample is thin, a small number of profiles overlap in the image plane and the measured thickness values are a good approximation to real ones, i.e., the distance between the average levels of the interfaces. In Fig. 16.2, this

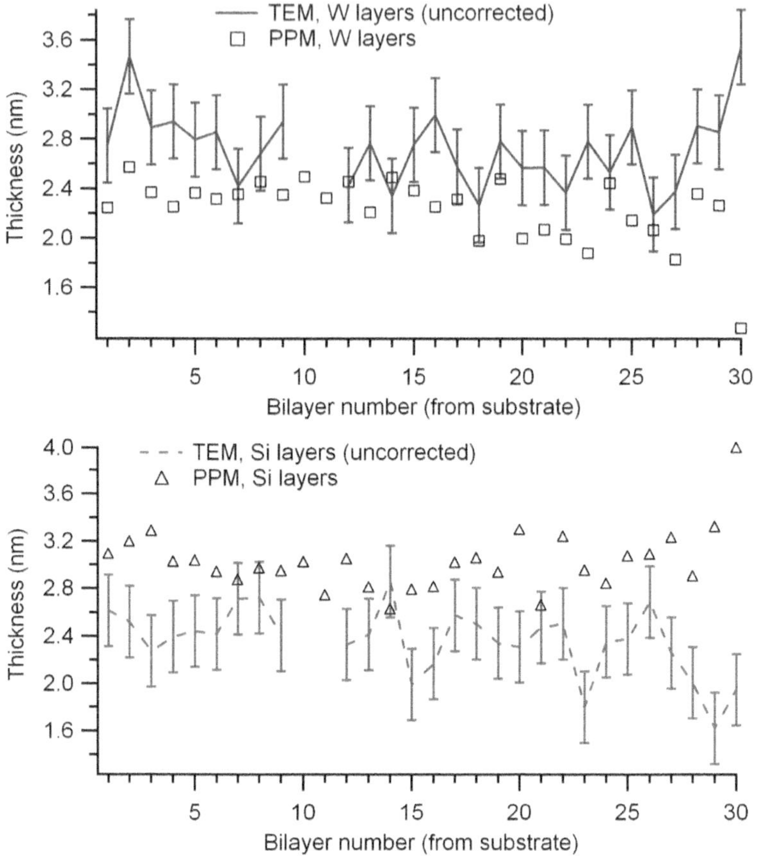

Fig. 16.9. The thickness distribution in W/Si multilayer as computed by PPM (*marks*) compared with TEM findings (*lines*). The error bars are the uncertainties of the TEM measurement

occurs near the edge of the sample, where the section is very thin. Far from the edge, a very large number of rough profiles are integrated with random phases along the line of sight, which is perpendicular to the page. This is schematically depicted in Fig. 16.10. Note that the obscuration equals roughly twice the maximum amplitude of the high-Z layers. Since the X-ray reflection occurs at the average level of the interfaces (the dashed lines in Fig. 16.10), the measured values by PPM will differ from those measured with TEM by the apparent "broadening" of dark layers.

TEM Artifacts Correction: Comparison of d-Spacings

To estimate quantitatively the correction of TEM data, we can suppose the interfaces to be isotropic. Therefore, the roughness PSD $P(f)$ along the

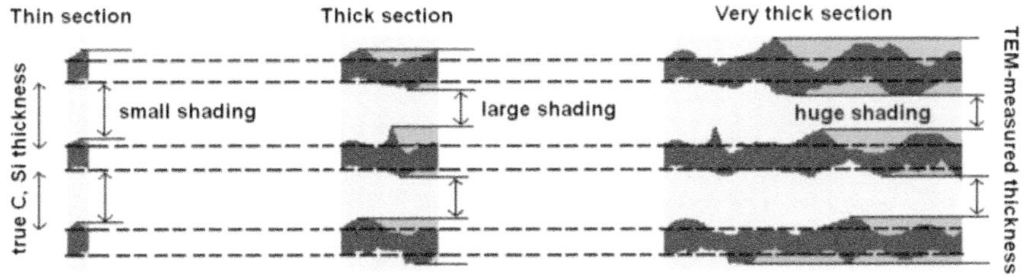

Fig. 16.10. Artifacts in TEM images of multilayers. If the TEM section is thin (*left*) the projection of the high-Z material layers approximately equals the average thickness (delimited by the dashed lines). If the thickness of the section is increased (*center*) the low-Z layers will start being "shaded" by the irregularities of the high-Z layers. Thus, the measured thickness with TEM (*vertical arrows*) will be lower and lower (*right*)

line-of-sight (perpendicular to the page) approximately equals the measured one in the TEM image plane (parallel to the page), and for a given thickness of the TEM section τ we can compute the broadening Δz of dark layers *as twice the peak value of the roughness*:

$$\Delta z \approx 2\sqrt{2}\left(\int_{1/2\tau}^{+\infty} P(f)\mathrm{d}f\right)^{1/2} . \qquad (16.9)$$

In (16.9) the factor $\sqrt{2}$ is the peak to rms ratio. The root of the integral of the PSD is the rms roughness and $1/2\tau$ is the minimum frequency being integrated, in other words, the minimum frequency with a maximum of the oscillation in a length τ. The increase of Δz with τ is partly due to the enlargement of the frequency band and partly due to the rapid increase of $P(f)$ for decreasing frequencies. Furthermore, τ in the TEM section is not constant, but it is larger in the upper part of the image. This makes the evaluation of Δz quite difficult.

Some constraints can be set on the value of τ by noting that we *do* observe profile undulations in the upper part of the TEM image (see Fig. 16.8). Then, under the reasonable assumption that the surface topography is isotropic, τ should be of the order of the average period of the observed oscillations (some 20 nm). The $P(f)$ function was not measured at such a short spatial scale; hence, we cannot compute directly Δz. We can, however, at least state that the root of the integral in (16.9) should be *much less* than the σ roughness value inferred from the XRR analysis, which usually refers to all the spatial periods below some microns. Such a value for σ could be used if the TEM sample were much thicker, i.e., $\tau \approx 1\,\mu\mathrm{m}$, and the correction Δz would amount (on average) to $2\sqrt{2} \times 0.4\,\mathrm{nm} \approx 1.1\,\mathrm{nm}$. Conversely, if the TEM sample is much thinner than $1\,\mu\mathrm{m}$, $\Delta z < 1.1\,\mathrm{nm}$.

In fact, if we assume $\Delta z = 0.5\,\mathrm{nm}$, the main discrepancy between TEM data and PPM findings is eliminated. To account for the projection effect

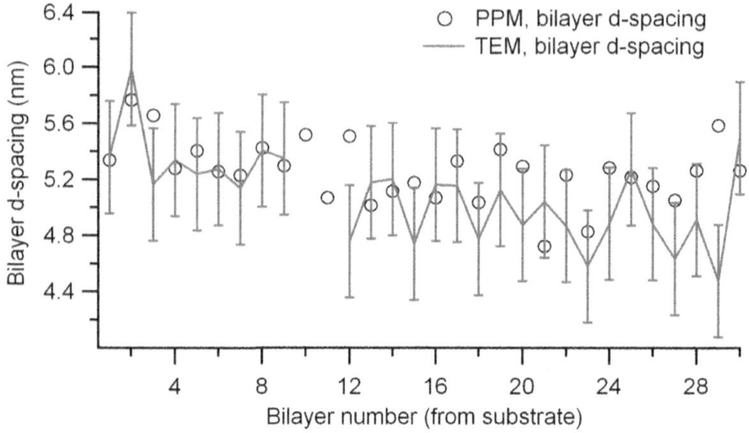

Fig. 16.11. Comparison of d-spacings as extracted from the XRR curve (*marks*) and TEM (*lines + bars*). The agreement is satisfactory

mentioned, this amount has to be subtracted from W thickness values as measured from TEM and to be added to the Si layers. The residual discrepancies can be due to errors in localization of the interfaces in the TEM image. The evolution of the roughness and the increase of τ can be the cause of the larger discrepancy in the last bilayers.

A confirmation of the assumed interpretation comes from the comparison of bilayer d-spacings obtained by XRR and TEM. They should not be affected by the projection effect in TEM image because the correction for W and Si have opposite signs and cancel out. The d-spacings comparison is shown in Fig. 16.11. Almost all the d-spacings, as measured with PPM, are in agreement with TEM to within the TEM error bars (0.5 nm). This is a confirmation of the correctness of the analysis performed with PPM.

16.3.4 Fitting a Multilayer with Several Free Parameters

We also applied PPM to fit the XRR curve of a graded multilayer deposited by DC magnetron sputtering at the *Harvard-Smithsonian Center for Astrophysics* (Boston, USA), using a deposition facility suitable for the production of multilayer coated hard and soft X-ray mirror shells [27]. The substrate used was a superpolished fused silica sample ($\sigma \sim 0.1$ nm). The multilayer is described by two power-law distributions of d-spacings: the outermost 20 bilayers are thicker and reflect soft X-rays, the innermost 75 bilayers are thinner and reflect the hardest X-rays.

The reflectivity of the sample was measured using the BEDE-D1 diffractometer at *INAF/Osservatorio Astronomico di Brera* up to 15,000 arcsec grazing incidence at the photon energy of 8.05 keV. In this case, because of the very large number of bilayers, we could not take all the thickness value as free parameters. To get around this, we initially modeled the thickness trends

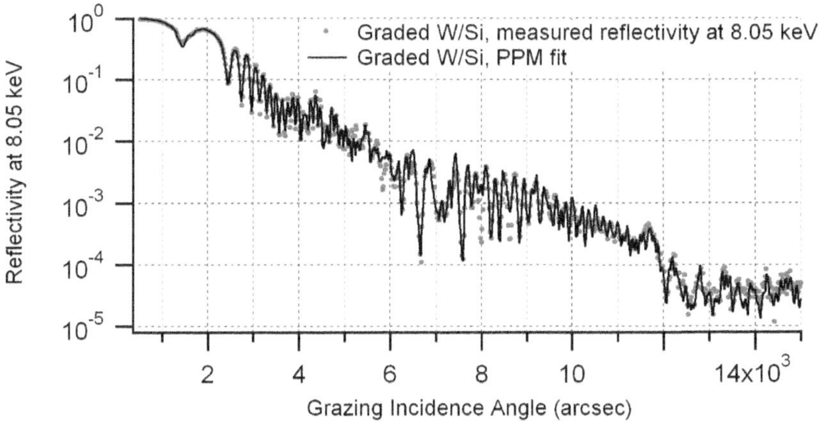

Fig. 16.12. PPM fit of a graded W/Si multilayer deposited at the *Harvard-Smithsonian Center for Astrophysics*. Measured data (*dots*) and PPM fit (*line*)

following two power laws, assuming as free parameters a, b, c, as in (16.4). To account for a drift of the Γ factor (see Sect. 16.2) throughout the stack, we assumed that the trend of the parameters for W and Si trend were independent. The roughness was free to drift according to second-order polynomial trends, one for each stack. The search for the best fit with PPM enabled us to determine the parameters a, b, c, for the two stacks.

In a second step, to refine the fit, the layers thickness values were treated as free variables, starting with the values found in the previous step and allowing small variations in them (± 0.3 nm). The very detailed fit results are shown in Fig. 16.12.

16.4 Conclusions

The methodology described here highlights the potential of the XRR scan analysis for investigations of the internal structure of nearly periodic and graded multilayers. The problem of multi-parametric XRR curve fitting can be solved by means of computer codes based on several algorithms. In particular, PPM has proven to be very effective in fitting structured XRR curves, and the structure inferred is in good agreement with TEM results after the correction of TEM artifacts.

XRR analysis with PPM yields a reliable description of the multilayer structure, on condition that the experimental curves are fitted very accurately. To do this, the fitting strategies can be summarized as follows:

1. For multilayers with 30 bilayers or less, assume each layer thickness value to be a free variable. Then let them vary within a wide variability range ($1 \div 2$ nm) in order to explore a *wide* parameter space region. If the multilayer

has a nearly constant d-spacing, assume a constant thickness throughout the stack as an initial guess.
2. For multilayers with more than 30 bilayers, start the fit by first assuming a continuous thickness drift in the stack. Then, let all layers vary freely within *small* (0.4 nm) limits around the values found in the previous step of the fit procedure. The fit can be restarted several times, adjusting the limits each time, until a satisfactory fit is reached.
3. If the experimental curve exhibits high reflectivity peaks, a preliminary computation can be done, assigning weights to data proportional to the reflectance values in order to approximately fit the primary reflectance peaks. The parameters should then be refined in a successive PPM run without weights.
4. If the actual density values are uncertain, they can be assumed as fit variables within small limits.
5. The roughness values can be considered as constant throughout the stack only for multilayers with less than 20 bilayers, otherwise a drift of the roughness should be included.
6. Possible angular offsets in the experimental curves, instrumental noise, and the angular resolution of the measurement should be accurately evaluated and included in the calculations.

Finally, when comparing XRR analysis and TEM results, correct the thickness values obtained from TEM according to (16.9). These fitting methodologies were tested on several multilayer samples [26] in addition to the example provided in the present work.

Further tests are foreseen in order to establish the reliability of PPM as a diagnostic tool for multilayers. If confirmed, the systematic use of the XRR scan analysis with PPM will be an important diagnostic tool in the development of multilayer mirrors for several applications, such as the wideband reflective coating of future soft and hard X-ray telescopes.

Acknowledgments

The author gratefully acknowledges A. Mirone, C. Ferrero, M. Sanchez del Rio (*European Synchrotron Radiation Facility*, Grenoble, France) for the development of the program PPM, L. Lazzarini and C. Ferrari (*IMEM-CNR*, Parma, Italy) for the TEM analysis of the samples and G. Nocerino (*Media-Lario technologies*, Bosisio Parini, Italy) for the support to this work. The valuable collaboration of G. Valsecchi, G. Grisoni, M. Cassanelli (*Media-Lario technologies*) is kindly acknowledged.

The author also thanks G. Pareschi, D. Vernani, V. Cotroneo, R. Canestrari (*INAF/Osservatorio Astronomico di Brera*) for support and useful suggestions and S. Romaine, P. Gorenstein, R. Bruni (*Harvard-Smithsonian CfA*, Boston, USA) for providing us with the graded multilayer sample referred to in Fig. 16.12. The graded multilayer reported in Fig. 16.1 was designed by V. Cotroneo (*INAF/Osservatorio Astronomico di Brera*).

The author is indebted with *MIUR* (the Italian Ministry for Universities) for the COFIN grant that provided support for the development of multilayer coatings for X-ray telescopes, and the *European Science Foundation – COST cooperation, action P7* for the grant that made the PPM training at the ESRF possible.

PPM is open-source software; it can be freely downloaded from ftp://www.esrf.fr/pub/scisoft/ESRF_sw/linux_i386_03/.

For PPM installation support, please contact the author of the present chapter.

References

1. G. Pareschi, P. Ferrando, Exp. Astron. **20**, 139 (2006)
2. P. Gorenstein, A. Ivan, R.J. Bruni, S.E. Romaine, F. Mazzoleni, G. Pareschi, M. Ghigo, O. Citterio, SPIE Proc. **4138**, 10 (2000)
3. A. Parmar, M. Arnaud, X. Barcons, J.M. Bleeker, G. Hasinger, H. Inoue, P. Palumbo, M.J. Turner, SPIE Proc. **5488**, 388 (2004)
4. L.G. Parrat, Phys. Rev. **95**(2), 359 (1954)
5. K.D. Joensen, P. Voutov, A. Szentgyorgyi, J. Roll, P. Gorenstein, P. Høghøj, F.E. Christensen, Appl. Opt. **34**(34), 7934 (1995)
6. G. Pareschi, V. Cotroneo, D. Spiga, M. Barbera, M.A. Artale, A. Collura, S. Varisco, G. Grisoni, G. Valsecchi, SPIE Proc. **5488**, 481 (2004)
7. Y. Tawara, K. Yamashita, H. Kunieda, K. Tamura, A. Furuzawa, K. Haga, N. Nakajo, T. Okajima, H. Takata, et al., SPIE Proc. **3444**, 569 (1998)
8. L. Névot, P. Croce, Revue Phys. Appl. **15**, 761 (1980)
9. G. Pareschi, O. Citterio, S. Basso, M. Ghigo, F. Mazzoleni, D. Spiga, W. Burkert, M. Freyberg, G.D. Hartner, G. Conti, E. Mattaini, G. Grisoni, G. Valsecchi, B. Negri, G. Parodi, S. Marzorati, P. Dell'Acqua, SPIE Proc. **5900**, 47 (2005)
10. D.G. Stearns, Appl. Opt. Lett. **62**(1515), 1745 (1993)
11. D.L. Windt, Comput. Phys. **12**, 360 (1998)
12. M. Wormington, C. Panaccione, K.M. Matney, D.K. Bowen, Phil. Trans. R. Soc. London **357**, 2827 (1999)
13. J.A. Nelder, R. Mead, Comput. J. **7**(4), 308 (1965)
14. V. Cotroneo, G. Pareschi, SPIE Proc. **5536**, 49 (2004)
15. K. Levenberg, Quart. Appl. Math. **2**, 164 (1944)
16. D. Marquardt, SIAM J. Appl. Math. **11**, 431 (1963)
17. I.V. Kozhevnikov, I.N. Bukreeva, E. Ziegler, NIM-A **460**, 424 (2001)
18. E. Ziegler, C. Morawe, I.V. Kozhevnikov, T. Bigault, C. Ferrero, A. Tallandier, SPIE Proc. **4782**, 169 (2002)
19. M. Sanchez del Rio, G. Pareschi, SPIE Proc. **4145**, 88 (2001)
20. D. Šimek, D. Rafaja, J. Kub, Mat. Struct. Krystalografická společnost **8**(1), 16 (2001)
21. N.A. Metropolis, A.W. Rosenbluth, M.N. Rosenbluth, A. Teller, E. Teller, J. Chem. Phys. **21**(6), 1087 (1953)
22. V. Cerny, in *Comenius University Report* (Bratislava, Slovakia, 1982)
23. S. Kirkpatrick, C.D. Gelatt, M.P. Vecchi, Science **220**, 671 (1983)

24. E. Ziegler, C. Ferrero, F. Lamy, C. Chapron, C. Morawe, CPDS 2002 **45**, 45 (2002)
25. S.D. Brown, L. Bouchenoire, A. Mirone, V. Cotroneo, in *Progress Report ESRF* (Grenoble, France, 2004)
26. D. Spiga, A. Mirone, G. Pareschi, R. Canestrari, V. Cotroneo, C. Ferrari, C. Ferrero, L. Lazzarini, D. Vernani, SPIE Proc. **6266**, 346 (2006)
27. S. Romaine, S. Basso, R. Bruni, W. Burkert, O. Citterio, G. Conti, D. Engelhaupt, M. Freyberg, M. Ghigo, P. Gorenstein, M. Gubarev, G.D. Hartner, F. Mazzoleni, S. O'Dell, G. Pareschi, B. Ramsey, C. Speegle, D. Spiga, SPIE Proc. **5900**, 225 (2005)

Part III

Refection/Refraction Optics

17

Hard X-Ray Microoptics

A. Snigirev and I. Snigireva

Abstract. This chapter presents a summary of micro-focusing optics and methods for X-rays in the energy range 4–100 keV, as provided by synchrotron radiation sources. The advent of third generation storage rings such as the ESRF, the APS and Spring-8 with X-ray beams of high brilliance, low divergence and high coherence has made possible efficient X-ray focusing and imaging. The main emphasis is on those methods which aim to produce submicrometre and nanometre spatial resolutions in imaging applications. These methods fall into three broad categories: reflective, refractive and diffractive optics. The basic principles and recent achievements are discussed for optical devices in each of these categories.

17.1 Introduction

A summary of microfocusing optics and methods for hard X-rays is presented. The hard X-ray region is taken as extending from about several keV (∼4 keV) to gamma rays with several hundreds keV (∼100 keV) provided by synchrotron radiation sources. The advent of third generation storage rings like ESRF, APS, and SPring-8 with radiation beams of high brilliance, low divergence, and high coherence makes possible efficient X-ray focusing and imaging. X-ray microscopy techniques are presented first. The main emphasis will be put on those methods that aim to produce nanometer resolution. These methods fall into three broad categories: reflective, refractive, and diffractive optics. The basic principles and recent achievements will be discussed for all optical devices. The report covers the latest status of reflective optics, including mirrors and multilayers, capillaries and waveguides. Special attention will be given for successful development of Kirkpatrick–Baez (KB) systems providing nanometer focusing in two dimensions. The basic principles and the state of the art of diffractive optics such as Fresnel zone plates are reviewed. Improvement of the spatial resolution without loss of efficiency is difficult and incremental due to the fabrication challenges posed by the combination of small outermost zone width and high aspect ratios. Particular attention will be given to recent invention of refractive optics. Refractive optics is a rapidly emerging option for focusing high energy synchrotron radiation from micrometer to nanometer dimensions. These devices are simple to align, offer a good working distance between the optics and the sample, and are expected to become standard elements in synchrotron beamlines instrumentation in general and in high energy X-ray microscopy in particular.

17.2 X-Ray Microscopy

The history of X-ray microscopy goes back to 1896, the year following the discovery of X-rays by Roentgen. The method used to study the structural details of biological objects by enlargement of X-ray radiographs was called by P. Goby as microradiography in 1913 [1]. Beginning in the late 1940s, X-ray microscopy with grazing incidence mirror optics was proposed by P. Kirkpatrick in order to surpass the optical microscope in resolution [2].

As a branch of earlier developments in electron microscopy, projection microscopy was proposed by Cosslett and Nixon [3] and it became very popular since the 1950s. In the early 1970s, several groups started new technological developments of X-ray optics, in particular, Fresnel zone plates, and the modern era of X-ray microscopy started. In 1974, Schmahl and collaborators built a full-field transmission microscope at DESY (Deutsches Elektronen Synchrotron) in Germany [4]. Kirz and Rarback at NSLS (National Synchrotron Light Source) at Brookhaven National Laboratory in USA built the first scanning transmission microscope using a zone plate objective in 1982 [5]. Traditionally, this type of X-ray microscopy deals with rather soft X-ray energies (100–2,000 eV), in particular, in the so-called water window region between the K-shell X-ray absorption edges of carbon and oxygen at 4.4 and 2.3 nm, where organic materials show strong absorption and phase contrast while water is relatively nonabsorbing. This enables imaging of specimens up to ~10 μm thickness, with high intrinsic contrast using X-rays with a lateral resolution down to 15 nm [6].

In recent years, considerable progress has been made in X-ray microscopy in the hard X-ray regime ($E > 4$ keV), as a result of the development of high brilliance, high energy X-ray sources coupled with advances in manufacturing technologies of focusing optics. One of the key strengths of hard X-ray microscopy is the large penetration depth of hard X-rays into the matter around 1 mm, allowing one to probe the inner structure of an object without the need for destructive sample preparation. Resolution of the order of 100 nm was reached with photon energies up to 30 keV.

Lens-based X-ray microscopy can be divided into two classes: full-field microscopy and scanning microscopy (Fig. 17.1). The full-field transmission X-ray microscope (TXRM) uses the same optical arrangement as conventional light and transmission electron microscopes. Such types of microscopes use optical elements like Fresnel zone plates or refractive optics as objective

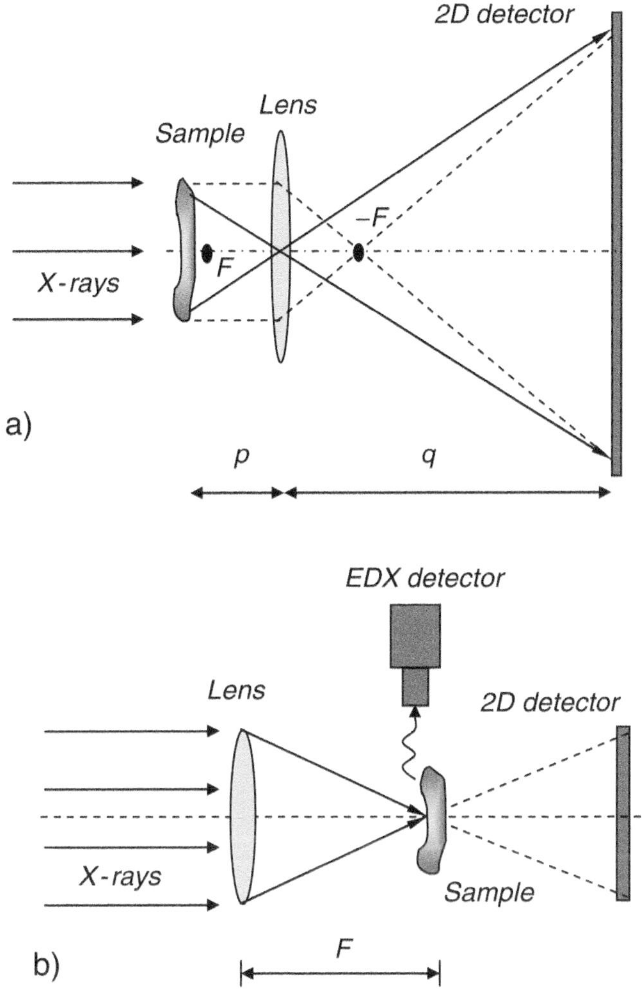

Fig. 17.1. Optical schematic of (**a**) transmission and (**b**) scanning-transmission X-ray microscopes

lenses for high-resolution imaging. Scanning microscopes usually use a focusing optics to form a finely focused spot or microprobe through which the specimen is rastered. The microscope of choice is generally determined by the specimen and observation to be made on it. Table 17.1 summarizes different types of X-ray microscopes available nowadays [7–11].

The full-field transmission microscope is illustrated schematically in Fig. 17.1a. The basic approach is to use a high quality imaging optics as a microscope objective to create a magnified image of the object. When the object is placed slightly outside the focal distance (p is slightly larger than F), then a strongly magnified image is generated at a distance $q = pF/(p - F)$

Table 17.1. Comparison of the features of different types of X-ray microscopy

Type of microscopy			Source	Energy	Mode of contrast	Optics	Resolution (nm)
Full field	Micro-radiography	Contact	SR, Lab	>0.5 keV	Absorption	–	100–1,000
		In-line imaging	SR	>1 keV	Absorption/phase XANES/EXAFS	–	500–1,000
		Magnified projection (no optics)	Lab	>1 keV	Absorption/phase	–	100–500
	Soft X-ray microscopy		SR, Lab	<4 keV	Amplitude/phase XANES	FZP	20–50
	Hard X-ray microscopy		SR, Lab	>4 keV	Amplitude/phase XANES/EXAFS possible	FZP, CRL	100–500
Scanning	Soft X-ray scanning microscopy		SR	<4 keV	Absorption, fluorescence, XANES/EXAFS photoemission	FZP	20–50
	Hard X-ray microprobe		SR, Lab	>4 keV	Absorption, fluorescence, XANES/EXAFS, diffraction	FZP, CRL, KB	100–1,000

Abbreviation: SR synchrotron radiation sources, Lab laboratory based sources, XANES X-ray absorption near edge spectroscopy, EXAFS Extended X-ray absorption fine spectroscopy

with magnification $M = q/p = F/(p-F)$. This technique offers the possibility of dynamic imaging and is well suited for phase contrast imaging.

The basic soft X-ray microscope approach is to use a condenser zone plate to monochromatize and to concentrate X-rays onto the sample. The micro-objective zone plate forms a magnified real image of the sample at roughly 1,000$\times$ magnification. The enlarged image is recorded on a backside-thinned CCD camera. The sample is mounted in an environmental chamber where it can be prepared between two thin foils in its natural state, including when wet. The object chamber is under atmospheric pressure, while the major parts of the X-ray optical path are in vacuum. Image exposure times range from under a second with dry specimens to a few seconds for wet specimens, depending on the zone plate used. Phase contrast in TXRM as a full analogue of Zernike phase contrast in visible light microscopy has been realized. A removal ring aperture is placed near the condenser and a phase-ring is permanently mounted in the conjugate plane. The improved contrast allowed for reduced illumination and therefore reduced radiation dose to the specimen. A cryogenic object chamber has been developed and implemented at the TXRM. The cryogenic method allows imaging of chemically unfixed samples with resolution 20–30 nm. Furthermore, the high stability of frozen-hydrated samples allows taking multiple images for tomography.

For hard X-rays, two types of optics are available today with sufficient quality to be used in the development of full-field imaging: Fresnel zone plates (FZPs) and compound refractive lenses (CRLs). Both have in common that their focal length for X-rays with photon energies around 10 keV and higher is in the order of meters, and that their apertures are limited to several hundreds of micrometers. The length of the setup will increase as M increasingly differs from unity. The length of the entire microscope can be of the order of 10–25 m. Another implication of a long focal length is a very small numerical aperture, which limits the resolution of the imaging setup. The resolution of about 100–300 nm is routine now.

X-ray microscopy in the scanning mode is illustrated in Fig. 17.1b. The basic idea in scanning microscopy is to form a microprobe across which the specimen is mechanically scanned. A proportional counter is used to detect the transmitted X-rays and the image is built up pixel (picture element) by pixel. The focused X-ray probe can also be used to excite other processes such as photoelectrons and fluorescent X-rays. The spatial resolution is limited by the focusing optics. Scanning transmission X-ray microscopes (STXRM) for soft X-rays use Fresnel zone plates as high resolution objectives to form a focused spot. Scanning microscopes require coherent illumination and must be used with high brightness sources such as undulators at synchrotron storage rings. Scanning microscopes generally have exposure times of minutes; each pixel can be formed in a few milliseconds using synchrotron radiation. Such microscopes impact 5–10 times less radiation dose onto the specimen and have an energy resolution of 0.1 eV. Scanning transmission X-ray microscopes are better suited to spectromicroscopy. Two main types of scanning microscopes exist today: STXRMs equipped with photon detectors and scanning photoemission microscopes equipped with electron detectors.

In the case of hard X-rays a microfocus is generated by KB mirrors, a Fresnel zone plate, or a parabolic CRL. The high-energy microfocusing optics benefits from longer focal lengths and a larger depth of field, up to a few millimeters, that is advantageous for the use of specific sample environment like furnaces or high-pressure cells. Shorter wavelengths are favorable for diffraction studies including wide and small angle scattering. In a microprobe, the strategy is to scan the beam over the sample and to measure a signal in diffraction, in fluorescence, or in absorption (XANES, EXAFS) for each beam position. When combining scanning microscopy with tomographic techniques the inner structure of a sample can be reconstructed, including the distribution of different atomic species and even of the valences of atoms.

17.3 X-Ray Optics

The small source size and low divergence of third generation synchrotron radiation sources gave rise to tremendous advances in the development of different types of microfocusing optics based on reflection, diffraction, and refraction phenomena. All available microfocusing devices for hard X rays are presented in the Table 17.2. The best resolution for the optical systems presented exceeds the 100 nm limit. From their principles of operation these optical systems are subdivided into the three broad groups.

17.3.1 Reflective Optics

Kirkpatrick–Baez Systems

Conventional mirrors, as used for visible wavelength, at normal incidence cannot be used, because the reflectivity is too low and typically less then one hundred thousand X-ray photons will be reflected. A high reflectivity can, however, be obtained at grazing incidence angles $\left(\theta < \sqrt{\delta}\right)$. To reduce the astigmatism, Kirkpatrick and Baez proposed the use of two spherical or cylindrical mirrors in a crossed configuration [2]. Figure 17.2 shows a geometrical arrangement of such so-called Kirkpatrick–Baez (KB) system. To enhance the reflectivity, multilayer mirrors, where the refractive index varies periodically with depth, can be used. To build such mirrors, alternative layers with a thickness $\lambda/4$ are evaporated or spattered onto a blank one with a relatively high refractive index and the other with a relatively low refractive index. The weak radiation reflected at the interfaces of the multiplayer is superimposed coherently and in phase, and can give a considerably increased reflectivity. The reflecting focusing systems can be either static, with mirrors polished according to the proper figure optimized for a given incidence angle and focus, or

Table 17.2. Microfocusing optics for hard X-rays

	Reflective					Diffractive	Refractive
	Kirkpatrick–Baez systems		Capillaries		Waveguides	Fresnel zone plates	Refractive lenses
	Mirrors	Multilayers	Multi-bounce	One-bounce	Feng et al., 1993	Baez, 1952	Snigirev et al., 1996
	Kirkpatrik and Baez, 1948	Undewood and Barbee, 1986	Kreger, 1948	Balaic, 1995			
	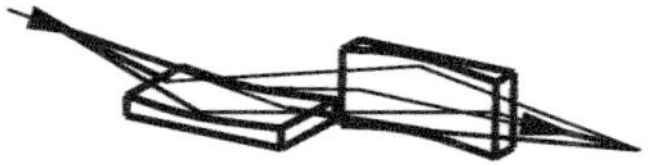				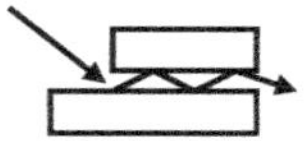	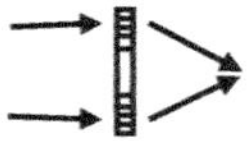	
E	<20 keV	<80 keV	<20 keV		<20 keV	<30 keV	<1 MeV
$\Delta E/E$	White beam	10^{-2}	White beam		10^{-3}	10^{-3}	10^{-3}
Δ	36 nm [12]	45 nm [13]	50 nm [14]	250 nm [15]	$40 \times 25\,nm^2$ [16]	30 nm [17]	50 nm [18]

E, energy range; $\Delta E/E$, energy bandwidth; Δ, best measured resolution

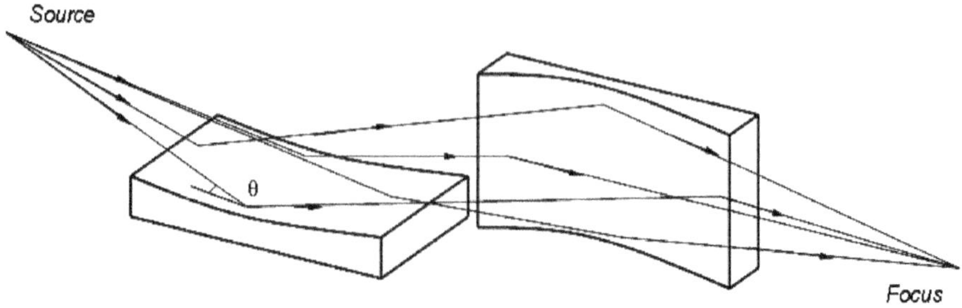

Fig. 17.2. Schematic view of the two-mirror Kirkpatrick–Baez system. θ is the mirror incidence angle

dynamic, with actuators bending flat mirrors into the elliptic shapes required by the experiment. Today, as a result of improved techniques for developing highly finished and perfect surfaces, ellipsoidal and paraboloidal mirrors can be manufactured to a high degree of perfection. Focusing down to 100 nm spot size is achievable with KB systems.

The reflecting multilayer supermirrors are able to focus X-rays over large energy range, whereas Bragg reflecting multilayer mirrors focus only a small energy bandwidth (~1–2%). As, within a given energy interval, the optical properties for total reflecting optics are independent of the X-ray wavelength, the focal spot is retained while tuning the energy. Therefore, experiments requiring energy tuning, like spectroscopy, can be performed without any readjustment of the optics. However, the disadvantage of grazing incidence optic is its inherent low acceptance and it is mostly used for scanning microscopes.

KB systems at the ESRF are based on dynamical bending systems where an initially flat, superpolished plate is used [13,19–21]. The obvious advantage of dynamical bending is that it permits one to tune the focusing conditions. Different mechanical benders are available at the ESRF. One can easily choose the most suitable system according to the experimental requirements, such as energy and energy range, the focusing parameters such as magnification and focusing distance. In addition, optimal substrate coatings, i.e., single layer mirrors or multilayers, can be used. The typical geometrical characteristics of the ESRF KB systems are focal distances ranging from 0.1 to 3 m and circular or elliptical bending radii varying from 20 to 1,000 m. Ideally KB systems require easy alignment procedure and they must preserve coherence or the wavefront of the reflected beam. Newly developed surface preparation techniques and a computer control alignment procedure allow a spot size of about 100 × 100 nm to be achieved. It is of great importance that the bending radii of the mirrors remain constant during several hours and even days of experimental measurements. Hence, the mirror mounting technique is a crucial point. Furthermore, vibrations should not deteriorate the spot by more than the mirror slope errors. At present, a typical value of the mirror local slope error corresponds to less than 0.3–0.4 μrad after polishing. To preserve high reflectivity, a good level of vacuum is required to avoid surface contamination. Normally, a KB prealignment operation is employed, consisting of a laser self-alignment mirror setup inserted temporarily in front of the KB system. Final alignment is done with X-ray beam.

Most ESRF beamlines are equipped with in-house developed KB systems. More than 50 mirrors have been built and installed for microfocusing applications. Some of them are mounted in pairs to make up KB systems with different dynamical technologies. Many substrates are multilayer-coated according to energy, energy tunability, acceptance, and reflectivity requirements. The developed technology allows to achieve resolution below 100 nm, and 45 nm was measured recently at ID19 using 24 keV [13].

Although KB mirrors with benders enable the focal length to be varied, the benders are bulky and the stability achieved is not sufficient to support long-time scanning experiments on nanofocusing systems. The efficiency for microfocusing is affected by dynamic stability. In their actual usage of X-ray KB optics, the ease of setup is a very important factor. As was recognized at the other facilities, the procedures involved in mirror bending to form an exact ellipse from a flat surface are so complicated that a long time is needed to set up and optimize the system.

For these reasons static KB systems were chosen at the SPring-8 [12,22–28] and APS [29–35] facilities. For KB mirror fabrication at SPring-8, plasma chemical vaporization machining (PCVM) is employed for rough figuring, and elastic emission machining (EEM) is used for final figuring and surface super smoothing. These methods do not cause mechanical damage to the processed surfaces because their removal mechanism is based on chemical reactions. By combining these two methods, surfaces of unprecedented accuracy can be created with 0.1 nm controllability and 0.3 mm spatial resolution. To control the surface accuracy, the combined technique of microstitching interferometry and relative-angle-determinable stitching interferometry (RADSI) is used. This technique makes possible the measurement of elliptical surface profiles with a spatial resolution of 0.03 mm and measurement height accuracy close to 1 nm. Typically, after the final figuring the surfaces were coated by a 50 nm Pt layer using an electron evaporation method. The KB system was tested at the 1-km-long beamline BL29XUL at SPring-8 using a 15 keV monochromatic beam. A focal spot of 36 nm (horizontal) and 48 nm (vertical) was measured [12]. This proposed KB device employing precisely figured mirrors is promising, not only on the basis of focusing performances, but also as a user-friendly instrumentation.

Elliptical surfaces can be obtained by differentially coating a flat or a spherical mirror with a continuously variable amount of gold. This idea was proposed more than 60 years ago and was recently realized at the APS [29–35]. Profile coatings have been applied on both cylindrical and flat Si substrates to make the desired elliptical shape. In a profile-coating process, the sputter source power is kept constant, while the substrate is passed over a contoured mask at a constant speed to obtain the desired profile along the direction perpendicular to the substrate-moving direction. A KB mirror pair, made using Au as the coating material, was recently tested at the APS, and X-ray results showed a focused spot size of $80 \times 70\,\text{nm}^2$ [35].

Capillaries

Similar to reflective mirrors, monocapillary optics rely on total external reflection of the X-rays from the internal surface of the tube to transport X-rays. Typical glass materials that have been used to fabricate capillary optics are borosilicate, lead based, and silica glasses. For hard X-rays the typical values for critical angles are 0.1–10 mrad. The simplest form of a capillary is straight (cylindrical) glass tube; these have first been used in the 1950s to guide X-rays from the source point to a distant sample. This leads to an effective reduction of the source sample distance and thus to an increase in flux on the sample. The next step was to use tapered capillaries not only to guide but moreover to squeeze the X-rays to a very small spot [36, 37] (see Fig. 17.3a). The first experiments were performed with conventional X-ray tubes. In the late 1980s, conical tapered capillaries were tested at synchrotron facilities [14, 36–38].

Tapered capillaries were shown to have nanofocusing capabilities more than 10 years ago [14]. However, no appreciable progress of using such simple devices for nanobeams has been made since that time. For multibounce capillaries, the problems are well understood and consist of significant losses in the multireflection process and almost zero-working distance, which substantially limits practical applications of capillaries in X-ray microscopy. On the other hand, proposed single-bounce capillaries [39] show a great potential in developing nanofocusing devices (Fig. 17.3b). Parabolic and elliptical capillaries have a large focal distance and very high reflectivity [40–43]. In a typical process the predetermined capillary profile is achieved by "pulling" an originally straight uniform glass tube with accurate control of mechanical movements and heating parameters. Unfortunately, because of some unavoidable factors, i.e., surface slope errors of the original glass tube and nonuniform mechanical movement, it is difficult to reliably achieve elliptical or parabolic shapes with the desired low-figure errors. State-of-the-art technology allows one to produce a capillary with 70 μrad slope errors [41]. Obviously such slope errors limit the resolution of the optics to a size of about 10 μm.

To overcome these problems, the use of a small elliptical capillary made by a stationary pulling technique has been proposed, where the elliptical shape is

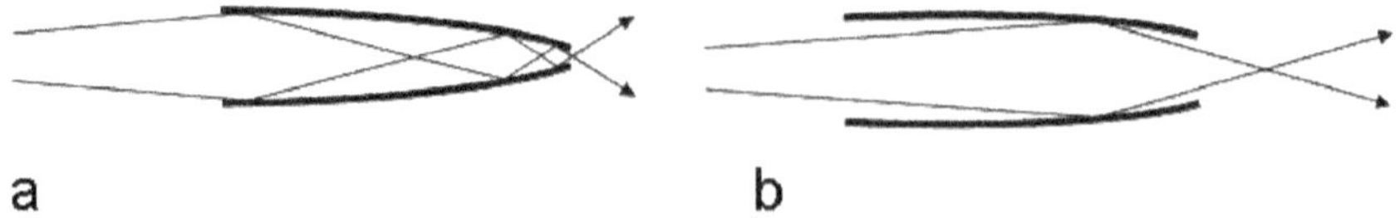

Fig. 17.3. Tapered multibounce (**a**) and single bounce (**b**) capillaries

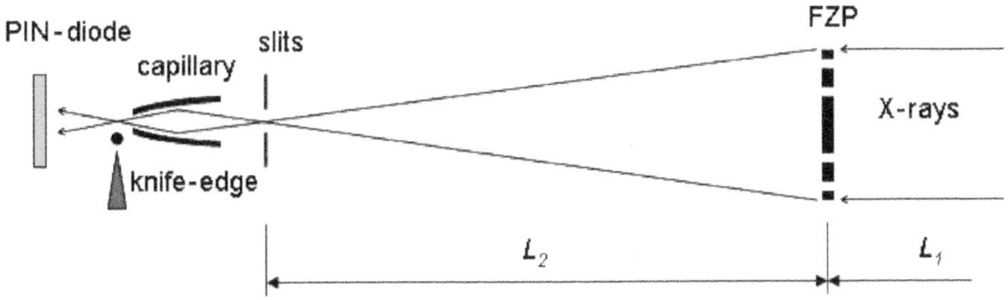

Fig. 17.4. Two-step focusing based on a Fresnel zone plate with a capillary

achieved by stretching an air bubble inside a glass fiber with constant velocity [44]. It appears that this approach might show a better quality surface as compared to the computer-controlled glass tube pulling process with a variable speed. For best performance of the ellipsoidal capillary a Fresnel zone plate was used to generate a secondary source at the first ellipse focus [15] (Fig. 17.4). The FZP serves as a first microfocusing element to produce a demagnified micrometer image of the source and then the elliptical capillary makes a final compression of the beam down to 250–500 nm. In general, the two-step focusing setup provides three important benefits: it significantly improves the flux by increasing the overall acceptance of the optical system; it makes possible optimal aberration-free focusing by the elliptical capillary surface; and it considerably minimizes the influence of slope errors. For practical applications, off-axis illumination of the capillary by a small prefocused beam eliminates the beam transmitted through the exit aperture and makes for an easy implementation of a beamstop. By accurate scanning of the tiny beam along the capillary surface one can control the demagnification factor and zoom the focal spot size.

The proposed approach allows one to shorten a small capillary further down to 5–10 mm, making it easy to align and operate. Use of short capillaries might open the possibility of coating the inside of capillaries with smooth films of the desirable materials such as platinum or gold. In this connection metal capillary optics look very attractive [45]. Finally, the single-bounce ellipsoidal capillaries are very attractive for X-ray nanobeam techniques because they are so simple and can potentially be reproduced inexpensively.

Waveguides

An X-ray planar waveguide is a thin film resonator in which a low absorbing material is enclosed between two metal layers with a smaller refractive index [46]. For particular grazing incidence angles, a resonator effect takes place inside the resonator film. Schematic representation of a thin-film waveguide is shown in Fig. 17.5. The beam is compressed in one direction and the trapped wave emerges from the end of the waveguide with enhanced intensity. In the direction of beam compression, the beam leaves the waveguide with the

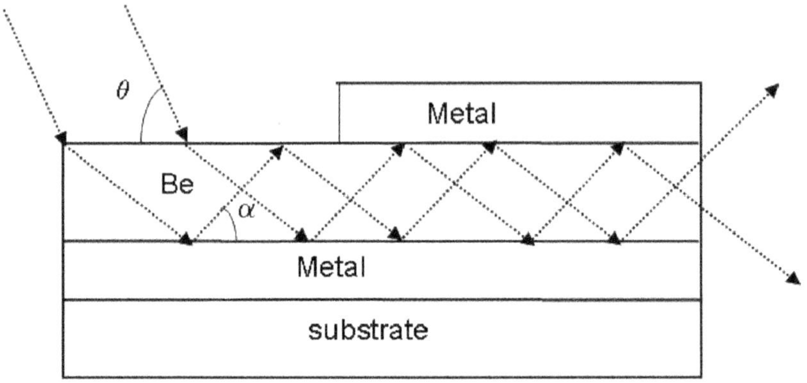

Fig. 17.5. The waveguide structure. θ is the incidence angle on the waveguide surface; incidence angle α on the interior metal surface of the waveguide differs from θ due to the refraction on the film

vertical size limited by the resonator layer thickness, which can be as small as 100 nm [47–49]. The cover layer and under layer can be constituted of the same material. To date, X-ray waveguide optics have been exclusively one-dimensional, whereas most nanobeam applications require two-dimensional point beams. Recently, the first proof of principle that resonant beam coupling can be realized in two dimensions was reported [16, 50]. The hard X-ray beam with a cross section of $25 \times 47\,\mathrm{nm}^2$ (FWHM) is impressive. Nevertheless, many technological improvements will be required before this device can be used as an efficient X-ray point source.

17.3.2 Fresnel Zone Plates

The focusing properties of zone plates were first discussed in the latter part of the nineteenth century, and Baez originally suggested their use as X-ray optical elements in 1952 [51]. In their most common form, this is circular diffraction grating that works as a lens for monochromatic light. A Fresnel zone plate consists of a series of concentric rings of radius $r_n^2 = n\lambda F$. The rings become narrower with increase in radius until the last, narrowest zone of width Δr_n is reached (Fig. 17.6). Linear, square, and elliptical zone plates have also been considered, but only circular and, to a lesser extent, linear and elliptical forms have generally been used. The focusing capability is based on constructive interference of the wavefront modified by passage through the zone plate. The wavefront modification is obtained through the introduction of a relative change in amplitude or phase in the beams emerging from two neighboring zones. A zone plate is called an amplitude zone plate if the focusing results from different absorptions between two neighboring zones. It is called a phase zone plate if the phase change upon transmission through a zone is the mechanism for focusing.

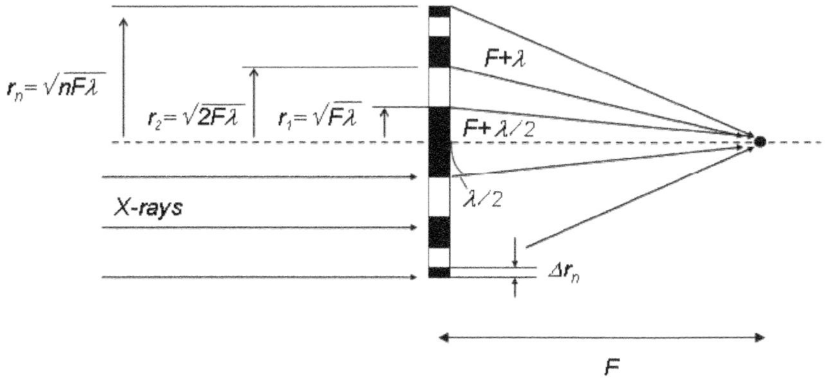

Fig. 17.6. Fresnel zone plate geometry

Just as the Rayleigh resolution of a light microscope lens is determined by its numerical aperture (NA), the Rayleigh resolution of a zone plate is determined by its maximum diffraction angle NA $= \lambda/(2\Delta r_n)$, so that the Rayleigh resolution is $1.22\Delta r_n$. Using the state-of-the-art lithographic technologies, zone plates with an outermost ring width of less than 20 nm can be fabricated. If illuminated with an X-ray beam whose spatial coherence length is equal to or greater than the diameter of the zone plate, a diffraction-limited focus can be obtained. The efficiency of zone plates, i.e., the fraction of the incident photons diffracted into the focal spot, depends on the phase shift and attenuation introduced by the FZP structures. In the soft X-ray region, zone plate efficiency is limited to about 15% due to photoelectric absorption. For X-ray energies greater than 4 keV, it is in principle possible to produce phase zone plates with focusing efficiencies close to 40%.

The efficiency of a binary FZP reaches its maximum value when the structure height is chosen to introduce a phase shift of π. As a phase shift generally decreases with increase in photon energy, higher and higher structures are required to provide useful efficiency values for harder X-rays. For example, the appropriate zone thickness for 10 keV X-rays is estimated to be about 2 μm even for high density materials like gold or tantalum. The resulting extreme aspect ratios (height/width of finest zone) are the reason why FZPs for hard X-rays cannot be made with a zone width as small as the ones for soft X-rays. In fact, FZPs have only very rarely been used for energies beyond 12 keV. It should be noted that APS [17, 52–68] and SPring-8 [69–87] have very strong in-house FZP long-term development programs for hard X-ray applications. Metal based FZPs are now commercially available from X-Radia (USA) [88] and NTT (Japan) [89].

As mentioned earlier, the spatial resolution of a zone plate is determined by the outermost zone width, Δr_n. Hence, manufacturing techniques must be capable of delivering small line widths over large areas, to give apertures as large as possible, with zone thicknesses of correct values to give good diffraction efficiencies. The accuracy of the technique must be sufficient to yield zone boundaries within about one-third of the outer zone width. To date, basically two main approaches are used to achieve this. The mostly used technique for FZP manufacturing is based on semiconductor MEMS (micro-electro-mechanical systems) technology, which is similar to the technique employed in the manufacture of microcircuits. This technique is based on lithography methods with consecutive deep pattern transfer. Since optical lithography is limited to 0.2 μm resolution, mainly electron beam lithography is used for manufacturing high resolution zone plates. Currently, the best performing zone plate lenses for multi-kiloelectronvolt X-rays in terms of resolution and efficiency are fabricated by means of a deep pattern transfer process.

To reproduce the zone plate structure in a suitable material the lithographically based techniques can be subdivided into two main transfer methods: (a) wet or reactive ion etching and (b) electrochemical deposition (electroplating). The first method can be applied to metals and semiconductors and is used for example by NTT in Japan (Ta FZP) [72–76, 81, 84–86] and by PSI (Si FZP) [90–94]. Electroplating can be applied only for metals and typically uses gold and nickel [88, 95–99]. To date, typical aspect ratios achieved using e-beam lithography and consecutive pattern transfer for 100 nm structures are of the order of 10:1–15:1: for wet etching Si – 15:1 [100]; for reactive ion etching Ni, W – 10:1 [89]; for electroplating, Au – 16:1 [88].

Recently, microfabrication MEMS technology was successfully used to produce Si-FZP for hard X-rays, and using reactive ion etching of Si an aspect-ratio of 40–50 has been demonstrated for modest resolution zone plates with a 400 nm outermost zone [101, 102]. Compared to other materials commonly used for FZPs (Ni, Ta, Au), the absorption losses are negligible in Si. The phase shifting property of Si is also comparatively weak. The structure height required for optimum efficiency at 24 keV, for example, is almost 30 μm (Fig. 17.7). The advantage of Si MEMS technology is that Si chips with a number of different diffraction optical elements (DOE-chips) can be made. At ESRF, three types of DOE-chips are used by different beamlines as medium and long focal distance optics (see Table 17.3).

To overcome the aspect ratio limitation inherent in the lithography technique, the sputtered-sliced FZP (ss-FZP) method was proposed in 1982 [103]. In this technique, two different materials of heavy and light elements are alternately deposited on a rotating gold wire core to give a concentric multilayer structure. The ss-FZP is then produced from the multilayer wire sample by making slices perpendicular to the wire axis and then thinning and polishing the slice down to the required thickness (10–200 μm). It has been found that for these ss-FZPs, imperfections in the wire and defects during deposition lead to accumulative errors in the fine outermost zones, which are deposited last. Furthermore, it proved very difficult to slice the wire without deforming the delicate zone plate structure. In principle, this method is capable of producing zone plates with 10–50 nm resolution, and strong activity is

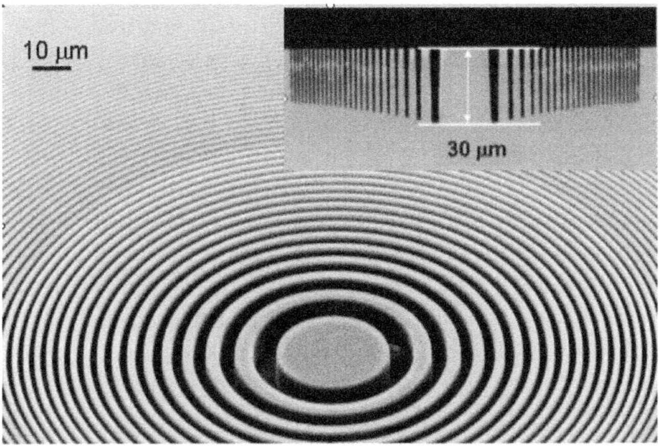

Fig. 17.7. SEM image of the SI Fresnel zone plate. Insert shows the cross section of FZP Chip 3 optimized for 23 keV (see Table 4.1.3)

Table 17.3. Chip specification

	T_{m}/h (µm)	E_{range} (keV)	η_{max} (%)	E_{max} (keV)
Chip1	12/9	6–12	30	7.5
Chip2	80/16	11–21	26	14
Chip3	90/30	17–40	32	23

T_{m}, thickness of silicon membrane; h, maximum height of the zone phase profile; E_{range}, energy range with focusing efficiency higher than 20%; η_{max}, maximum focusing efficiency (achieved at energy E_{max}); E_{max}, energy at which maximum efficiency η_{max} is achieved

going on at SPring-8 in collaboration with the Photonic Research Institute AIST (Osaka) [77–79, 85].

Another way of solving the aspect-ratio problem proposed recently by a PSI group is to apply an inclined geometry for a linear zone plate. Linear Si lenses were produced by electron-beam lithography and anisotropic wet etching. Two linear FZPs in a crossed geometry similar to a KB mirror system allowed one to produce a focal spot of 170 nm at 12.7 keV [100]. In addition, the device, consisting of two crossed linear lenses, can be matched with the asymmetric shape of synchrotron X-ray sources. This results in a better performance in terms of focal spot shape, spot size, and flux within the spot. Obviously, using two linear zone plates in series has the disadvantage that the losses in two optical components have to be considered.

It should be noted that achieving the utmost resolution is much easier in a magnified imaging mode as compared with a microbeam or microprobe mode. The microbeam size is a result of the convolution of the demagnified source size and the resolution of the optics, whereas in the imaging mode the source size is not important and the resolution of the image depends only on the intrinsic resolution of the optical element. As a most illustrative example, a 60 nm resolution has been obtained using a laboratory source with 5.4 keV X-rays by an X-Radia full-field microscope equipped with Au FZPs [104].

It is clear, however, that the focusing properties of all binary FZPs developed for X-ray applications are far weaker than those of an ideal lens because a significant fraction of incident photons is delivered to the undiffracted zero-order beam and to diffraction orders other than the primary, first order. For instance, although in principle the undiffracted zero-order beam for an ideal phase zone plate is completely reduced to zero, about 60% of the incident beam is still delivered to other unwanted orders. Improving the focusing efficiency of a zone plate in the X-ray spectral region by optimizing the zone profiles was studied in detail in a theoretical article by Tatchyn in 1990 [105], and the first zone plate with a blazed zone profile was manufactured at Wisconsin and tested at the APS using 8 keV X-rays showing 45% efficiency for linear and 39% efficiency for circular ZPs, respectively [57]. Later 55% efficiency for the nickel circular zone plate at 7 keV was achieved at the ESRF [95]. This measured focusing efficiency is the highest value that has been demonstrated in the X-ray region for zone plate optics. In practice, the production of high-resolution zone plates with a blazed profile is very complicated and expensive and a resolution better than 0.5 μm was never realized for these devices.

A novel approach for high-resolution X-ray focusing, a Multilayer Laue Lens (MLL), was proposed at the APS [17, 62, 63, 68]. The MLL concept is a system of two crossed linear zone plates, manufactured by deposition techniques. The approach involves the deposition of a multilayer with a graded period, sectioning it to the appropriate thickness, assembling the sections at the optimum angle, and using it in Laue geometry for focusing. The approach is particularly well suited for high-resolution focusing optics for use at high photon energies (30 nm resolution and 70% efficiency).

To overcome the aspect-ratio limitation to produce thicker FZPs, one can attempt a multiple zone plate setup. Two stacked zone plates made of gold were tested at 50 keV [59, 60], but did not show the theoretically predicted efficiency and resolution for the following reasons: the zone plates were not exactly identical; the alignment procedure was not accurate enough; the stability was not sufficient; and a 3 mm separation of zone plates was far more than that is allowable.

Recently, the Microoptics test bench at BM5 at the ESRF was used for the FZP stacking technique, using Si FZP chips described earlier [101, 102]. The first FZP chip was mounted on the optics stage with all necessary angular and linear movements. The second chip was mounted on the slit support frame with two X-Y linear translations. The precise and smooth vertical movement of the Microoptics test bench enabled one to check the alignment of different FZPs on the chip. Achieving the exactly aligned condition was easily done by a straightforward X-ray phase contrast imaging technique by looking at the X-ray CCD camera image of the beam. When you look through one FZP at

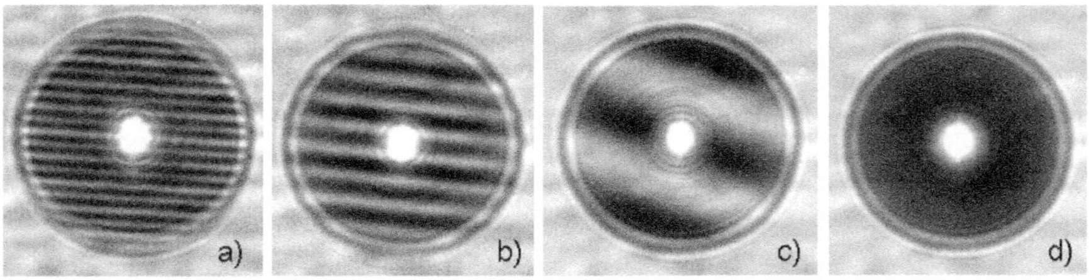

Fig. 17.8. Experimental moiré patterns recorded at 12 keV at 76 cm, with 6 μm (**a**), 3 μm (**b**), 1 μm (**c**), and 0 μm (**d**) lateral displacement

another, a *moiré pattern* always appears as the two repetitive patterns overlap. A slight motion of one of the objects creates large-scale changes in the moiré pattern (Fig. 17.8). Two chips with zone plates were joined together in a way that the front side of the first chip faces the backside of the other one. Therefore, the corresponding zone plates are separated from each other by a substrate thickness of 500 μm. For two zone plates to behave as one, their relative transverse positions must be adjusted properly to within a hundred nanometers, so as to directly line up the zones of the two elements. A compact micromechanical motion system (piezo-based Y-Z stage) was used to execute such precise alignment with 50 nm step size. When two zone plates are nearly perfectly aligned, the image shows no interference fringe patterns – the fringe spacing becoming infinite as the stacking alignment becomes perfect. To provide the desired stability, zone plates are bonded together with epoxy glue with slow solidification speed for the on-line alignment correction.

Two identical Si Chips3 were used for stacking in order to focus 50 keV X-rays. Each chip contains nine different FZP elements (five circular and four linear) with a depth of 30 μm, providing a focusing efficiency of about 34% at 23 keV. Two closely juxtaposed zone plates focus 50 keV X-rays with 35% efficiency [100–102]. Thus, there is a significant improvement in going from one to two such elements. It should be noted that such systems provide a focused beam with more than 20% efficiency in the energy range from 40 to 100 keV and can be used now for microdiffraction/scattering experiments. One can envisage some new imaging and interferometry techniques using FZP systems at higher energies. The technique developed will be used for future Fresnel optics with higher resolution and can be extended to "thin" nanofocusing FZPs.

17.3.3 Refractive Optics

Refractive lenses made of glass are among the most widely used optical components for visible light, with the wide spectrum of applications in focusing and imaging. Refractive lenses for X-rays were considered unfeasible for a long time due to the weak refraction and strong absorption. However, in 1996 it

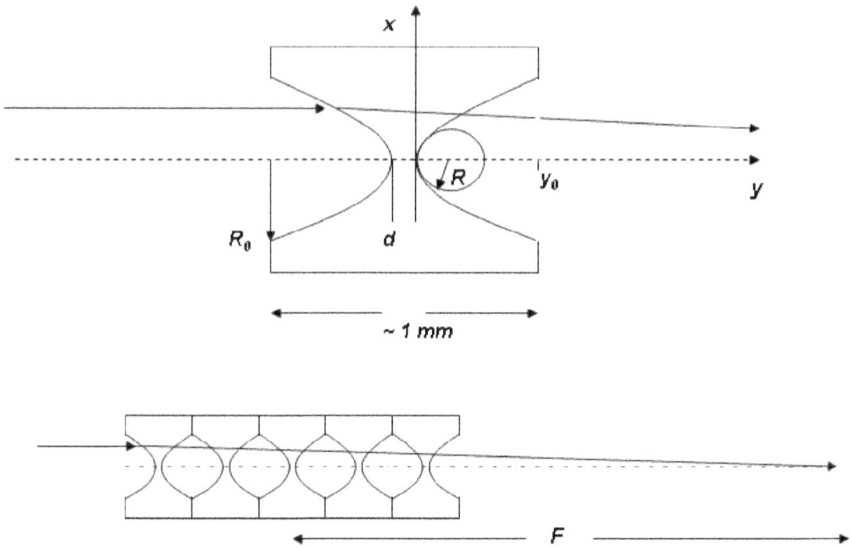

Fig. 17.9. Parabolic compound refractive lens (CRL). The individual lenses (**a**) and stacked behind one another to form a CRL

was shown that focusing by X-ray lenses is possible [106]. Since the $(1 - \delta)$ in the index of refraction is smaller than 1, lenses must have a concave shape [106–108]. To obtain a focal length F in the range of 1 m, many single lenses have to be stacked behind each other to form a compound refractive lens (CRL) as shown in Fig. 17.9. Fabricating the lenses from low-Z materials like Li, Be, B, C, and Al minimizes the problems associated with absorption. The focal length of such CRL with a parabolic profile $x^2 = 2Ry$ and N individual biconcave lenses is $F = R/2N\delta$, where R is the radius of curvature at the apex of parabola. A lens with thickness $2y_0 + d$ has an aperture $2R_0 = 2\sqrt{2Ry_0}$.

Refractive lenses act as a conventional lens and one can apply the Gauss lens formula, which relates the source distance p, the image distance q, and the focal distance F via $q = Fp\,(p - F)$. The diffraction-limited resolution of the lens Δ is defined by an effective aperture: $\Delta = 0.75\lambda/2\mathrm{NA}$, where the numerical aperture is $\mathrm{NA} = A_{\mathrm{eff}}/2q$. A_{eff} is the effective aperture of the lens, reduced by photon absorption and scattering, compared with the geometrical aperture $2R_0$.

The first lenses consisted of a row of holes, about 1 mm in diameter, drilled in a material such as Al or Be [107]. Two of these lenses in crossed geometry are able to focus an X-ray beam to a spot size of a few microns. Soon after this first successful experimental demonstration it was understood that refractive lenses can be used as a condensers or collimators with relatively long focal distances. Be, Graphite, and Al lenses were installed at the front-ends (FE) about 24 m from the source at various beamlines [109–112]. The typical FE

Table 17.4. Typical parameters of the FE CRL lenses at the ESRF

Lens	N	E for $L_2 = 41\,\text{m}$ (keV)	E for collimation (keV)	f for $E = 20\,\text{keV}$ (m)
1	7	12.7	15.1	42
2	16	17.0	22.9	18.4
3	28	22.8	30.2	10.5
4	44	28.2	37.9	6.7
5	64	35	45.7	4.6

CRL consists of a series of cylindrical holes drilled into a material. By varying the number of holes and their radius, it is possible to fine-tune the focal length of the lenses, making them a very useful device not only to focus but also to collimate a divergent X-ray beam: by choosing $F = p$ one obtains $q = \infty$, and the beam after the lens will be parallel [113–115].

Table 17.4 gives an overview for typical focal lengths of the FE lenses, collimation energies, and energies used to image the source size with a camera in the second experimental hutch.

Nowadays some ESRF beamlines (ID2, ID16, and ID18) are equipped with cylindrical CRL installed in optics hutches. For example, at ID18, in addition to FE lenses there is a CRL with 120 Al holes installed up-stream of the high-heatload (HHL) monochromator at ID18 to meet the acceptance of the Si (111) reflection for energies above 30 keV. At 64 keV this lens collimates the beam from 15 to 1.5 μrad and improves the resolution from about 10–1 eV while keeping the integral flux. Down-stream of the HHL monochromator another CRL is also installed to match the beam divergence to the acceptance of the first crystal of the high resolution monochromator. At 14.4 keV the intrinsic divergence of the X-ray beam of about 20 μrad has been decreased to 6 μrad, improving the throughput by a factor of two and the resolution from 0.82 to 0.65 meV [113].

In the meantime, Al and Be parabolic refractive lenses have been developed in collaboration with Aachen University [116–124]. They focus in both directions and are free of spherical aberrations and other distortions. Parabolic refractive lenses can be used to focus hard X-rays in both directions in the range from about 5 keV to about 200 keV. They are compact, robust, and easy to align and to operate. They can be used like glass lenses used for visible light and provide a resolution on the order of 300–500 nm, the main difference being that numerical aperture is much smaller than 1 [116]. Their main applications are in micro- and nanofocusing and in imaging by absorption and phase contrast [121]. In combination with tomography, 3D imaging of opaque media with sub-micrometer resolution is possible [123]. The Be and Al lenses for two-dimensional focusing are now used extensively as a standard tool in experiments. Table 17.5 shows the ESRF beamlines equipped by Al and Be parabolic refractive lenses made by RWTH in Aachen.

Table 17.5. Parabolic CRLs from RWTH Aachen used at the ESRF beamlines

Beam line	Material	Energy (keV)	Number of indiv. lenses	Source distance (m)	Focal distance (m)
ID1	Be	6–9	20	42	0.5–1.5
ID10A	Be	6–20	20	40/55	0.5–3
ID10B	Be	7–20	40	35/40	0.5–3
ID11	Be	15–80	61	30/55	1–10
	Al	20–100	254	55/100	0.5–10
ID11/ID15	Al	20–200	500	40/50/60	0.5–10
ID13	Be	12–14	24	1.5	1.5
ID22	Al	6–10	200	40/50	0.5–1
	Be	8–60	100	40/50	10–30
ID18F	Al	15–30	200	20	20
ID32	Be	8–23	15	34	6–13
ID14	Be	14	30	40	0.5–1.5
MOTB/BM5	Be	7–30	25	40/54	1
	Al	15–60	200	40/54	0.5–50
MPI/ID15	Al	50–90	300	60	3–6

In recent years a significant demand for focusing of hard X-rays above 40 keV has developed. A number of new applications such as surface and interface scattering, high pressure Compton magnetic scattering, and depth strain analysis using powder microdiffraction are under extensive development [125–127]. The Max–Planck Institute (MPI) end-station for surface and interface scattering, which has been recently installed at ID15, is a nice example of such a development.

Recently, microelectronics planar fabrication technology has been applied to create silicon-based devices [128–135]. One-dimensionally focusing parabolic refractive lenses have been manufactured in collaboration with the Institute of Microelectronics Technology (Chernogolovka, Russia) and Dortmund University using lithography and highly anisotropic plasma etching techniques. This type of planar lens is well suited for high-resolution diffraction experiments, including standing wave techniques [133–135]. It is possible to make a composite lens consisting of a set of parallel parabolas with different focal distances. To change the focal distance or the desirable working energy, one can switch from one array to another by moving the composite lens. Driven by the requirements of new 100-m-long beamlines at the ESRF, Si planar parabolic lenses were designed and fabricated (Fig. 17.10). They have a short focal distance in the energy range of 10 and 100 keV. The optical test of the new planar lenses was performed at the ESRF beamlines BM5 and ID15. The resolution below 200 nm was measured in the energy region of 15–80 keV. The best resolution of 150 nm was demonstrated at 50 keV energy. Using the same approach

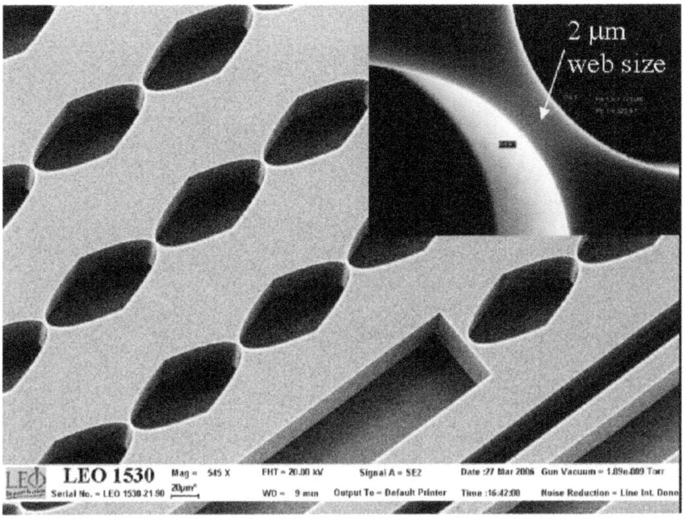

Fig. 17.10. SEM image of a Si planar refractive lens. The insert shows the 2 μm web size

of the Si-planar technology, nanofocusing lenses were developed by the Aachen group [136–138]. They have a focal distance in the range of a few millimeters at hard X-ray energies. In a crossed geometry, two lenses were used at ID13 to generate a nanobeam with a lateral size of 115 nm by 160 nm at 15.2 keV, and in December 2004 a focus spot of about 50 nm was achieved [18]. The planar lens technology is being transferred to materials like diamond that has low X-ray absorption, low thermal expansion, and high heat conductivity [139, 140]. These lenses are mechanically robust and can withstand the high heat load of the white beam produced by the ESRF in vacuum undulators and from future X-ray free electron lasers.

The applicability of Al lenses for microbeam analysis at energies above 100 keV is limited by the physical size of the lens assembly, because the number of individual lenses required to produce a reasonable focal distance grows quickly with energy. Using denser lens materials, such as nickel, the number of lenses that are needed can be drastically reduced. While the absorption in nickel is still tolerable, its density and thus its refraction are higher compared to the low Z materials used. Nickel is the most promising since it is radiation and corrosion stable and, what is more important, it is one of the best materials for electroplating. LIGA technology including deep X-ray lithography and electroplating has been widely used in the last ten years for the fabrication of various microstructures in Ni. These techniques make possible the formation of planar lens arrays with a wide range of parameters. Lens apertures can range from a few microns to a few millimeters. Structures up to few millimeters in depth can be realized. Their focal distances can range from a few

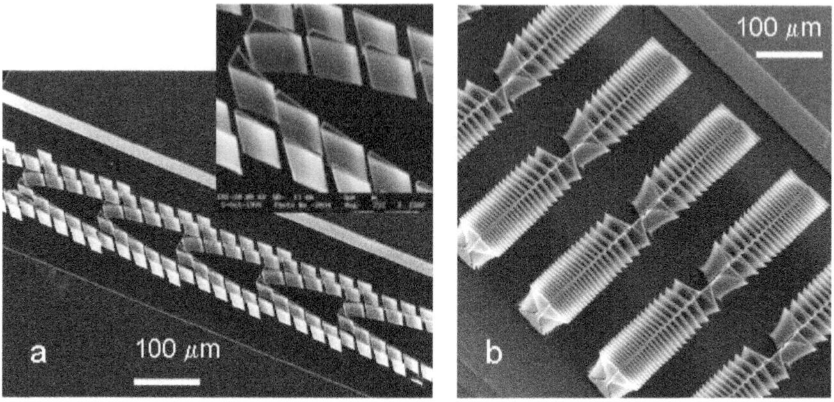

Fig. 17.11. SEM images of two different types of kinoform lenses made in Si (see text) [128] (**a**); [130] (**b**)

millimeters to tens of meters. Ni planar refractive lenses have been manufactured by deep X-ray lithography and LIGA techniques. The optical properties of lenses were determined at the ESRF ID15 beamline at energies from 40 to 220 keV. One- and two-dimensional focusing was performed. Sub-micrometer focusing was measured in the energy range from 40 to 150 keV [141, 142].

Recently, holographic or kinoform optical elements (Fig. 17.11) with a combination of refractive and diffractive properties were manufactured [128, 130]. In these refractive lenses passive parts of the material that cause multiples of 2π in phase shift are removed thereby reducing absorption. With this method drawbacks of purely diffractive or refractive elements are eliminated and advantages such as high transmission, absence of zero-order, high efficiency are combined. Recently, Ni kinoform lenses made by LIGA focused 212 keV X-rays to a focal line 5 μm wide with a tenfold gain [141, 142]. The ability to manipulate the local amplitude and phase of the incoming wave opens the perspective to make a new class of beamshaping X-ray optics for coherent synchrotron radiation.

17.4 Concluding Remarks

The foregoing overview shows the tremendous development in the possibilities for X-ray focusing that now makes possible the construction of powerful instruments for microscopy at synchrotron radiation beamlines.

In conclusion we compare the different focusing systems. First, we should mention that reflective, diffractive, and refractive microoptics have the following features in common:

- All three types are under intensive development at all three big hard X-ray facilities
- They are becoming commercially available
- They are used as a standard instrumentation at the beamlines
- All three types show nanofocusing capabilities

KB mirrors have an intrinsic advantage over the other focusing elements, such as Fresnel zone plates and refractive optics: nondispersive or broadband focusing. In the case of dynamic KB systems sophisticated bending techniques have been developed to bend mirrors to the desired elliptical shape for micro- and nanofocusing. The vibration level has to be controlled to within a few microradians and the figure accuracy of the elliptical mirrors to within a few nanometers. This is technologically challenging. The reflected beam is deflected with respect to the incoming beam. These constraints can all be managed, but have to be taken into account when selecting the most appropriate microfocusing technology. Mirrors with benders can provide an adjustable focal length, but the benders are bulky. Monolithic or static KB systems are much easier to use if the desired elliptical surface profile can be fabricated.

FZP and refractive optics being in-line optics have certain advantages over KB systems:

- On-axis optics do not change the beam direction
- They provide easy alignment and operation
- They can be easily implemented at any beamline (including nonspecific beamlines)
- In the case of nanofocusing geometries FZP and CRL should have greater distance from the optics to the sample

FZP elements have attractive features in that they are very compact and easy to use. The alignment mechanics requires basically only two orthogonal translations (XZ) and therefore they can be easily used at any nonspecific beamline [143, 144]. Si FZPs are compatible with ML and "pink" beams because of high radiation and temperature stability.

The advantages of CRLs are the following: they are very robust and small, the focal length and size are adjustable by adding or removing individual lenses, and the lenses can withstand a high heatload. The lens aperture can range from a few microns to a few millimeters. Their focal distance can range from a few millimeters to tens of meters. What is more, CRLs can cover the energy range from 4 to 200 keV and higher.

Compared to mirrors, refractive lenses are about a factor of 1,000 less sensitive to surface roughness. This is an important aspect in the production process of the lenses. Surface roughness plays no role in imaging by refractive X-ray lenses.

For comparing different optics, it is important to consider the physical limits to the efficient focusing of hard X-rays. It was found that mirrors and waveguides have a numerical aperture, which is limited by the critical angle of total reflection. The ultimate resolution limit is 10 nm [145], while for refractive optics this limit is slightly lower and 2 nm may be achievable [146]. Unlike reflective and refractive optics, zone plates can focus X-rays below 1 nm [147,148]. In this case, complex multilayer zone plates have to be manufactured and Bragg conditions have to be fulfilled for the outermost zones. As for conventional zone plates, there is a simple pathway to achieve sub-10 nm resolution X-ray imaging by using a higher diffraction order, such as the third diffraction order of a currently available zone plate. While progress in the fabrication of hard X-ray zone plates has significantly advanced within the last few years, the pattern transfer fabrication process may reach a practical limit very soon. As the polymer structures of the electroplating mold become smaller and smaller in width they lose strength and tend to collapse during the fabrication process. Also, the directionality of the reactive ion etch may impose practical limits to the achievable sidewall angle in the resist, limiting the achievable width of features that can be fabricated. From current fabrication data it can be estimated that the practical limit for hard X-ray zone plates using current pattern transfer technology is 20–30 nm (structures height ≅1 μm). It is believed that by using higher diffraction orders, such as the third diffraction order, it would be possible to achieve sub-10 nm resolution X-ray imaging.

To discuss the applicability of one or another type of focusing systems for nanofocusing applications, let us consider the conditions for new 100-m-long beamlines at the ESRF. To obtain a resolution about 50 nm in the vertical direction we need to apply a demagnification factor of ×1,000 for the vertical source size of 50 μm. Therefore, a microoptics device placed at the source-to-optics distance $p = 100$ m must have a focal length (distance to detector/sample distance) $q = 0.1$ m. We consider the following three optical systems:

- KB mirror system (Pt coated) with 40-mm-long mirrors and 30 mm working distance
- Fresnel zone plates with the outermost zone width 40 nm
- Planar refractive lenses made of Si, Be, and C (diamond)

The graph in Fig. 17.12 shows effective apertures or acceptances of the KB-mirrors, FZP and CRLs. The FZP effective aperture in the graph is normalized to the FZP efficiency ε: $A_{\mathrm{eff}} = A_{\mathrm{fzp}}\varepsilon$. We optimistically assume that the FZP is made with optimal thickness providing a phase shift π and an efficiency not less than 30% over the entire energy range. As can be seen from this graph, the FZP elements can be applied up to 20 keV energy, whereas KB-mirror systems look competitive up to 40–50 keV. Si nanofocusing lenses can easily beat FZP after 20–25 keV and become competitive with KB-m after 40 keV.

Use of microoptics and exploiting the high brilliance and the spatial coherence of the X-ray beam provided by the third generation synchrotron radiation

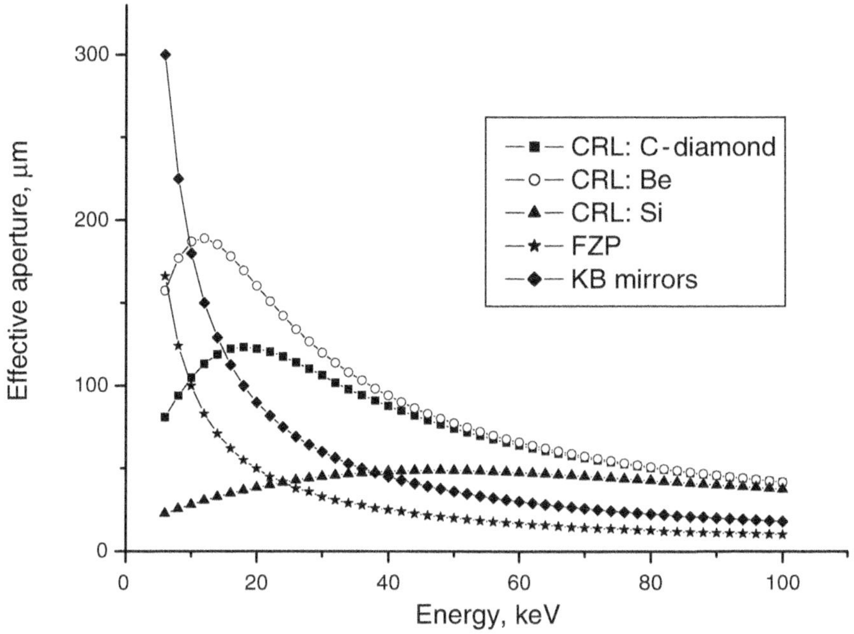

Fig. 17.12. Graph showing the effective aperture dependence vs. energy for different optical elements

sources makes possible high energy X-ray microscopy as a combination of diffraction, fluorescence, and imaging techniques. The availability of these techniques is opening up research opportunities for a broad range of disciplines, including material science, biology, environmental, and geosciences.

References

1. P. Goby, Comptes Rendue de l'Academie des Sciences Paris **156**, 686 (1913)
2. P. Kirkpatrick, A.V. Baez, J. Opt. Soc. Am. **38**, 766 (1948)
3. V.E. Cosslett, W.C. Nixon, Nature **168**, 24 (1951)
4. B. Niemann, D. Rudolph, G. Schmahl, Opt. Commun. **12**, 160 (1974)
5. J. Kirz, H. Rarback, Rev. Sci. Instrum. **56**, 1 (1985)
6. W. Chao, B.D. Harteneck, J.A. Liddle, E.H. Andersen, D.T. Attwood, Nature **435**, 1210 (2005)
7. A.G. Michette, *Optical Systems for Soft X-rays* (Plenum Press, New York, 1986)
8. A. Michette, S. Pfauntsch (eds.), *X-rays: The First Hundred Years* (Wiley, New York, 1997)
9. J. Thieme, G. Schmahl, D. Rudolph, E. Umbach (eds.), in *X-ray Microscopy and Spectroscopy.* Status Report from the Fifth International Conference, Wurzburg, August 19–23, 1996 (Springer-Verlag, Berlin Heidelberg, New York, 1998)

10. W. Meyer-Ilse, T. Warwick, D. Attwood (eds.), in *X-ray Microscopy Proceedings of the Sixth International Conference*, Berkeley, CA 1999. AIP Conference Proceedings 507 (American Institute of Physics, Melville, New York, 2000)
11. J. Susini, D. Joyeux, F. Polack (eds.), *X-ray Microscopy 2002, 7th International Conference on X-ray microscopy, ESRF*, Grenoble, France, July 28–August 2, 2002. EDP Sciences, Journal de Physique IV, vol. 104 (2003)
12. H. Mimura, S. Matsuyama, H. Yumoto, H. Hara, K. Yamamura, Y. Sano, M. Shibahara, K. Endo, Y. Mori, Y. Nishino, K. Tamasku, M. Yabashi, T. Ishikawa, K. Yamauchi, Jpn. J. Appl. Phys. **44**, L539 (2005)
13. O. Hignette, P. Cloetens, C. Morawe, C. Borel, W. Ludwig, P. Bernard, A. Rommeveaux, S. Bohic, AIP Conf. Proc. **879**, 792 (2007)
14. D.H. Bilderback, S.A. Hoffman, D.J. Thiel, Science **263**, 201 (1994)
15. A. Snigirev, A. Bjeoumikhov, A. Erko, I. Snigireva, M. Grigoriev, M. Erko, S. Bjeoumikhova, J. Synchrotron Radiat. **14**, 227 (2007)
16. A. Jarre, C. Fuhse, C. Ollinger, J. Seeger, R. Tucoulou, T. Salditt, Phys. Rev. Lett. **94**, 074801 (2005)
17. H.C. Kang, J. Maser, G.B. Stephenson, C. Liu, R. Conley, A.T. Macrander, S. Vogt, Phys. Rev. Lett. **96**, 127401 (2006)
18. C.G. Schroer, O. Kurapova, J. Patommel, P. Boye, J. Feldkamp, B. Lengeler, M. Burghammer, C. Riekel, L. Vincze, A. van der Hart, M. Kuchler, Appl. Phys. Lett. **87**, 124103 (2005)
19. Y. Dabin, G. Rostaing, O. Hignette, A. Rommeveaux, A. Freund, Proc. SPIE **4782**, 235 (2002)
20. O. Hignette, P. Cloetens, W.K. Lee, W. Ludwig, G. Rostaing, J. Phys. IY France **104**, 231 (2003)
21. O. Hignette, P. Cloetens, G. Rostaing, P. Bernard, C. Morawe, Rev. Sci. Instrum. **76**, 063709 (2005)
22. Y. Mori, K. Yamauchi, H. Mimura, Y. Sano, A. Saito, K. Ueno, K. Endo, A. Souvorov, M. Yabashi, K. Tamasaku, T. Ishikawa, Proc. SPIE **4782**, 58 (2002)
23. K. Yamauchi, K. Yamamura, H. Mimura, Y. Sano, A. Saito, A. Souvorov, M. Yabashi, K. Tamasaku, T. Ishikawa, Y. Mori, J. Synchrotron Radiat. **9**, 313 (2002)
24. K. Yamauchi, K. Yamamura, H. Mimura, Y. Sano, A. Saito, K. Endo, A. Souvorov, M. Yabashi, K. Tamasaku, T. Ishikawa, Y. Mori, Jpn. J. Appl. Phys. **42**, 7129 (2003)
25. K. Yamamura, K. Yamauchi, H. Mimura, Y. Sano, A. Saito, K. Endo, A. Souvorov, M. Yabashi, K. Tamasaku, T. Ishikawa, Y. Mori, Rev. Sci. Instrum. **74**, 4549 (2003)
26. H. Yumoto, H. Mimura, S. Matsuyama, H. Hara, K. Yamamura, Y. Sano, K. Ueno, K. Endo, Y. Mori, M. Yabashi, Y. Nishino, K. Tamasaku, T. Ishikawa, K. Yamauchi, Rev. Sci. Instrum. **76**, 063708 (2005)
27. S. Matsuyama, H. Mimura, H. Yumoto, K. Yamamura, Y. Sano, K. Endo, Y. Mori, Y. Nishino, K. Tamasku, T. Ishikawa, M. Yabashi, K. Yamauchi, Rev. Sci. Instrum. **76**, 083114 (2005)
28. H. Yumoto, H. Mimura, S. Matsuyama, S. Handa, Y. Sano, M. Yabashi, Y. Nishino, K. Tamasaku, T. Ishikawa, K. Yamauchi, Rev. Sci. Instrum. **77**, 063712 (2006)
29. G.E. Ice, J.S. Chung, J. Tischler, A. Lunt, L. Assoufid, Rev. Sci. Instrum. **71**, 2635 (2000)

30. A. Khounsary, G. Ice, P. Eng, Proc. SPIE **4782**, 65 (2002)
31. C. Liu, L. Assoufid, A.T. Macrander, G.E. Ice, J.Z. Tischler, Proc. SPIE **4782**, 104 (2002)
32. C. Liu, L. Assoufid, R. Conley, A.T. Macrander, G.E. Ice, J.Z. Tischler, Opt. Eng. **42**, 3622 (2003)
33. C. Liu, R. Conley, L. Assoufid, Z. Cai, J. Qian, A.T. Macrander, AIP Conf. Proc. **705**, 704 (2004)
34. G.E. Ice, E.D. Specht, J.Z. Tischler, A.M. Khounsary, L. Assoufid, C. Liu, Proc. SPIE **5347**, 1 (2005)
35. W. Liu, G.E. Ice, Z. Tischler, A. Khonsary, C. Liu, L. Assoufid, A.T. Macrander, Rev. Sci. Instrum. **76**, 113701 (2005)
36. D.R. Kreger, Recl. trav. Bot. Neerlandais **41**, 603 (1948)
37. E.A. Stern, Z. Kalman, A. Lewis, K. Lieberman, Appl. Opt. **27**, 5135 (1988)
38. P. Engstrøm, S. Larrson, A. Rindby, A. Buttkewitz, S. Garbe, G. Gaul, A. Knöchel, F. Lechtenberg, Nucl. Instrum. Methods **A302**, 547 (1991)
39. D.X. Balaic, K.A. Nugent, Z. Barnea, R. Garret, S.W. Wilkins, J. Synchrotron Radiat. **2**, 296 (1995)
40. R. Huang, D. Bilderback, AIP Conf. Proc. **705**, 712 (2003)
41. R. Huang, D. Bilderback, J. Synchrotron Radiat. **13**, 74 (2006)
42. A. Bjeoumikhov, S. Bjeoumikhova, R. Wedell, Part. Part. Syst. Charact. **22**, 384 (2005)
43. J. Bartoll, S. Rohrs, A. Erko, A. Firsov, A. Bjeoumikhov, N. Langhoff, Spectrochim. Acta **B59**, 1587 (2004)
44. A. Knochel, G. Gaul, F. Lechtenberg, German Patent DE 44441092C2, 1994
45. G. Hirsch, J. X-ray Spectrom. **32**, 229 (2003)
46. Y.P. Feng, S.K. Sinha, H.W. Deckman, J.B. Hastings, D.P. Siddons, Phys. Rev. Lett. **71**, 537 (1993)
47. S. Di Fonzo, W. Jark, S. Lagomarsin, C. Giannini, L. De Caro, A. Cedola, M. Muller, Nature **403**, 638 (2000)
48. W. Jark, A. Cedola, S. Di Fonzo, M. Fiordelisi, S. Lagomarsino, N.P. Konovalenko, V.A. Chernov, Appl. Phys. Lett. **78**, 1192 (2001)
49. A. Cedola, S. Lagomarsino, Synchrotron Radiat. News **17**, 30 (2004)
50. F. Pfeiffer, C. David, M. Burghammer, C. Riekel, T. Salditt, Science **297**, 230 (2002)
51. A.V. Baez, J. Opt. Soc. Am. **51**, 405 (1961)
52. B. Lai, W. Yun, D. Legnini, Y. Xiao, J. Chrzas, P.J. Viccaro, V. White, S. Bajikar, D. Denton, F. Cerrina, E. Di. Fabrizio, M. Gentilli, L. Grella, M. Baciocchi, Appl. Phys. Lett. **61**, 1877 (1992)
53. E. Di Fabrizio, M. Gentili, L. Grella, N. Baciocchi, A. Krasniperova, F. Cerina, W. Yun, B. Lai, E. Gluskin, J. Vasc. Sci. Technol. **B12**, 3979 (1994)
54. B. Lai, W. Yun, Y. Xiao, L. Yang, D. Legnini, Z. Cai, A. Krasnoperova, F. Cerrina, E. Di Fabrizio, M. Gentilli, Rev. Sci. Instrum. **66**, 2287 (1995)
55. Z. Chen, Y. Vladimirsky, M. Brown, Q. Leonard, O. Vladimirsky, F. Moore, F. Cerina, B. Lai, W. Yun, E. Gluskin, J. Vasc. Sci. Technol. **B15**, 2522 (1997)
56. W. Yun, S.T. Pratt, R.M. Miller, Z. Cai, D.B. Hunter, A.G. Jarstfer, K.M. Kemner, B. Lai, H.R. Lee, D.G. Legnini, W. Rodrigues, C.I. Smith, J. Synchrotron Radiat. **5**, 1390 (1998)
57. W. Yun, B. Lai, A.A. Krasnoperova, E. Di Fabrizio, Z. Cai, F. Cerina, Z. Chen, M. Gentili, E. Gluskin, Rev. Sci. Instrum. **70**, 3537 (1999)

58. X. Su, C. Stagarescu, G. Xu, D.E. Eastman, I. McNulty, S.P. Frogo, Y. Wang, C.C. Retsch, I.C. Noyan, C.K. Xu, Appl. Phys. Lett. **77**, 3465 (2000)
59. S.D. Shastri, J.M. Maser, B. Lai, J. Tys, Opt. Commun. **197**, 9 (2001)
60. Z. Cai, B. Lai, Y. Xiao, S. Xu, J. Phys. IY France **104**, 17 (2003)
61. K.M. Kemner, S.D. Kelly, B. Lai, J. Maser, E.J. O'Loughlin, D. Sholto-Douglas, Z. Cai, M.A. Schneegurt, C.F. Kulpa, Jr., K.H. Nealson, Science **306**, 686 (2004)
62. J. Maser, G.B. Stephenson, D. Shu, B. Lai, S. Vogt, A. Khounsary, Y. Li, C. Benson, G. Schneider, AIP Conf. Proc. **705**, 470 (2004)
63. L.E. Ocola, J. Maser, S. Vogt, B. Lai, R. Divan, G.B. Stephenson, Proc. SPIE **5539**, 165 (2004)
64. T. Buonassisi, A.A. Istratov, M. Heuer, M.A. Marcus, R. Jonczyk, J. Isenberg, B. Lai, Z. Cai, S. Heald, W. Warta, R. Schindler, G. Willeke, E.R. Weber, J. Appl. Phys. **97**, 074901 (2005)
65. C.C. Abnet, B. Lai, Y.L. Qiao, S. Vogt, X.M. Luo, P.R. Taylor, Z.W. Dong, S.D. Mark, S.M. Dawsey, J. Natl. Cancer Inst. **97**, 301 (2005)
66. T. Buonassisi, M.A. Marcus, A.A. Istratov, M. Heuer, T.F. Ciszek, B. Lai, Z. Cai, E.R. Weber, J. Appl. Phys. **97**, 063503 (2005)
67. Y. Xiao, Z. Cai, Z.L. Wang, B. Lai, Y.S. Chu, J. Synchrotron Radiat. **12**, 124 (2005)
68. H.C. Kang, G.B. Stephenson, C. Liu, R. Conley, A.T. Macrander, J. Maser, S. Bajt, H.N. Chapman, Appl. Phys. Lett. **86**, 151109 (2005)
69. Y. Suzuki, N. Kamijo, S. Tamura, K. Handa, A. Takeuchi, S. Yamamoto, H. Sugiyma, K. Ohsumi, M. Ando, J. Synchrotron Radiat. **4**, 60 (1997)
70. N. Kamijo, S. Tamura, Y. Suzuki, K. Handa, A. Takeuchi, S. Yamamoto, M. Ando, K. Ohsumi, H. Kihara, Rev. Sci. Instrum. **68**, 14 (1997)
71. M. Koike, I.H. Suzuki, S. Komiya, Y. Amemiya, J. Synchrotron Radiat. **5**, 794 (1998)
72. Y. Kagoshima, T. Ibuki, K. Takai, Y. Yokoyama, N. Miyamoto, Y. Tsusaka, J. Matsui, Jpn. J. Appl. Phys. **39**, L433 (2000)
73. Y. Kagoshima, T. Ibuki, Y. Yokoyama, Y. Tsusaka, J. Matsui, K. Takai, M. Aino, Jpn. J. Appl. Phys. **40**, L1190 (2001)
74. Y. Suzuki, A. Takeuchi, H. Takano, T. Ohigashi, H. Takenaka, Jpn. J. Appl. Phys. **40**, 1508 (2001)
75. Y. Kagoshima, T. Ibuki, Y. Yokoyama, K. Takai, Y. Tsusaka, J. Matsui, Jpn. J. Appl. Phys. **41**, 412 (2002)
76. A. Takeuchi, Y. Suzuki, H. Takano, J. Synchrotron Radiat. **9**, 115 (2002)
77. M. Awaji, Y. Suzuki, A. Takeuchi, H. Takano, N. Kamijo, S. Tamura, M. Yasumoto, J. Synchrotron Radiat. **9**, 125 (2002)
78. S. Tamura, M. Yasumoto, N. Kamijo, Y. Suzuki, M. Awaji, A. Takeuchi, H. Takano, K. Handa, J. Synchrotron Radiat. **9**, 154 (2002)
79. N. Kamijo, Y. Suzuki, M. Awaji, A. Takeuchi, H. Takano, T. Ninomiya, S. Tamura, M. Yasumoto, J. Synchrotron Radiat. **9**, 182 (2002)
80. A. Takeuchi, K. Uesugi, H. Takano, Y. Suzuki, Rev. Sci. Instrum. **73**, 4246 (2002)
81. Y. Kagoshima, Y. Yokoyama, T. Ibuki, T. Niimi, Y. Tsusaka, K. Takai, J. Matsui, J, Synchrotron Radiat. **9**, 132 (2002)
82. M. Awaji, A. Takeuchi, H. Takano, N. Kamijo, M. Yasamoto, Y. Terado, S. Tamura, Rev. Sci. Instrum. **74**, 4948 (2003)

83. N. Kamijo, Y. Suzuki, H. Takamo, M. Yasumoto, A. Takeuchi, M. Awaji, Rev. Sci. Instrum. **74**, 5101 (2003)
84. Y. Kagoshima, Y. Yokoyama, T. Niimi, T. Koyama, Y. Tsusaka, J. Matsui, K. Takai, J. Phys. IY France **104**, 49 (2003)
85. Y. Suzuki, M. Awaji, A. Takeuchi, H. Takano, K. Uesugi, Y. Komura, N. Kamijo, M. Yasumoto, S. Tamura, J. Phys. IY France **104**, 35 (2003)
86. Y. Kagoshima, T. Koyama, I. Wada, T. Niimi, Y. Tsusaka, J. Matsui, S. Kimura, M. Kotera, K. Takai, AIP Conf. Proc. **705**, 1263 (2004)
87. Y. Suzuki, A. Takeuchi, H. Takano, K. Uesugi, T. Oka, K. Inoue, Rev. Sci. Instrum. **75**, 1155 (2004)
88. X-Radia website: http://www.xradia.com/
89. NTT website: http://www.ntt-at.com/products_e/x-ray_optics/
90. C. David, B. Kaulich, R. Barrett, M. Solome, J. Susini, Appl. Phys. Lett. **77**, 3851 (2000)
91. C. David, B. Nohammer, E. Ziegler, Microelectron. Eng. **61–62**, 987 (2002)
92. C. David, E. Ziegler, B. Nohammer, J. Synchrotron Radiat. **8**, 1054 (2001)
93. C. David, B. Nohammer, E. Ziegler, Appl. Phys. Lett. **79**, 1088 (2001)
94. H. Solak, C. David, J. Gobrecht, Appl. Phys. Lett. **85**, 2700 (2004)
95. E. Di Fabrizio, F. Romanato, M. Gentili, S. Cabrini, B. Kaulich, J. Susini, R. Barrett, Nature **401**, 895 (1999)
96. B. Kaulich, S. Ostereich, M. Salome, R. Barrett, J. Susini, T. Wilheim, E. Di Fabrizio, M. Gentili, P. Charalambous, Appl. Phys. Lett. **75**, 4061 (1999)
97. T. Wilheim, B. Kaulich, E. Di Fabrizio, S. Cabrini, J. Susini, Appl. Phys. Lett. **78**, 2082 (2001)
98. B. Kaulich, R. Barrett, M. Solome, S. Ostereich, J. Suzini, ESRF Newsletter **34**, 27 (2000)
99. E. Di Fabrizio, D. Cojoc, S. Cabrini, B. Kaulich, J. Susini, P. Facci, T. Wilheim, Opt. Express **11**, 2278 (2003)
100. B. Nohammer, C. David, M. Burghammer, C. Riekel, Appl. Phys. Lett. **86**, 163104 (2005)
101. I. Snigireva, A. Snigirev, G. Vaughan, M. Di Michiel, V. Kohn, V. Yunkin, M. Grigoriev, AIP Conf. Proc. **879**, 998 (2007)
102. I. Snigireva, A. Snigirev, V. Kohn, V. Yunkin, M. Grigoriev, S. Kuznetsov, G. Vaughan, M. Di Michiel, Phys. Stat. Sol. A, to be published
103. D. Rudolph, G. Schmahl, B. Niemann, Proc. SPIE **316**, 103 (1982)
104. W. Yun, M. Feser, A. Lyon, F. Duewer, Y. Wang, Proc. SPIE **5539**, 133 (2004)
105. O. Tatchyn, in *X-ray Microscopy*, ed. by G. Schmahl, D. Rudolph, Springer Series in Optical Sciences, vol. 43 (Springer, Berlin Heidelberg New York, 1990), p. 40
106. A. Snigirev, V. Kohn, I. Snigireva, B. Lengeler, Nature **384**, 49 (1996)
107. A. Snigirev, V. Kohn, I. Snigireva, A. Souvorov, B. Lengeler, Appl. Opt. **37**, 653 (1998)
108. B. Lengeler, J. Tummler, A. Snigirev, I. Snigireva, C. Raven, J. Appl. Phys. **84**, 5855 (1998)
109. P. Elleaume, ESRF Newsletter **28**, 33 (1997)
110. A. Snigirev, B. Filseth, P. Elleaume, Th. Kolocke, V. Kohn, B. Lengeler, I. Snigireva, A. Souvorov, J. Tummler, Proc. SPIE **3151A**, 164 (1997)
111. P. Elleaume, J. Synchrotron Radiat. **5**, 1 (1998)
112. P. Elleaume, Nucl. Instrum. Methods **A412**, 483 (1998)

113. A. Chumakov, R. Ruffer, O. Leupold, A. Barla, H. Thiess, T. Asthalter, P. Doyle, A. Snigirev, A. Baron, Appl. Phys. Lett. **77**, 31 (2000)
114. A.Q. Baron, Y. Kohmura, Y. Ohishi, T. Ishikawa, Appl. Phys. Lett. **74**, 1492 (1999)
115. A.Q.R. Baron, Y. Kohmura, V.V. Krishnamurthy, Yu.V. Shvyd'ko, T. Ishikawa, J. Synchrotron Radiat. **6**, 953 (1999)
116. B. Lengeler, C.G. Schroer, M. Richwin, J. Tummler, M. Drakopoulos, A. Snigirev, I. Snigireva, Appl. Phys. Lett. **74**, 3924 (1999)
117. B. Lengeler, C. Schroer, J. Tummler, B. Benner, M. Richwin, A. Snigirev, I. Snigireva, M. Drakopoulos, J. Synchrotron Radiat. **6**, 1153 (1999)
118. B. Lengeler, C. Schroer, J. Tummler, B. Benner, M. Richwin, A. Snigirev, I. Snigireva, M. Drakopoulos, Synchrotron Radiat. News **12**(5), 45 (1999)
119. C.G. Schroer, J. Tummler, F. Gunzler, B. Lengeler, W.H. Schroder, A.J. Kuhn, A. Simionovici, A. Snigirev, I. Snigireva, Proc. SPIE **4142**, 287 (2000)
120. B. Lengeler, C. Schroer, B. Benner, F. Gunsler, M. Kuhlmann, J. Tummler, A. Simionovici, M. Drakopoulos, A. Snigirev, I. Snigireva, Nucl. Instrum. Methods A **467–468**, 944 (2001)
121. C. Schroer, F. Gunsler, B. Benner, M. Kuhlmann, J. Tummler, B. Lengeler, C. Rau, T. Weitkamp, A. Snigirev, I. Snigireva, Nucl. Instrum. Methods A **467–468**, 966 (2001)
122. C. Schroer, B. Benner, F. Gunzler, M. Kuhlmann, C. Zimprich, B. Lengeler, C. Rau, T. Weitkamp, A. Snigirev, I. Snigireva, J. Appenzeller, Rev. Sci. Instrum. **73**, 1640 (2002)
123. C.G. Schroer, J. Meyer, M. Kuhlmann, B. Benner, T.F. Gunsler, B. Lengeler, C. Rau, T. Weitkamp, A. Snigirev, I. Snigireva, Appl. Phys. Lett. **81**, 1527 (2002)
124. C.G. Schroer, M. Kuhlmann, B. Lengeler, T.F. Gunsler, O. Kurapova, B. Benner, C. Rau, A.S. Simionovici, A. Snigirev, I. Snigireva, Proc. SPIE **4783**, 10 (2002)
125. H. Reichert, V. Honkimaki, A. Snigirev, S. Engemann, H. Dosch, Physica B **336**, 46 (2003)
126. S. Engemann, H. Reichert, H. Dosch, J. Bilgram, V. Honimaki, A. Snigirev, Phys. Rev. Lett. **92**, 205701 (2004)
127. C. Mocuta, H. Reichert, K. Mecke, H. Dosch, M. Drakopoulos, Science **308**, 1287 (2005)
128. V. Aristov, M. Grigoriev, S. Kuznetsov, L. Shabelnikov, V. Yunkin, T. Weitkamp, C. Rau, I. Snigireva, A. Snigirev, M. Hoffmann, E. Voges, Appl. Phys. Lett. **77**, 4058 (2000)
129. V. Aristov, M. Grigoriev, S. Kuznetsov, L. Shabelnikov, V. Yunkin, C. Rau, A. Snigirev, I. Snigireva, T. Weitkamp, M. Hoffmann, E. Voges, Proc. SPIE **4145**, 285 (2001)
130. I. Snigireva, A. Snigirev, C. Rau, T. Weitkamp, V. Aristov, M. Grigoriev, S. Kuznetsov, L. Shabelnikov, V. Yunkin, M. Hoffmann, E. Voges, Nucl. Instrum. Methods A **467–468**, 982 (2001)
131. M. Grigoriev, L. Shabelnikov, V. Yunkin, A. Snigirev, I. Snigireva, M. Di Michiel, S. Kuznetsov, M. Hoffmann, E. Voges, Proc. SPIE **4501**, 185 (2001)
132. I. Snigireva, M. Grigoriev, L. Shabelnikov, V. Yunkin, A. Snigirev, S. Kuznetsov, M. Di Michiel, M. Hoffmann, E. Voges, Proc. SPIE. **4783**, 19 (2002)

133. M. Drakopoulos, J. Zegenhagen, A. Snigirev, I. Snigireva, M. Hauser, K. Eberl, V. Aristov, L. Shabelnikov, V. Yunkin, Appl. Phys. Lett. **81**, 2279 (2002)
134. M. Drakopoulos, J. Zegenhagen, T.-L. Lee, A. Snigirev, I. Snigireva, V. Cimalla, O. Ambacher, J. Phys. D: Appl. Phys. **36**, A214 (2003)
135. M. Drakopoulos, J. Zegenhagen, A. Snigirev, I. Snigireva, Synchrotron Radiat. News **17**(3), 37 (2004)
136. C.G. Schroer, M. Kuhlmann, U.T. Hunger, T.F. Gunsler, O. Kurapova, S. Feste, F. Frehse, B. Lengeler, M. Drakopoulos, A. Somogyi, A.S. Simionovici, A. Snigirev, I. Snigireva, C. Schug, W.H. Schroder, Appl. Phys. Lett. **82**, 1485 (2003)
137. C.G. Schroer, B. Benner, T.F. Günzler, M. Kuhlmann, J. Patommel, B. Lengeler, A. Somogyi, T. Weitkamp, C. Rau, A. Snigirev, I. Snigireva, Proc. SPIE **5535**, 701 (2004)
138. C.G. Schroer, T.F. Günzler, M. Kuhlmann, O. Kurapova, S. Feste, M. Schweitzer, B. Lengeler, W.H. Schröder, A. Somogyi, A.S. Simionovici, A. Snigirev, I. Snigireva, Proc. SPIE **5535**, 162 (2004)
139. A. Snigirev, V. Yunkin, I. Snigireva, M. Di Michiel, M. Drakopoulos, S. Kouznetsov, L. Shabelnikov, M. Grigoriev, V. Ralchenko, I. Sycho, M. Hoffmann, E. Voges, Proc. SPIE **4783**, 1 (2002)
140. B. Nohamme, J. Hoszowska, A.K. Freund, C. David, J. Synchrotron Radiat. **10**, 168 (2003)
141. V. Nazmov, E. Reznikova, M. Boerner, J. Mohr, V. Saile, A. Snigirev, I. Snigireva, M. Di Michie, M. Drakopoulos, R. Simon, M. Grigoriev, AIP Conf. Proc. **705**, 752 (2004)
142. A. Snigirev, I. Snigireva, M. Di Michiel, V. Honkimaki, M. Grigoriev, V. Nazmov, E. Reznikova, J. Mohr, V. Saile, Proc. SPIE **5539**, 244 (2004)
143. A. Mazuelas, A. Snigirev, I. Snigireva, C. David, P. Boesecke, H. Djazouli, T.H. Metzger, Proc. SPIE **5539**, 259 (2004)
144. B. Struth, A. Snigirev, O. Konovalov, A. Otten, R. Gauggel, T. Phohl, AIP Conf. Proc. **705**, 804 (2004)
145. C. Bergemann, H. Keymeulen, J.F. Van der Veen, Phys. Rev. Lett. **91**, 204801 (2003)
146. C. Schroer, B. Lengeler, Phys. Rev. Lett. **94**, 054802 (2005)
147. C.G. Schroer, Phys. Rev. B **74**, 033405 (2006)
148. F. Pfeiffer, C. David, J.F. van der Veen, C. Bergemann, Phys. Rev. B **73**, 245331 (2006)

18

Capillary Optics for X-Rays

A. Bjeoumikhov and S. Bjeoumikhova

Abstract. Capillary optics have relatively recently been introduced for the control of X-ray beams. Such optics have also been widely applied in X-ray analysis. In this chapter the basic ideas of capillary systems are presented, along with a discussion of the technological processes involved in their practical realization. Particular examples of parameter optimization for specific applications are included, and several applications of capillary optics in X-ray diffractometry, micro-X-ray fluorescence analysis and absorption spectroscopy are presented.

18.1 Introduction

Traditionally X-ray optical elements consist of crystals or different kinds of diffraction elements such as multilayer mirrors, zone plates, etc. In the past 10 years, a new type of optical element has been developed – capillary optics.

Capillary optics is based on the principle of total external reflection (TR) on smooth surfaces in capillary systems. These elements are mainly characterized by the critical angle of TR, which is relatively small. Nevertheless, optical elements using single and multiple reflections are relatively efficient and are, in some cases, unique instruments for beam shaping of X-rays from laboratory sources as well as of synchrotron radiation. In spite of the sceptical attitude of many specialists concerning the use of capillary optics, these elements have already found many applications, and new scientific and industrial instruments and methods have been developed based on capillary optics.

X-ray microprobes are widely used in combination with the most modern experimental techniques, such as the X-ray absorption fine structure (XAFS) and X-ray fluorescence analysis (XRF). XAFS is generally divided into two parts: a near-edge region within some tens of eV above an absorption edge and an extended region well above the edge. Near-edge X-ray absorption fine structure (NEXAFS), also known as X-ray absorption near-edge structure (XANES), requires high-energy resolution within an energy scan range on the order of 100 eV. The extended X-ray absorption fine structure (EXAFS) technique consists of relatively weak and broad wiggles, requires a lower energy resolution and an energy scan range on the order of >1,000 eV. Therefore, X-ray polychromatic focusing optics is required to cover the necessary energy range.

In this chapter, the physical basics of capillary optics have been described, their efficiency is shown, and application examples for different analytical methods are presented.

18.2 Physical Basics of Capillary Optics

It is well known that the absolute value of the refraction index for X-rays is generally lower than 1, which is quite different from the behavior of the corresponding refraction index for visible light. Therefore, at grazing incidence on the planes between air or vacuum and a solid, total external reflection is realized. Similarly, inside of glass fibers, the effect of internal total reflection leads to the transport of visible light. The refraction index for X-rays depends on the wavelength and on the energy of the X-ray photons, so that the following simplified formula for the critical angle (TR) can be found:

$$\theta_{\mathrm{cr}} = \frac{0.02\sqrt{\rho}}{E}, \qquad (18.1)$$

where ρ is the density of the reflecting material in g cm^{-3} and E is the energy of the X-ray photons in keV.

The reflection process is influenced by the surface roughness, the latter giving rise to a diffusely scattered component in the reflected beam. Unlike the specularly reflected Fresnel component, this part shows a dependence on the statistics of the roughness, yielding a defined angular distribution. The mathematical description of the scattering process on rough surfaces and its consideration for the modeling of capillary optical elements is a relative difficult problem. The diffuse scattering plays a negative role and leads to additional intensity losses in optical elements.

Capillary optical elements may be divided into two types: elements based on single reflections and elements based on multiple reflections.

18.2.1 Optical Elements Based on Single Reflections

Optical elements based on single reflections are monocapillaries of elliptical or parabolic shape. Using an elliptical capillary, X-rays emitted from a point source can be focused in a focal spot. In this case, the source must be positioned in the first focal point of the ellipse and the focal spot appears then in the second focal point. A parabolic capillary can focus a parallel beam and collimate divergent radiation emitted by a point source. These optical elements belong to the class of "imaging" systems. Therefore, the size of a focal

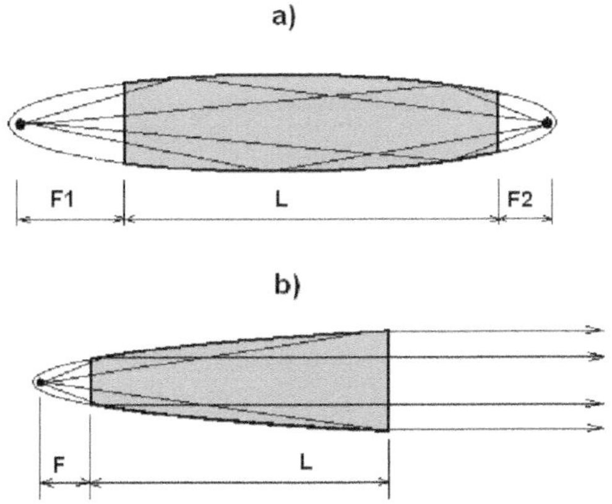

Fig. 18.1. Scheme of the optical principle by an elliptical (**a**) and a parabolic (**b**) capillary

spot will be determined not only by the accuracy of the form of the optical element but also by the properties of the source.

Figure 18.1 shows schematically the principle of focusing by means of elliptical and parabolic capillaries. The main parameters of the ellipse – the semimajor and semiminor axes – must be optimally fitted for every concrete case. For focusing, only a part of the ellipsoid is used. In this region, it should be guaranteed that the reflection angles are always below the critical angle of total external reflection for the radiation in the energy interval, which has to be focused by this optical element. If a magnification of the source is required, that part of the ellipsoid will be used which is positioned closer to the source. In the case of demagnification, that part of the ellipsoid will be used which is closer to the image point. By such an arrangement, the necessary asymmetry of focusing will be guaranteed as required by the principles of geometric optics.

18.2.2 Optical Elements Based on Multiple Reflections

A cylindrical capillary can transport X-rays by means of multiple reflections. This principle was already applied for X-rays transport in the end of the 1920s of the last century [1]. The intensity of the radiation at a defined distance from the source can be increased by such an optical element in comparison to a pinhole without a capillary. If the capillary is slightly bent (see Fig. 18.2), radiation will be still transported but the intensity losses are higher than in the case of a straight capillary. However, the transmission degree may be acceptable for not too small radii of curvature. By using a capillary, not only the radiation transport can be realized, but also the direction of radiation propagation can be changed.

If a parallel beam enters the entrance cross section of a bent capillary, then it follows from the geometry that only for $d < d_{cr}$ the entire amount of

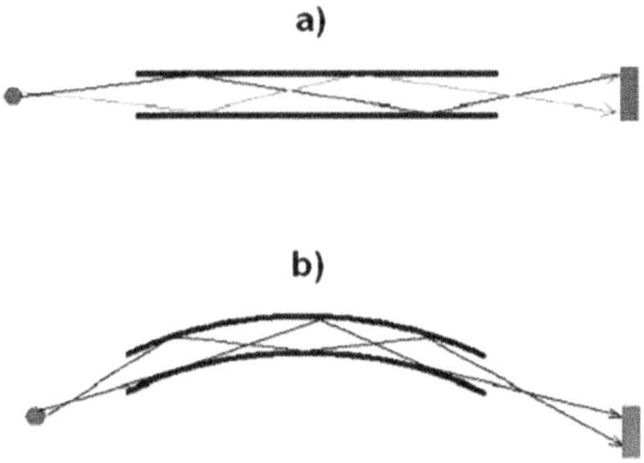

Fig. 18.2. Principle of radiation transport in a straight (**a**) and a bent (**b**) capillary

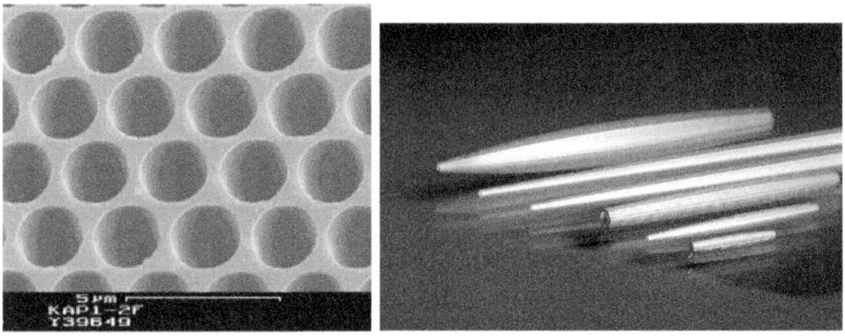

Fig. 18.3. An SEM image of the cross section of a lens structure (*left*) and photo of polycapillary lens (*right*)

incoming radiation will be captured and transported. The critical diameter depends on the critical angle (18.1) and the radius of curvature R as follows:

$$d_{\mathrm{cr}} = \frac{R\theta_{\mathrm{cr}}^2}{2}. \tag{18.2}$$

This simple formula enables one to optimize the parameters for an efficient radiation transport at a given energy.

A system of many bent capillaries, which are directed to the source at the entrance side and to the focal spot at the exit side, is able to concentrate effectively X-rays. In this way, the gain effect of single capillaries is increased by many times. Such a system, called a *polycapillary lens*, was first proposed in [2]. Although such a system does not represent a lens in a traditional sense, this name has been used permanently in the international literature.

A polycapillary lens is a monolithic system made of microstructured glass consisting of a huge number of capillary channels (in some cases several millions). Figure 18.3 shows a scanning electron microscope (SEM) image of the cross section of a lens structure and photo of polycapillary lenses as an

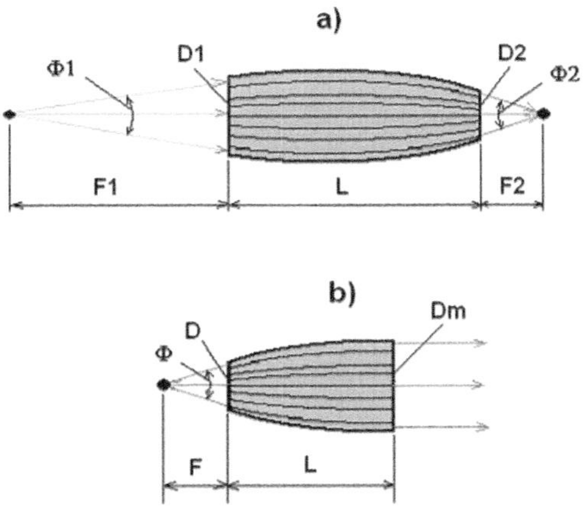

Fig. 18.4. Practical representation of two capillary lens types: (**a**) focusing lens and (**b**) collimating semilens

example. The dimensions of the capillary channels in lenses in use today lie between 1 and 100 μm.

Figure 18.4 presents schematically the working principle of two types of polycapillary lenses. In the first case (Fig. 18.4a), a full lens focuses the radiation from a point source into a focal spot of small size. In the second case (Fig. 18.4b), the so-called semilens transforms a divergent beam into a parallel beam or focuses a parallel beam into a focal spot.

Capillary lenses have the following main parameters: first focal distance, length, second focal distance, and capturing angle. The efficiency of focusing is characterized by the following parameters: focal spot size and intensity gain in comparison with the direct beam. These parameters depend on the energy of photons.

The optimization of the lens can be performed starting with the source size and the energy interval, on the one side, and taking into account the focal spot size to be reached, on the other side. The optimum inner diameter of the capillary channels depends also on the energy interval of the radiation. It can be estimated by means of formula (18.2) if the other geometric parameters of the lens are fixed.

The relations between the first focal distance and the size of the given source and between the second focal distance and the required focal spot size are given by the following formulas

$$\Delta X_{\mathrm{s}} = 2\theta_{\mathrm{cr}} f_1 + D_{\mathrm{c}}, \tag{18.3}$$

$$\Delta X_{\mathrm{f}} = 2\theta_{\mathrm{cr}} f_2 + D_{\mathrm{c}}, \tag{18.4}$$

where ΔX_{s} is the source size, f_1 is the first focal distance, D_{c} is the inner diameter of the capillary channels, ΔX_{f} is the focal spot size, and f_2 is the second focal distance. In practice, the inner diameter of capillary channels can

Table 18.1. Geometrical parameters of the lens 51MLS02

f_1 (mm)	47.5 ± 0.3
f_2 (mm)	15.3 ± 0.2
L (mm)	101.5
Input lens size (mm)	4.6
Input capillary size (μm)	6.7
Maximum lens size (mm)	6.9
Maximum capillary size (μm)	10
Output lens size (mm)	2.5
Output capillary size (μm)	3.6

Table 18.2. Physical parameters of the lens 51MLS02

E (keV)	3–5	5–7.5	7.5–10	10–15	15–20	20–25	25–30
Focal spot size FWHM (μm)	38	40	41	41	35	31	33
Intensity gain	2,892	7,078	7,929	8,070	8,502	5,631	1,550

be neglected in most cases relative to the part containing the critical angle $2\theta_{\mathrm{cr}} f_{1,2}$. From these formulas, the following important relation between the main lens parameters is found

$$\frac{\Delta X_{\mathrm{f}}}{\Delta X_{\mathrm{s}}} = \frac{f_2}{f_1}. \tag{18.5}$$

The set of formulas (18.1)–(18.5) allows practically to optimize all lens parameters taking into account the given energy interval, the source size, and the required focal spot size.

As an example, experimental results of test measurements of a lens are presented in Table 18.2 for the geometrical parameters given in Table 18.1. In this case, the lens is asymmetric; the entrance focal distance is three times longer than the exit one. Therefore, a focal spot size smaller than the source size will be expected.

The measurement procedure of the focal spot of an X-ray focusing system is nontrivial. There are known different methods with advantages and disadvantages. Several authors use the method of a knife edge (see, e.g., [3]). This method provides a good energy resolution and a high spatial resolution. However, this method is not applicable to high intensities of the collimated beam and does not yield two-dimensional intensity distributions in the focal plane.

In the present chapter, the method employing a small pinhole was used to obtain two-dimensional distributions in the focal plane and to avoid detector problems at high count rates. This pinhole had a conical shape instead of a

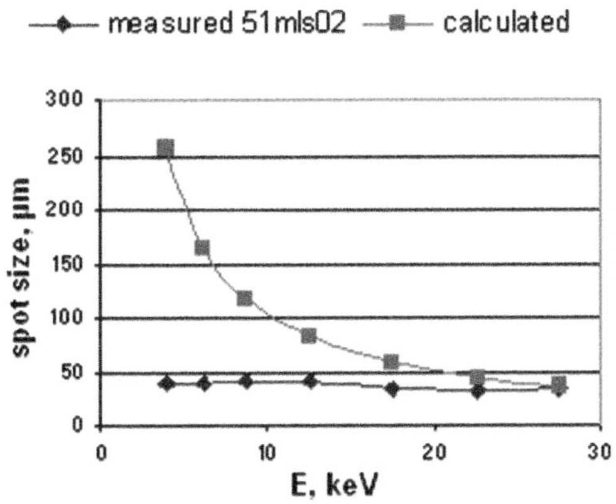

Fig. 18.5. Calculated and measured (pinhole scan) dependence of the focal spot size on energy for the lens 51MLS02

cylindrical one to take into account the beam convergence especially for lower photon energies at the exit of the measured capillary.

Table 18.2 shows the measured physical parameters of the given lens – the focal spot size and the gain factor. A microfocus X-ray tube with a molybdenum anode was used as source. The anode spot size of the tube is 50 µm. For detection of X-rays, an SDD detector (manufacturer BRUKER-AXS, MA) was used. The focal spot size at different energies was measured by scanning over it with a pinhole of 5 µm in diameter situated in front of the detector entrance window. The scan with a pinhole enables one to measure the intensity distribution in the focal spot as a function of energy. The intensity gain factor was also measured using this pinhole and determined as the ratio of the intensity in the focal point with the lens and without the lens.

Figure 18.5 shows the radiation energy dependence of the focal spot size as measured and as calculated by (18.4). As considered above, the focal spot size should rise strongly with decreasing radiation energy. However, the measured values depend only weakly on energy and below 4 keV they decrease with decreasing energy. Such a behavior at low energies is shown also by computer simulations [3]. Near 25 keV, the measured value approaches the calculated one.

This result can be explained by the fact that, in the propagation of radiation in the capillary channels, the absorption for rays with relative large incidence angles is higher which leads to a self-collimation of the beam. This effect is especially related to the low-energy part of the spectrum. Despite a relatively high critical angle of total reflection at low energies, only photons moving under small angles relative to the capillary axis have favored conditions for propagation within the capillary channels. A small increase of the focal spot size in the energy interval 25–30 keV in comparison with 20–25 keV is obviously connected with edge effects at the pinhole. The effective size of the

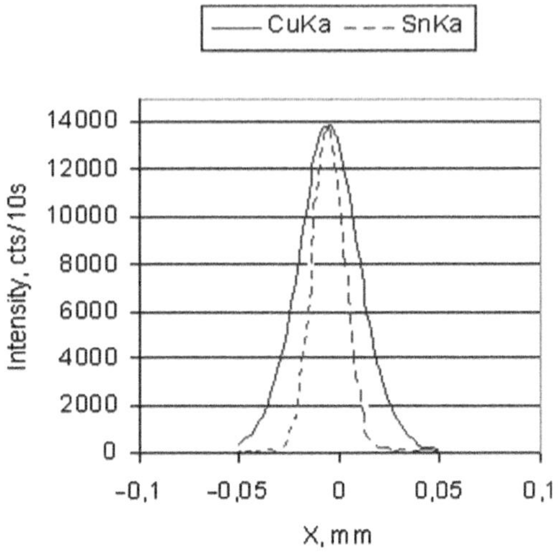

Fig. 18.6. Intensity distributions of Cu Kα and Sn Kα made by scanning with a wire of 10 µm in diameter (anode high-voltage $U_a = 50$ keV)

pinhole increases due to the decrease of radiation absorption at higher photon energies in the pinhole material. To support this assumption, a scan with a wire of 10 µm in diameter containing 94% Cu and 6% Sn was carried out (see Fig. 18.6). This experiment was important to show whether the so-called *halo effect* appears or not. In many publications, the existence of a transmission halo around the main sharp focus of polycapillary lenses is mentioned. This effect occurs in the high-energy region and deteriorates the spatial resolution in the analysis of heavy elements /18/.

As seen from the Sn Kα distribution in Fig. 18.6, there is no "halo effect." This means that the focal spot is not blurred for radiation energies above the K absorption edge of Sn (>29.2 keV) and the spatial resolution will not be deteriorated. This result confirms the assumption made above that in the pinhole method the edge effect influences the measured value of the focal spot size. Therefore, in the energy region above 29 keV, the correct size of the focal spot is not larger than 19 µm.

Figure 18.7 shows the intensity distribution in the focus of the lens in the energy interval 15–20 keV. The measurement was performed by making a two-dimensional scan with a pinhole of 5 µm in diameter in the focal plane of the lens. As seen, the focal spot is relatively symmetric without any distortions.

In this section, different methods of characterization of polycapillary lenses were demonstrated. In the high-energy region for the same lens, the experimentally determined parameters differ in dependence on the measuring procedure used. Therefore, for a correct characterization of the optics, a combination of different methods is required.

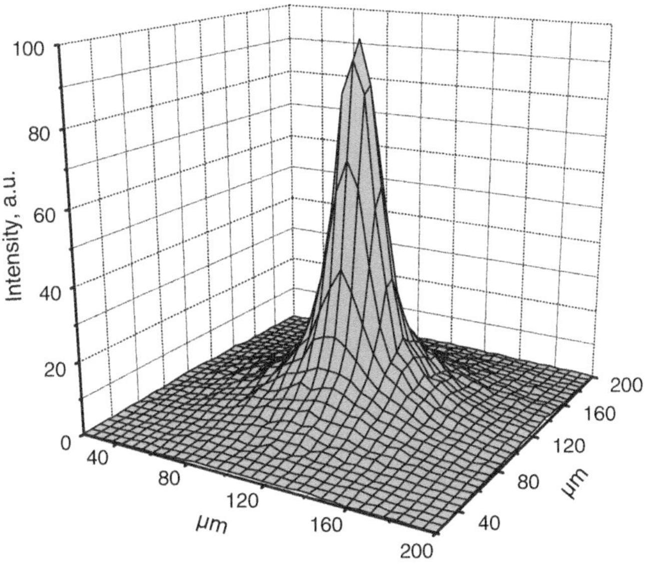

Fig. 18.7. Two-dimensional intensity distribution in the focal plane of the lens 51MLS02

18.3 Application Examples for Capillary Optics

18.3.1 X-Ray Fluorescence Analysis with Lateral Resolution

Measurement of the Elemental Distribution at the Sample Surface

As shown in Sect. 18.2.2, a polycapillary lens focuses X-ray efficiently in an energy interval from several keV to several tens of keV. This corresponds to the energy interval required for the XRF. Due to the obtainable sizes of focal spots, polycapillary lenses are very useful for XRF analysis with lateral resolution. Using such lenses, a high density of the primary excitation radiation in a small focal spot can be reached at the sample surface. By this way, an intensity of the fluorescence radiation was obtained that was not only high enough to carry out explicit analysis of elements in the measured point, but also to determine the two-dimensional distribution of elements across the sample surface in a reasonable time.

Many papers are devoted to the realization of micro-XRF using monocapillary systems and polycapillary lenses in combination with laboratory and synchrotron X-ray sources [4–9]. The lateral resolution obtained for monocapillaries at a synchrotron source amounts to about 1 µm. Polycapillary lenses allow one to reach a minimum lateral resolution of about 15 µm. In this case, the element detection limit amounts to several ppm.

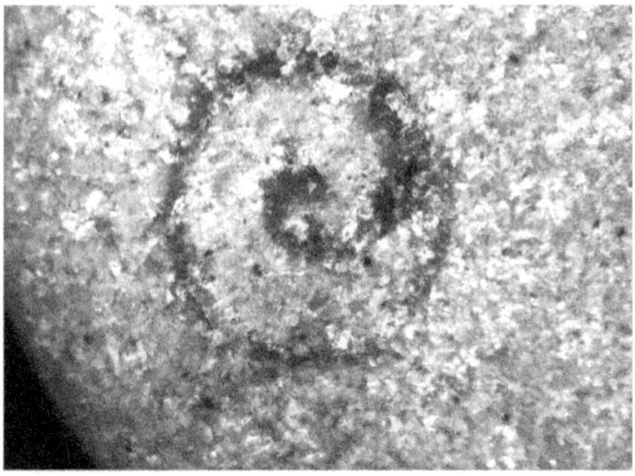

Fig. 18.8. Microscopic image of a fossil snail

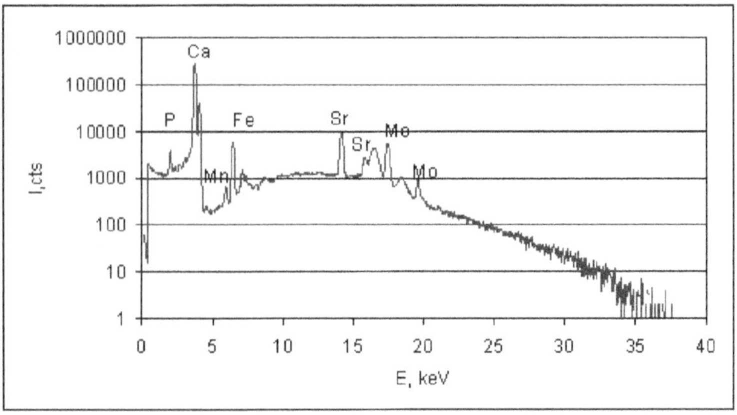

Fig. 18.9. XRF spectrum of the stone sample

Another example for the application of micro-XRF is the analysis of a stone with an included fossil snail. This stone was found on the coast of the Baltic Sea on the island Ruegen. The sample shows a strong inhomogeneity with crystalline inclusions and a rough surface. In Fig. 18.8, an image of the area of interest taken by an optical microscope is shown. Crystalline blocks, different inclusions, etc., are clearly observed. The size of the snail amounts to ca. 3 mm in diameter, and it has a darker color than the general background. The surface of the stone was placed in the focal plane of the lens. Spectra taken from different points are quite different. Therefore, in a first step, a spectrum was accumulated from a surface area of about $4 \times 4\,\mathrm{mm}^2$ for the determination of the general elemental composition of the sample. The accumulation time was 500 s.

Figure 18.9 shows the spectrum obtained in this way. As seen from this spectrum, the sample contains mainly the elements P, Ca, Mn, Fe, and Sr. The lines are situated on a background of scattered primary bremsstrahlung.

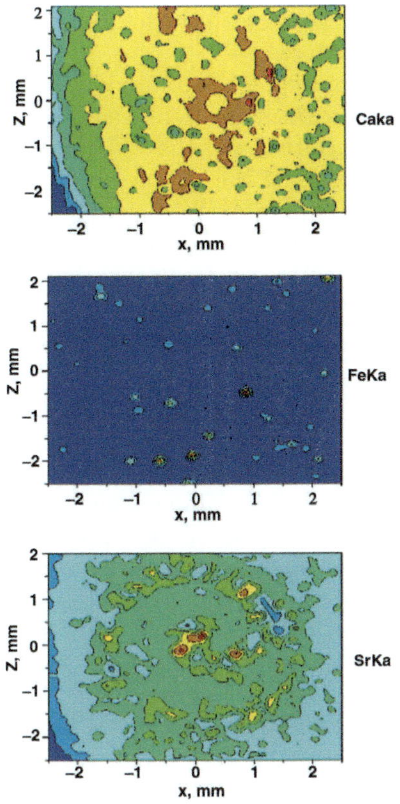

Fig. 18.10. Two-dimensional distributions of the elements Ca, Fe, and Sr in the stone shown in Fig. 18.8

The basic element is Calcium, its line intensity is more than one order of magnitude higher than the other lines. Two-dimensional distributions of the main elements Ca, Fe, and Sr were measured at an X-ray tube high voltage of 40 kV and 700 μA tube current. The accumulation lifetime for every point was 5 s, the dead time was about 15%, i.e., 0.75 s. For changing the position from one point to another, 0.4 s was needed. An area of $4.5 \times 5\,\text{mm}^2$ was scanned with a step size of 0.05 mm. Therefore, a general scanning time of 10 h was required. Figure 18.10 shows the two-dimensional distributions of the above-mentioned three elements.

As seen from the picture, Ca has a more or less homogeneous distribution, Fe forms inclusions, and Sr is contained only in the snail, i.e., its distribution follows the snail's silhouette.

In [10], the development of a special X-ray source for the application of micro-XRF in SEMs has been described. As is well known, the sensitivity of microanalysis excited by an electron beam is lower than that excited by an X-ray beam. Therefore, a combination of the high spatial resolution of an SEM image of the surface morphology with the high elemental detection sensitivity of micro-XRF is of great interest.

A further interesting solution is the combination of micro-XRF and Raman spectrometry within one instrument. Here, the micro-XRF gives information about the elemental composition and the Raman spectrometer delivers data about the molecular structure of the sample.

The lateral resolution for micro-XRF amounts to about 25 μm and for Raman spectrometry about 5 μm. The simultaneous analysis with both methods to determine the distribution of elements and molecules can be carried out at specific points or by scanning the sample surface. Such an instrument has been developed within the framework of the European project PRAXIS [11].

Measurement of the Depth Profile of an Elemental Distribution

In micro-XRF, the elemental composition of a small sample volume in a layer near the surface has been studied. This volume is approximately a cylinder whose axis corresponds to the axis of the primary beam. The diameter of this cylinder is determined by the beam cross section and depends only on the focusing properties of the optical element. The depth of the analyzed layer will be determined by the emission depth of the secondary fluorescence radiation and depends on the emission angle and the absorption in the sample material.

Recently, a 3D micro-XRF arrangement has been realized which is based on the principle of a selective depth-resolved detection of the secondary fluorescence radiation. For this arrangement, a so-called polycapillary conical collimator (poly-CCC) [12–14] or a polycapillary semilens [15] is used. In this way, an additional depth resolution of micro-XRF can be reached within the limits of the emission depth for the secondary fluorescence radiation. These methods have been realized by using a synchrotron source, because a rather high intensity is required. However, recently it was shown that, besides the use of powerful synchrotron sources, a combination of a microfocus X-ray tube with an efficient polycapillary lens provides enough intensity for the realization of a tabletop 3D micro-XRF setup [16, 17] under laboratory conditions. However, the sensitivity of the laboratory arrangement was found to be one to two orders of magnitude lower than the synchrotron-based setup.

In [18], another method for the determination of the depth profile of elemental distributions in samples has been proposed. As before, a polycapillary lens is used for focusing the primary beam from a laboratory X-ray source. The depth resolution is achieved by the knife-edge method on the detector side. The sharp edge of an absorber, which is positioned between the sample and the detector, generates a transition "light–shadow" at the detector surface. This transition region contains information about the intensity of fluorescence radiation emitted from different depths of the sample. The intensity distribution in the transition region has been measured by shifting a slit in front of the detector. In the arrangement described, a depth resolution of

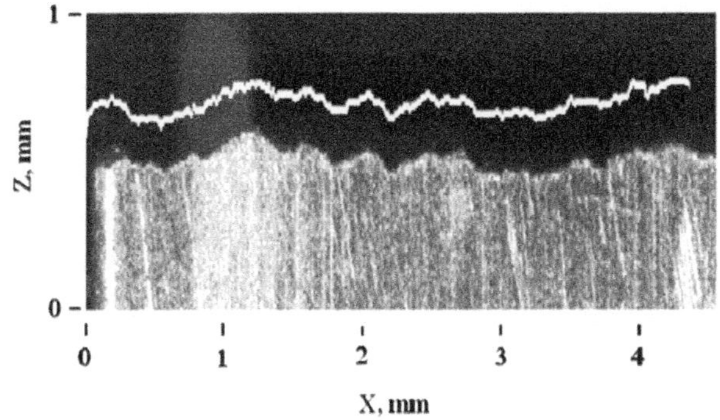

Fig. 18.11. Microscopic image of the edge of a brass sample and the corresponding profile measured by the X-ray method

24 µm was achieved with a focal spot size of 40 µm. Using this method, not only the depth profile of an elemental distribution but also the surface profile of the sample can be determined. Here a depth resolution of 7 µm was obtained. Figure 18.11 shows the microscopic image of the surface profile of a brass sample and the corresponding X-ray measurement result. Unlike optical profile measurements, the procedure described allows one to also observe profiles which are covered by surface layers or also boundary profiles of different materials.

The 3D micro-XRF is a relatively informative method and it will find applications in different fields, such as in analysis of biomedical materials, of objects of arts with a layered structure or of multilayered microelectronic circuits, etc. The depth resolution becomes important, if the layered structure consisting of different elements is unknown and especially, if this structure is laterally nonhomogeneous so that the usual quantitative analysis becomes impossible because an appropriate model is not available.

18.3.2 X-Ray Diffractometry

X-Ray Structure Analysis (Single Crystal Diffractometry)

It is important in X-ray structure analysis that the primary radiation directed on a sample consisting of an isolated single crystal of small dimensions (typical 100–500 µm) has a high intensity and is well collimated. In the conventional method, pinholes are used for collimation of the divergent primary beam. In this case, only a very small part of the radiation intensity emitted by the source can be used.

The simplest method to increase the efficiency of the source is to use a cylindrical capillary between the source and the crystal. Dependent upon the geometry, such an optical element will increase the radiation intensity at the crystal by a factor of 3–10 due to the radiation transport inside the capillary. The time of measurement is correspondingly decreased. This simple procedure has been shown to be very useful in practice and is applied today in many diffractometers for a variety of purposes [19]. Due to the radiation transport through the capillary, an exit beam is formed with the diameter of the capillary and a divergence of the order of magnitude of the critical angle of total external reflection.

Microdiffractometry

The principle described above is also useful if a locally high resolution during the investigation of massive samples is required. The inner diameter of a capillary can amount to 10 µm. Using such an optical element, irradiation of a small volume element of the polycrystal or an individually selected single crystal becomes possible.

A further increase of the irradiation intensity for local analysis can be achieved by using a capillary of ellipsoidal form. Such an optical element realizes not only the radiation transport but also its concentration, so that a higher intensity gain can be reached. For example, in [20], stress analysis with a spatial resolution of 70 µm has been performed.

Local diffractometric analyses can be carried out by using polycapillary lenses. In this case, the polycapillary lens focuses the primary radiation from the source onto the sample surface. In this way, in [21], the method of energy-dispersive microdiffractometry has been realized. Accordingly, the secondary radiation has been registered by an energy-dispersive detector. Here, phase and texture in a selected small surface area can be estimated by the position and the intensity relations of the Laue diffraction peaks. By scanning the sample surface, changes in the texture can be measured. Figure 18.12 shows the two-dimensional intensity distribution of the Laue lines (420) and (521) of a molybdenum sample. It was observed that an intensity decrease of the (420) line corresponds to an intensity increase of the (521) line.

Analysis of Polycrystals and Powders

Powder diffractometry has been realized by means of a polycapillary semilens. The primary radiation beam formed by such a lens is directed onto the massive sample. Using a Soller slit or a plane secondary monochromator, the diffracted secondary radiation beam will be separated from the incident beam [22]. By this way diffractometry is realized in a collimated beam.

Using a focusing polycapillary lens, phase analysis can be carried out. In this case, a relatively large second focal distance is used. The sample, in form of a foil or a powder, is held between two X-ray transparent foils and is positioned directly at the exit of the lens. The diffractogram is registered in the focal plane of the lens. For this purpose, a single channel detector with a two-Theta scan [23] or a bent position-sensitive detector can be used. In these

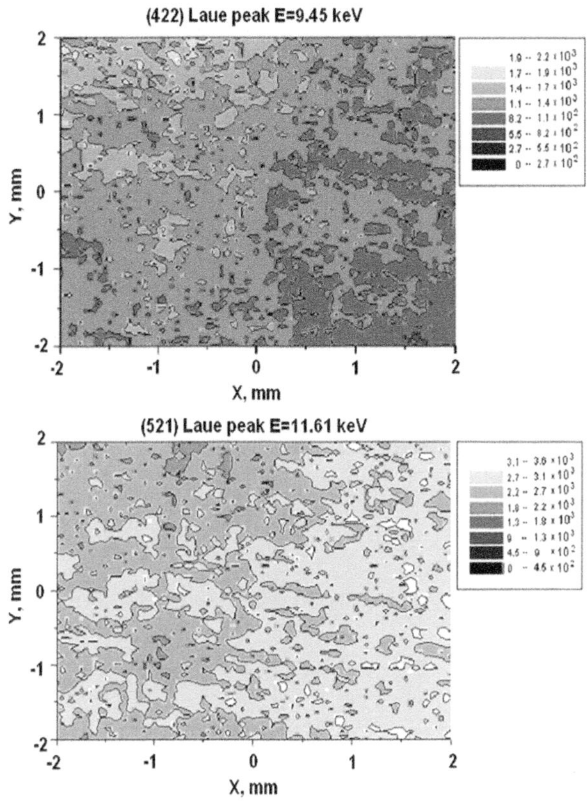

Fig. 18.12. Two-dimensional intensity distributions of Laue diffraction peaks (420) and (521) from a molybdenum sample

methods, high intensities are achieved due to the relatively large aperture of the polycapillary lenses.

At a synchrotron beamline, a method of measurement has been realized, where the probe volume was imaged by detecting the small angle scattering of the primary synchrotron beam [24]. To separate the primary beam and the diffracted beam, a parallel polycapillary structure was positioned between the sample and a CCD detector.

Real-Time Diffractometry

The principle described in the second part of the last section can also be used to record diffractograms in real time. In the same way, focusing in a transmission arrangement is applied and the diffractograms are recorded in the focal plane of the polycapillary lens. For detection, an X-ray-sensitive CCD camera was used which can record the diffraction pattern in real time [25]. Figure 18.13 shows the diffractogram of a zirconium foil which is a single

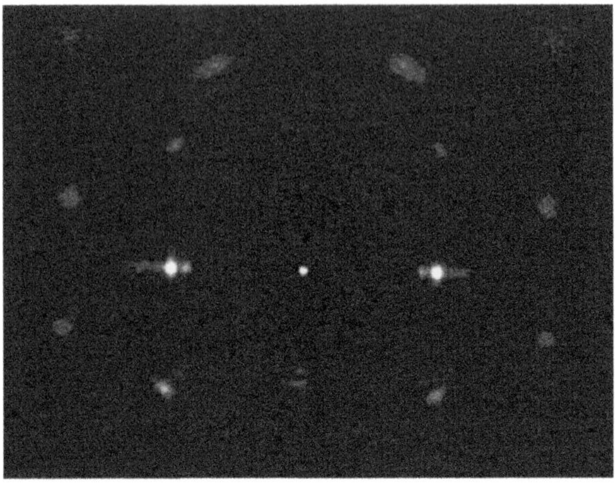

Fig. 18.13. Diffractogram of a zirconium foil in form of a single half-frame captured from a live video sequence

half-frame captured from a live video sequence. The appearance of a texture is clearly seen.

18.4 Capillary Optics for Synchrotron Radiation

Capillary optics in synchrotron beamlines for focusing the radiation onto a spot of micron and submicron size employs tapered monocapillaries [26]. Development of new methods of glass treatment makes the fabrication of elliptical monocapillaries with micrometer spatial resolution possible. Such monocapillaries can be placed into the preliminary focused beam at the end of conventional beamline. This provides a fine focusing of an X-ray beam into a micrometer spot size.

The work deals with the experimental results on the beam focusing and μXFA, μXANES, application in archeometry at BESSY GmbH (Berlin) [27]. The beam, in the energy range of 3.5–15 keV, is monochromatized by a double-graded-crystal monochromator and focused onto the sample with a spot size of 150 μm (horizontal) by 700 μm (vertical). The total flux in the spot is in the range 10^9–10^{10} photons s^{-1} per 100 mA. The capillary is placed close to the focal position of the refocusing mirror of the beamline with a beam cross section of 400 μm (horizontal) by 800 μm (vertical). The optical scheme of the experimental setup is shown in Fig. 18.14.

The parameters of the inner elliptical surface of the glass monocapillary are shown in Table 18.3. A capillary was fabricated by "bubble" technology from the borosilicate glass [28]. In the first stage of technology, a bubble is injected into the molten glass. In the second stage, the glass is drawn with a given pulling factor.

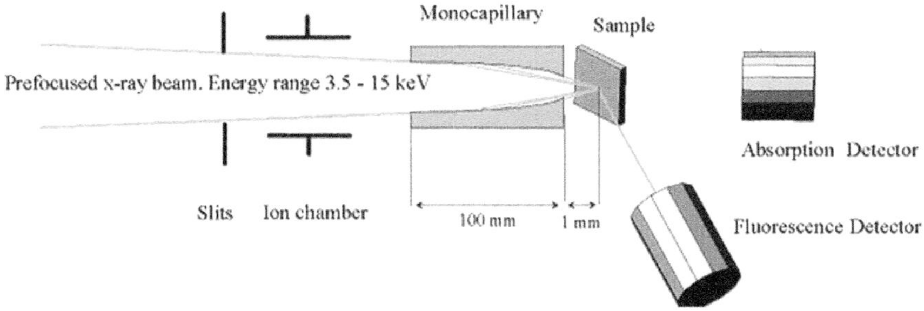

Fig. 18.14. Optical scheme of the microprobe experimental setup

Table 18.3. The parameters of the elliptical glass capillary

Large half-axis (mm)	Small half-axis (μm)	Number of input aperture (μm)	Number of output aperture (μm)	Focal length (mm)	Experimental intensity gain	Focal size (μm)
100	175	350	15	0.8	40–50	∼5

The final parameters of the ellipse are defined by the diameter of the bubble and the pulling factor. The form of a capillary was optimized to the parameters of the exit beam on the given beamline. The capillary material contains very few amounts of heavy elements and does not affect on the fluorescence spectrum in the above-mentioned energy range.

The microprobe beamline provides an experimental setup for μEXAFS and μXANES measurements in air or He in the energy range of 3.6–15 keV. The detector system consists of a Si-PIN photodiode for fluorescence measurements, three ion chambers, a scintillation counter, and an energy-dispersive detector.

A sample stage and detectors are mounted on a two-rotation axis goniometer. Precise X–Y–Z translations provide a 3D scan with 1 μm reproducibility of the sample position.

Knife-edge tests have been done to measure the size of the focal spot in transmission. Sharp-edge platinum blades were used for these measurements. A two-dimensional mapping of a focal spot was performed using a platinum pinhole of 5 μm in diameter and a scintillation counter. The results of measurements are shown in Fig. 18.15a, b. A flux density gain was measured in the capillary focal plane using the same pinhole at the energies 5, 10, and 14 keV. The flux through the pinhole was compared with the direct flux without the capillary.

Comparative XANES and EXAFS spectra measurements of test objects consisting of microparticles of Sn, Mn, Fe, and Ge microcrystals were carried out using primary and microfocus beams. The results are completely

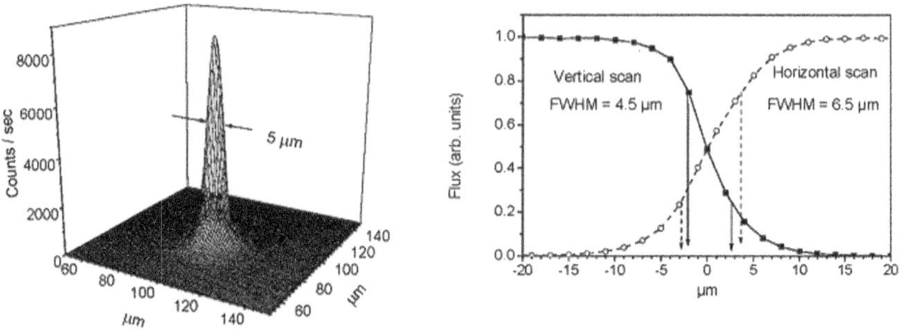

a) b)

Fig. 18.15. (**a**) Flux distribution in the focus of a monocapillary measured with a 5 μm Pt pinhole at 12 keV. (**b**) Knife-edge test at 9 keV. Vertical spot size is 4.5 μm FWHM and horizontal is 6.5 μm FWHM

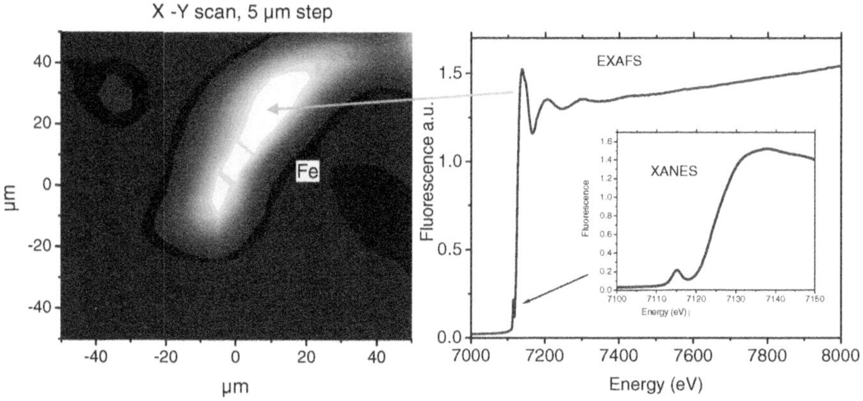

a) b)

Fig. 18.16. (**a**) Two-dimensional scan of a Fe microparticle on a test object measured in the fluorescence mode at 6.4 keV. (**b**) EXAFS and XANES spectra of the iron oxide microparticle

identical, verifying the significance of the method. A two-dimensional map of a Fe microparticle is shown in Fig. 18.16a. An EXAFS (XANES) spectrum of this particle, measured with a microbeam of 5 μm diameter, is represented in Fig. 18.16b.

In addition, in the work presented, the investigation of corrosion processes of amalgam layers on a microscopic scale is shown using μXANES [26]. Sn–Hg alloys, Hg and SnO_2, and possibly SnO could be identified within the corrosion areas studied.

18.5 Concluding Remarks

As shown, capillary optics can be used for X-ray diffractometry and X-ray fluorescence analysis. Efficient beam focusing allows it to realize micro-XRF and micro-XRD, where the intensity of secondary radiation is high enough not only for a scan across the sample surface but also for a measurement of depth profiles. Further development of capillary optics is planned with the goal of achieving an increased efficiency and a decrease in the focal spot size. Combinations of sources with a high emission density with capillary optical elements will find additional applications in the future.

References

1. F. Jentzsch, Phys. Zeitschrift **30**, 268 (1929)
2. M.A. Kumakhov, *Radiation of Channeled Particles in Crystals* (Energoatomizdat, Moscow, 1986) (in Russian)
3. N. Gao, K. Janssens, in *X-Ray Spectrometry: Recent Technological Advances*, ed. by K. Tsuji, J. Injuk, R. Van Grieken (Wiley, Chichester, 2004), pp. 89–110
4. A. Rindby, Nucl. Instrum. Meth. Phys. Res. Sect. A **249**, 536 (1986)
5. N. Gao, I.Yu. Ponomarev, Q.F. Xiao, W.M. Gibson, D.A. Carpenter, Appl. Phys. Lett. **71**, 3441 (1997)
6. H. Bronk, S. Röhrs, A. Bjeoumikhov, N. Langhoff, J. Schmalz, R. Wedell, H.-E. Gorny, A. Herold, U. Waldschläger, Fresen. J. Anal. Chem. **371**, 307 (2001)
7. A. Bjeoumikhov, N. Langhoff, R. Wedell, V. Beloglazov, N. Lebedev, N. Skibina, X-Ray Spectrom. **32**, 172 (2003)
8. C.A. MacDonald, W.M. Gibson, X-Ray Spectrom. **32**, 258 (2003)
9. A. Bjeoumikhov, S. Bjeoumikhova, N. Langhoff, R. Wedell, Rev. Sci. Instrum. **76**, 063115 (2005)
10. A. Bjeoumikhov, V. Arkadiev, F. Eggert, V.-D. Hodoroaba, N. Langhoff, M. Procop, J. Rabe, R. Wedell, X-Ray Spectrom. **34**, 493 (2005)
11. A. Bjeoumikhov, N. Langhoff, J. Schmalz, J. Tilgner, R. Wedell, Kompakter mobiler Messkopf für die Element- und Molekülspektrometrie durch die Nutzung der Methodenkopplung (RF- und Raman-Spektroskopie) (Compact Mobile Measuring Head for Elemental and Molecular Spectroscopy by Using Method Coupling (XRF and Raman Spectroscopy)). *11. Anwendertreffen Röntgenfluoreszenz- und Funkenemissionsspektrometrie*, 8–9 März 2004, Universität Duisburg-Essen (Report published on CD)
12. A.A. Bzhaumikhov, N. Langhoff, J. Schmalz, R. Wedell, V.I. Beloglazov, N.F. Lebedev, Proc. SPIE **3444**, 430 (1998)
13. C. Fiorini, A. Longoni, A. Bjeoumikhov, IEEE Trans. Nucl. Sci. **48**, 268 (2001)
14. B. Kanngießer, W. Malzer, I. Reiche, Nucl. Instrum. Meth. Phys. Res. Sect. B **211**, 259 (2003)
15. L. Vincze, B. Vekemans, F.E. Brenker, G. Falkenberg, K. Rickers, A. Somogyi, M. Kersten, F. Adams, Anal. Chem. **76**, 6786 (2004)
16. B. Kanngießer, W. Malzer, A.F. Rodriguez, I. Reiche, Spectrochim. Acta B **60**, 41 (2005)

17. W. Malzer, *3D Mikro-Röntgenfluoreszenzanalyse (3D Micro-XRF Analysis)*, Ph.D. Thesis, Universität Bremen, 2004
18. A. Bjeoumikhov, S. Bjeoumikhova, R. Wedell, R. Gubzhokov, Z. Margushev, Nucl. Instrum. Meth. Phys. Res. Sect. B **248**, 142 (2006)
19. R. Wedell, Phys. Blätter **52**, 1134 (1996)
20. B. Eigenmann, N. Langhoff, A. Bjeoumikhov, A. Hase, R. Stabenow, Mater. Sci. Forum **404–407**, 303 (2002)
21. A. Bjeoumikhov, S. Bjeoumikhova, N. Langhoff, R. Wedell, Appl. Phys. Lett. **86**, 144102 (2005)
22. M. Gubarev, E. Ciszak, I. Ponomarev, W. Gibson, M. Joy, J. Appl. Crystallogr. **33**, 882 (2000)
23. A. Bjeoumikhov, N. Langhoff, J. Rabe, R. Wedell, X-Ray Spectrom. **33**, 312 (2004)
24. T. Wroblewski, A. Bjeoumikhov, Nucl. Instrum. Meth. Phys. Res. Sect. A **521**, 571 (2004)
25. A. Bjeoumikhov, S. Bjeoumikhova, N. Langhoff, H. Wolff, J. X-Ray Sci. Technol. **13**, 185 (2005)
26. C. Riekel, Nucl. Instrum. Meth. Phys. Res. Sect. B **199**, 106 (2003)
27. J. Bartoll, S. Röhrs, A. Erko, A. Firsov, A. Bjeoumikhov, N. Langhoff, Spectrochim. Acta B **59**, 1587 (2004)
28. A. Knöchel, G. Gaul, F. Lechtenberg, German Patent DE 4444102C2 (1994)

19

Reflective Optical Arrays

S. Lagomarsino, I. Bukreeva, A. Surpi, A.G. Michette, S.J. Pfauntsch, and A.K. Powell

Abstract. In this chapter the prospects of using grazing incidence reflection from custom built nested mirrors or reflector arrays are discussed. The aim is to provide a high gain in focused intensity from either laboratory or synchrotron sources. The calculated performances of such systems are presented, taking into account manufacturing tolerances and surface roughness. These calculations indicate that the improvement of roughness is of primary importance, and ways of addressing this during possible manufacturing processes are discussed.

19.1 Introduction

X-ray reflective mirrors are widely used in both synchrotron radiation and laboratory sources to collimate or focus X-ray beams with high efficiency. Improvements to fabrication technologies now allow manufacture of long X-ray mirrors with slope errors less than 1 μrad and root-mean-square (rms) roughnesses less than 0.1 nm [1–3]. In the Kirkpatrick–Baez geometry, such mirrors can provide significantly sub-micrometre focal spots. However, these systems are very expensive and, in general, only large-scale facilities can afford high precision mirrors. To improve flexibility, adaptive mirrors using benders to give the required shapes have been developed [1] but alignment, stability and optimisation of the curvature are critical issues.

If a given spot size is desired, i.e., if a given demagnification factor at a certain distance from the source is required with high focusing efficiency, the length of the mirror should not exceed an optimum value [4]. The flux density of photons in the focal spot can be increased significantly compared to that from a single surface if a system of confocal mirrors, similar to those used in astronomy, is employed [5].

Mono- and poly-capillaries [6, 7] and microchannel plate arrays [8, 9] have also been used for X-ray optics. These work by grazing incidence reflections along many small diameter channels, up to about 10^6 for poly-capillaries and microchannel plates. They can have large apertures and bandpasses, with transmission efficiencies of several tens of percent. If there are two reflections from the same wall of each channel, they can approximately satisfy the Abbe sine condition [10], thereby reducing coma. If there are more than two reflections, correspondence between object and image points may be lost. Current manufacturing methods, in which optical fibres undergo several stages of pulling [7, 11], limit channel diameters to ~0.5 μm, arrays to a few millimetres square and focal spots to ~10 μm. Performances are limited by channel tilting, curvature errors, waviness, diffraction, uncontrolled substrate bending and defects, i.e., misplaced channels [12]. In addition, microchannel plates are primarily made for another purpose, i.e., X-ray detection, so that they are not optimised for optical performance; for example, the channel wall roughnesses are not a major concern for X-ray detection. Such optics can also have very poor point spread functions [13]; in principle, it is possible to improve their performances by using specific designs – microstructured optical arrays (MOAs) [14]. These may also allow adaptivity and controllable focal length [15].

In the following sections, the performances of such optics will be presented, along with discussions of fabrication routes to compact systems adaptable to a large range of photon energies and X-ray sources. These systems can be tailored to be focusing, collimating or condensing devices.

19.2 Nested Mirror Systems

Nested mirror systems use arrays of confocal reflecting surfaces such as shown in Fig. 19.1. In this example of elliptical mirrors, rays emanating from one focus F_1, common to all the mirrors, converge after reflection to the second common focus F_2. It is worth noting that, in general, the mirrors do not have to be elliptical; the shape depends on the application.

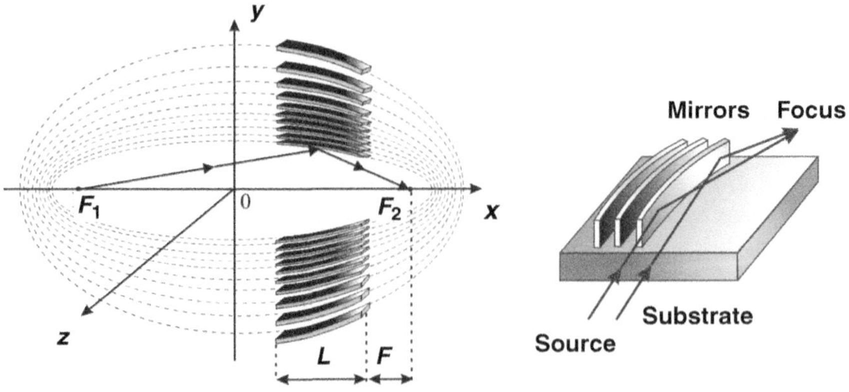

Fig. 19.1. Nest of confocal elliptical mirrors

19.2.1 Computer Simulations

The proper choices of system configuration and material for the reflecting surfaces are very important. For optimisation of the relevant parameters and evaluation of the fabrication tolerances, a computer ray-tracing code allowing parallel treatment of the optical elements has been developed [16, 17]. This code gives information about the intensity distribution in the focal spot, the system efficiency and the gain. Gain is defined as the ratio between the output and input flux densities, and corresponds to the enhancement in intensity with respect to a hypothetical aperture of the same size as the focal spot placed in the focal position. The code has been used to evaluate nested mirror systems for a synchrotron radiation source and for a laboratory-scale source.

For the synchrotron, a bending magnet source of the European Synchrotron Radiation Facility, namely the optical beamline BM5, was used as an example. A geometry which focuses the beam in the horizontal plane was considered, with a horizontal source size of 270 μm, a divergence of 2.4 mrad, a source-focus distance of 40 m and a wavelength of 0.1 nm. The focusing system (Fig. 19.2) has four nickel mirrors with length of 25 mm, height of 300 μm and wall thicknesses of 37 μm at the entrance and 20 μm at the exit. The simulations indicate that this system can focus the incident beam into a spot of ≈0.14 μm full width at half maximum (FWHM) at 1 cm from the exit edge of the system, with a gain of around 980.

For the laboratory-scale microfocus source, a focused electron spot size of 15 μm, a wavelength of 0.154 nm and a source-focus distance of 150 mm were assumed. The optimised system in this case has two symmetrical nickel mirrors of length 25 mm, height 300 μm and wall thickness 100 μm (Fig. 19.3). This provides focusing into a spot of around 0.8 μm FWHM, 5 mm from the exit edge of the system, and a gain of around 90.

A variety of characteristic distortions in technological processing and errors that are inevitably introduced during manufacture can both lower the

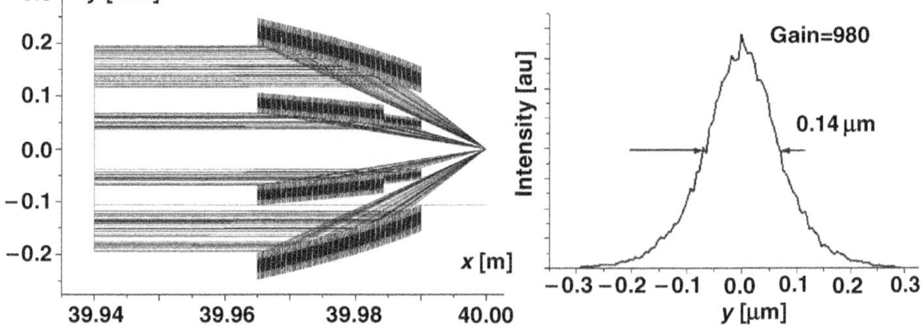

Fig. 19.2. Results of optimisation: (*left*) an optical system for a synchrotron radiation source and (*right*) the distribution of intensity in the focal spot

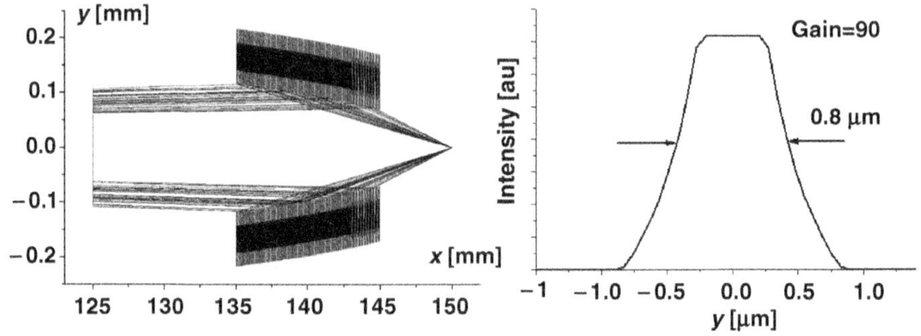

Fig. 19.3. Results of optimisation: (*left*) an optical system for a laboratory-scale microfocus source and (*right*) the distribution of intensity in the focal spot

reflectivity and broaden the focal spot leading to significant reduction in the gain. Simulations have therefore been carried out to take into account various deviations from the ideal design, including slope error, relative shift of the mirrors and poor wall verticality, i.e., the angular inclination of the walls with respect to the axis normal to the surface. To minimise deviations which cause blurring at the scale of the focal spot size, for the synchrotron radiation source, the slope error should be less than 2 μrad, relative shifts of the mirrors should not exceed 0.16 μm and the wall verticality should be better than 1.5 mrad. For the microfocus source, the fabrication tolerances are not so tight but are still quite severe. The slope error must be better than 30 μrad, relative shifts of the mirrors must be smaller than 1 μm and wall verticality must be better than 6 mrad.

The surface roughness required for efficient mirrors can be estimated from the amount of power reflected in the specular direction. Using the simple Debye–Waller model, the total intensity I in the geometrically focused beam is given approximately by

$$I = I_0 \exp\left(-\left(\frac{4\pi\sigma \sin\theta}{\lambda}\right)^2\right), \tag{19.1}$$

where I_0 is the intensity in the absence of roughness, σ is the rms surface roughness and θ is the grazing incidence angle. To obtain 50% power into the geometrical image and with $\theta \sim 3$ mrad, the surface roughness must satisfy $\sigma < 2$ nm for photon energies around 12 keV.

19.2.2 Mirror Fabrication Procedures

As discussed in Sect. 19.2.1, the tolerance limits are rather tight, especially for synchrotron radiation sources. Thus, control of the fabrication process is critical and advanced procedures are required. Moreover, in addition to the accuracy requirements in shape and positioning, etc., it is important to have

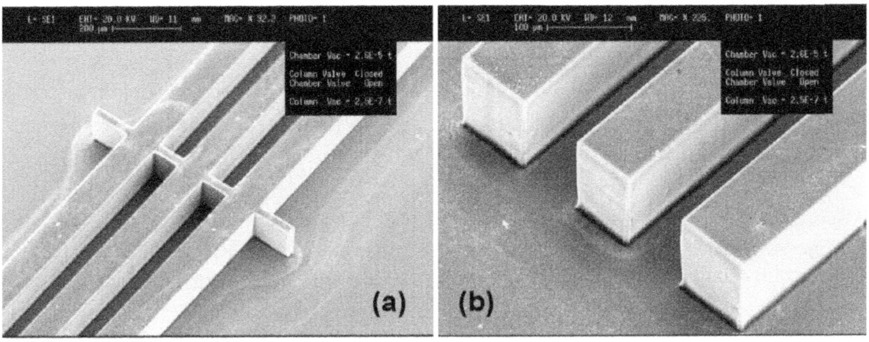

Fig. 19.4. (**a**) A mirror system in SU8 resist fabricated using process steps involving electron-beam and optical lithography. A bridge aimed at mechanically reinforcing the structure is visible. (**b**) A detailed view of the system shown in (**a**)

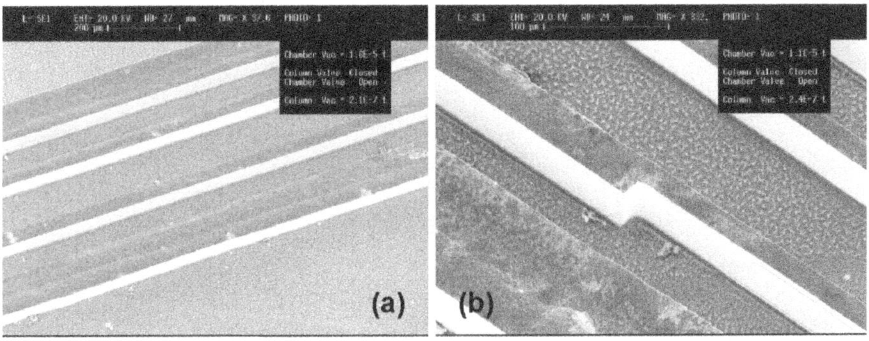

Fig. 19.5. (**a**) A nickel mirror system made using electron-beam and deep X-ray lithography and electro-deposition. (**b**) An enlarged view of the mirror system shown in (**a**)

long structures in the direction of the optical axis, at least in the hundreds of micrometres range, to allow significant acceptance of radiation, as suggested by Fig. 19.1. Thus, if the system is composed of more than two symmetrical mirrors as, e.g., in Fig. 19.2, the aspect ratio must be high, typically greater than 10. If only two mirrors are used, as in the case of the system for laboratory sources shown in Fig. 19.3, large structures and low or moderate aspect ratios are sufficient.

To satisfy these requirements, several fabrication routes can be followed, involving one or more of electron-beam lithography (always necessary to define the shape of the reflecting surface with sufficient precision), optical lithography, X-ray lithography and electro-galvanic growth of material [17]. Figure 19.4 shows a prototype made directly in SU8 photoresist by optical lithography using a mask made by electron-beam lithography. A second prototype is shown in Fig. 19.5; this was made in electro-deposited nickel after electron-beam and X-ray lithography steps.

In both cases, the mirrors follow the design shape, and the overall structures are satisfactory. The main problem is the wall roughness which presently exceeds the acceptable limits. Very recent results on post-fabrication treatment have given interesting indications of possible routes to reduce wall roughness significantly. This will be discussed in Sect. 19.3.2.

Despite the technological problems related to fabrication issues, such systems are very promising, with the great advantages of flexibility and compactness. Using modern microfabrication tools, mirrors of virtually any shape can be manufactured, so that focusing, collimating or condenser systems can be easily designed for different types of source.

Here, simulations involving just one reflection have been presented, but systems working with more reflections (see also Sect. 19.3), including whispering galleries [18], can be conceived. Since an entire system has lateral dimensions of the order of 0.5 mm, several systems with different characteristics can be arranged side by side, allowing easily interchangeable optics. Another aspect of flexibility is provided by the wide operating energy range of such reflective optics.

19.3 Microstructured Optical Arrays

MOAs [14,15], as shown schematically in Fig. 19.6, work on the same principles as poly-capillary and microchannel plate optics. Now, however, the reflecting channels are made to specific designs and are, for example, etched into thin silicon (100–200 μm) which can be flexed to provide adaptivity and/or focal length control. To reduce aberrations such as coma, two reflections are needed in such systems; since the channel lengths are small (due to the thin silicon), reflections from two successive components must be used. In Fig. 19.6, the channel widths are shown as increasing radially outwards, to compensate for the increased grazing incidence angle. In practice, to date, this has not been necessary as achievable aspect ratios (i.e., the ratio of channel length to channel width), means that most X-rays pass straight through without reflecting,

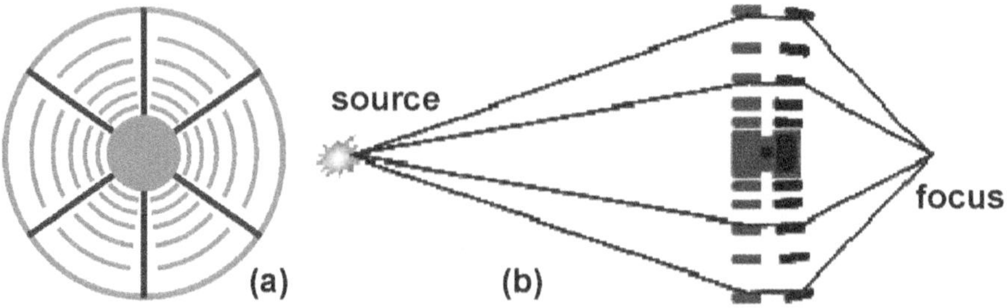

Fig. 19.6. (**a**) A microstructured optical array, face on, and (**b**) arrangement of two MOAs, in which either or both can be flexed to change the position of focus

Fig. 19.7. An MOA consisting of an arrangement of 1D strips to give a 2D focus

whatever the radial distance. Additionally, in practice, many more channels would be used compared to the number shown in the schematic diagram; typically channels would be ~10 μm wide, with walls of comparable thickness, over areas of a few square millimetres.

Flexing may be carried out either mechanically or by coating piezo material on, for example, the spokes shown in Fig. 19.6a. By controlling each piece of piezo independently, the X-ray beam could be further manipulated, for example to reduce aberrations in an adaptive or active way.

The two-dimensional focusing capabilities of such arrays could be simulated by making a series of 1D strips, as shown schematically in Fig. 19.7. As well as being technically less challenging to manufacture each strip could also be flexed independently. Other arrangements, designed for specific applications, would also be possible.

19.3.1 Computer Simulations

Modelling the performances of MOAs, even in the simplest way, is challenging. It requires both finite element analysis (FEA), to determine the effect of flexing on the channel walls, and ray tracing to characterise the optical performances. More sophisticated analyses will require wavefront propagation and studies of the effects of diffraction.

FEA and ray tracing are both complicated for such optics, as the effects of many channels have to be taken into account. For FEA, this means that the number of elements to be analysed is very large, leading to problems with mesh sizes, while ray tracing has to be carried out non-sequentially as at most two optical surfaces out of many hundreds will be encountered by an individual array. So far, only rudimentary FEA studies have been carried out, but many characteristics of the optical performances have been investigated using the optical design software ZEMAX©. Recently, ray-tracing analysis has been carried out using the much more flexible (and user-friendly) "Q" software developed at the University of Leicester (UK) [19].

As an example, a silicon MOA designed for X-rays of energy 4.5 keV (Ti K_α) has been modelled using ZEMAX©. This type of optic will be suitable for irradiating cells, in studies related to cancer research, using an X-ray

Table 19.1. Parameters of the prototype MOA for the Gray Cancer Institute microprobe

Source size	5 μm
Source to optic distance	160 mm
Diameter of optic	2 mm
Separation of optic components	1 mm
Length of channels	200 μm
Width of channels	10 μm
Bending radius of first component	∞
Bending radius of second component	100 mm
Focal distance	73 mm

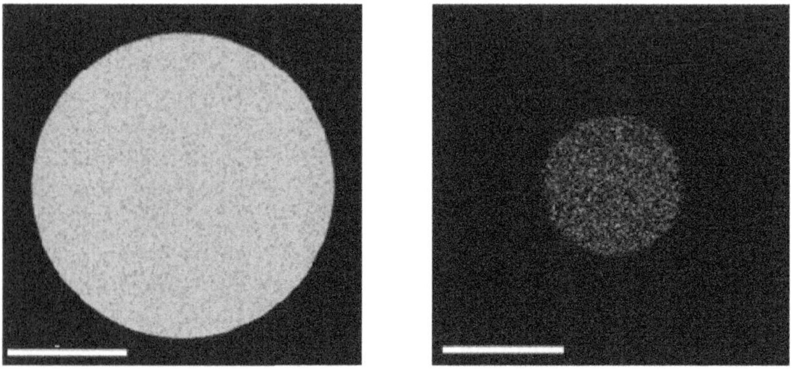

Fig. 19.8. Results of ray tracing the MOA for the Gray Cancer Institute microprobe. The source (*left*) emitted 4.5×10^6 keV photons, of which (*right*) 8,600 were doubly reflected and brought to focus. The scale bars are 2 μm

microprobe at the Gray Cancer Institute (UK) [20]. The parameters are shown in Table 19.1.

The bending radius of the second component (the first was unbent) was chosen to give a focal spot size of about 2 μm, assuming a 5 μm diameter source. Although zone plates can give smaller focal spots than this, the intention was, in the first instance, to aim for something experimentally feasible at an early stage, while providing a focal spot size useful for studies using the microprobe. Smaller spot sizes could be achieved by using a smaller bending radius, or by bending both components. Although this would mean that fewer X-rays would pass straight through without reflection, the effects of roughness would be more pronounced and a detailed analysis needs to be carried out to determine the optimum configuration.

The ray tracing took into account the efficiency of each reflection, which decreases radially outwards as the grazing incidence angle increases, as well as the channel wall roughness. Results for zero roughness (Fig. 19.8) indicate that the configuration of Table 19.1 results in a focusing efficiency of slightly under 1% (primarily since most X-rays pass straight through). However, the focused flux is some two orders of magnitude higher than that which could be achieved by a state-of-the-art zone plate with a diameter ~100 μm. A channel wall roughness of 10 nm, the effects of which were modelled using (19.1), reduces this gain by a factor of about 3, suggesting that a roughness of a few nanometres is acceptable at energies of a few keV. An additional advantage of MOAs, over zone plates, is that the focal length is independent of energy, so that (unless energy-dependent effects are being studied) the bremsstrahlung as well as the characteristic radiation could be used, enhancing the gain in useful flux. To date, using zone plates, all studies using the microprobe have been concerned with cell death [21], rather than the much more important phenomenon of mutation which occurs at a rate several orders of magnitude lower; hence the need for increased focused flux.

19.3.2 Manufacture of Microstructured Optical Arrays

Because of the necessity for high aspect ratios, techniques such as the Bosch process [22] of deep etching in silicon are required to manufacture MOAs. The Bosch process utilises successive etch/passivate stages to create the channels while preventing side-wall etching. Until recently, the applications of such manufacture did not require tight tolerances on wall roughness, and so values of the order of micrometres were acceptable. MOAs require improvements of around three orders of magnitude over this, and so new procedures have had to be devised. By shortening the etch/passivate cycle time, the Scottish Microelectronics Centre at the University of Edinburgh has shown that channel wall roughnesses of less than 20 nm are possible. Subsequent coating with 100 nm of silicon dioxide improved this further to less than about 10 nm [23], which suggests that the ultimate goal of roughnesses of a few nanometres is achievable.

19.4 Conclusions

The nested and array systems presented here show promising capabilities as future generation X-ray optics. Some technological challenges, including roughness, appear close to being overcome, while others, e.g., control of surface shapes for adaptive systems, must still be addressed in detail.

Recent experiments involving coating of nested mirror systems with silica sol-gel showed very promising wall roughness reduction, and 50% reflectivity at Cu line from SU8 walls coated with sol-gel.

Acknowledgments

In addition to the European Science Foundation support, the work on Microstructured Optical Arrays is part of that being carried out by the UK Smart X-Ray Optics (SXO) consortium funded by the Basic Technology Programme of Research Councils UK (grant code EP/D04880X/1). The members of the SXO consortium are University College London (including the Mullard Space Science Laboratory), King's College London, the Gray Cancer Institute, the Scottish Microelectronics Centre at the University of Edinburgh, the University of Birmingham, the University of Leicester and STFC Daresbury Laboratory. Silson Ltd is an associate member. IFN acknowledges partial financial support from National project SPARX.

References

1. O. Hignette, G. Rostaing, P. Cloetens, A. Rommeveaux, V. Ludwig, A. Freund, Proc. SPIE **4499**, 105 (2001)
2. K. Yamamura, K. Yamauchi, H. Mimura, Y. Sano, A. Saito, K. Endo, A. Souvorov, M. Yabashi, K. Tamasaku, T. Ishikawa, Y. Mori, Rev. Sci. Instrum. **74**, 4549 (2003)
3. Y. Mori, K. Yamauchi, K. Yamamura, H. Mimura, Y. Sano, A. Saito, K. Ueno, K. Endo, A. Souvorov, M. Yabashi, K. Tamasaku, T. Ishikawa, Proc. SPIE **4782**, 58 (2002)
4. I.N. Bukreeva, S.B. Dabagov, S. Lagomarsino, Appl. Opt. **43**, 6270 (2004)
5. C.G. Cheng, R.K. Heilmann, P.T. Konkola, O. Mongrard, G.P. Monnely, M.L. Schattenburg, J. Vac. Sci. Technol. B **18**, 3272 (2000)
6. H.N. Chapman, K.A. Nugent, S.W. Wilkins, Rev. Sci. Instrum. **62**, 1542 (1991)
7. M.A. Kumakhov, Proc. SPIE **3444**, 424 (1998)
8. S.W. Wilkins, A.W. Stevenson, K.A. Nugent, H. Chapman, S. Steenstrup, Rev. Sci. Instrum. **60**, 1026 (1989)
9. G.W. Fraser, J.E. Lees, J.F. Pearson, M.R. Sims, K. Roxburgh, Proc. SPIE **1546**, 41 (1992)
10. J.F. McGee, A Catoptric X-Ray Optical System (for Use in Laser-Fusion Diagnostics). Final Technical Report, Saint Louis University, Missouri, 1982
11. J.L. Wiza, Nucl. Instrum. Methods **162**, 587 (1979)
12. M.V. Gubarev, C.D. Bankston, M.K. Joy, J.J. Kolodziejczak, C.E. McDonald, C.H. Russell, W.M. Gibson, Proc. SPIE **3444**, 467 (1998)
13. A.N. Brunton, G.W. Fraser, J.E. Lees, I.C.E. Turcu, Appl. Opt. **36**, 5461 (1997)
14. P.D. Prewett, A.G. Michette, Proc. SPIE **4145**, 180 (2000)
15. A.G. Michette, P.D. Prewett, A.K. Powell, S.J. Pfauntsch, K.D. Atkinson, B. Boonliang, J. Phys. IV France **104**, 277 (2003)
16. I. Bukreeva, A. Gerardino, A. Surpi, A. Cedola, S. Dabagov, S. Lagomarsino, Proc. SPIE **5974**, 59741D (2005)
17. I. Bukreeva, A. Surpi, A. Gerardino, S. Lagomarsino, F. Perennes, M. Altissimo, S. Cabrini, A. Carpentiero, A. Vicenzo, P. Cavallotti, Opt. Commun. **259**, 366 (2006)
18. V.I. Ostashev, V.E. Asadchikov, I.N. Bukreeva, O.N. Gilev, N.A. Havronin, I.V. Kozhevnikov, S.I. Sagitov, Opt. Commun. **155**, 17 (1998)
19. R. Willingale, http://www.star.le.ac.uk/~rw/(last accessed 25 April 2007)

20. K.D. Atkinson, M. Folkard, B. Vojnovic, G. Schettino, K.M. Prise, B.D. Michael, A.G. Michette, Radiat. Res. **161**, 103 (2004)
21. M. Folkard, K.M. Prise, C. Shao, S. Gilchrist, A.G. Michette, B. Vojnovic, Acta Phys. Pol. A **109**, 257 (2006)
22. A.A. Ayón, X. Zhang, R. Khanna, Sens. Actuators A **91**, 381 (2001)
23. W. Parkes, Private communication (2006)

20

Reflective Optical Structures and Imaging Detector Systems

L. Pina

Abstract. New types of grazing incidence X-ray mirror systems based on single reflections have been studied, including modelling of optical performance, effects of surface figure errors and micro-roughness, actual performances and astronomical and laboratory applications. Ray-tracing simulations of multi-foil reflective optics for focusing radiation from a gas puff plasma source have been studied in detail for soft X-rays in the wavelength range 3–20 nm. Such sources are debris free because of the use of noble gases as the working medium. The ray-tracing was performed for both point and extended sources. The optics consist of two orthogonal stacks of ellipsoidal mirrors with gold reflecting surfaces forming a double focusing device. Unlike multilayer optics, grazing incidence optics are efficient at focusing soft X-rays over a wide wavelength range. Optics designed for collecting solid angle of 0.1sr were manufactured and tested in the visible and EUV regions. It has been demonstrated that multi-foil optics are a good candidate for concentrators of EUV radiation in applications such as lithography.

High resolution imaging screens and detector systems with thin YAG:Ce and other monocrystal scintillator screens have been designed and tested. Camera systems based on monocrystal scintillator, optics and CCD detector were built and used for testing of scintillators. Screens with thicknesses down to 5 µm and a fast, high resolution, cooled 16 bit CCD camera have been used to achieve resolutions of 1 µm and comparative studies of sensitivity (for YAG:Ce and fine grain Gadox screen) were carried out. Scintillator screens and systems with sub-micrometre resolution have also been studied.

20.1 Introduction

Recent progress in high-intensity microfocused EUV beam generation is presented in this chapter. Ellipsoidal thin glass foils were used in multifoil optical systems for focusing the radiation in a 50 to 150 eV energy band from a gas–puff laser plasma source. A multifoil optical (MFO) condenser was designed and tested for applications with an Xe laser plasma gas–puff source. A high intensity EUV beam focal spot was recorded, analyzed, and compared with theoretical results from computer ray tracing. Direct EUV lithography using radiation-induced decomposition and ablation of TEFLON was studied.

EUV sources are considered as the sources for lithography working with the wavelength of 13.5 nm, i.e., 92 eV. Two working media as a laser target, Xe and Sn, were used in our particular case to obtain a high efficiency of laser energy conversion into the radiation in this wavelength range. The EUV sources based on such media emit radiation in relatively wide wavelength range. The applications of such sources include proximity X-ray lithography, soft X-ray contact microscopy, or micromachining of polymers by direct photoetching.

Grazing incidence X-ray optics can be used to collect the radiation in a wide wavelength range. However, X-rays are reflected only at grazing incidence angles and, thus, only a relatively small collecting angle can be used. Different solutions can be applied to enlarge the collecting angle such as a polycapillary [1], nested Wolter type optics [2], or multifoil optics (MFO) [3]. While grazing incidence optics are commonly used in space X-ray telescopes, they can be also successfully used for laboratory imaging as well as for collecting X-rays from the laboratory sources. The critical angle is relatively large in the case of EUV radiation. It can be up to 15° for gold coated mirrors with surface microroughness below 1 nm and radiation wavelength around 10 nm. The design, ray-tracing X-ray tests and recently obtained results for the multifoil condenser for the laser plasma EUV source presented in this chapter resulted from the cooperation of three laboratories. The gas-puff laser plasma EUV source [4] is operated at the WAT Institute of Optoelectronics, Warsaw, where EUV experiments were done. Ray-tracing calculations were done at the Czech Technical University and design was done at Reflex, in Prague, where the multifoil optics technology was developed. The condenser collects photons from the source located in the source chamber and directs them onto a plane in the experimental chamber used for experiments on the interaction of high-intensity EUV radiation with matter and for lithography.

There were several requirements for the condenser:

- Working energy range $E = 80$–120 eV
- Focal length $f = 440$ mm to fit into the existing vacuum chamber
- Source diameter = 100–500 μm
- Focal spot diameter = 500–1,000 μm
- Restriction for the front area in order to fit into the existing vacuum chamber, aperture size 80 mm.

There are two contradictory requirements. First, there is a large field of view (FOV) and a large solid angle of collection to obtain the best possible effectivity of the condensing system. Second, there is restriction on the focal spot size because it is more difficult to control the distortions and imperfections for larger optics. Wolter type optics and Lobster Eye in the Schmidt [5] arrangement can be considered. The Wolter optical system is the proven technology, and it is successfully used in variety of applications. Wolter optics consists of a number of nested axially symmetric hyperbolic/parabolic/ellipsoidal mirrors. The actual shape depends on the specific application. However, manufacturing mirrors with the accuracy required and nesting them precisely into the optical system is rather difficult and expensive. In the case of a condenser, imaging quality is not so important and a Lobster Eye optics can be sufficient.

20.2 Design

The first concept was based on the use of a Lobster Eye (LE) in the Schmidt arrangement. The design consists of two orthogonal sets of reflecting mirrors. However, the simple Schmidt design was shown to be impractical after the first calculations and simulations, because very short reflecting surfaces had to be used. Otherwise, it was not possible to meet the required focal spot size with a given source diameter and to simultaneously utilize the incoming radiation optimally. Initial calculations suggest using curved mirrors in order to increase the focusing power of the Lobster Eye. As long as the problem covers, only the single-point-to-single-point focusing system, the effect of severe image distortions for off-axis sources can be neglected, because the source will always effectively be on the optical axis. The system for focusing from point to point is needed and thus the mirror shape should be elliptical. To optimally cover the FOV of the system in this case, each mirror not only has to have a different curvature but also has to be in the different distance from its neighbor, i.e., the mirrors cannot be equally spaced.

Several iterations have been done. The mirror length was changed to have the mirrors as long as possible in order to reduce the number of them. These iterations ended up with the final design consisting of 4 cm long, 8 cm wide, and 300 μm thick ellipsoidal mirrors. Half of the profile is shown in Fig. 20.1. Note the changing distances between the mirrors, which were calculated to allow illumination of the entire surface of each of the mirrors. The necessary number of mirrors is optimized.

The final design can be described as a multifoil, bifacial Kirkpatrick–Baez system. Term "bifacial" is used to stress that the reflecting mirrors are on

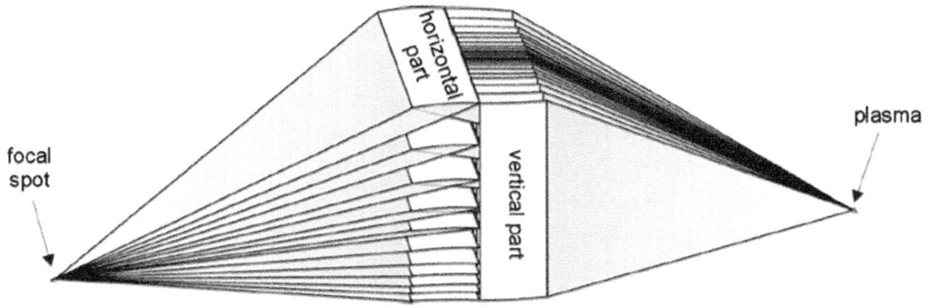

Fig. 20.1. One half of the multifoil (MFO) EUV bifacial Kirkpatrick–Baez condenser

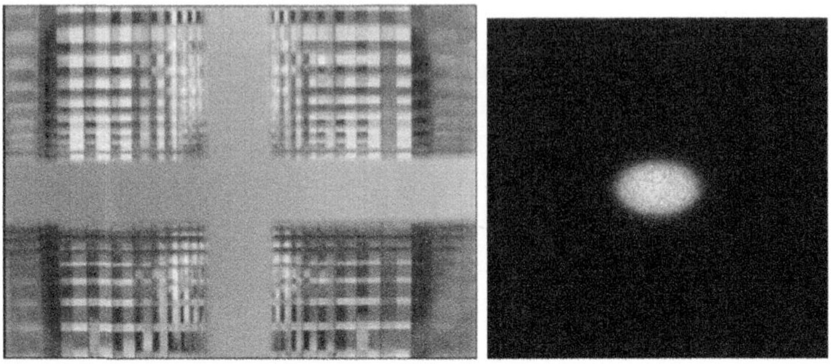

Fig. 20.2. Front view of the multifoil (MFO) EUV bifacial Kirkpatrick–Baez condenser (*left*). The ray-tracing simulation of a 0.5 mm size EUV source focus (*right*). Vertical and horizontal intensity profiles give a peak FWHM of 0.45 and 0.65 mm, respectively

both sides of the optical axis unlike the case of the classic KB system, where the optics are asymmetric.

A number of ray-tracing simulations of the selected design, as well as of each of the particular designs during the iterative design process were performed. The ray-tracing simulation of the final system design for a flat circular photon source with diameter of 0.5 mm is plotted in Fig. 20.2 together with the front view of the condenser. Mirrors with varying distances between them are clearly depicted. The intermirror distance increases for larger mirror off-axis distances. The gray central cross is part of the optics holding structure. It additionally shields the central part of the optics, where no reflection is possible, against the direct beam.

The optics has different magnifications in two perpendicular directions. This is because the reflection occurs at different distances from the source for different directions and an asymmetry is thus introduced. If the mirrors could penetrate each other to form the channels, i.e., if all the mirrors were at the same distance from the source, no differences in magnification would be visible.

Simulated peak widths strongly indicated that the proposed condenser is feasible. Distortions due to the design itself are acceptable. Decay of the intensity with the distance from the optical axis was also theoretically studied. An extremely extended uniform source has been simulated in this case. It can be seen that the FWHM FOV (i.e., the intensity falls to 1/2 at the edges of the FOV) is $5 \times 5\,\text{cm}^2$, i.e., $13 \times 13\,\text{deg}^2$. This means that although not originally designed to be an imaging device, this MFO still have some imaging power in relatively large focal area.

The ratio of the number of photons gathered inside the central peak to the number of photons blocked by the central support structure characterizes the MFO compared to the LE design. The ratio of all the photons in the entire focal cross without the central peak to the photons inside the central peak is

0.2%. The length of the cross bar is about 10 mm. This means that the cross structure is strongly reduced if compared to standard Schmidt LE because of the curved and optimized mirrors. While still present, it plays a negligible role in the experiments described.

The solid angle from which the photons are collected in case of the MFO condenser and the maximal solid angle which can be reached with an axially symmetric condenser have also been studied in order to compare the efficiency of the systems. The simulation indicates that the solid angle which is covered by the MFO condenser is about 0.09 sr. This number includes correction on the gaps between the mirrors and shows how many photons are reflected at least once. The solid angle which can be reached by the nested shells of a corresponding axially symmetric condenser is about 0.18 sr at maximum. In fact, this number should be smaller because only a finite length of the mirror can be used, or, equivalently, only a limited number of shells are feasible. Therefore, the ratio between the MFO condenser and the ideal condenser with the same outer dimensions is about 0.5. The realistic ratio including various technical aspects is somewhere between 0.5 and 0.9.

20.3 MFO

The exact parameters of each of the reflecting mirrors were calculated. Some problems were encountered during the manufacturing, however (unclear). All the previously created devices of the comparable size and complexity employed only flat mirrors, which was not the case of the proposed condenser. Hence, the technology of shaping the mirrors with the desired quality has had to be developed.

In the framework of alternative proposals, thermal shaping of glass mirrors was studied. Thermally shaped mirrors keep their proper shape after the process, and the internal stress is minimized. Therefore, the probability of glass damage and/or cracks is substantially reduced even under extreme conditions, such as strong vibrations and temperature changes.

Although the technology has promising results, its economy is a great disadvantage for the condenser. As mentioned earlier, each mirror of the condenser has a different shape. If the thermal forming technology is used, a number of different forms would be required, one for each mirror.

A different approach was developed to avoid this problem. The key features are as follows:

- A modular concept of multifoil optics with a large number of mirrors is used to facilitate assembling them into complex optical systems
- Relatively low-cost methods to create many differently shaped mirrors in one module are used
- The accuracy of shaping is lower compared to the thermal shaping, however, there is still room for further development

Finally, the optics using this new technology was successfully manufactured. The module uses glass sheets with thin layers as reflecting optical elements deposited on the surface and with a metal support structure.

20.4 Experiments

20.4.1 Experiments in VIS Region

First tests were performed using visible light. The optics was illuminated by a 500 μm light source and the focal spot was imaged with a scientific grade CCD camera. A sample image is shown in Fig. 20.3. The FWHM of the detected peak is about $0.7 \times 1.1\,\text{mm}^2$. FWHM values are up to about 70% larger than the simulated ones. This is probably mainly caused by misalignments of the reflecting surfaces and mirror profile errors.

Broadening would be relatively more apparent if a power spectrum was displayed. The focal spot enlargement is caused mainly by the optics imperfections in this case, not by the source size. However, several additional reasons for the focal spot broadening exist. All of them are due to the measurement with visible light:

- Effects coming from diffraction on a number of thin reflecting foils, which does not occur in case of X-rays due to extremely short wavelengths.
- Glass foils work as optical waveguides for visible light multiple reflections between reflecting surfaces, which cannot be neglected in case of the nonplanar geometry and shaped surfaces.

According to our experience, any kind of multifoil X-ray optics, which was already manufactured and tested in both visible light and in X-rays, exhibits the worse peak blurring when tested with visible light as compared with testing with X-rays. However, if the measurement in visible light is consistent with

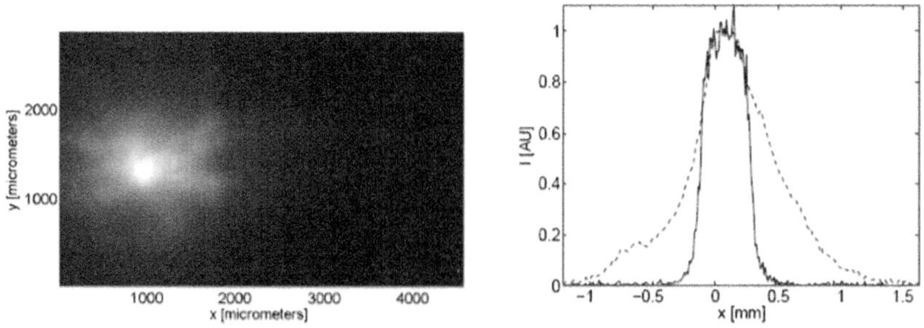

Fig. 20.3. Optical test results. The image of the focal spot from the 0.5 mm source is plotted on the *left*. Comparison between the simulation for soft X-rays (*solid line*) and the profile of the measured focal spot in visible light (*dashed line*) is on the *right*. Broadening of the measured VIS peak relative to the EUV simulation is discussed in the text

the desired requirements, one can expect that it will be even more favorable for X-rays.

20.4.2 Experiments in EUV Region

During the design phase a Gaussian profile for the source with a FWHM of 0.5 mm was assumed. However, the true source size has to be determined in future experiments. Several methods for characterization of the focal spot profile have been used. A schematic view of the three different experimental setups is in Fig. 20.4. The last method used the optical system to create a pattern on a sample surface by EUV lithography and by then analyzing the depth of the grooves formed.

First, the focal spot was imaged using the Wolter type X-ray microscope. The EUV focal spot was positioned on the Sn target. Radiation emitted under an angle of 60° with respect to the condenser optical axis passed through aluminum filter and was focused by the Wolter optics and detected by a back illuminated CCD camera. The image obtained is shown in Fig. 20.5. The image is not corrected for the geometrical inclination or for any possible change of intensity with the inclination angle.

Second, the profile of the focal spot was mapped for two different gases, Xe and Kr, using a scanned pinhole inside the source chamber as plotted in Fig. 20.6. The pinhole was placed in a particular position and the flux was measured. After measurement of sufficient number of distinct pinhole positions, the interpolation was used to create the profile. Because of the nature of the probing process, each point in the resulting image is based on a completely different set of shots. Thus, an averaging of a large number of shots exactly as in case of the measurements with Wolter optics was incorporated. However, the detection was done in the direction parallel to the optical axis of the condenser.

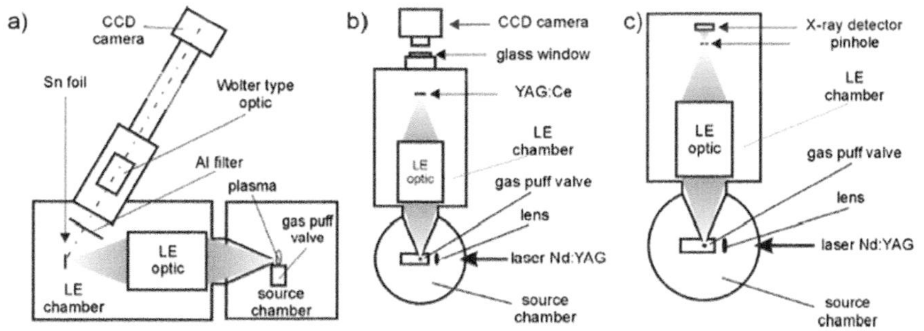

Fig. 20.4. Experimental arrangements for the intensity measurements in the focal spot: (**a**) EUV imaging with the Wolter type microscope, (**b**) 2D imaging with a use of YAG:Ce crystal scintillator, (**c**) scanning with pinhole and AXUV 100 detector covered by Zr/C filter

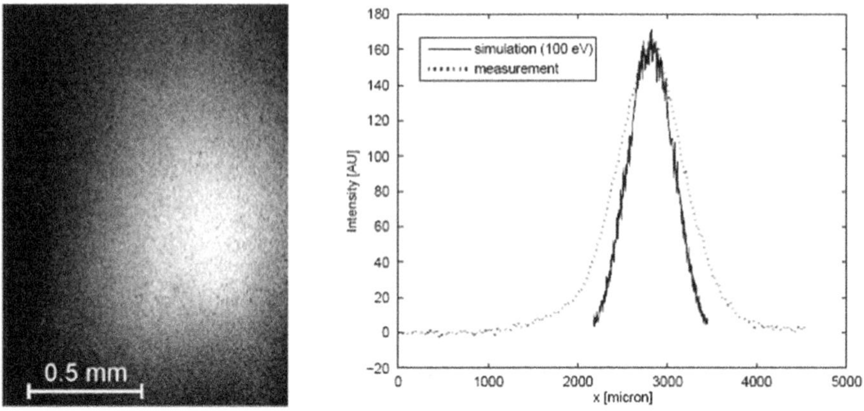

Fig. 20.5. The focal spot obtained by the Wolter X-ray optics after 200 shots. The FWHM of the peak is 0.6 mm and 0.8 mm, respectively

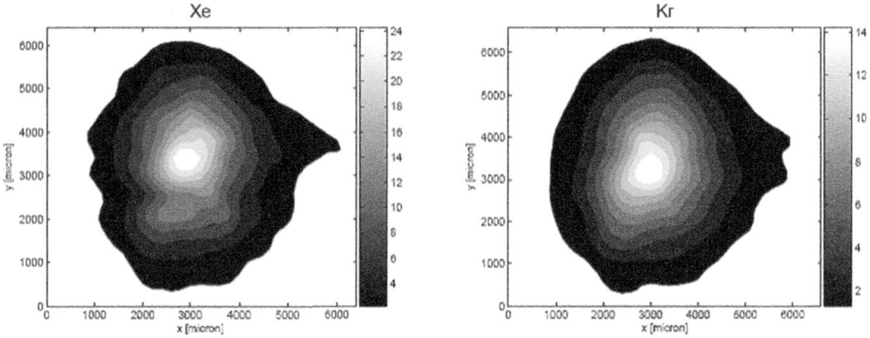

Fig. 20.6. Results of absolute measurements of EUV intensity distributions in the focal plane obtained with the use of the calibrated pinhole coupled to the AXUV 100Zr/C detector

The condenser tends to focus a circular source into a slightly asymmetric one, as seen in Fig. 20.2. Asymmetry of the focal spot was studied and a simulation for the circular source was compared with the measured data. The ratio between the small and the large axis for a simulated 0.5 mm source is about 0.67. The measurement in Fig. 20.5 gives the ratio 0.7. This means that the source either has a symmetric profile or the asymmetry is relatively large and it can switch the longer semiaxis from one direction to the other one, which is less probable. However, this assumes that the condenser optical transfer function resembles the simulated one at least in the shape, if not in the overall size. The apparent circular source shape is a very probable case, because an extremely large number of shots had to be used to generate the focal spot map, and this results in a statistically integrated source shape which

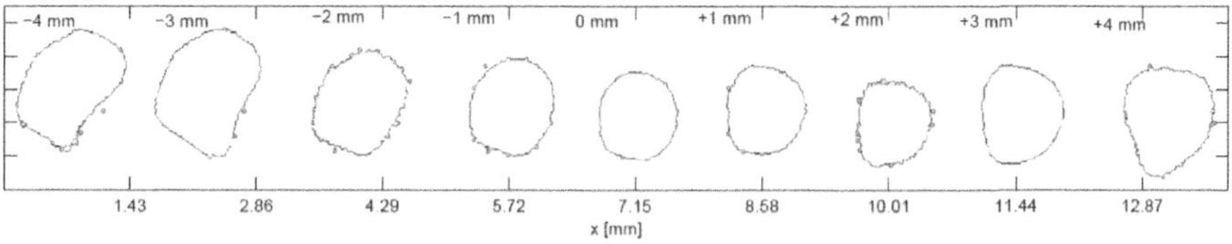

Fig. 20.7. Isointensity lines plotted from X-ray CCD images obtained at the FWHM level. Shift from the focal position is indicated above each profile

resembles the circular Gaussian profile, even if the focus shape in case of a single shot is completely different.

Third, the last measurement was based on the use of a 20 μm YAG:Ce crystal scintillator plate combined with a CCD camera. This allows for much faster measurements of the focal spot size change with the distance from the optimal focal plane. Visible light emitted from the crystal under EUV irradiation was registered by a CCD camera, which was mounted outside the experimental chamber behind a glass window. To avoid possible influence of the visible light emitted from the EUV source, a 200-nm thick aluminum filter was placed in front of the crystal. The radiation from the source had to be decreased substantially and a number of shots had to be performed to obtain the images. Stronger EUV radiation leads to crystal saturation. This effect is the subject of further study aimed at exploiting maximum intensity. The crystal was mounted on a micropositioning stage enabling it to be moved along the axis of the condenser. Isodensity profiles of a measured focal spot are plotted in Fig. 20.7. These results were compared with the simulation for the Gaussian source with FWHM = 0.5 mm. Comparison showed a not negligible difference between simulations and experimental results in focus shape, but a smaller difference in the focus size. This strongly indicates that the optical transfer function of the condenser is more complex than expected and that the imperfections in system manufacturing have to be further studied. On the other hand, a multifoil bifacial Kirkpatrick–Baez EUV condenser has considerably better focusing characteristics than does relevant classic Lobster Eye optics. This MFO system, as designed and tested, has proven to be a good compromise between true high quality focusing optics with precise solid parts and LE optics based on flat foils. In spite of higher figure error of elliptically shaped foil mirrors in a MFO system, the optical characteristics of the system are quite promising for condenser applications. Other applications should be possible with an improved figure error.

The fourth and only complementary method is for the focal spot profile estimation which is useful in lithography applications. The image in Fig. 20.8 shows the material illuminated by EUV radiation from the optics through the lithographic matrix. The depth of the grooves is proportional to the incoming beam fluence allowing one to measure the profile of the beam in this way. The image was processed to obtain the profile.

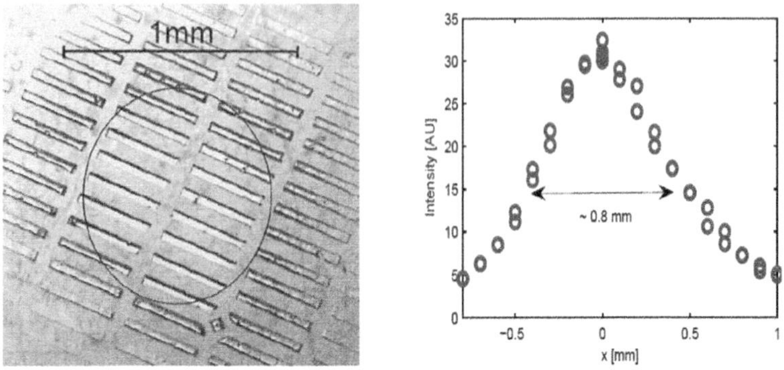

Fig. 20.8. TEFLON dry etched with EUV radiation passing through a matrix (*left*). The circle shows the FWHM of the spot obtained from the changing height of the lithographically created pattern. The line profile is on the *right*

20.4.3 Future Experiments with MFO

There are two sets of topics that should be studied in further experiments. First, optics themselves should be studied with inclusion of all relevant measurements. This will provide more information needed for further optical designs and manufacturing. Second, set of experiments should be dedicated to study differences in focal spot sizes as obtained by pinhole camera mapping and with a YAG:Ce scintillator. Experiments should be done to image the source under given conditions with an absolutely calibrated pinhole camera. Imaging is needed for integration times between 0.2 and 20.0 s.

Imaging the source by the MFO optics rotated along the optical axis by 90° should reveal whether the focal spot asymmetry is due to source asymmetry or due to optics themselves. This test is useful especially if the previous tests are incomplete or impossible.

Comparison of focal spot size measurements done with the YAG:Ce screen and with a scanned pinhole show a larger focal spot in the case of the YAG scintillator. Subsequently, a systematic study confirmed this result which, however, remained unexplained. Dedicated experiments with scintillator focal spot enlargement will be necessary to explain the behavior of the scintillator under the influence of intense EUV radiation. YAG:Ce screen behavior should also be studied separately in general terms of saturation, superluminiscence, and light signal propagation in the screen.

20.5 Conclusions

We have studied, designed and manufactured a novel soft X-ray condenser for EUV radiation based on multifoil optics (MFO) technology. It has a broadband response from 50 up to 120 eV with a collecting solid angle of about

0.1 sr, which is approaching the value attainable by axially symmetric optical designs. Tests of the optical system were performed with a visible light source and with an EUV source using several imaging methods. Measured focal spot sizes and measured focal lengths were compared with theoretical values obtained from computer modeling. The feasibility of the MFO for EUV/SXR radiation collection was demonstrated. New design tools and new manufacturing technologies were developed during the project.

Further experiments have to be completed in order to investigate the optical properties of the system and scintillator crystal in more details. The high intensity of focused EUV beams opens new possibilities in various fields of science and technology. Higher photon fluxes make possible studies of, for example, scintillation physics on wider group of materials, radiation modification of biocompatible materials, and direct molecular decomposition with applications in lithography.

References

1. M.A. Kumakhov, in *Nonimaging Optics and Efficient Illumination Systems II*, ed, by R. Winston, J. Koshel,, Proceedings of the SPIE, Volume 5943, p. 102 (2005), M.A. Kumakhov, R.B. Hoover, eds., p. 102, July 2005
2. O. Citterio, M. Ghigo, F. Mazzoleni, G. Pareschi, B. Aschenbach, H. Braeuninger, P. Friedrich, G. Hasinger, G. Parodi, Adv. Space Res. **34**, 2637 (2004)
3. R. Hudec, L. Pina, A. Inneman, L. Sveda, V. Semencova, M. Skulinova, V. Brozek, M. Mika, R. Kačerovský, J. Sik, in *Optics for EUV, X-Ray, and Gamma-Ray Astronomy II*, ed. by O. Citterio, S.L. O'Dell, Proceedings of the SPIE, Volume 5900, p. 276 (2005), O. Citterio, S.L. O'Dell, eds., p. 276, Aug. 2005
4. A. Bartnik, H. Fiedorowicz, R. Jarocki, J. Kostecki, R. Rakowski, M. Sczurek, "Wide band laser-plasma soft X-ray source using a gas puff target for direct photo-etching of polymers," in *Infrared Photoelectronics*, ed. by A. Rogalski, E.L. Dereniak, F.F. Sizov, Proceedings of the SPIE, Volume 5958, p. 279 (2005), K.M. Abramski, A. Lapucci, E.F. Plinski, eds., p. 279, Sept. 2005
5. W.K.H. Schmidt, Nucl. Instrum. Methods **127**, 285 (1975)

21

CLESSIDRA: Focusing Hard X-Rays Efficiently with Small Prism Arrays

W. Jark, F. Pérennès, M. Matteucci, and L. De Caro

Abstract. CLESSIDRA is Italian for hourglass, which is a good description of the appearance of the optical components discussed in this chapter. In these optics, many small prisms are arranged to form two large prisms, which almost touch each other at their tips, giving rise to the focusing of X-rays in one direction. From the optical point of view, CLESSIDRA is a type of Fresnel transmission lens, containing less absorbing material than normal transmission lenses. Its aperture can thus be larger. In this contribution experimental data on the focusing properties will be compared to the predicted characteristics.

21.1 Introduction

The phenomena of refraction and reflection have been exploited over the centuries for redirecting or manipulating light beams. In principle, a light beam can be focused by both phenomena, i.e., by use of transmission lenses and by use of curved mirrors. The latter are almost always oriented for normal incidence of the arriving radiation and they can then be radially symmetric. The important properties of these objects depend only on a few parameters, namely on the radius of curvature R of the surfaces on the center axis and eventually on the refractive index, n, of the optical component. The focal length f of a concave reflecting surface is then simply (the discussion in this first chapter follows textbooks on optics [1])

$$f_{\text{mirr}} = \frac{R}{2}, \tag{21.1}$$

while it is

$$f_{\text{lens}} = \frac{R}{2(n-1)} \tag{21.2}$$

for an axially symmetric lens as shown in Fig. 21.1a with two identically curved surfaces operated in air, for which we assume $n_{\text{air}} = 1$.

The main parameters for transmission lenses are explained in Fig. 21.1a for the case of a bi-concave lens. The sign convention is such that a positive

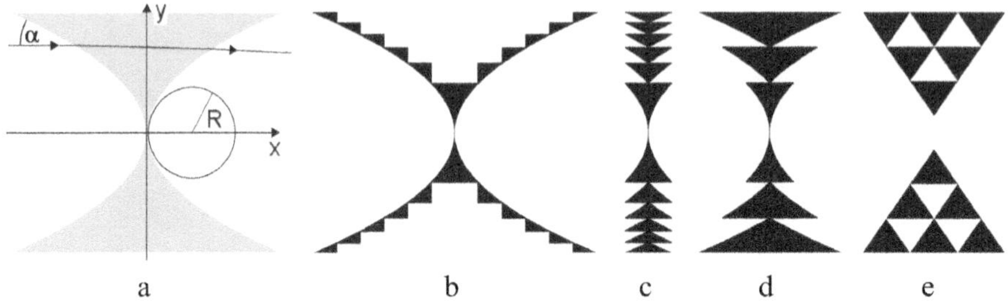

Fig. 21.1. Cross-sectional view of different transmission devices for the focusing of X-rays with the trajectory along the x-axis. In the symmetric bi-concave parabolic lens in (**a**) the radius of curvature on the optical axis of the curved lens surface is R and α is the local angle of grazing incidence. The lens in (**b**) is obtained by modeling the lens surface from (**a**) with curved prism-like objects of equal base width. It is more compact in the structure (**c**). The lens in (**d**) has external segments of equal height. These outer segments can be subdivided into smaller identical structures as shown in E. All drawings have a common scale and identical focal length

radius of curvature $R > 0$ describes a convex lens surface, and a component with a positive focal length $f > 0$ produces a real image downstream of it at image distances $p \geq f$. For $f < 0$ a virtual source is obtained between the source and the optical component. Thus, the concave surface in Fig. 21.1a has $R < 0$. According to (21.2) focusing by the use of lenses can then be achieved with the combinations $R > 0,\ n > 1$ and $R < 0,\ n < 0$.

A single mirror focuses only with a concave surface.

The distance p between the lens/mirror and the plane with the sharpest image can be calculated using

$$\frac{1}{f} = \frac{1}{p} + \frac{1}{q}, \tag{21.3}$$

where q is the distance of the lens from the real object, e.g., a radiation source. For a series of focusing optical components, the resulting focal length is given by

$$\frac{1}{f} = \sum_i \frac{1}{f_i}. \tag{21.4}$$

The reflectivity, r, of and the transmission, t, through any surface are given by

$$r = \left(\frac{n-1}{n+1}\right)^2 \quad \text{and} \quad t = 1 - r = \frac{4n}{(n+1)^2}. \tag{21.5}$$

The spatial resolution, s, obtainable for these systems for a particular wavelength λ is given as

$$s = 1.22\frac{\lambda}{2\,[\mathrm{NA}]}, \tag{21.6}$$

where [NA] is the numerical aperture of the system, which is half of the sinus of the angular extent Φ under which the geometrical aperture A of the optical component, if operated in air or vacuum, is seen from the focus position, i.e.,

$$[\mathrm{NA}] = \sin\frac{\Phi}{2} = \frac{A}{2f}. \tag{21.7}$$

To achieve this spatial resolution, the Rayleigh criterion [1] requires the transmitted or reflected wavefronts to be distorted by not more than a quarter of their separation, i.e., by $\lambda/4$. This puts the following requirements onto the permitted displacements, $\Delta\nu$, of small segments in the optical surfaces from the ideal position:

$$\frac{\lambda}{4} = 2\Delta v_{\mathrm{mirr}} \tag{21.8}$$

and

$$\frac{\lambda}{4} = (\Delta v_{\mathrm{lens},1} + \Delta v_{\mathrm{lens},2})\,(n-1). \tag{21.9}$$

A typical refractive index for transparent materials in the visible spectral range is $n \approx 1.5$. Then, according to (21.2), lenses can have a focal length, f, and also apertures, A, identical to their radii of curvature, R. The situation is similar for mirrors, which according to (21.1) can have an even shorter focal length, f. According to (21.5) the lens surfaces have transmissions of $t = 0.96$, while typical reflection coefficients for metal mirrors are of the order of $r \approx 0.9$. Ultimately for both objects, [NA] in (21.7) can be $[\mathrm{NA}] \geq 0.5$, which permits in (21.6) $s \approx \lambda$, i.e., spatial resolutions of the order of $s \approx 0.5\,\mu\mathrm{m}$. To achieve this performance according to (21.8), the mirror surface needs to be figured accurately to the perfect shape with an error not exceeding $\Delta v_{\mathrm{mirr}} = \frac{\lambda}{8}$. On the other hand, in the worst case for additive errors $\Delta v_{\mathrm{lens},1} = \Delta v_{\mathrm{lens},2}$, the required tolerance in (21.9) for the lens surfaces is more relaxed with $\Delta v_{\mathrm{lens}} = \frac{\lambda}{4}$.

These numbers can be easily controlled by interferometry with visible light [1]. Thus, neither lenses nor mirrors present a particular problem for their production. These considerations are valid if one wants to achieve the theoretically best possible spatial resolution. If a lower resolution is acceptable, a lens can also be made more compact, i.e., segmented, with the strategy of Fresnel, applied originally to lighthouse lenses, and now more frequently found in overhead projectors, in which the projection lens can be kept almost flat.

21.2 Historical Development of X-Ray Transmission Lenses

All the equations presented above are also valid for the X-ray range, for which transmission lenses were thought to be impossible for a long time. Where did this idea originate? Actually for X-rays the refractive index for all materials is slightly smaller than unity. Consequently, it is more convenient to use the refractive index decrement from unity δ defined by

$$n = 1 - \delta. \tag{21.10}$$

In addition, one now has to consider the unavoidable beam attenuation in matter. A sample of thickness d will have a transmission given by

$$t = \exp(-\frac{d}{L}), \tag{21.11}$$

where L is the attenuation length of the material. Typical numbers for the index δ are of the order of 10^{-5} to 10^{-6}. Then according to (21.5), X-rays are not reflected appreciably anymore at normal incidence. A mirror must be used in the total external reflection regime below the critical angle, which is given by [1]

$$\theta_{\mathrm{crit}} = \sqrt{2\delta}. \tag{21.12}$$

At this angle the focal length for focusing in the plane of incidence, i.e., in the plane containing the incident and the reflected beam, is

$$f_{\mathrm{mirr}} = \frac{R \sin(\theta_{\mathrm{crit}})}{2} \tag{21.13}$$

for a rotationally symmetric mirror. A lens can still be used at normal incidence for which (21.2) is valid and is now better written as

$$f_{\mathrm{lens}} = -\frac{R}{2\delta}. \tag{21.14}$$

The latter equation shows that the refraction is very small and that focusing is achieved for $\delta > 0$ by the use of a concave lens ($R < 0$) as shown in Fig. 21.1a. Such a lens has a natural limitation for its aperture as the absorption increases away from the optical axis. Kirkpatrick and Baez [2] discuss this argument just when synchrotron radiation had been seen for the first time [3]. However, they intended to use an X-ray tube. Thus, the focal lengths, which they derived for X-ray lenses, were of the order of $f \approx 100$ m, and hence, impractical. Instead, they introduced for the focusing, what is now known as Kirkpatrick–Baez mirrors, i.e., a crossed pair of concave mirrors. Was their choice responsible for the fact that transmission lenses for X-rays were not considered anymore for a while or was it the unverified idea that the lens surfaces might have to be too precisely polished with errors of the order of the operating wavelength? If this latter argument would have been looked at earlier, lens development would certainly not have been postponed so long. For the lens the surface perfection requirement is still given by (21.9), and leads, for an X-ray lens in the worst case, to

$$\Delta v_{\mathrm{lens}} = \frac{\lambda}{8\delta}. \tag{21.15}$$

A lens in plexiglass or polymethylmethacrylate (PMMA), which is $C_5H_8O_2$ with density 1.19 g cm^{-3}, has $\delta = 4.13 \times 10^{-6}$ [4] for copper K_α radiation with $\lambda = 0.154$ nm (photon energy, $E = 8.05$ keV), and has a surprisingly large tolerance with $\Delta v_{\text{lens}} = 5\,\mu$m. One can use a stack of lenses in order to reduce the focal length according to (21.4) and still deal with objects that can have surface errors larger than the wavelength of visible light with roughly $0.4\,\mu\text{m} < \lambda_{\text{vis}} < 0.8\,\mu\text{m}$.

For a mirror one has now

$$\Delta v_{\text{mirr}} = \frac{\lambda}{8\sin(\theta_{\text{crit}})}, \tag{21.16}$$

which for a hypothetical mirror in plexiglass requires the surface errors to be smaller than $\Delta v_{\text{mirr}} = 7$ nm, i.e., ca. $\lambda_{\text{vis}}/100$. Such a small number requires specialised instruments for tests and a careful and time consuming production process.

The interesting point in the large tolerance for the surface errors in X-ray transmission lenses is the fact that these lenses, at these tolerances, do not even require polishing. Indeed the tolerance is even within reach of standard workshop tools. Consequently, Suehiro et al. [5] resume the discussion of X-ray transmission lenses when a larger focal length was practical, e.g., at a relatively great distance from small synchrotron radiation sources. They propose lightening the lens for reduced absorption with the strategy of Fresnel now applied to concave lenses, and they propose using a lathe for the lens production. Tomie [6] subsequently mentions methods to produce concave Fresnel lenses, in particular by drilling holes into metal substrates. The feasibility of this attempt was verified soon after in experiments by Snigirev et al. [7], who call such objects *"compound refractive lenses"* (CRL). The latter collaboration then concentrated on the development of rotationally symmetric concave CRLs [8] as shown in Fig. 21.1a.

However, one does not necessarily have to maintain the classical lens shape for the focusing objectives. Workshop tools allow one to prepare other forms with the same tolerances. Combs or LP (long playing record) lenses [9] or alligator lenses [10, 11] are dented arrays, which appear from the side like a slightly opened alligator mouth. They approximate a concave lens stepwise and can easily be prepared on a milling machine [9, 11] as can be a master die for the rapid stamping of the form into soft and low-absorbing but rapidly deteriorating lithium [10].

Lithographic methods can easily achieve the tolerance for the X-ray lenses in the side walls of the masks and of the replicated structures [12]. With this technique even more freedom for the shape of the objects than workshop tools is possible. Lithography offers many advantages for the production of X-ray lenses [13]. First of all, the production can profit from further progress in the lithography process, which is driven by the microprocessor industry. Second, lithography for mask production and mask replication is possible in mass production with constant quality. Lithographically produced X-ray lenses are economically feasible and could eventually become disposable objects, when operated with reduced life time in high heat load beams, as will be found in future free electron lasers [14].

At this point the question arises whether more unconventional lens shapes, possible with lithography, can be used for the focusing of X-rays. Up to now only two attempts have been made in this direction by Jark et al. [15, 17] and Cederstroem et al. [16]. In both cases, prisms or prism-like objects are arranged in structures, which have no equivalent in classical or visible light optics.

21.3 Optimization of X-Ray Lenses with Reduced Absorption

The following discussion will be limited to parameters and structures, which are already state of the art in lithography. First of all, shaping in only one dimension will be assumed. Bi-dimensional focusing requires the use of two lenses in series orthogonally oriented. Consequently, the lens apertures and lens depths have to match. Second, we will consider that the chemical developing process will attack sharp edges and will round them. State of the art for the minimum radius of curvature of tips is about $R \approx 1\,\mu\text{m}$ [18]. Any shaped structure needs to have minimum dimensions of the order of 5–10 μm. It should be noted that this is up to three orders of magnitude larger than the minimum structure sizes which have already been realized in zone-plates [19] and which are proposed for high resolution refractive lenses [20].

A lens can focus an incident plane wave if throughout the lens aperture one has

$$f_{\text{lens}} = \frac{y}{\gamma}, \tag{21.17}$$

where y is the distance from the lens center and γ is the local refraction angle. As long as the local angle of grazing incidence, α, onto the lens in Fig. 21.1a is larger than the critical angle, θ_{crit}, of the lens material, the refraction angle γ in such a thin lens can be derived as

$$\gamma = \frac{2\delta}{\tan(\alpha)}. \tag{21.18}$$

The $\tan(\alpha)$ can be substituted with the local variation of the path, Δl, in material, i.e.,

$$\tan(\alpha) = \frac{\Delta y}{\Delta x} = \frac{2\Delta y}{\Delta l} \tag{21.19}$$

for the bi-concave lens. The refraction angle is given as

$$\gamma = \delta \frac{\Delta l}{\Delta y} \tag{21.20}$$

and the focal length in (21.17) can be written as

$$f = y \frac{\Delta y}{\Delta l} \frac{1}{\delta}. \tag{21.21}$$

This form for the lens-maker's equation is now valid generally, even if both lens surfaces have completely uncorrelated curvatures.

Obviously f is constant throughout the lens aperture when the local derivative of the material distribution $\Delta l / \Delta y$ increases linearly with increasing y. This latter requirement describes a parabolic lens profile, which the group of Lengeler is exclusively using for their lens stacks [8].

In this case the transmission function is of Gaussian shape [9]

$$t = \exp(-\frac{y^2}{2\delta f L}). \tag{21.22}$$

For comparison purposes we will define here an optimum aperture, which is given by

$$A_{\mathrm{CRL,opt}} = 2\sqrt{2\delta f L}. \tag{21.23}$$

This geometrical aperture transmits 84% of the flux transmittable through an unlimited aperture and has an average transmission of about $\bar{t} \approx 0.75$. The size of a pinhole collecting the photon flux, which is passing a lens of a given geometrical aperture, was defined by Lengeler et al. [21] as the effective aperture of an optical system and it is then for the CRL with unlimited aperture

$$A_{\mathrm{CRL,eff}} = \frac{0.75}{0.84} A_{\mathrm{CRL,opt}} = 1.786\sqrt{2\delta f L}. \tag{21.24}$$

According to (21.7), the numerical aperture of this lens will increase with decreasing focal length, and thus according to (21.6) the diffraction limited spatial resolution will improve towards shorter focal lengths.

Up to this point, we have used an incident plane wave. In reality we will always find a finite distance of the lens from a source. According to (21.3) this will move the image plane away from the focus position. In addition, an elliptically shaped lens would be required for imaging as was pointed out by Evans-Lutterodt et al. [22]. However, with a parabolic lens the difference become significant only far away from the optical axis. Thus, it will be ignored here.

We see in (21.21) that it is only the curved surface that contributes to the beam refraction. We could divide the curved lens surfaces in Fig. 21.1a into small prisms or prism-like objects and remove all other unavoidably absorbing material as shown, for example, in the object in Fig. 21.1b. This raises the question whether this strategy can be freely applied or whether some boundary conditions have to be respected for the size of the prisms. In the X-ray range away from absorption edges we find $\delta/\lambda^2 = \text{const}$ [4], which, according to (21.21), makes these lenses chromatic. The related image blurring, when refractive X-ray lenses were used for microspot production, was avoided by operating the lenses in sufficiently monochromatized beams.

21.3.1 Focusing Spatially Incoherent Radiation

In case the incident radiation is completely incoherent, the radiation from two adjacent prism rows does not interfere and will thus not produce a diffraction pattern. Nevertheless, the radiation will be diffracted at the borders of each row. After passing through prisms of height h, the beam will produce a blurred image in the lens focal plane with a size of approximately

$$a = \frac{\lambda}{h} f. \tag{21.25}$$

To make all incoherent images of the same size, the prism height needs to be constant throughout the lens aperture. This is the case for the lens structure in Fig. 21.1e, which is equivalent to the lens in Fig. 21.1d. In lens 21.1e the prism closest to the optical axis focuses into the common focus for $\gamma f = h$. With a prism base length in beam direction of d this leads in (21.18) to

$$\gamma = \delta \frac{\Delta l}{\Delta y} = \delta \frac{d}{h} \tag{21.26}$$

and thus finally to

$$a = \frac{\lambda}{\delta} \frac{h}{d}. \tag{21.27}$$

For the plexiglass lens from above with $h = d = 10\,\mu\text{m}$ this gives $a = 38.5\,\mu\text{m}$. This is larger than h. However, it can be reduced by increasing d. In any case, in external rows the accumulated prism base width is not to exceed the attenuation length, L, of the material, which, for plexiglass, at 8.05 keV is $L = 1.36\,\text{mm}$ [4]. The base, d, of a single prism has thus to fulfill $d << L$. As a result, with reasonable numbers for d (<0.2 mm) the prism lens 21.1E will not be able to produce an image of submicron size if illuminated spatially incoherently at 8.05 keV photon energy.

It goes beyond the scope of this report to discuss alternative applications for the prism array in Fig. 21.1e in incoherent light, e.g., in combination with an X-ray tube.

21.3.2 Focusing Spatially Coherent Radiation

If submicron image size is the goal, one has to use the constructive interference of the beam passing in adjacent rows to sharpen the image. The incident beam has to be spatially coherent over more than a single prism height. The transmitted wavefronts are not to be disturbed beyond the Rayleigh limit of $\lambda/4$. However, the wavefield periodicity allows us to introduce distortions, which are integer multiples of the wavelength [5]. The question then is whether the amount of material in the beam direction, which will shift the incident wave by its wavelength (or by a phase of 2π) or a multiple of it, is compatible with the capabilities of lithography.

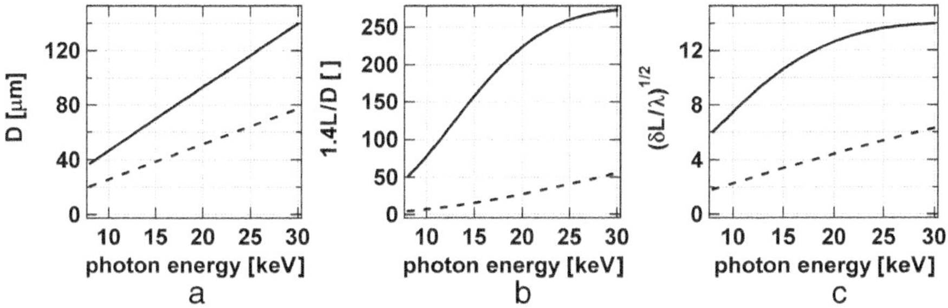

Fig. 21.2. From left to right are presented for the materials plexiglass (PMMA) as a solid line and for silicon as a dashed line the following parameters depending on photon energy: D in (**a**) is the amount of material that shifts the phase of a passing wave by 2π according to (21.28). $1.4L/D$ in (**b**) is the number of 2π phase shifts, which occur in a slab with a transmission of $t = 0.5$ according to (21.32). $(\delta L/\lambda)^{1/2}$ in (**c**) is the factor in (21.34) by which the optimum aperture of a clessidra lens with $m = 1$ can be larger than the optimum aperture of the standard concave lens (CRL) of parabolic shape in the same material

The phase shift is proportional to δ and to the material thickness d and amounts to 2π in the length

$$D = \frac{\lambda}{\delta}. \tag{21.28}$$

For minimum absorption the prisms need to be produced in a light material. Lithography concentrated mostly on photoresists, like plexiglass used in this study and on Si. It should be noted that the negative resist SU8 [23] is more radiation resistant, and has already been used successfully for X-ray lenses [15, 18], as the exposed areas become insoluble to the solvent and they will be hardened even further when focusing X-ray beams, while plexiglass will be slowly destroyed. For plexiglass and for silicon D, as required according to (21.28), is shown in Fig. 21.2a. These numbers are feasible with $D > 10\,\mu\text{m}$ for both materials for photon energies from about 8 keV to beyond 30 keV. This fixes the dimensions for the prism base width but not yet for the prism heights. It tells us that with $\Delta l = D$ in (21.21), for constant focal length $\Delta y = h_y$ needs to decrease reciprocally with increasing distance y from the optical axis. This leads to the type of Fresnel lens suggested previously [5,6,21] and shown in Fig. 21.1b. It was first produced and tested by Aristov et al. [24]. Among other solutions it can be compacted to the form in Fig. 21.1c.

Here we will look onto the other alternative of constant prism height $\Delta y = h$. According to (21.21) in such a lens Δl needs to increase linearly with distance y from the optical axis. This concept is shown in Fig. 21.1d in a compacted version of such a lens. So while the first Fresnel lens (Fig. 21.1b, c) has mostly constant average absorption throughout the lens aperture, in the lens in Fig. 21.1d the average absorption increases approximately linearly with increasing y. In the present study the structure in Fig. 21.1d was made more regular by subdividing the prisms with larger base width into many smaller identical prisms, which fulfill separately $d = mD$, where m is an integer, which can be regarded as an order number. This structure is shown in Fig. 21.1e and received for its appearance the name clessidra, which is the Italian translation for hour glass [17]. As far as the vertical direction is concerned both structures (Fig. 21.1d, e) are linear diffraction gratings with special blazing. They will create line patterns at the diffractive focal distance

$$f_{\text{diff}} = \frac{h^2}{m\lambda}, \tag{21.29}$$

which does not depend on material properties, like δ.

The refractive focal distance can be derived from (21.17) and (21.18) for the lens structure in Fig. 21.1e as

$$f_{\text{ref}} = \frac{h\tan(\alpha)}{2\delta}. \tag{21.30}$$

The intensity is then refracted mostly into only one of the diffraction peaks if both focal lengths are identical, i.e., for

$$\lambda = h\frac{2\delta}{m\tan(\alpha)}. \tag{21.31}$$

For given h, m, and α this allows for only one wavelength λ for the lens operation, as δ varies approximately with the square of λ. The relative uncertainty in the knowledge of δ produces an identical relative uncertainty for the correct operation wavelength setting. Achieving optimal alignment can thus be a lengthy procedure as it also requires finding the optimum wavelength setting. Away from the correct wavelength setting the refracted intensity will be distributed in the diffractive focal plane over several peaks.

For this lens we will consider the optimum aperture to have an average transmission of $\bar{t} \approx 0.75$, as was also used for the CRL. This requires the average transmission of the last prism row to be approximately $t = 0.5$. Then, the number of prisms to be used for the focusing is limited to

$$N = \frac{1.4L}{mD}. \tag{21.32}$$

This number is shown in Fig. 21.2b for plexiglass and for silicon for $m = 1$. The optimum clessidra aperture is then given by

$$A_{\text{Cle,opt}} = (2N + 1)h. \tag{21.33}$$

Then we see from Fig. 21.2b that PMMA has to be favored over silicon for the focusing at smaller photon energies. By the use of (21.24), (21.28), (21.29), and (21.32) the aperture in (21.33) can also be written for $N >> 1$ approximately as

$$A_{\text{Cle,opt}} \approx 2\sqrt{\frac{\delta L}{m\lambda}}\sqrt{2\delta f L} = \sqrt{\frac{\delta L}{m\lambda}} A_{\text{CRL,opt}}. \tag{21.34}$$

The prefactor $\sqrt{\frac{\delta L}{m\lambda}}$ is the advantage factor for the clessidra optimum aperture compared with the same aperture of the CRL. It is shown in Fig. 21.2c for $m = 1$. We see that the clessidra design in plexiglass can provide even more than an order of magnitude larger aperture than the normal concave lens of parabolic shape in the same material. The advantage in silicon is smaller at the smaller photon energies but increases towards higher photon energy. A more detailed comparison of the related behavior in lenses of different shapes and in more materials is presented in [17]. This report discusses only perfect prisms and thus a stepwise approximation of the parabolic lens shape. It can be shown that one can give the structure the correct material gradient $\Delta l/\Delta y$ everywhere by curving identically one or two of the prism side walls per row [15].

The focal length of a state-of-the-art clessidra lens in plexiglass with $h = 10\,\mu\mathrm{m}$ for $\lambda = 0.154\,\mathrm{nm}$ would be $f = 650\,\mathrm{mm}$ according to (21.29). Its optimum aperture, according to (21.33), is $A_{\mathrm{Cle,opt}} = 1.04\,\mathrm{mm}$ and its effective aperture is $A_{\mathrm{Cle,eff}} = 0.78\,\mathrm{mm}$. The spatial resolution obtainable with the optimum aperture would be $s = 120\,\mathrm{nm}$ according to (21.6). By use of (21.4) the long focal length can be reduced in a stack of several lenses. The combination of two clessidra lenses is shown in Fig. 21.3a, which can be converted to the very elegant alternative in Fig. 21.3b as proposed by Cederstroem et al. [16]. Both lenses have the same focal length and approximately also the same average transmission function. However, the lens of Cederstroem approximates the parabolic profile now with step heights of $\Delta y = h/2$ instead of h and thus with a smaller variation of the transmission within this height. Finally, the Cederstroem concept is more flexible as the vertical shift of the prism columns is not limited only to integer fractions of h [16]. In both cases, the average of the transmission function falls off linearly towards the lens border, and both lenses will now have a focal length of $f = 325\,\mathrm{mm}$. The optimum and effective apertures will be $A_{\mathrm{opt}} = 0.52\,\mathrm{mm}$ and $A_{\mathrm{eff}} = 0.39\,\mathrm{mm}$, respectively. In this

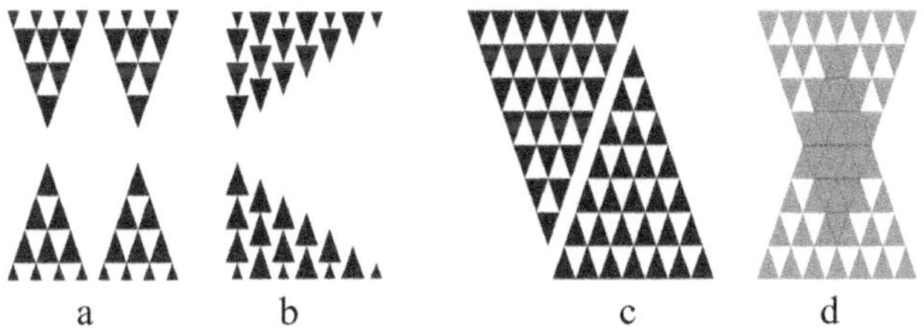

Fig. 21.3. Possible designs for focal length reduction in prism array lenses. In (**a**) two clessidra lenses provide half the focal length of a single lens. The same is achieved in the lens in (**b**) proposed by Cederstroem et al. [16] by arranging the prisms differently and more independently. The lens halves in (**c**) achieve the same focal length reduction, but now with a constant average transmission function. The number of single elements in this lens can be reduced and the structure can be compacted to the lens in (**d**)

lens concept the stacking procedure shortens the focal length and reduces the aperture by the same factor. Consequently, neither the numerical aperture in (21.7) nor the spatial resolution in (21.6) is changing. By use of the prism array lenses presented, an improvement of the spatial resolution at shorter focal lengths can only be achieved with smaller prism heights. In fact, for a fixed wavelength, by use of (21.29) and (21.33), the spatial resolution in (21.6) depends linearly on the prism height, h.

If a more constant transmission function is desired, we can reduce the focal length by half by making a lower lens half of clessidra overlap with its upper counterpart as shown in Fig. 21.3c. Then, the average value of the transmission function is constant. However, in any row some prisms will always cancel their beam refraction. The number of refracting interfaces could be minimized as is done in the lens at the right in Fig. 21.3d. The effective aperture is a maximum in this design if the lens has an average transmission of $\bar{t} \approx 0.5$. The focal length and the optimum aperture do not change as compared to the clessidra lens stack in Fig. 21.3a. However, the effective aperture is slightly reduced to $A_{\mathrm{eff}} = 0.26\,\mathrm{mm}$.

21.4 Discussion of Experimental Data

This report will now deal exclusively with the properties of the clessidra lens from Fig. 21.1e fabricated in plexiglass, which can exceed the corresponding aperture of a concave lens in the same material according to Fig. 21.2c by about sixfold for 8.05 keV photon energy. A more general comparison of the properties of this lens with those of first-order Fresnel lenses and of ideal parabolic lenses is presented in [17].

21.4.1 Parameters of the Clessidra Lens

The lens production is described in detail in [25]. The mask production and the mask replication were made at the deep X-ray lithography beamline of ELETTRA [26]. Out of a number of lenses with different h and m, we chose for this study the parameter set $h = 25.67\,\mu\mathrm{m}$, $\alpha = 35°$, and $m = 2$, which leads to a diffractive focal length of $f_{\mathrm{diff}} = 2.14\,\mathrm{m}$ for $\lambda = 0.154\,\mathrm{nm}$. The refractive focal length would be identical for $\delta = 4.2 \times 10^{-6}$. The lens has $N = 29$ prisms in the last row, which is slightly larger than the optimum of $N = 26$ according to (21.32). Its geometric aperture is $A_{\mathrm{geo}} = 1.51\,\mathrm{mm}$, i.e., 3.4 times larger than the optimum aperture of the corresponding concave lens in the same material, and the diffraction limited spatial resolution could be as high as $s = 266\,\mathrm{nm}$. Lenses with the indicated parameters were produced with consistent performance [17].

21.4.2 Properties of the Radiation Source

For the lens test one of the problems is the accurate wavelength determination as discussed above and the other is the beam coherence. Fully coherent X-ray sources are not yet operational, and thus one will find in a radiation beam spatially coherent radiation only in a full-width-at-half-maximum (FWHM) aperture, which is given by [27]

$$A = 0.44\frac{\lambda q}{S}, \tag{21.35}$$

if S is the FWHM source size.

Experimental stations for the test of optical components are available at beamline BM05 [28] at the ESRF and at the SYRMEP beamline [29] at ELETTRA. The test objects are mounted at BM05 at ca. 55 m from a source with verified FWHM dimensions of (vertical $\times$ horizontal) $80 \times 270\,\mu m^2$ [28]. At SYRMEP the lenses can be mounted at 22.5 m from a source with theoretical FWHM dimensions of $66 \times 330\,\mu m^2$ for 2.0 GeV ring energy. For $\lambda = 0.154$ nm the spatially coherently illuminated area according to (21.35) measures then (vertically $\times$ horizontally) $22 \times 4.5\,\mu m$ FWHM at SYRMEP and $45 \times 13\,\mu m^2$ FWHM at BM05. As far as the present lens is concerned, with nominally $h = 25.67\,\mu m$, one finds the following:

Not even one prism row is spatially coherently illuminated at 8 keV photon energy. The only exception is in the vertical direction at BM05, where almost two rows are illuminated spatially coherently. For spatially coherent illumination of the whole lens, one needs to go further away from the source, even at smaller sources. So, under the experimental conditions chosen, we expect to find in the lens image plane mostly an incoherent superposition of the beams passing through the different prism rows.

21.4.3 Beam Diffraction in the Clessidra Structure

The dependence of the lens performance on the size of the spatially coherently illuminated area of the lens was tested at BM05 and at SYRMEP by rotating the lens around its optical axis. With image plane distances of ca. $p = 2.2$ m, the demagnified source image of the BM05 source should measure then $3.4 \times 11.3\,\mu m^2$ fwhm, which is much larger than the diffraction limit for the lens. For the cases where the illumination would be considered as spatially incoherent, the blurred image according to (21.27) would measure at least $13\,\mu m$ FWHM. This latter number is thus to be expected for the horizontal source image. A possibly visible diffraction line structure in the vertical direction should have a peak separation of about $13\,\mu m$. The demagnified source image at SYRMEP instead should measure $8 \times 40\,\mu m^2$ FWHM and the insufficient spatial coherence in the vertical direction should widen the corresponding image size to $13\,\mu m$.

At the latter source the expected large horizontal image size is in agreement with the experimental finding. However, the measured vertical image size is always significantly larger than expected. The experimental data from BM05 show more clearly the effect of the changing spatial coherence when the lens is rotated around its optical axis. For equal diffractive and refractive focal distances we expect in vertical focusing in the image plane an intense central peak with width 3.4 μm FWHM surrounded by smaller secondary peaks. If the diffractive and the refractive focal lengths are different, then several intense narrow peaks should be observed in a line structure with 13 μm separation in the diffractive image plane. For horizontal focusing with now insufficient spatial coherence in the incident beam, we expect to find only a single peak and no line structure independent of the distance from the lens. Its width is expected to be about 13 μm FWHM in the refractive image plane. It will be larger in the diffractive image plane if the latter does not coincide with the refractive image plane.

The experimental data in Fig. 21.4 mostly confirm these expectations. The data presented are for a lens with partly curved prisms, and are essentially identical to those measured for lenses with perfect prisms. To explain the observed discrepancy between the position of the diffractive and the refractive images, we need to assume that the refractive index is ca. 1.5% smaller than predicted [4]. Then, the diffractive and refractive focal lengths would coincide according to (21.31) for $\lambda = 0.159\,\mathrm{nm}$. The experiment covered only wavelengths $\lambda < 0.155\,\mathrm{nm}$, at which the refractive focus always lies downstream of the diffractive focus. In fact, in the best diffractive image position at $p = 2.21\,\mathrm{m}$ for $\lambda = 0.155\,\mathrm{nm}$, the diffraction patterns measured independently from both lens halves have the peaks at almost identical positions as shown in Fig. 21.4a. The line image had a constant width over more than 0.35 mm of structure depth and was registered with an X-ray CCD camera with 0.645 μm pixel size. For the present analysis the signal was integrated over at least 50 pixels in the direction perpendicular to the focusing direction. The peak separation is constant with the expected value of 13 μm.

The peak width is around 5 μm FWHM and the peak positions remain fixed even if we move a smaller sub-aperture of 0.1 mm opening vertically through the lens aperture. Only the prisms close to the optical axis refract the beam mainly into the central peak, while the outer areas refract more into the adjacent peaks. When one moves away from the best diffractive image planes, the resultant intensity distribution becomes more structured with changing peak separations and peak widths as is presented in Fig. 21.4b. The solid curve for $p = 2.27\,\mathrm{m}$ in Fig. 21.4b was registered close to the refractive focus in (21.30) and its FWHM is significantly larger than expected. In the best diffractive image position for $\lambda = 0.153\,\mathrm{nm}$ at $p = 2.24\,\mathrm{m}$ in Fig. 21.4c, the upper lens half (dotted curve) directs the radiation mostly into the first secondary peak in vertical focusing. This remains so, even after the rotation of the lens for focusing more spatially incoherent radiation. As expected from the different source sizes the peak width increases from 5 μm to ca. 13 μm

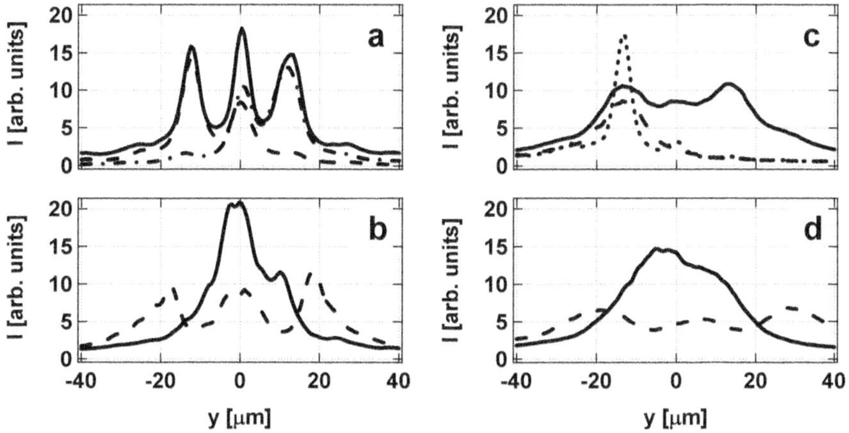

Fig. 21.4. Intensity distribution in different image planes for vertical focusing (left column **a, b**) with a larger spatially coherently illuminated area and for horizontal focusing (right column **c, d**) with a smaller spatially coherently illuminated area. (**a**) The intensity distribution in the diffractive image plane at $p = 2.21\,\mathrm{m}$ for $E = 8.0\,\mathrm{keV}$ ($\lambda = 0.155\,\mathrm{nm}$). The solid curve is for full illumination of the lens, the dashed and the dashed-dotted curves are for illumination of only the lower and the upper halves of the lens, respectively. (**c**) The intensity distribution in the diffractive image plane at $p = 2.24\,\mathrm{m}$ for $E = 8.1\,\mathrm{keV}$ ($\lambda = 0.153\,\mathrm{nm}$) with a smaller spatially coherently illuminated area at the lens. The solid curve is for full illumination of the lens and the dashed curve is for illumination of only the left part of the lens (this corresponds to the lower half in the left plot). The dotted curve presents the data for vertical focusing under the same conditions with only the lower lens half illuminated. (**b, d**) The intensity distributions in the image planes are shown as dashed lines, which are 6 cm closer to the lens. The solid line in (**b**) is measured 6 cm further from the lens, while in (**d**) it is only 4 cm further from the lens. Both solid curves are then closer to the refractive image planes. The plotted signals still contain dark counts of about 0.5 arbitrary units, while the incident intensity is about 0.6. The intensity gain would thus be approximately the numbers at the y-axis multiplied by 1.6

FWHM. The peak width for the completely illuminated lens is larger (solid curve), which indicates a different distance for the best refractive image. This peak is still structured with 13 μm periodicity, which should not have been observed for illumination with spatially completely incoherent radiation, for which we would have expected at this position a single rather wide peak. As before at other image distances the intensity distributions in Fig. 21.4d are more complicated and they do not converge towards a value as small as 13 μm for the incoherently created image in the best refractive image position according to (21.30). As a consequence, obviously, in order to characterize the focusing properties of the lens more precisely one would have to make the analysis for coincidence of the diffractive and refractive images.

21.4.4 Refraction Efficiency in the Clessidra Structure

Since the relevant operating wavelength is difficult to locate, we decided to test the lens properties by other means. The important properties are the refraction efficiency, the refraction angle, and the scattering, depending on the position in the lens and on the position in the single prism rows. The quantitative results from normalized radiographs, as shown in Fig. 21.5, taken at 8.5 keV photon energy just downstream of the clessidra lens, confirmed an average transmission t consistent with the material distribution in the lens according to Fig. 21.1e [17]. In a structure of perfect prisms the local refraction efficiency e would be identical to the local transmission.

The refraction efficiency integrated over four prism rows was measured by scanning a slit of width 100 µm vertically through the lens aperture upstream of it and registering the intensity in the incoherent image in the best refractive image plane [17]. These results, poorer than expected, are also presented in Fig. 21.5 as filled circles.

The refraction in the single rows was investigated by reducing the aperture to a width of 12 µm. Data for the refraction efficiency, the refraction angle, and the beam scattering could then be measured by use of a high resolution X-ray camera. However, the data can be measured much more precisely if the radiation passing the slit is dispersed in a diffraction crystal mounted downstream of the lens. In this configuration a rocking curve could be measured for any position of the slit. The experiment was performed for a photon energy of 8.5 keV at SYRMEP for which the FWHM rocking curve width including the diffraction of the incident beam at the aperture was 41 µrad. The expected beam deflection according to (21.18) is 10.6 µrad per prism. The results are

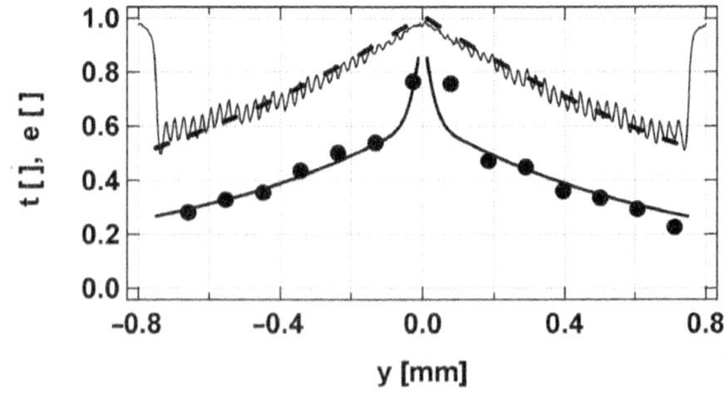

Fig. 21.5. Transmission, t, and refraction efficiency, e, for the clessidra lens tested depending on the distance from the optical axis y. The lens transmission, t, measured right behind the lens with an X-ray CCD camera is presented as a thin line. The dashed line is the calculated average transmission of the prism rows for the projected material distribution. The filled circles present the measured refraction efficiency, e, averaged over four rows. The thick line is the simulation of the averaged refraction efficiency based on the findings of this study. Note that a transmission of $t = 1$ is measured outside the lens for $y < -0.75$ mm and $y > 0.75$ mm

very encouraging. In fact, as long as the small beam does not hit any prism border the beam is refracted as expected, i.e., the refraction efficiency is identical to the local transmission and the beam deflection is consistent with (21.18) and increases with the expected step size between different rows. Even after the passage through 28 prisms at the lens border, no appreciable scattering, i.e., no widening of the rocking curve, is observed. The situation changes significantly when we approach the prism tips and finally illuminate the prism borders. Then, the divergence of the transmitted beam increases significantly. Close to the prism tips some intensity passes even completely unrefracted. The data presented for the refraction efficiency with the 100 μm slit in Fig. 21.5 as filled circles and measured data for the 12 μm slit can consistently be explained (solid curve in Fig. 21.5) when we assume that 0.63 of any prism row height refracts as expected. This applies to the 16.2 μm forming the thicker part of the prisms next to the prism base. The remaining 9.5 μm towards the prism tips are more transparent, but do not refract the beam into the FWHM image size. Only the single prism in the lens center is assumed to refract perfectly. The inefficient areas of the prisms then account for the missing efficiency of almost 50%. Obviously, the zone with the erroneous material gradient next to the prism tips is more extended than one would expect from the optical micrograph of the top surface of the lens structure in Fig. 21.6.

That these zones remain dark for the purpose of focusing can also be made evident by magnifying the lens transmission function locally by use of an asymmetrically cut crystal [30]. A Si(111) crystal with an asymmetry angle of 10° provided a magnification of 6.8 in the vertical direction for 8.5 keV photon energy ($\lambda = 0.146\,\text{nm}$) for the pictures in Fig. 21.7. This setup with an increased rocking curve width of 91 μrad is then less sensitive for the accurate determination of a beam divergence increase. The FWHM of the rocking curve covers the refraction from about 8.5 prism rows, as the difference in the refraction angle in adjacent rows is 10.6 μrad. The crystal was rotated with a step size of 175 μrad, which is almost twice the rocking curve width, and thus it was centered on refraction in the rows with 17 and with hypothetical 33 prisms, respectively. With the rocking curve centered on the incident beam the refracted intensity from rows 13–17 (Fig. 21.7b) and 26–29 (Fig. 21.7d) should not be transmitted appreciably at the crystal. Nevertheless, one observes some intensity in the corresponding points at the CCD camera, which was obviously not refracted as expected. In fact, the same points remain consistently dark in

Fig. 21.6. Optical micrograph of the present clessidra lens. The contrast was adjusted such that the resist surface is particularly revealed

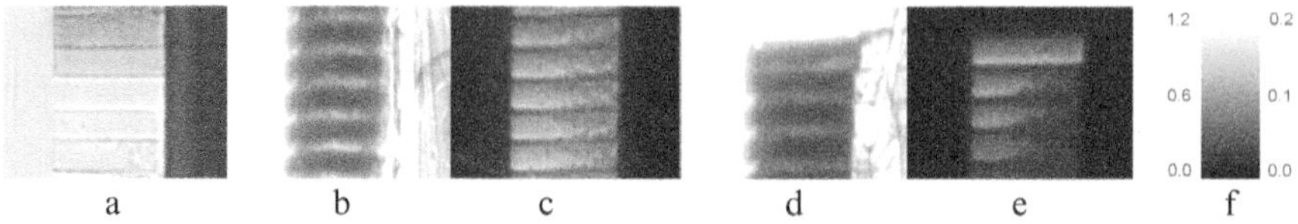

Fig. 21.7. Signal (normalized to incident beam) registered with a CCD camera at 300 mm from the lens in different regions of the lens. The pictures are taken downstream of an asymmetric analyzer crystal, which is used in magnifying configuration with ×6.8 magnification applied only to the vertical direction. The pictures cover 0.125 mm in this direction. (**a**) The first two prisms in the two lens halves adjacent the central air gap. In (**b**) and (**c**) rows 13–17 from the upper lens half and in (**d**) and (**e**) the last four rows 26–29 are shown. In (**a**), (**b**), and (**d**) are obtained with the maximum of the crystal rocking curve set to the incident beam. In (**c**) and (**e**) the crystal was inclined by 0.175 mrad and by 0.35 mrad with respect to (**a**), respectively. The grey scale in (**f**) has always black for a signal of 0, while white is a signal of 1.2 (scale to the left of the bar) in (**a**), (**c**), and (**e**), and it is 0.2 (scale to the right of the bar) in (**b**) and (**d**)

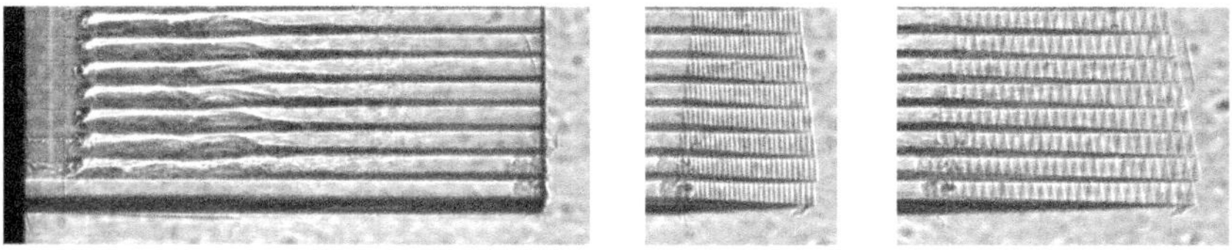

Fig. 21.8. Radiographs from a clessidra lens well aligned in tilt and roll angle at different positions in yaw taken with a photon energy of 15 keV in the optical test setup at BM05 at the ESRF. The pictures are registered with an X-ray CCD camera at 100 mm from the lens and cover 0.25 mm in the vertical direction. The left picture has the rays pass tangentially to the lens substrate. The center and the right picture are rotated off of the best yaw alignment by 3.5° and 7°, respectively. These latter orientations enable one to see the structure of the resist surface in projection. The prisms seen towards the right in these radiographs are the farthest from the CCD camera. Note that the two prisms at the edges of the last row with nominally 29 prisms are detached. One of them can be identified along the diagonal in the right radiograph

Fig. 21.7c, d, when the crystal rocking curve covers their refraction angle. In Fig. 21.7a the tips of the first prism in each lens half are particularly apparent due to phase contrast [31].

The inappropriately refracting zones can also be observed when we illuminate the lens with the much higher photon energy of 15 keV. Note: in this photon energy range the present lens would focus with $m = 1$ and a focal length of about $f \approx 8$ m. The lens can be rotated in yaw around the vertical axis such that we should also see the top surface of the prisms in projection. The related radiographs are shown in Fig. 21.8. The radiation is still sufficiently spatially coherent in the vertical direction that we expect to see the prisms and the prism rows mostly because of phase contrast [31]. The radiographs from a perfect structure taken in this condition should then show identical prisms everywhere and division lines between the different prism rows, which are periodically structured along their length. However, the width of these lines should be independent of the number of prisms through which the beam has traversed. In fact this is observed in the center and the right picture in Fig. 21.8 towards the right, where the radiation passes only a few prisms. However, to the left, after passing more prisms, the width of the division line increases until it covers about 1/3 of the prism height.

Obviously, the radiation entering the lens close to the prism tips is accumulating misdeflections along its trajectory until it is deflected out of this region. Finally, it is not refracted anymore into the image size. There is now complete consistency among the different experiments. It should be noted that this explanation is not valid at the lens border. Here we see simply a dark wedge, which is the expected shadow between the radiation passing unrefracted at the lens border and the radiation being refracted in an increasing number of prisms in the last prism row.

21.5 Conclusion

We have shown that prism arrays, which hardly look like focusing objects, can be used as transmission lenses for the focusing of X-rays. Even though the focusing in this case requires the passage through many prism surfaces, the transmission losses caused by surface imperfections are negligible. Instead the losses have to be assigned to production problems leaving imperfections in the prism tips, where the etching does not succeed in removing all the material from the corners, where the prisms touch each other with a shallow angle of only 35°. Nevertheless, the average refraction efficiency in Fig. 21.5 is 0.4 for an expected average efficiency of 0.75. Thus, the lens presented has a refraction efficiency of more than half of that theoretically expected. This lens, with a geometrical aperture of $A_{\text{geo}} = 1.51\,\text{mm}$, refracts approximately the same photon flux to the focus, which would pass through a perfect aperture of $1.51\,\text{mm} \times 0.4 = 0.6\,\text{mm}$ size.

Acknowledgments

We thank very much the following for the help with the lens production, during the experiments and for discussions: B. Marmiroli, L. Mancini, L. Rigon and R.H. Menk from ELETTRA and A. Snigirev and I. Snigereva from ESRF.

References

1. See textbooks on optics, e.g. M. Born, E. Wolf, *Principle of Optics*, 6th edn (Pergamon, New York, 1980)
2. P. Kirkpatrick, A. Baez, J. Opt. Soc. Am. **38**, 766 (1948)

3. F.R. Elder, A.M. Gurewitsch, R.V. Langmuir, H.C. Pollock, Phys. Rev. **71**, 829 (1947)
4. B.L. Henke, E.M. Gullickson, J.C. Davis, At. Data Nucl. Data Tables **54**, 181 (1993). http://www-cxro.lbl.gov/optical_constants/
5. S. Suehiro, H. Miyaji, H. Hayashi, Nature **352**, 385 (1991)
6. T. Tomie, Japanese Patent 6-045288, 18 Feb 1994; U.S. Patent 5,594,773, 1997; German Patent DE1995019505433, 1998
7. A. Snigirev, V. Kohn, I. Snigireva, B. Lengeler, Nature **384**, 49 (1996)
8. B. Lengeler, C.G. Schroer, M. Richwin, J. Tuemmler, M. Drakopoulos, A. Snigirev, I. Snigireva, Appl. Phys. Lett. **74**, 3924 (1999)
9. B. Cederström, R.N. Cahn, M. Danielsson, M. Lundqvist, D.R. Nygren, Nature **404**, 951 (2000)
10. E.M. Dufresne, D.A. Arms, R. Clarke, N.R. Pereira, S.B. Dierker, Appl. Phys. Lett. **79**, 4085 (2001)
11. W. Jark, X-Ray Spectrom. **33**, 455 (2004)
12. F.J. Pantenburg, J. Mohr, Nucl. Instrum. Methods A **467–468**, 1269 (2001)
13. A. Stein, C. Jacobsen, K. Evans-Lutterodt, D.M. Tennant, G. Bogart, F. Klemens, L.E. Ocola, B.J. Choi, S.V. Sreenivasan, J. Vasc. Sci. Technol. B **21**, 214 (2003)
14. J. Andruszkow, et al. Phys. Rev. Lett. **85**, 3825 (2000)
15. W. Jark, F. Pérennès, M. Matteucci, L. Mancini, F. Montanari, L. Rigon, G. Tromba, A. Somogyi, R. Tucoulou, S. Bohic, J. Synchrotron Radiat. **11**, 248 (2004)
16. B. Cederström, C. Ribbing, M. Lundqvist, J. Synchrotron Radiat. **12**, 340 (2005)
17. W. Jark, F. Pérennès, M. Matteucci, J. Synchrotron Radiat. **13**, 239 (2006)
18. V. Nazmov, L. Shabel_nikov, F.-J. Pantenburg, J. Mohr, E. Reznikova, A. Snigirev, I. Snigireva, S. Kouznetsov, M. DiMichiel, Nucl. Instrum. Methods **B217**, 409 (2004)
19. W. Chao, B.D. Harteneck, J.A. Liddle, E.H. Anderson, D.T. Attwood, Nature **435**, 1210 (2005)
20. C.G. Schroer, B. Lengeler, Phys. Rev. Lett. **94**, 054802 (2005)
21. B. Lengeler, C.G. Schroer, J. Tuemmler, B. Benner, M. Richwin, A. Snigirev I. Snigireva, M. Drakopoulos, J. Synchrotron Radiat. **6**, 1153 (1999)
22. K. Evans-Lutterodt, J.M. Ablett, A. Stein, C.-C. Kao, D.M. Tennant, F. Klemens, A. Taylor, C. Jacobsen, P.L. Gammel, H. Huggins, S. Ustin, G. Bogart, L. Ocola, Opt. Express **11**, 919 (2003)
23. C. Cremers, F. Bouramrane, L. Singleton, R. Schenk, Microsyst. Technol. **7**, 11 (2001)
24. V. Aristov, M. Grigoriev, S. Kuznetsov, L. Shabelnikov, V. Yunkin, T. Weitkamp, C. Rau, I. Snigireva, A. Snigirev, M. Hoffmann, E. Voges, Appl. Phys. Lett. **77**, 4058 (2000)
25. F. Pérennès, M. Matteucci, W. Jark, B. Marmiroli, Microelectron. Eng. **78–79**, 79 (2005)
26. F. Pérennès, F. De Bona, F.J. Pantenburg, Nucl. Instrum. Methods A **467–468**, 1274 (2001)
27. D. Attwood, *Soft X-Rays and Extreme Ultraviolet Radiation: Principles and Applications*, chap. 8 (Cambridge University Press, Cambridge, 1999)
28. BM05 X-ray Imaging Group http://www.esrf.eu/UsersAndScience/Experiments/Imaging/BM05/

29. SYRMEP: BL 6.1 R, SYnchrotron Radiation for MEdical Physics (SYRMEP) http://www.elettra.trieste.it/experiments/beamlines/syrmep/index.html
30. K. Kohra, in *Proceedings of the Sixth International Conference on X-Ray Optics and Microanalysis*, ed. by G. Shinoda, K. Kohra, T. Ichinokawa (University of Tokyo Press, Tokyo, 1972), pp. 35–45
31. T.J. Davis, D. Gao, T.E. Gureyev, A.W. Stevenson, S.W. Wilkins, Nature **373**, 595 (1995)

Part IV

Multilayer Optics Developments

22

Neutron Supermirror Development

Th. Krist, A. Teichert, R. Kovács-Mezei, and L. Rosta

Abstract. Neutron supermirrors are interference systems which reflect at small glancing angles. They consist of layers with thicknesses of a few nanometres. While in the 20 years after the such systems were proposed the number of layers was limited to 100, recent improvements in the control of layer growth and process stability now allows the production of supermirrors with over 1000 layers. For certain material combinations such supermirrors can be used to polarize the reflected neutrons.

22.1 Introduction

Supermirrors are aperiodic multilayers of two materials, which enable neutrons to be reflected at small grazing angles. They are used mainly for the transport of neutrons through guides or to separate polarization components.

One of the most important properties of a supermirror is its critical angle, Θ_c^{SM}, the angle up to which it reflects. By convention it is measured in multiples, m, of the critical angle of natural nickel, Θ_c^{Ni}, which has the largest critical angle of all naturally occurring elements:

$$\Theta_c^{SM} = m\Theta_c^{Ni}. \tag{22.1}$$

The critical angle is determined by the thickness of the thinnest layers. The reflectivity of a bilayer is roughly proportional to the fourth power of its thickness and to the square of the number of layers. Thus, the number of layers needed to reach the same reflectivity for a given bilayer thickness increases quadratically with decreasing thickness. From the number of layers needed for a supermirror with a certain m-number a simple approximation can be derived. The number of layers is am^4, with $a = 4.5$ for Ni–Ti, and $a = 6.5$ for Fe–Si supermirrors.

Since supermirrors were first proposed, it took about 20 years to master the production of $m = 2$ mirrors which have a total thickness of 1 μm and a smallest layer thickness of 7 nm. To increase the m-value two problems had to be solved: the interface roughness and the stress, both of which tend to increase with the number of layers and must, therefore, be kept constant or reduced. Also the stability of the sputtering machine used to produce the mirrors has to be sufficient.

The first part of the article describes the improvements of Ni–Ti supermirrors for guides in the Budapest laboratories, which were achieved by relating the crystalline structure of supermirrors to their reflectivity. Furthermore, the behavior of supermirrors under heat load and irradiation was characterized and the homogeneity of the coating process improved. This made it possible to produce Ni–Ti supermirrors with $m = 4$.

In the second part, from the Berlin laboratory, the polarization of neutrons is introduced and several methods to prepare polarized neutron beams are described with the emphasis on polarizing supermirrors. They reflect only one-spin component and transmit the other one. The main challenge today is to increase the number of layers, which requires improved interfaces between the layers and reduced stress in the layers. In the past few years, it has been possible to increase the number of layers from 150 to 1,000 and the m-value from $m = 2.3$ to 3.4.

22.2 Development and Investigation of Ni/Ti Multilayer Supermirrors for Neutron Guides

22.2.1 Neutron Guides

Neutron guides are used to transport neutrons to the instruments, which may be situated up to 100 m from the source. Mirror reflection is the basis of this transmission process. Neutron supermirrors, which were proposed and built by Mezei in the 1970s [1, 2], are nonperiodical multilayer systems. They are designed so that their angle of total reflection for neutrons is larger than from a pure material, by exploiting interference effects. As the scattering contrast for neutrons between Ni and Ti is large, a high reflectivity can be expected when using them for supermirrors. Therefore, they are most frequently used in nonpolarizing neutron optics.

Figure 22.1 shows the neutron reflectivity curve of some Ni/Ti supermirrors which have different m values. From the preparation viewpoint it is interesting to know that an $m = 2$ supermirror has about 100 layers, while an $m = 3$ mirror has about 500 layers.

Due to the low primary flux of neutrons, the mirrors used, as well as the other optical devices, have to be highly efficient. That is, the reflectivity of the mirrors should be as large as possible. To achieve good reflectivity, it is extremely important to produce smooth layers. Not only the substrate roughness should be as low as possible, but the interface roughness as well. The larger the number of layers the more critical the interface roughness is, since it can grow layer by layer. This means that developing good quality supermirrors with a higher critical angle requires appreciable effort. To produce smooth

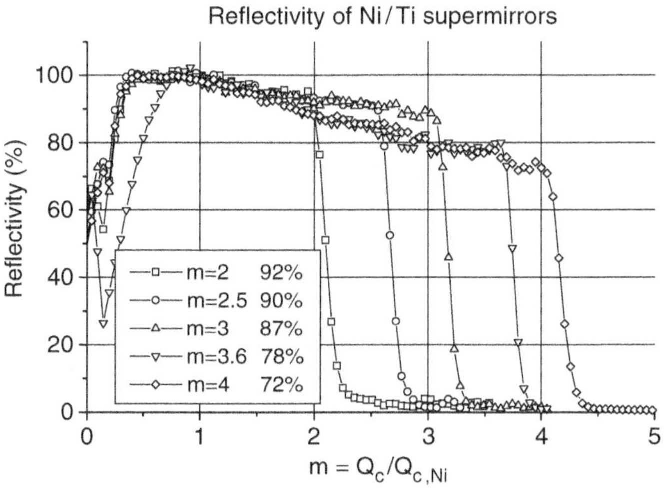

Fig. 22.1. Neutron reflectivity curves of Ni/Ti supermirrors produced by Mirrotron Ltd.

layers by our two DC magnetron sputtering machines – based on the extensive research of Peter Böni and his group [3, 4] – we use the reactive sputtering method, with which we could reproducibly prepare high quality mirrors.

A standard neutron guide is constructed from coated glass plates assembled to a rectangular cross section, the dimensions of which are generally about 150 mm high and 50 mm wide. However, there is also a need to produce guides with larger cross sections. With our high capacity sputtering machine we aimed to develop large surface mirrors.

Neutron guides are mounted with tolerances of less than 10 μm since small misalignments result in a considerable loss of neutrons after some tens of meters of guide. In addition, at the neutron facilities one would like to use a guide for at least 10 years without changing its parts. Therefore, the stability of the coated substrate under extended irradiation and heat load caused by irradiation is a very crucial point.

In the framework of the COST P7 2002–2006 project we aimed at improving the quality and variety of Ni/Ti multilayers by determining the crystalline structure of the supermirrors produced and their stability.

22.2.2 Relation Between Crystalline Structure of Layers in a Multilayer Structure and its Reflectivity

The structures of several Ni/Ti supermirrors with $m = 3$, produced by an optimized reactive DC sputtering method in various qualities on various substrates were investigated by means of wide angle X-ray diffraction (Philips, Cu K_{α}). In the case of a multilayer system reflections of crystalline planes can be seen that are nearly parallel to the surface. We characterized the crystal orientation by the ratio of the area under the respective peaks. From the change in the position of the peaks related to the bulk value, we determined the lattice distortion. As far as we know, no publication in the literature has dealt so far with XRD measurements on Ni/Ti supermirrors.

In all samples, both Ni and Ti show polycrystalline structure. Generally four maxima occur in the spectra, which can be identified as face-centered cubic (*fcc*) Ni lattice (Ni*fcc*(200), Ni*fcc*(111)) and hexagonal close-packed (*hcp*) Ti lattice (Ti*hcp*(011), Ti*hcp*(002)).

For the best quality mirror – prepared on a smooth substrate – the XRD spectra only show one Ni and one Ti peak. This means that both materials crystallize with a preferred orientation with the Ni(200) planes and Ti(011) planes parallel to the substrate surface.

The medium quality mirror was prepared on a similarly smooth substrate; however, the quality is not the same. Again a similar orientation is preferred, but not to the same extent, even though Ni(111) and Ti(002) reflections appear.

The worst mirror was prepared on a quite rough substrate. We can see that all four peaks appear with no preferred orientation in this case. Summarizing we can say that there are two ways of orientation: Ni(200) is connected to Ti(011), and Ni(111) to Ti(002) [5]. On a rough substrate both orientations are present and the reflectivity is low ($R \sim 50\%$). In this case $\sigma \sim 1.2$ nm, where σ is the rms microroughness measured by X-ray reflectivity. On a smooth substrate ($\sigma \sim 0.4$ nm) the Ni(200)/Ti(011) orientation is preferred. The more this orientation is preferred, the better is the quality of the mirror (reflectivity $R > 80\%$), as can be seen in Fig. 22.2.

In Fig. 22.3 one can see the extent of orientation depending on the reflectivity of supermirrors with $m = 3$. Circles represent the ratio of the area under the maxima Ni(200) and Ni(111). Triangles are for the ratio of the area under the maxima Ti(011) and Ti(002). Where no triangles are shown only the Ti(011) peak was present. Thus, we can find a relation between the quality of the mirror and the crystal orientation.

To explain this correspondence TEM pictures were taken (Fig. 22.4). In the good quality mirror there are smooth parallel layers. In the second, less good mirror the layers are not parallel at some parts. The size of the deflection is about 1,000 A. The crystalline orientation is likely to be the same in both cases, but due to the deflection of the layers, the layers are not parallel to the surface and the reflections from the other lattice planes appear as well in the XRD spectra. The deflections start from the substrate, and grow almost straight upward. Their origin might be some inhomogeneity of the substrate surface on a mesoscale, which does not change the observed roughness.

The Ni(200) lattice spacing is in all cases larger by 0.001–0.003 nm than in the bulk. In accordance with the literature, during sputtering in a reactive atmosphere, nickel crystals grow such that Ni(200) planes are parallel to the surface, because gas atoms can be incorporated easily into the lattice in that

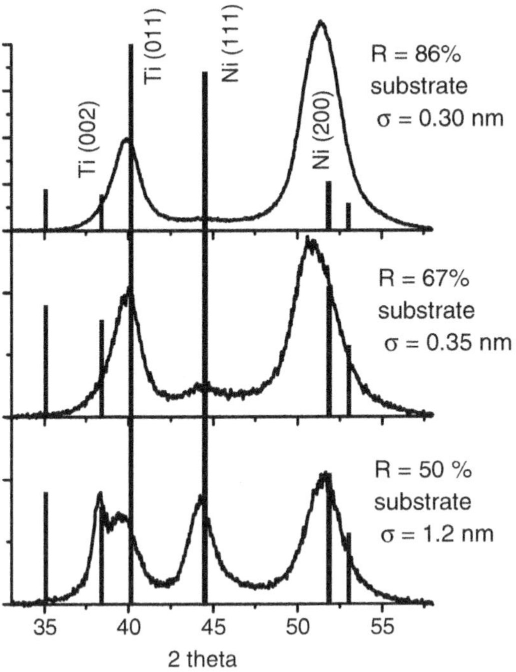

Fig. 22.2. XRD spectra of Ni/Ti supermirrors

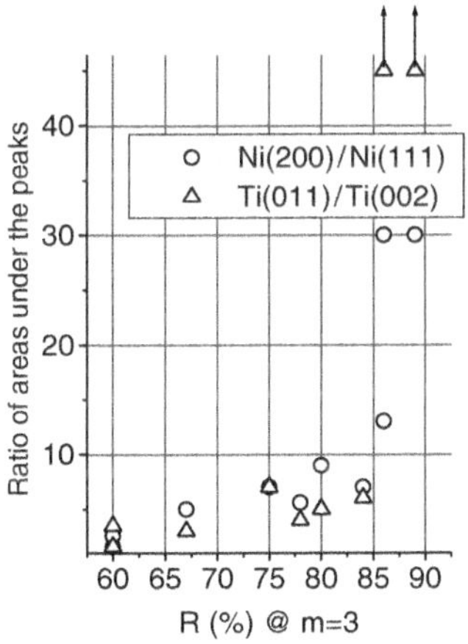

Fig. 22.3. Extent of crystalline orientation in relation to the reflectivity of the supermirrrors with $m = 3$

orientation. The increased lattice spacing also shows this incorporation. Moreover, the dilatation of 0.001–0.0015 nm in the Ti(011) direction indicates the diffusion of gas atoms through the Ni/Ti interface. The Ni(111) peak position is the same as in the bulk. It is related to the presence of small pure Ni phases.

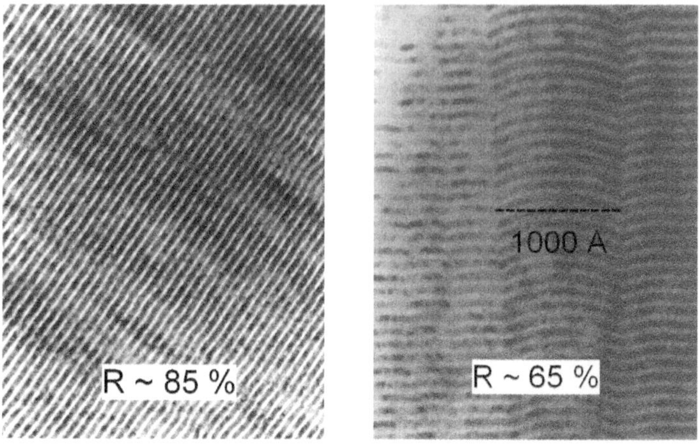

Fig. 22.4. TEM picture taken on a high and a low quality Ni/Ti supermirror

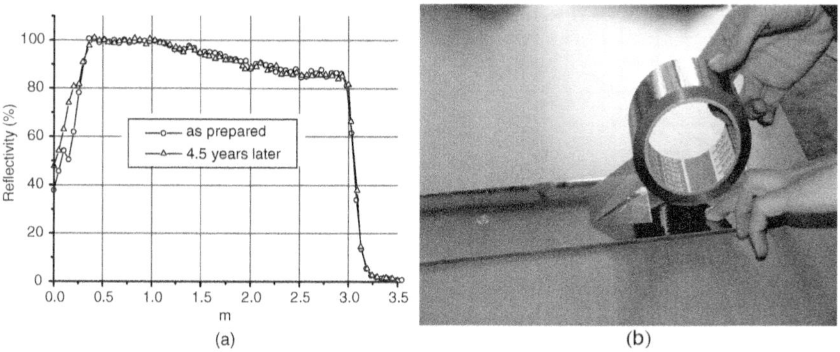

Fig. 22.5. (**a**) Reflectivity curves of a Ni/Ti supermirror before and after extended storage. (**b**) Adherence check

22.2.3 Stability of Supermirrors

Extended Storage

We have performed tests on unused supermirrors 4.5 years after their production. The reflectivity was found to be the same as at the time they were produced. The adherence checked by strong tesa tapes also meets the requirements (Fig. 22.5).

Stability Under Heat Load

We studied the structural changes during heating by X-ray diffraction with a heatable vacuum chamber as sample holder, to be able to perform in situ measurements during heating under low pressure. We simulated the same

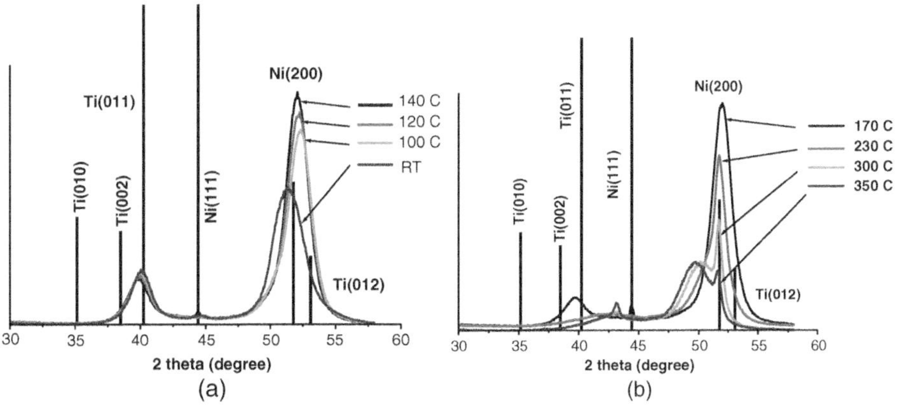

Fig. 22.6. (**a, b**) XRD spectra of a Ni/Ti supermirror taken in situ during heating over two temperature ranges

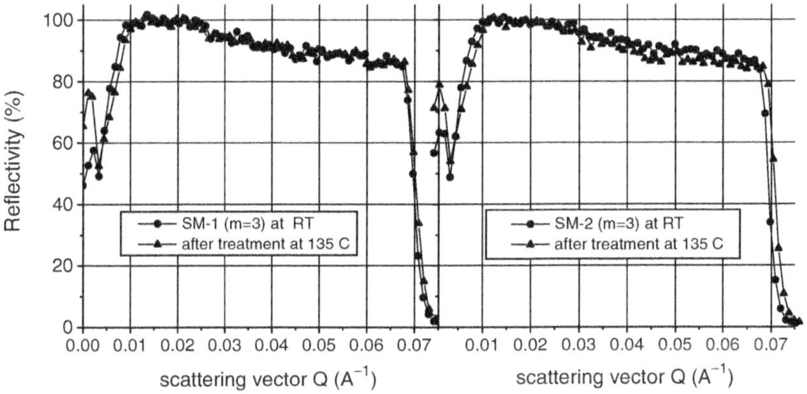

Fig. 22.7. Neutron reflectivity curves of two heated Ni/Ti supermirrors

circumstances as in the real neutron guides, where there is a pressure of about 10^{-4} bar.

Figure 22.6a shows the change of the XRD spectrum up to 140°C. The only change is the sharpening and slight shifting of the Ni(200) maximum. Figure 22.6b shows the further changes up to 350°C. In this temperature region fundamental structural transformations occur.

Based on these results we expect that up to about 140°C the supermirror structure will be stable, and its reflectivity does not change. Experiments on the changes of reflectivity have been made by heat treatment of supermirrors at 100, 120, and 135°C for 50 min in a vacuum chamber The reflectivity does not change after the treatment at 100 and 120°C. However, after the 135°C treatment the critical scattering vector may increase and the reflectivity decreases with 2–3% above $m = 1.21$ (Fig. 22.7).

The detailed process of structural changing during heating is as follows:

As described above in good mirrors at room temperature the polycrystalline Ni and Ti layers show a preferred orientation of Ni *fcc*(200)/Ti*hcp*(011) with some dilatation. When heating to 100 °C, peak broadening and further dilatation can be observed, the extent of which is higher than can be explained by simple thermal expansion. This process is likely to be due to gas atoms bound in the Ni layers during the sputtering in reactive atmosphere. They start to diffuse at the interfaces into the very good getter Ti. Up to 230°C, Ti gradually becomes amorphous and/or forms an amorphous compound.

At the same time at 100°C the Ni(200) peak sharpens and shifts unexpectedly toward higher angles. That is, the lattice spacing decreases even below the bulk value and the crystallite size seems to increase. The reason for this change is not yet clear: it may be ascribed to a rearrangement of the Ni lattice due to the diffusion of the gas atoms. By further heating, a slight thermal expansion occurs. At 140°C the Ni(111) peak arises. Above 170°C, Ni gradually transforms to a yet unidentified compound characterized by a broad maximum at $d \sim 0.180$nm, and partly to NiO.

Stability Under Irradiation

One of the currently used substrates for supermirrors in neutron guides is Borofloat 33 (13 wt% B_2O_3) glass produced by the company Schott. This material has the advantages of low surface roughness due to the float technology and the absorption of neutrons coming through the multilayers, providing shielding for the guide system via the following reaction

$$n + {}^{10}B \Rightarrow \alpha + {}^{7}Li + \gamma \; (1.47\,\mathrm{MeV}).$$

However the effect of this process on the glass and the coating is not yet clear. It is possible that this process or the energy released may cause damage if the neutron dose is large enough.

The question is how large is the onset of neutron dose damage that determines the lifetime of a borofloat guide piece? Recently at the ILL, Grenoble it was found that at the first part of the out-of-pile guide the coating from a borofloat substrate pealed off after three years irradiation and at the same time the glass surface was destroyed. During this period the total incoming dose is estimated to be 3×10^{16} n cm^{-2}. The other mirrors on normal float (no B content) or polished Borkron glass (prepared without float-technology) were found to be stable at the ILL. In Gatchina, and in Budapest, however, they found no damage at in-pile borofloat guides exposed to similar doses.

This indicates an emerging need for the detailed examination of radiation damage of guide substrates and coatings. In the framework of COST action we have performed several irradiation tests in the reactor water in various neutron channels at the 10MW BNC reactor. The samples were packed in an Al capsule, and wrapped in Al foil to transmit the cooling effect of the surrounding water.

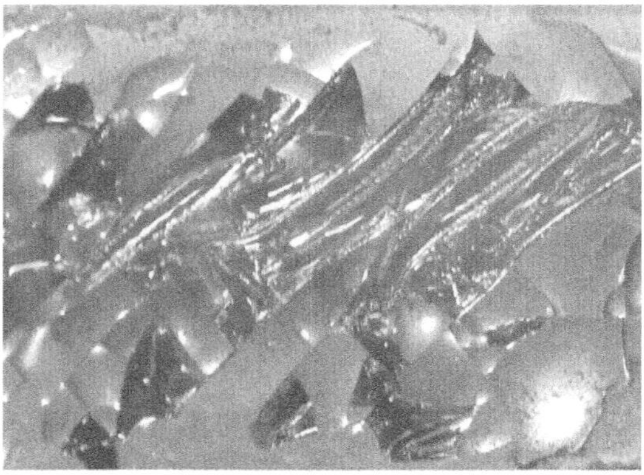

Fig. 22.8. Destruction of the Ni/Ti supermirror coating and the glass surface of borofloat glass after various irradiation doses larger than 1.5×10^{17} n cm^{-2} in channel 11/2

Irradiation of uncoated glass substrates was performed in the 69/3 channel for 72 h. Borofloat and normal float glass alone were not damaged at a dose of 5×10^{19} n cm^{-2} in reactor water. However, the color of borofloat glass changed to brownish.

Irradiation of NiTi supermirror coatings on several glass substrates was performed in different channels of the reactor. The dose applied varied between 10^{17} and 10^{19} n cm^{-2}. In the channel 11/2 during the irradiation of borofloat glass with a supermirror coating, we found coating and the glass surface destruction and discoloration (Fig. 22.8).

However, the noncoated side of the glass remained intact. Float glass with a supermirror coating remained intact in all cases. We have performed tests in another channel, channel 17, where a smaller fast neutron flux and less gamma radiation is expected. Here, as you can see in Fig. 22.9, even the coated side of the borofloat glass remained intact under a dose of 4×10^{18} n cm^{-2}.

Based on these results we can conclude that the borofloat glass without a metal layer is stable under irradiation. The destruction of coated borofloat glass under irradiation does not depend on the thermal flux of neutrons according our experience. The destroying factor is probably the thermal effect of high gamma radiation on the metal layer, which causes the glass surface destruction as well.

By irradiating NiTi coatings on Si substrates, which absorb neutrons only to a small extent, we can investigate the stability of the supermirror coating itself. Moreover, in that case we have the opportunity to check the reflectivity as well. We have applied about 10^{19} n cm^{-2} on coated Si wafers and measured the reflectivity curves before and after irradiation. We found that the reflectivity above $m = 1.8$ is degraded by about 3%. The slope of the reflectivity curve above the maximum value, $m = 3.2$, is somewhat less steep compared to the curve obtained before irradiation.

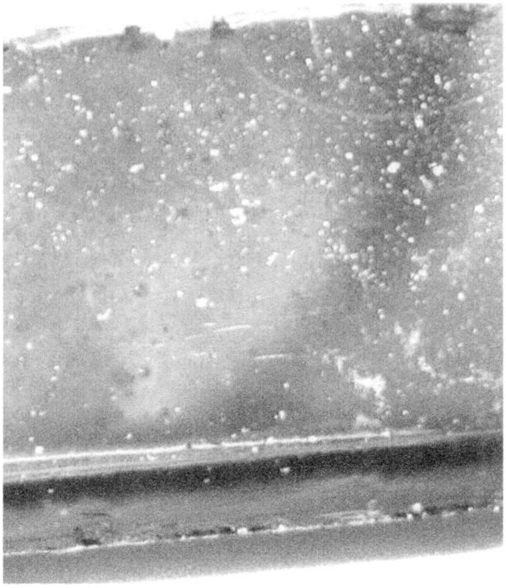

Fig. 22.9. Ni/Ti supermirror coating on a borofloat glass surface after an irradiation dose larger than 4×10^{18} n cm^{-2} in channel 17

22.2.4 Development of $m = 4$ Supermirror Technology

Subsequent to the development of the $m = 3.65$ Ni/Ti neutron supermirror, the task of realization of the considerably more difficult $m = 4$ multilayer system has been taken on.

For this purpose the multilayer system to be sputtered has been extended, namely instead of the 900 layers for the $m = 3.65$ mirror, a system of at least 1,600 layers is needed for reaching $m = 4$, with decreasing layer thickness. The layer system has been optimized by means of the code for reflectivity computation (REFLEX) [6].

To achieve high quality mirrors, appropriate substrate is needed. The substrate quality has been assessed by X-ray reflectometry. Finally the Schott Borofloat glass has been chosen (roughness <0.4 nm, lower density surface layer <1.5 nm), allowing excellent, reproducible $m = 3$ coating.

At $m = 4$, a reflectivity of 72% has been obtained, the expected result after the 76% obtained for $m = 3.65$ (see Fig. 22.1). Further experiments are planned using higher number of layers in order to improve the reflectivity. Today users (e.g., Spallation National Source, USA) require about 60% reflectivity for $m = 4$ supermirrors. Thus the quality of the produced supermirror exceeds the internationally expected quality level.

22.2.5 Increase of Homogeneity Over Large Substrate Sizes

In some new neutron sources there is a need for using large cross-section guides. For that one has to produce supermirrors on substrates with a width larger than the typical 50–100 mm, namely 200 or even 300 mm. We performed tests to determine whether the coating prepared on substrates with various widths is of the same quality. The length was in all cases the usual 500 mm. Substrates were coated in the same sputtering machine, under the same conditions and the neutron reflectivity curves were measured. We have found a larger critical angle for supermirrors deposited on larger substrates. On the basis of this change we can conclude that the integral thickness of the deposited layers is 3–4% smaller for substrates of 200 × 500 mm and 7–8% smaller for substrates of 300 × 500 mm, with respect to substrates of 50 × 500 mm size. The cause of this difference in the deposition process is not yet fully understood. Some electrical charging can be supposed which can be dependent on the substrate size. It was also concluded that this thickness variation of the layers does not influence the mirror quality (interface roughness, adherence) because after compensating for the difference in deposition rate we obtained the same reflectivity for each substrate size.

22.3 Polarizing Supermirrors

22.3.1 Neutron Polarization

In a magnetic field the neutron energy has an additional energy term, the Zeeman term $\pm\mu B$ [7,8]. The magnetic moment of the neutron, μ, has the value 61 neV T^{-1} and B is the magnetic field, which in Fe, for example, has a value of 2.2 T. The sign refers to the orientation of the neutron spin, which is either parallel or antiparallel to the magnetic field direction. The spin of a neutron is antiparallel to its magnetic moment. In ferromagnetic materials the Zeeman term has the same order of magnitude as the nuclear interaction.

The refractive index, n, of a magnetic material for neutrons including nuclear and magnetic interactions is given by:

$$n = 1 - \lambda^2 N(b \pm p)/2\pi \tag{22.2}$$

with λ, the neutron wavelength and N the atomic density. The magnetic scattering length, p, is given by

$$p = 2\mu m_n M/\hbar^2 N \tag{22.3}$$

with m_n, the neutron mass, M, the magnetization in the material and $\hbar$ Planck's constant divided by 2π.

This refractive index gives rise to two critical angles for the total reflection for the two different spin components:

$$\sin\Theta^{\pm} = \lambda\sqrt{(N(b \pm p)/\pi} \tag{22.4}$$

The product $N(b \pm p)$ is called scattering length density (SLD).

Two quantities are used to characterize how well the spin components of a neutron beam have been separated. In terms of the number of neutrons in the two spin states, n^+ and n^-, the polarization, P, and the flip ratio, f_r, are defined as:

$$P = (n^+ - n^-)/(n^+ + n^-) \tag{22.5}$$

$$f_r = n^+/n^-. \tag{22.6}$$

The polarization of a sample is determined by using a neutron beam and a spin analyzer of known polarization. Methods of calibrating spin analyzers are discussed in [7].

22.3.2 Neutron Polarizers

Nowadays mainly three methods are used to polarize neutrons: by the use of Heusler alloys, by ^{3}He spin filters and by polarizing supermirrors. Heusler alloys like Cu_2MnAl can simultaneously monochromatize and polarize a neutron beam [9]. The cross section for Bragg reflection for a magnetic field perpendicular to the scattering plane is given by the square of the sum of the nuclear and the magnetic atomic structure factor. If both have the same value, a high polarization can be achieved. In practice polarization values of 95% for reflected intensities of 90% can be achieved. Heusler alloys are expensive and not easily available on the market. They are mostly used for neutrons with wavelengths below 0.2 nm.

^{3}He spin filters exploit the spin-dependent absorption cross section of ^{3}He atoms for neutrons [10]. The cross section amounts at a neutron wavelength of 0.18 nm to 5,333 barn for antiparallel and to 5 barn for parallel spins. The ^{3}He atoms are kept in a cell with specially prepared walls to reduce polarization losses during wall reflections and are polarized either by spin exchange or by metastable optical pumping. The polarization efficiency, P, for neutrons depends on the polarization, P_{He}, of the ^{3}He atoms and the so-called opacity, O, of the gas:

$$P = \tan\ h(P_{\mathrm{He}}O), \tag{22.7}$$

with

$$O = p\ \mathrm{bar}^{-1}\ l\ \mathrm{cm}^{-1}\lambda\ \text{Å}^{-1}, \tag{22.8}$$

with p, the helium pressure in the cell, and l, the flight path of the neutron in the ^{3}He gas.

The transmission of neutrons through the gas is given by

$$T = \cos\ h(P_{\mathrm{He}}O)T_0 \exp(-O), \tag{22.9}$$

with T_0, the absorption of the cell.

Thus, the degree of neutron polarization can be chosen at the expense of the transmitted intensity. A polarization of 90% at a transmission of 30% of the incoming unpolarized beam is presently a reasonable compromise between maximum transmission and maximum polarization and can be reliably reached.

^{3}He spin filters need an expensive infrastructure and permanent maintenance and the technology is still strongly improving. Magnetic field gradients larger than 10^{-4} reduce their polarizing efficiency. Their advantages are the absence of any small angle scattering and any sensitivity to the angles under which neutrons pass the filter. They enable a very high degree of polarization to be achieved if a corresponding reduction of the transmitted intensity is acceptable.

Supermirrors in general were introduced above. Polarizing supermirrors exploit the fact that ferromagnetic materials have two strongly different scattering length densities (SLD) for the two spin components. After choosing two materials which exhibit the same SLD for one spin component, the supermirror sequence is calculated from the contrast of the two materials for the other spin component. Such a system reflects only the second spin component and transmits the first one. Polarizations up to 98% can be reached for intensities of 30–40% of the nonpolarized beam.

Historically the first mirrors were made from the material pairs Fe–Ag [2] and Co–Ti [11]. Nowadays two groups of combinations are used: Fe-SiN_x [12] and $Fe_{89}Co_{11}$–Si [6] or Co–Ti, FeCo–TiZr [13], and $Fe_{50}Co_{48}V_2$–TiN_x [14].

The materials in the first group have an SLD of the spin-down component close to the SLD of Si. They are used for solid-state devices where the neutrons travel inside thin Si wafers and one spin state is reflected from the supermirror coating at the walls of the wafers. The other spin component is not reflected by the supermirror since there is no or only a very small contrast to Si.

The materials in the second group have an SLD close to or slightly below zero for the spin-down component. In this case no reflection of the spin-down neutrons occurs from the supermirror if the neutrons hit the supermirror in air. However, there are only two kinds of substrates which have the required small surface roughness and are available at reasonable prices for areas in the m^2 range: glass and Si wafers. From these substrates the spin-down component is reflected up to their critical angle. This amounts to $m = 0.5$ for Si and $m = 0.6$ for glass. To maximize the angular and wavelength range where neutrons are reflected with good polarization, an antireflecting layer is introduced between the supermirror and the substrate, this antireflecting layer absorbing the neutrons before they reach the substrate. Such layers are made from Gd or Gd alloys or multilayers of Gd and Ti [15].

Polarizing supermirrors are not sensitive to magnetic fields and the technology is quite mature. However, they show some small angle scattering, in some cases only if used in small magnetic fields, and they work only in an angular range on the order of 1°. They are most useful to polarize neutrons for wavelengths above 0.2 nm with a small angular divergence.

22.3.3 Increase of the Critical Angle

In order to increase the available angular range of polarizing neutron supermirrors and to facilitate the construction of polarizing devices, the critical angle of the supermirrors should be as large as possible. To reach higher m-values the neutron optics group at the Hahn-Meitner-Institut Berlin performed a detailed study on the development of interface layers and their relation to the sputter parameters argon pressure, sputter voltage, and sputter rate [16, 17]. The choice of certain sets of these parameters enabled the interface layer thickness to be reduced to about 1 nm, keeping it constant for an arbitrary number of layers. The results of a second study, which focused on the development of stress, are reported in another contribution to this chapter. These studies allowed one to choose the optimum parameters for the sputter process.

Additionally, the computer control system of the sputtering machine was improved by a new program, which allowed for the measurement and control of a much larger number of sputter parameters such as plasma potential and substrate bias during the sputtering process.

As a result of all these improvements it was possible to increase the number of layers from 150 to 1,000 and the critical angle from $m = 2.3$ to 3.4.

Figures 22.10 and 22.11 show the reflectivity curves of such mirrors, measured at the neutron reflectometer V14 at the Hahn-Meitner-Institut Berlin. The neutron beam had a wavelength of 0.48 nm, a divergence of 0.035°, and a polarization of 97%. The data points in the figures give the neutron reflectivity for the two spin components, which are calculated by subtracting the background of 3×10^{-4} of the direct beam and correcting for the polarization of the incoming beam. Additionally shown are the polarization and the flip ratio.

The reflectivity for both mirrors decreases to about 90% at $m = 2$ and in the second case to 83% at $m = 3.3$. The average flip ratios are for the first mirror 50 (polarization 96%) in the interval from $m = 1$ to 2 and for the

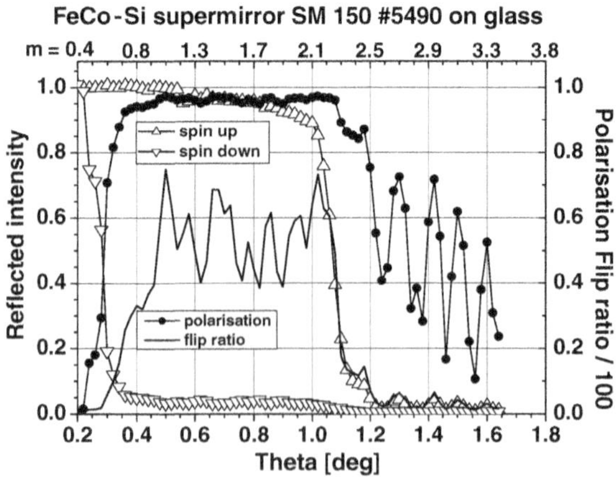

Fig. 22.10. FeCo–Si supermirror with 150 layers on a float glass substrate showing the reflectivity of both spin components of neutrons with a wavelength of 0.47 nm together with the polarization and the flip ratio

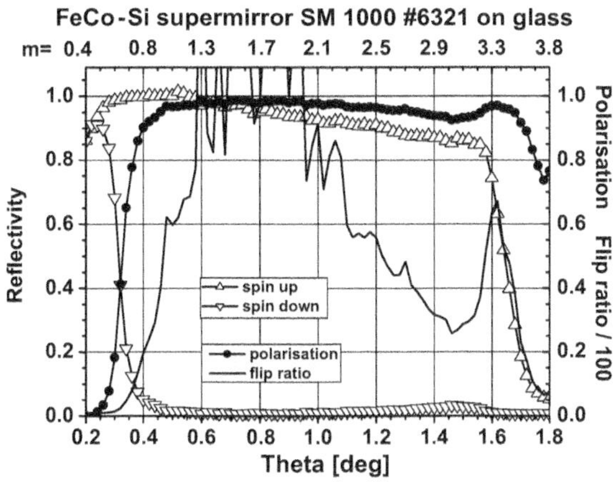

Fig. 22.11. FeCo–Si supermirror with 1,000 layers on a float glass substrate showing the reflectivity of both spin components of neutrons with a wavelength of 0.47 nm together with the polarization and the flip ratio

second mirror 80 (polarization 97.5%) in the interval from $m = 1$ to 2 and 50 in the interval from $m = 2$ to 3.3.

In conclusion it can be said that in the past 5 years the critical angle of polarizing and nonpolarizing supermirrors has been increased considerably, as has been the reliability of the sputtering machines. This was achieved by understanding the effects previously limiting the layer numbers to a few hundred, and finding ways to reduce them by improved sputtering processes.

In the near future, a further increase of the critical angle can be expected. In the case of nonpolarizing supermirrors this will increase the divergence and hence the flux of neutrons transmitted through neutron guides and make it possible to transport shorter wavelengths.

In the case of polarizing supermirrors this will lead to fewer restrictions in the construction of polarizing and analyzing elements and will enable one to handle larger beam divergences.

References

1. F. Mezei, Comm. Phys. **1**, 81 (1976)
2. F. Mezei, P.A. Dagleish, Comm. Phys. **2**, 41 (1977)
3. O. Elsenhans, P. Böni, H.P. Friedli, H. Grimmer, P. Buffat, K. Leifer, J. Söchtig, I.S. Anderson, Thin Solid Films **246**, 10 (1994)
4. P. Böni, D. Clemens, M. Senthil-Kumar, S. Tixier, Physica B **241–243**, 1060 (1998)
5. K. Makihara, J. Yang, J. Shi, M. Hasimoto, S. Maruyama, A. Barna, J. Vacuum Sci. Tech A, **19**, 2494 (2001)
6. D. Clemens, Ph.D. Thesis, Technische Universität, Berlin, 1993

7. W.G. Williams, *Polarized Neutrons* (Clarendon Press, Oxford, 1988)
8. V.F. Sears, *Neutron Optics* (Oxford University Press, New York, Oxford, 1989)
9. A. Freund, R. Pynn, W.G. Stirling, C.M.E. Zeyen, Physica B **120**, 86 (1983)
10. R. Surkau, Ph.D. thesis, Universität Mainz, Germany, 1995
11. O. Schärpf in *Symposium on Neutron Scattering*, Argonne, IL, USA, 12–14 Aug 1981, AIP Conference Proceedings, 1982, p. 182
12. P. Höghöj, R. Anderson, R. Siebrecht, W. Graf, B. Ben-Saidane, Physica B **267–268**, 355 (1999)
13. A.F. Schebetov, N.K. Pleshanov, V.M. Pusenkov, B.G. Peskov, G.E. Shmelev, W.H. Kraan, P.T. Por, M.Th. Rekveldt, V.E. Mikhailova, Nucl. Instrum. Methods B **94**, 575 (1994)
14. D. Clemens, P. Böni, H.P. Friedli, R. Göttel, C. Fermon, H. Grimmer, H. van Swygenhoven, J. Archer, F. Klose, Th. Krist, F. Mezei, P. Thomas, Physica B **213–214**, 942 (1995)
15. O. Schärpf, Physica B **156–157**, 631 (1989)
16. S.J. Cho, Th. Krist, F. Mezei, Thin Solid Films **434**, 136 (2003)
17. S.J. Cho, Th. Krist, F. Mezei, Thin Solid Films **444**, 158 (2003)

23

Stress Reduction in Multilayers Used for X-Ray and Neutron Optics

Th. Krist, A. Teichert, E. Meltchakov, V. Vidal, E. Zoethout, S. Müllender, and F. Bijkerk

Abstract. Multilayer systems have important applications in many areas of X-ray and neutron optics. For some applications the positions of the optical surfaces have to be controlled with accuracies in the sub-nanometre range. For neutron supermirrors with over a thousand layers, stresses above 1000 MPa can occur. In addition to bending the substrate such stresses can lead to the films peeling from the substrate, or even to the destruction of the substrate surface, and so must be avoided. After an introduction to stress, this chapter describes how stresses can be reduced to acceptable values and discusses two examples – FeCo/Si polarizing neutron supermirrors and Mo/Si multilayer mirrors for extreme ultraviolet lithography.

23.1 Introduction

Multilayer systems find important applications in many areas of X-ray and neutron optics. Besides the reflective properties of the multilayer film, the actual position of the optical surface has to be controlled with unprecedented accuracy. Often the position of the mirror surface has to be accurate in the sub-nanometer regime. This puts high demands on the allowable stress in the multilayer films. After giving an introduction to stress we will show how the stress can be reduced to acceptable values using the two examples of FeCo/Si polarizing neutron supermirrors and Mo/Si multilayer mirrors for extreme ultra violet (EUV) lithography.

In the case of neutron supermirrors with some hundred to some thousand layers and a total thickness up to 5 μm, values of stress above 1,000 MPa can be reached. Such stress values lead to a bending of the substrate, to the peeling off of the films from the substrate, or even to the destruction of the substrate surface. Most high reflectance Mo/Si multilayer mirrors reported so far have a stress value of −350 up to −450 MPa. They are usually produced by magnetron sputter deposition [1–3]. These stress values result in a surface deformation of several nanometers up to tens of nanometers, depending on the substrate dimensions and the material.

Relaxation of the film stress deforms the substrate at the film–substrate interface. The net result of the bending momentum of the film and the substrate determines the resulting deformation. Change of surface curvature can now be ascribed to the (multilayer) film on top of the substrate without knowing the elastic properties of this film. Stress can be divided into two types: compressive stress, which refers to the situation where the reflecting surface becomes more convex, and tensile stress, which refers to a more concave reflecting surface.

To reduce the stress to acceptable levels, several schemes can be applied. First of all the parameters in the deposition process can be optimized. Evaporation techniques, however, usually do not allow for enough process latitude to obtain sufficiently different layer compositions for stress optimization. Sputtering techniques can use a variation of the working gas pressure, although here limits are given by changes in surface structure, which lead to increased interface roughness. The additional use of ions during or after layer deposition offers more freedom. Either one or both of the layers can be treated with different polishing conditions, resulting in different layer smoothening and densification. Furthermore, it is known that a larger metal fraction of the bilayer thickness (Γ) shifts the stress from compressive to more tensile [1, 2]. Also the thermal stress can be used for compensation of stress by choosing an appropriate substrate temperature during deposition. Another option is to compensate stress by a secondary single layer or multilayer film underneath the high reflectance multilayer [2]. Finally, new materials can be introduced in thin layers in the multilayer system. In this approach an additional (buffer) layer is admitted, which is incorporated in the multilayer stack. To have no or only a negligible effect on the reflectivity, such a layer must be optically neutral. It should be able to induce an opposite stress and its thickness is chosen in order to compensate the original stress, thus making the total biaxial stress equal zero. These interlayers can influence the stress dramatically [4], but other film properties can change as well. So far, most methods show limitations in terms of the reflectivity achieved, and no experimental demonstration of an effective compensation method has been given without loss of reflectivity.

23.2 Origin, Description, and Measurement of Stress

Multilayer systems develop stress due to several reasons [5]: through defects in the crystal lattices, columnar instead of layer-by-layer growth, the formation of new phases at the interfaces or simply by different thermal expansion coefficients of the two materials.

Stress, σ_t, arising from the difference in thermal expansion, $\Delta\alpha$, between two materials is given by

$$\sigma_t = Y_f \Delta\alpha \, (T_d - T_m), \tag{23.1}$$

where T_d and T_m are the temperatures of the deposition and the measurement, respectively. Y_f is the biaxial elastic modulus of the film,

$$Y_\mathrm{f} = E_\mathrm{f}/(1 - \nu_\mathrm{f}), \tag{23.2}$$

where E_f is the Young's modulus and ν_f is Poisson's ratio. To control induced thermal stress, one has to keep the temperature of the substrate stable. In addition, a stress known as epitaxial may appear when a mismatch exists in the lattice parameters of the different materials. Although this origin of stress is typical for techniques of epitactic deposition and exclusively for crystalline layers, similar effects may be present in a sputtering process. The stress related to the interfaces can be induced by the configuration change of interfaces, for example, phase formation [6]. Deposition stress may result from a nonequilibrium growth of a film whose density undergoes variation, due to gas incorporation, formation of voids, etc.

The stress induced in films deposited by the sputtering process is directly dependent on the mobility of the adatoms [7–9]. For low values of mobility, a layer is generally of lower density or even porous. Indeed, in the presence of voids, the stress becomes compressive due to the atomic forces. On the other hand, for high values of the mobility the adatoms are able to reach lattice sites and fill the voids. The layer is denser and the stress is normally tensile. The value of the stress is a function of the mobility of adatoms and passes abruptly from tensile to compressive for increasing mobility. In turn, the mobility of adatoms is a function of the deposition conditions, particularly on the working gas pressure: the lower the pressure, the higher the mobility. Thus, the stress induced by gas pressure changes is different for different materials (Fig. 23.1).

The total stress in a multilayer results from the stress of each layer in the multilayer stack and from the interfacial stress within the structure.

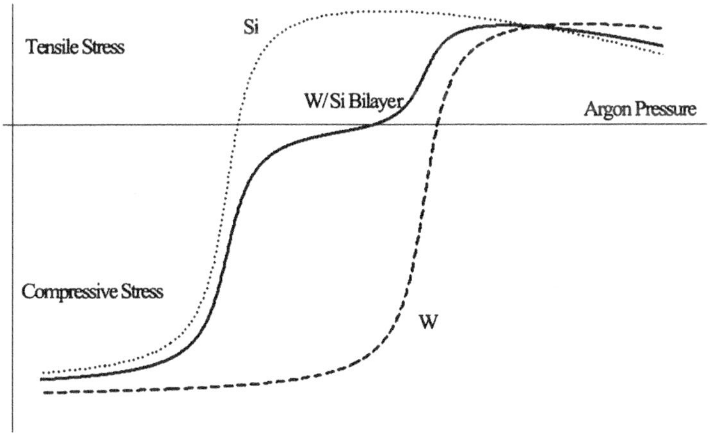

Fig. 23.1. Stress variation vs. argon pressure in single layer W (*dashed line*), single layer Si (*dotted line*), and bilayer Si/W (*solid line*) [5]

In the case of a periodic binary multilayer consisting of alternatively deposited materials, h and l, with respective thickness, d_h and d_l, and assuming that the interfacial stress (as for instance, in the system Mo/Si and W/Si) is negligible compared to the stress within the layers, the biaxial total stress, σ_{total}, is given by [10]

$$\sigma_{\text{total}} = (d_{\text{h}}\sigma_{\text{h}} + d_{\text{l}}\sigma_{\text{l}})/(d_{\text{h}} + d_{\text{l}}). \quad (23.3)$$

We suppose here that the different layers are isotropic in the plane of the layers, the interfaces between the different layers are abrupt, and that the stress of the first and last layer is negligible taking into consideration the large number of periods.

A great number of stress measurements in deposited layers are concerned with the determination of the substrate curvature. When considering small deviations from an almost flat substrate due to isotropic biaxial stress of a thin film, Stoney's equation is best suited to describe stress [11,12]:

$$\sigma = Y_{\text{f}} \frac{D^2}{6d} \left(\frac{1}{R_2} - \frac{1}{R_1} \right), \quad (23.4)$$

where D is the substrate thickness, R_2 and R_1 are the radii after and before the coating, d is the layer thickness. For Si Y_{f} has the value of 180 GPa.

According to Finot [13], for circular substrates it is convenient to introduce a stress parameter, A, defined as

$$A = \sigma\, d\, \frac{B^2}{D^3}, \quad (23.5)$$

where B is the substrate diameter.

There are three distinct curvature modes. A_{c} is defined as the value of the stress parameter for which the curvature undergoes a transition from axial to cylindrical symmetry. When A/A_{c} is lower than 0.2, the Stoney formula can be used. For a larger A/A_{c}, i.e., a larger deformation compared to the thickness of the substrate, the Stoney formula does not work any more. Then, if the ratio A/A_{c} is between 1 and 0.2, the deformation remains axially symmetric but is not spherical any more. Instead, it has an inhomogeneous curve, which is more important at the edge of the sample. The relation between the force on the surface and the curve becomes nonlinear. For a ratio A/A_{c} larger than 1, the curvature is not axially symmetric any more but becomes cylindrical.

The most often used experimental methods are those that measure the bending of a substrate. Various techniques and methods can be applied for this. A scheme of the curvature measurement, which uses a deflected laser beam, is presented in Fig. 23.2. A swiveling motorized mirror sends the laser beam to the sample via a lens. The sample reflects the beam and a detector records the position of the beam after it has again passed through the lens. By placing the detector and the motorized mirror in the focal plan, a variation of the position of the laser on the detector allows one to determine the radius of curvature. The sample can be placed horizontally in a furnace within a

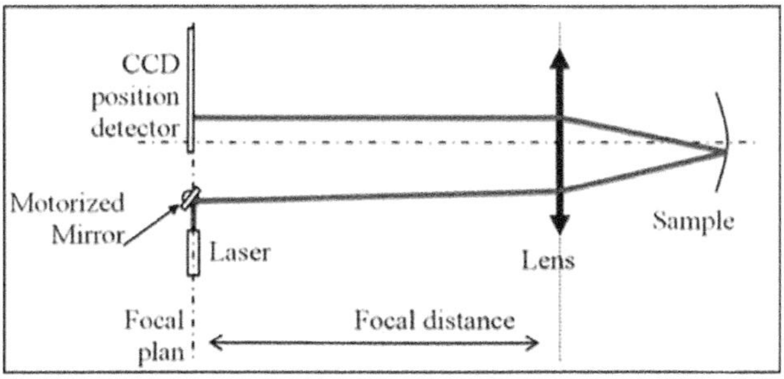

Fig. 23.2. The curvature measurement by deflection of a laser beam

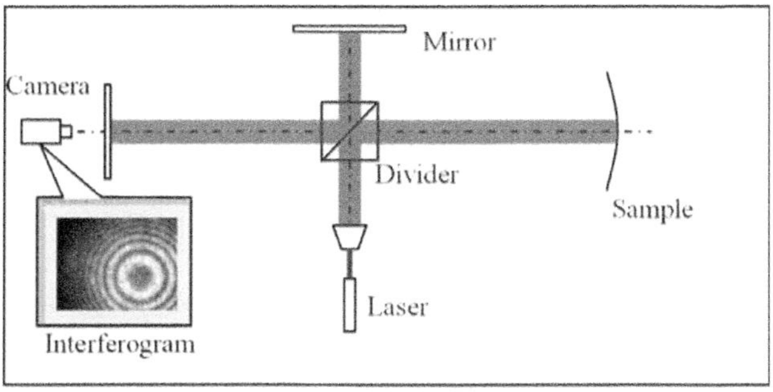

Fig. 23.3. A Michelson interferometer

vacuum chamber, which enables one to follow the evolution of the stress with the temperature while the sample is annealing.

A Michelson interferometer can also be used for the determination of the curvature. The schematic diagram is presented in Fig. 23.3. It is a traditional Michelson interferometer, which measures the wave front. It consists of a laser; a beam expander; a divider cube creating the two waves, which interfere; reference mirrors; the sample support; and the screen associated with the CCD camera. The return mirrors make it possible to position the sample horizontally in order to not introduce stress due to the sample fixation. With this device the two-dimensional surface of the sample plane can be determined in only one measurement. One can deduce the shape of multilayer optics with a precision of the order of the wavelength of the laser light.

Curvature measurements can also be performed by using a local probe microscope such as an atomic force microscope (AFM). A tip with a radius of curvature of the order of some nanometers comes in intermittent contact with a surface. A real-time feedback control loop using the piezoelectric transducer makes it possible to pull the cantilever with constant force and to scan the sample surface (Fig. 23.4).

This makes possible precise topography measurements of the stress in very thin layers such as those deposited onto the membranes. However, because of

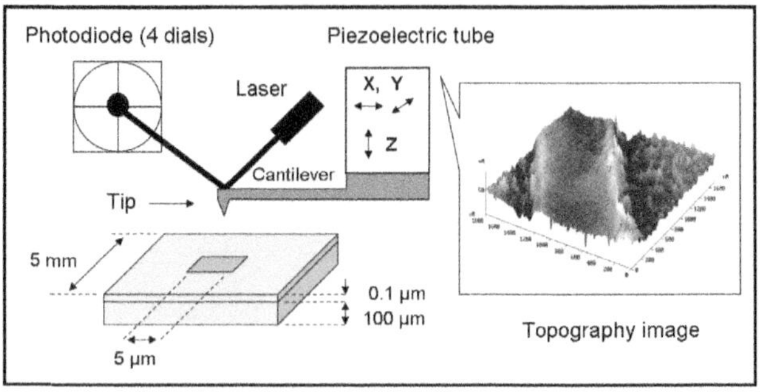

Fig. 23.4. Schematic of a membrane and measurement by atomic force microscopy (AFM)

the geometrical characteristics of the membranes, it becomes impossible to apply the Stoney formula. Numerical simulations are needed to obtain the value of the stress.

As a further method not relying on the curvature of a substrate, diffraction analysis can be applied to the stress measurements in order to calculate the constraint induced by the multilayer stack deposited on crystalline substrate. The stress induces a deformation field that modifies the diffraction pattern from the substrate. One can thus calculate the stress according to the angular displacement of the diffraction peaks.

23.3 FeCo/Si Polarizing Neutron Supermirrors

We report on two studies of the stress developing in $Si/Fe_{89}Co_{11}$ multilayer systems, which are used for polarizing neutron supermirrors (cf. Chap. 22). The stress was examined as a function of the thickness of the layers [14] and of the substrate bias voltage.

23.3.1 Experimental

The Si and $Fe_{89}Co_{11}$ (in the following text: FeCo) multilayers were produced in a triode sputter machine [15]. The pressure of the working gas Ar was 1.5×10^{-3} mbar and the sputter power was 240 W. In a first series of experiments the nominal layer thickness of one of the materials in a monochromator system was kept constant, while the thickness of the other material was varied from 5 to 25 nm. In a second series the bias potential of the substrate was varied from 30 to 60 V for monochromators and supermirrors.

The systems were simultaneously sputtered onto 3-mm-thick float glass substrates and onto two thin Si substrates with a thickness between 120 and 250 µm.

After the sputtering the multilayers were characterized by X-ray reflection with a wavelength of 0.154 nm and neutron reflection with a wavelength of 0.47 nm. In the case of X-rays, fits to the data using the program Parrat [16] enabled us to determine average values for the thickness and roughness of the two individual layers and one extra layer of a mixed material at each interface.

XRD measurements were employed to determine the grain size and the crystallinity, and finally polarized neutron reflection was used to determine the quality of the supermirrors. The magnetic measurements were accomplished on a SQUID magnetometer at 300 K.

The bending of the Si substrates was measured on a Dektak 3030 profilometer before and after coating them to calculate the stress using the Stoney equation (23.4).

23.3.2 Layer Thickness Variation

In the first set of experiments, monochromator systems of ten bilayers with a capping layer of Si with the same thickness as the other Si layers were used. Systems with 14 different FeCo layer thicknesses between 4 and 22.5 nm were produced.

The multilayers were then characterized by X-ray reflection to determine the layer sequence. As an example, Fig. 23.5 shows an X-ray measurement and a fit to the data using the program Parrat. The results for the average layer thicknesses of the two materials and two interface layers from the fit are $d_{\text{Si}} = 18.34$ nm, (roughness, 0.52 nm), $d_{\text{Si-FeCo}} = 1.23$ nm (0.65 nm), $d_{\text{FeCo}} = 7.11$ nm (1.56 nm), and $d_{\text{FeCo-Si}} = 1.69$ nm (0.86 nm).

For the complete set of the 14 samples, the average value for the Si layer thickness from the X-ray measurements was 18.55 ± 0.36 nm. The average thickness values for the interface layers were on top of the Si layer

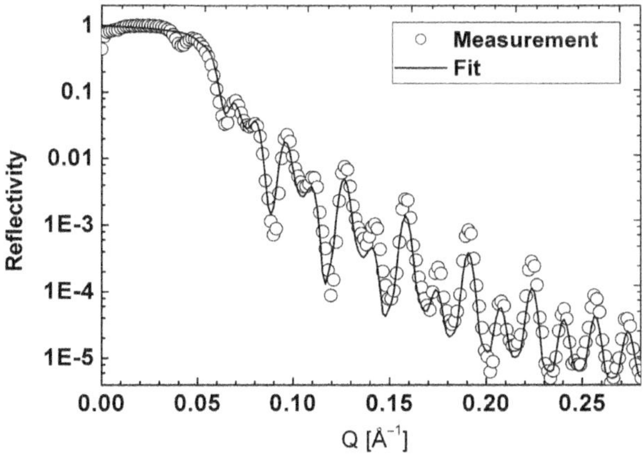

Fig. 23.5. X-ray reflection measurement on a FeCo–Si monochromator together with a fit using the program Parrat

1.23 ± 0.10 nm and on top of the FeCo layer 1.76 ± 0.17 nm. The average values for the roughness were 0.51 ± 0.03 nm for the Si layer, 0.62 ± 0.12 nm for the interface layer on top of the Si, 1.58 ± 0.17 nm for the FeCo layer, and 0.89 ± 0.14 nm for the other interface layer.

Figure 23.6 shows the increasing tensile stress in this series. It can be fitted by a linear function giving the relation between the total stress, σ, in the multilayer in MPa and the thickness, d, in nm of the n FeCo layers where the number n is constant:

$$\sigma = n0.023d - 463. \tag{23.6}$$

X-ray diffraction was used to determine crystallinity and – by use of the Scherrer formula – the grain size of the FeCo layers. Figure 23.7 shows that

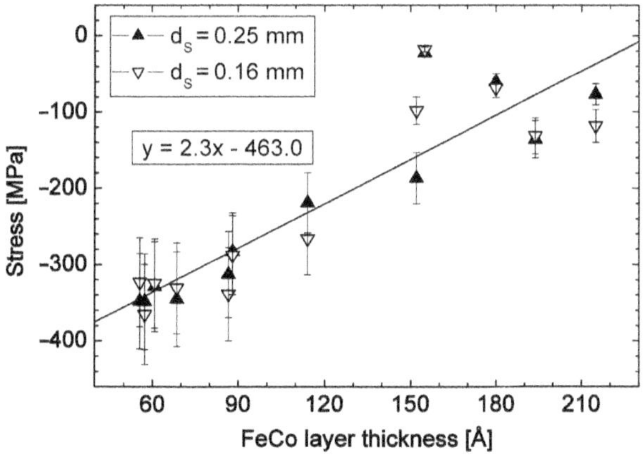

Fig. 23.6. Stress values for FeCo–Si monochromators with varying thickness of the FeCo layer together with a linear fit

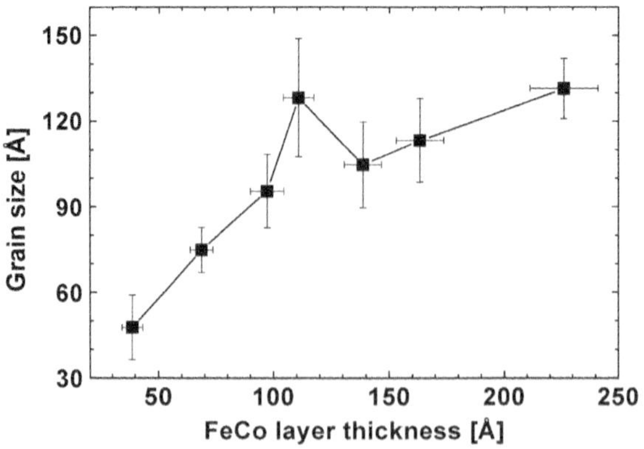

Fig. 23.7. Grain size in the FeCo layers depending on the layer thickness

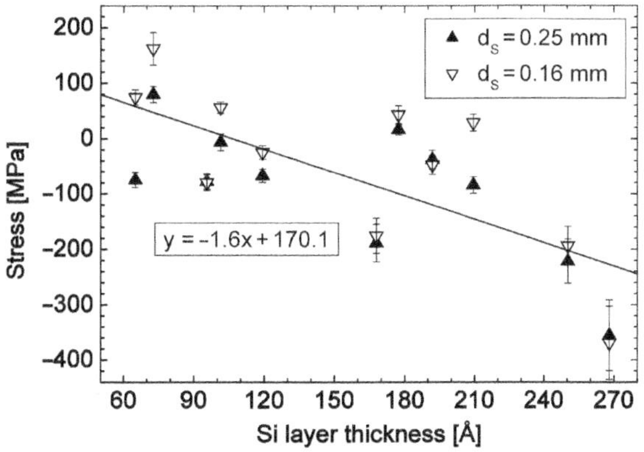

Fig. 23.8. Stress values for FeCo–Si monochromators with varying thickness of the Si layer together with a linear fit

the grain size of the FeCo crystallites equals the layer thickness from 4 to 10 nm and then only slowly increases to 12 nm at a layer thickness of 24 nm.

A similar set of samples was produced and characterized with a constant thickness of the FeCo layer and a variation of the thickness of the Si layer from 4 to 27 nm.

From the X-ray measurements, the average value for the FeCo layer thickness was 15.1 ± 0.7 nm. The respective values for the interface layers were on top of the Si layer 1.54 ± 0.16 nm and on top of the FeCo layer 1.9 ± 0.08 nm. The average values for the roughness were 0.53 ± 0.04 nm for the Si layer, 0.8 ± 0.29 nm for the interface layer on top of the Si, 1.57 ± 4 nm for the FeCo layer, and 1.02 ± 0.25 nm for the other interface layer.

Figure 23.8 shows the decreasing tensile stress due to the increase of the Si layer thickness. It can be fitted by a similar linear function as above:

$$\sigma = -n0.145d + 170. \tag{23.7}$$

The total stress of a multilayer with layer thicknesses in the examined ranges can be calculated. With the linear fits above the total stress of the monochromator σ_M is given by

$$\sigma_M = na_{FeCo}d_{FeCo} + na_{Si}d_{Si} + C, \tag{23.8}$$

where d_{FeCo} and d_{Si} are the thickness of the FeCo or Si layer, respectively. The constants are $a_{FeCo} = 2.3\,\text{MPa nm}^{-1}$, $a_{Si} = -1.45\,\text{MPa nm}^{-1}$, and $C = -172\,\text{MPa}$. With these data a Γ value can be deduced for any given bilayer thickness.

23.3.3 Substrate Bias Voltage

In a second study the bias potential of the substrate was varied from 30 to 60 V for three different multilayer systems: monochromators with 25 bilayers with

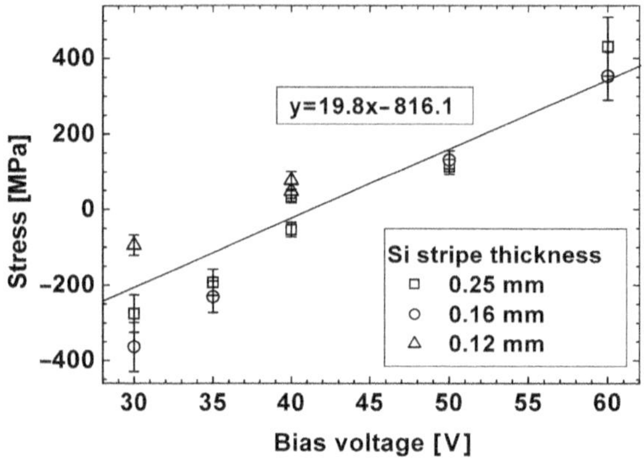

Fig. 23.9. Stress dependence on bias voltage in Si–FeCo monochromators with 51 layers, each with a nominal thickness of 15 nm

a nominal thickness of 15 nm and a Si capping layer of the same thickness, and polarizing supermirrors with 50 and with 330 layers. The self-bias potential of the substrate is 45 ± 3 V.

For the Si–FeCo monochromators, the total thickness remains constant within the error limits. The stress developing in the whole system for the different bias voltages is presented in Fig. 23.9. At 30 V the stress is compressive with -363 ± 65 MPa and at 60 V it is tensile with 431 ± 78 MPa. At a bias voltage of 41 ± 1 V the stress is zero. A linear fit to the experimental data shows that increasing the bias potential by 1 V results in an increase of the stress by 19.8 MPa for this system.

The influence of the bias voltage on the crystallinity of the FeCo layers was determined by X-ray diffraction measurements. Figure 23.10 shows the intensity of the (110) peak of the FeCo crystallites, which crystallize in the body-centered cubic (bcc) phase as also pure Fe does.

Table 23.1 gives some data derived from the curves in Fig. 23.10. The integral intensity has the highest value at 30 V and decreases with increasing voltage. Also, the grains in the FeCo layers become smaller. For a bias voltage below the self-bias value of 45 V Ar ions are attracted to the growing film and due to the additional energy they convey to the film the fraction of the crystalline area increases. This effect increases for decreasing bias voltage.

Further it can be seen that with increasing bias voltage the lattice parameter, d_{hkl}, decreases while the full width at half maximum and the center of gravity of the curves increase. This means that the lattice constant acquires values below and above the bulk value, which clearly is the final cause of stress: for the bulk value the stress is zero.

Additionally, the magnetic properties were measured by SQUID magnetometry. Like the stress, the remanence and coercivity were isotropic. The

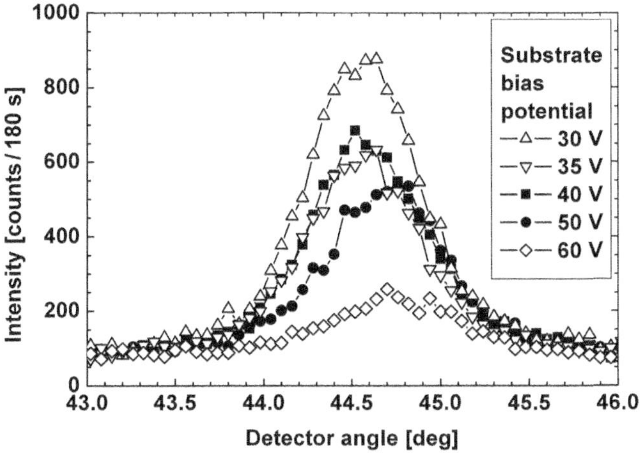

Fig. 23.10. XRD measurement of Si–FeCo monochromators on float glass with different bias voltages ($\lambda = 0.1541$ nm)

Table 23.1. Results of the X-ray diffraction measurements on Si–FeCo monochromators

Bias (V)	Intensity (counts)	2θ (deg)	FWHM (deg)	D (nm)	d_{hkl} (nm)
30	860	44.541	0.701	12.25	0.2033
35	575	44.540	0.714	12.03	0.2033
40	621	44.577	0.717	11.98	0.2032
51	486	44.696	0.732	11.74	0.2026
60	170	44.736	0.838	10.25	0.2025

remanence, with values of 50% ± 10%, does not depend on the bias voltage, while the coercivity increases slightly from 8 to 10.5 G.

In the next step, supermirrors with 50 and 330 layers were investigated. The total layer thicknesses for these systems were 620 ± 100 nm and 2,500 ± 200 nm.

As for the monochromators, for the supermirrors the stress becomes more tensile with increasing positive bias voltage. The stress vanishes at 32 ± 2 V for the supermirrors with 50 layers and at 39 ± 2 V for those with 330 layers. The results of the stress measurements for the supermirrors with 50 layers are given in Fig. 23.11. A change of 1 V leads to an additional tensile stress of 23.5 MPa. Here the change is larger than that for the monochromators, because the total layer thickness of FeCo is larger.

Figure 23.12 shows the reflectivity curves for two polarizing neutron supermirrors, which were grown for two different substrate bias potentials of 35 and 60 V. As can be seen, the reflectivity for the spin up state is the same within the error limits. The flip ratio, which is given by the ratio of spin up intensity

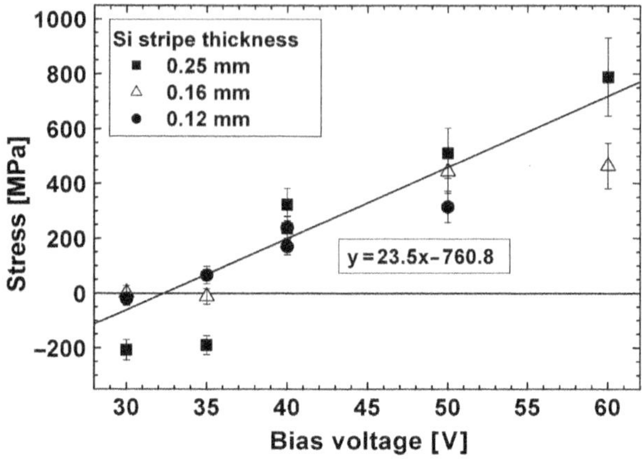

Fig. 23.11. Stress values as a function of the bias voltage in Si–FeCo supermirrors with 50 layers

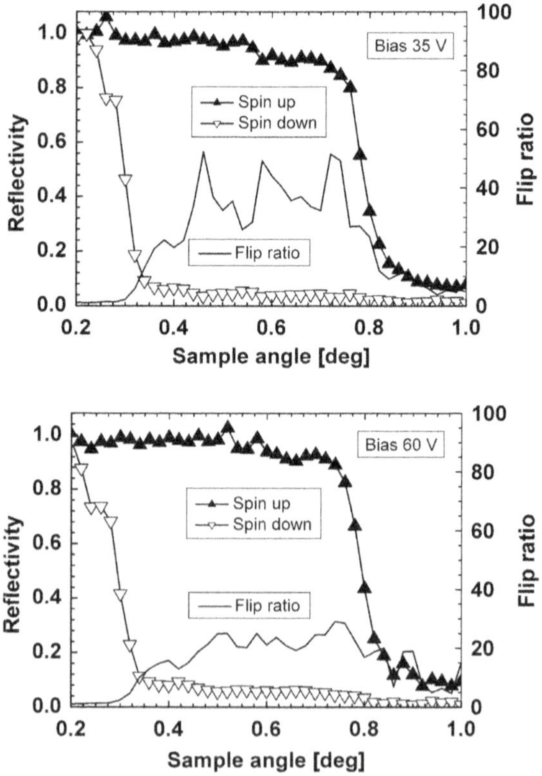

Fig. 23.12. Reflectivity of polarizing supermirrors with 50 layers, which were grown with substrate bias values of 35 and 60 V, for neutrons with a wavelength of 0.47 nm and for both spin components together with the flip ratio

divided by spin down intensity, is about 30% higher for the supermirror grown with 35 V. It can be concluded that, within the parameter range explored in this study, the variation in substrate bias influences the growth conditions in a way which only changes the stress and not the reflectivity.

In conclusion, it can be said that an increase in layer thickness leads to a linear increase of the tensile stress for FeCo layers and to a linear decrease of the tensile stress for Si layers. Increase in the bias voltage leads to a linear increase of the tensile stress in all systems. Its value depends on the particular system.

These results show that the bias potential is a very convenient parameter to influence the stress developing in a multilayer. For all systems tested it was possible to find a bias potential for which the total stress of the system was close to zero. This makes it possible to grow multilayer systems with a stress below 50 MPa, which is sufficient for the substrates presently used for neutron supermirrors.

23.4 Stress Mitigation in Mo/Si Multilayers for EUV Lithography

To accommodate the next generation of integrated circuits, the semiconductor industry is looking at soft X-rays as the source for lithographic processes. The so-called extreme ultraviolet (EUV) lithography development programs use reflecting Mo/Si multilayer optics, because the 13.5 nm wavelength of the soft X-rays is absorbed in any realistically thick lens optics. Another consequence of the short wavelength is that the exact position of the mirror surface, the surface figure, has to be controlled accurately. This puts strict demands not only on the substrate manufacturing, but also on the multilayer induced substrate deformation as well. In general, all thin films will try to relax the intrinsic film stress by macroscopic deformation of the substrate. Standard film stress values of −350 to −450 MPa, reported for Mo/Si systems produced by magnetron sputtering, induce an intolerable deformation of the surface figure of EUV optical components. At FOM Rijnhuizen, an extensive stress mitigation program has been carried out on multilayers produced by e-beam deposition and medium energy ion polishing. The resulting stress in standard, high reflectance Mo/Si multilayers is less than −200 MPa. Although e-beam deposition apparently halves the typical stress values obtained by sputter deposition, it is still above the allowable limit for the first lithographic system, the so-called Alpha Tool. To further reduce stress, the critical parameters in Mo/Si multilayers, the Mo fraction, the number of periods, and the multilayer period or d-spacing have been investigated. Different Mo fractions in e-beam deposited multilayers result in a similar dependency as reported for magnetron sputtered coatings, although at strongly reduced absolute values. Variation of the d-spacing has only a small influence on stress. The number of periods, however, has no influence on the stress value in the range from

20 to 50 periods. Applying stress mitigation techniques based on adjustment of the Mo fraction, a high reflectance at near normal incidence at 13.5 nm has been obtained for multilayers with a stress value of only −33 MPa. This has been achieved by using Mo and Si only. This stress value is sufficiently low to enable the first generation of EUVL optics to be prepared.

23.4.1 Experimental

All Mo/Si multilayer films in this paper have been deposited by e-beam deposition and ion polishing in an ultra high vacuum system [17–19]. The base pressure during deposition is typically 5×10^{-8} mbar. Growth and polishing of the Mo and Si layers are monitored in-situ by N-K radiation (3.16 nm) reflectometry. The development of film roughness is mitigated by medium energy Kr ion treatment. Furthermore, a quartz oscillator is used for additional growth monitoring. Super polished Si(001) substrates of 0.5 mm thickness are used and have been characterized for their curvature before and after deposition by interferometry. Film thickness is analyzed ex-situ by grazing incidence Cu-K radiation (0.154 nm). The near normal incidence reflectivity at 13.5 nm has been measured at the Physikalisch Technische Bundesanstalt (PTB) in Germany.

23.4.2 Results

The distinct difference of the e-beam deposition method applied here is that the atoms reach the substrate with low thermal energy. The layers can therefore grow differently from layers grown using the higher thermal energy atoms in the more popular magnetron sputter deposition. To obtain a stress value for a typical high reflectance Mo/Si multilayer produced by an e-beam, several 50 period multilayers have been produced on curvature characterized Si(001) substrates. These high reflectance multilayers show a stress value of −180 MPa. This is already an improvement of a factor of two compared to multilayers produced by magnetron sputtering. Nevertheless, the lower e-beam stress value still deforms the optical components by several nanometers.

To mitigate the film stress of the Mo/Si multilayer, the parameters of the polishing step in the e-beam deposition process have been a topic of investigation. Changing the ion treatment of Si, with minimal reduction of optimal EUVL performance, has only a minor influence on the film stress. This parameter is therefore not useful for stress mitigation.

Furthermore, different Γ of the multilayer structures, with Γ defined as the as-deposited Mo fraction of a single multilayer period, is known to influence film stress. Figure 23.13 is an overview of the stress values at different Γ for e-beam deposition and magnetron sputter deposition. The results shown in Fig. 23.13 in the black circles represent e-beam deposited multilayers consisting of 50 periods Mo/Si with a d-spacing of 7 nm and ion treatment of the Si layers only. Stress values increase nearly linearly with increasing Γ. The zero

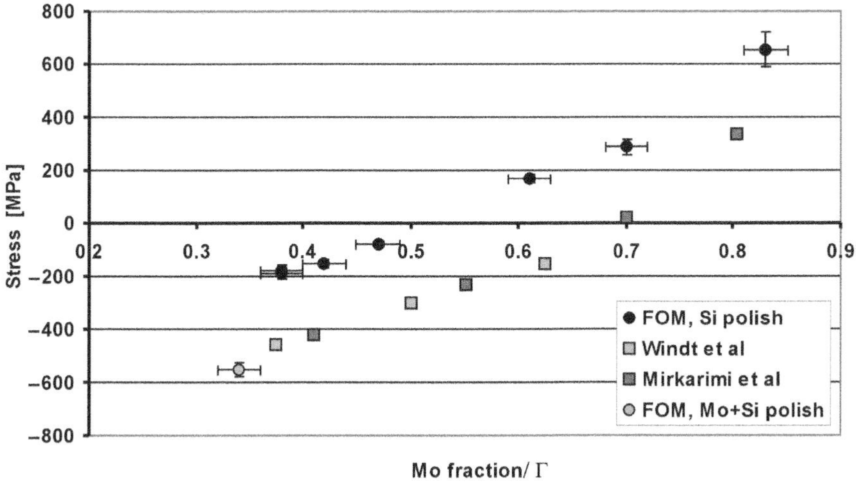

Fig. 23.13. Stress values of different Mo fractions (Γ) for e-beam deposition (FOM) and magnetron sputtering by Windt et al. [1] and Mirkarimi et al. [2]. The black circles have been produced with e-beam and ion treatment of the Si layers, the gray circle by e-beam and ion treatment of the Si and Mo layers, the squares have been produced by magnetron sputter deposition

stress crossover occurs at a Γ of about 0.5. This value divides the Γ range into two: below 0.5 there is a region where the Mo/Si multilayer has a compressive stress and above 0.5 a region has tensile stress. A similar linear dependency is observed for magnetron sputtering [1,2], though the zero stress crossover is at a Γ between 0.65 and 0.7. The difference between e-beam and magnetron sputter deposition appears to be the constant off-set of the linear dependency of stress with Γ.

Furthermore, Fig. 23.13 shows a high reflectance multilayer, produced by e-beam deposition, with both Mo and Si layers ion-treated (gray circle). The stress value of −550 MPa found for this multilayer is more than twice that found in the case of Si ion treatment. This result shows that, besides the Mo fraction (Γ), ion treatment of Mo has a significant effect on the film stress. A possible explanation for the observed shift between e-beam and magnetron sputter deposited multilayers can be found when not only the Si, but also the Mo layers are treated with ions. This result fits well on the magnetron sputter deposition curve. Apparently, conditions can be found where the ion treatment of the Mo and Si mimics the growth of magnetron sputtering. Polishing seems to provide the Mo layer with sufficient energy to reconstruct to a similar composition (e.g., smoothness, density, crystal sizes) as in magnetron sputtering.

The Γ-dependency has been investigated for a d-spacing around 7 nm and 50 periods. Stress values do not a priori have to be the same for different d-spacings nor for different numbers of periods. Especially in the nanometer

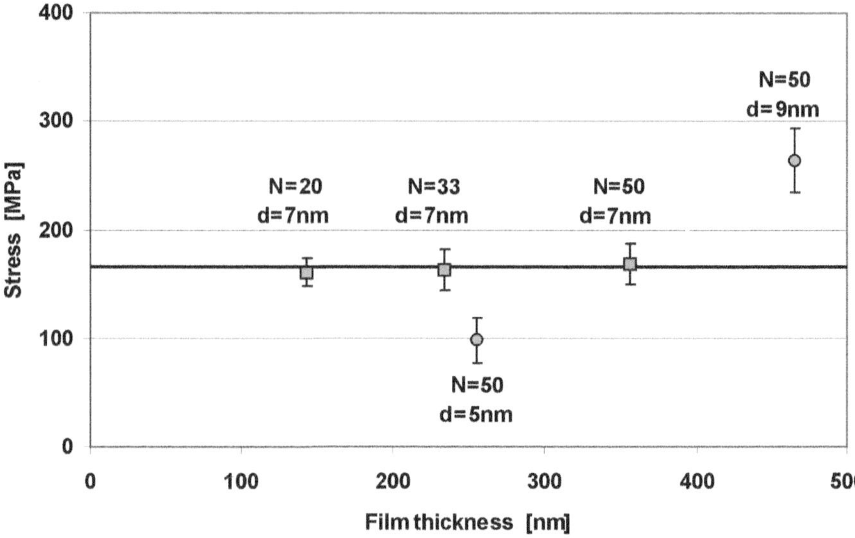

Fig. 23.14. Stress values at $\Gamma = 0.6$ for Mo/Si multilayers produced with e-beam deposition. The different number of periods (N) does not influence stress whereas different d-spacing (d) appears to

range, stress values are known to differ for different layer thickness [3]. Figure 23.14 shows the results for different d-spacings and for different numbers of periods, in all cases with a Γ of 0.6. The stress is not influenced by the number of periods, but it is influenced by the d-spacing. A smaller d-spacing results in a less tensile stress. Keeping in mind that the observed stress value is a result of the balance between film and substrate deformation, as defined by Stoney's equation, films with a similar composition should show a similar stress value per unit film thickness. The number of periods of a multilayer should therefore not be of any influence. This is indeed observed in Fig. 23.14. For different d-spacings, however, the stress values do change. This indicates a different multilayer composition, or layer thickness dependent stress relaxation.

The results so far have shown that a wide variety of stress values can be obtained. The aim is to produce a Mo/Si multilayer that meets the EUVL requirements, of which reflectance is of paramount importance. Using the results depicted in Fig. 23.13, the simplest solution would be to produce a multilayer with a Γ of 0.5, but the reflectance would not meet the requirement. Another approach that has been suggested is to deform the substrate in the opposite direction with a so-called anti-stress layer (ASL) [2]. E-beam deposited Mo/Si multilayers with Γ larger than 0.5 can serve as an ASL. Additional benefit of Mo and Si in an ASL is the use of the same deposition technique as for the high reflectance multilayer part. The ASL needs to compensate 50 periods of -180 MPa stress. Using the results from Fig. 23.13, we find that

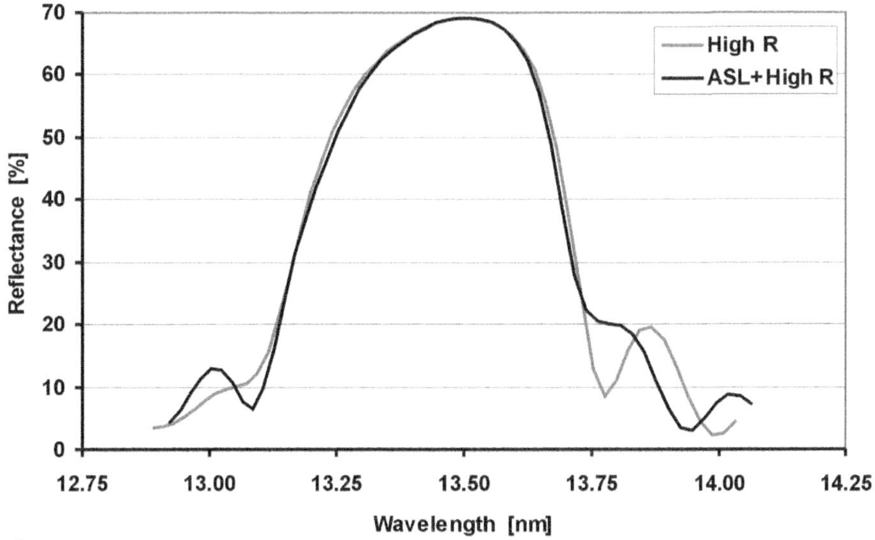

Fig. 23.15. Near normal incidence reflectance of a high reflectance multilayer and a stress mitigated high reflectance multilayer. The stress mitigation reduced the stress value to −33 MPa at a reflectance of 69% and a wavelength of 13.5 nm

a 50 period Mo/Si multilayer with Γ of 0.6 will compensate that stress. This combination, however, would require the deposition of two full stacks.

Figure 23.14 confirms that stress is independent of the number of periods, allowing one to balance the number of periods with the proper Γ to mitigate a selectable amount of the high reflectance part. An ASL of only 30 periods of Γ of 0.7 and 7 nm d-spacing has therefore been selected for experimental verification. Figure 23.15 shows the reflectance as a function of the wavelength for both a standard high reflectance multilayer and a stress compensated high reflectance multilayer. The standard high reflectance multilayer reflects 69% at near normal incidence at 13.5 nm, and has a stress value of −180 MPa. The stress mitigated high reflectance multilayer has the same high reflectance (69%), but shows a dramatically reduced stress value of −33 MPa.

We have reported the results on stress mitigation of e-beam deposited Mo/Si multilayers. Standard, high reflectance multilayers show a stress value of −180 MPa, about half the stress value of multilayers produced by magnetron sputter deposition. We found two different regimes when varying the Γ in the multilayer: if the Mo fraction is below 0.5 the resulting stress is compressive, while for larger Γ the stress is tensile, making possible the design of a stress compensating multilayer under the high reflectance coating. Following this approach, we demonstrated the first stress compensated Mo/Si multilayer with a reflectivity of 69% at 13.5 nm radiation for near-normal incidence. This reflectivity is identical to a coating without stress compensation. A residual stress value of −33 MPa has been achieved. This stress value is sufficiently low to enable the deposition of EUVL optics for the first generation of EUV wafer steppers. In this approach, the same material combination is used in the high reflection and in the stress mitigation part, greatly facilitating the multilayer fabrication process. It is expected that this technique will enable the production of stress free multilayer coatings.

Acknowledgments (only for part 23.4)

We acknowledge Frank Scholze and J. Tümmler for the reflectance measurements performed at the Radiometry Laboratory of the Physikalisch Technische Bundesanstalt (PTB, Berlin Germany), Claudia Junghans and Dr. Horst Feldermann for doing additional stress measurements. This research is part of the program of FOM, the Dutch organization for Fundamental Research on Matter, and has been carried out under contract with Carl Zeiss Semiconductor Manufacturing Technologies AG (Oberkochen, Germany).

References

1. D.L. Windt, W.L. Brown, C.A. Volkert, W.K. Waskiewicz, Appl. Phys. **78**, 2423 (1994)
2. P.B. Mirkarimi, C. Montcalm, SPIE **3331**, 133 (1998); Mirkarimi PB, Montcalm C, Opt. Eng. **38**, 1246 (1999)
3. J.M. Freitag, B.M. Clemens, Appl. Phys. Lett. **73**, 43 (1998)
4. M. Shiraishi, W. Ishiyama, T. Oshino, K. Murakimi, Jpn. J. Appl. Phys. **39**, 6810 (2000)
5. W.D. Nix, Metal. Trans. A **20**, 2217 (1989)
6. U. Welzel, J. Ligot, P. Lamparter, A. Vermeulen, E. Mittemeijer, J. Appl. Crystallogr. **38**, 1 (2005)
7. J.A. Thornton, J. Vac. Sci. Technol. **11**, 666 (1974)
8. G.S. Bales, A. Zangwill, J. Vac. Sci. Technol. A **9**, 145 (1991)
9. F.A. Dollack, R.W. Hoffmann, Thin solid films **12**, 71 (1972)
10. J.A. Ruud, A. Witrouv, F. Spaepen, J. Appl. Phys. **74**, 2517 (1993)
11. G.G. Stoney, Proc. Roy. Soc. Lon. Ser A **82**, 172 (1909)
12. M. Ohring, *The Material Science of Thin Films* (Academic Press, New York, 1992), p. 416 or similar text books
13. M. Finot, I.A. Blech, S. Surech, J. Appl. Phys. **81**, 3457 (1997)
14. A. Teichert, Th. Krist, F. Mezei, Phys. B **385–386**, 1262 (2006)
15. D. Clemens, Th. Krist, P. Schubert-Bischoff, J. Hoffmann, F. Mezei, Phys. Scr. **50**, 195 (1994)
16. Available on the webpage of BENSC: http://www.hmi.de/bensc/software/refl/parratt/parratt.html
17. E. Spiller, *Soft X-Ray Optics* (SPIE Optical Engineering Press, Bellingham, Washington, 1994), p. 170
18. E. Louis, A.E. Yakshin, P.C. Görts, S. Abdali, E.L.G. Maas, R. Struik, F. Bijkerk, *Reflectivity of Mo/Si multilayer systems for EUVL*, SPIE Conference 3676, Santa Clara (1999)
19. R. Schlatmann, S. Lu, J. Verhoeven, E.J. Puik, M.J. van der Wiel, Appl. Surf. Sci **78**, 147 (1994)

24

Multilayers with Ultra-Short Periods

M. Jergel, E. Majková, Ch. Borel, Ch. Morawe, and I. Maťko

Abstract. Multilayers with ultra-short periods (less than 2nm) are needed for particular applications such as mirrors for water-window X-ray microscopy and gamma-ray telescopes. Due to the extreme requirements on interface quality, manufacture is challenging even when realized with traditional material pairs. In this chapter, several types of ultra-short period multilayers, prepared by different deposition techniques from constituents with different miscibilities, are studied and compared in terms of the interface quality, structure and thermal stability. Specular/non-specular X-ray reflectometry, transmission electron microscopy and X-ray and electron diffraction were employed to obtain an insight into the interface phenomena. UHV e-beam deposition with optimized *in situ* substrate heating was tested successfully as a simpler and cheaper alternative to *in situ* ion beam polishing to deposit high quality Cu/Si and Ni/C multilayers. However, true ultra-short periods could not be achieved. Replacement of elemental by compound layers in Ni/B^4C multilayers using distributed electron cyclotron resonance sputtering allowed multilayer periods of less than 2 nm to be produced, with enhanced thermal stability up to 500° C compared to Ni/C multialyers. Similar excellent interface quality, ultra-short period and good thermal stability were obtained with Sc/Cr along with negligible miscibility. In both cases, the multilayer decay is controlled by the formation of a fine granular phase from amorphous layers. Vertical correlation of the interface profiles was found to be too weak to be detrimental effect to the specular imaging contrast. The results have direct implications for the targeted optimization of ultra-short period interference mirrors.

24.1 Introduction

An interference multilayer (ML) mirror is among the most attractive optical elements developed during the last two decades of the past century [1]. Flexibility in the material choice and the stack parameters result in the applicability in a wide wavelength range. This fact is of special importance when the refractive index is close to unity and the use of lenses and mirrors is not as straightforward as in visible optics, i.e. from the ultraviolet (UV) through

X-ray up to gamma regions. Many practical examples may be found particularly at synchrotron beam lines, in astronomy, plasma diagnostics, surface science instrumentation or X-ray diffractometry. Industrial applications are foreseen in the near future, Mo/Si mirrors for extreme UV (EUV) lithography at 13.4 nm being the best example. Here, differently shaped substrates may be used which is a further advantage of ML mirrors.

MLs are prepared by alternate deposition of two or more elements on a substrate. An interplay between extrinsic and intrinsic interface roughness induced by the substrate and growth process, respectively, controls the interface morphology while thermodynamic characteristics of the forming layers determine interface reactions such as mixing, interdiffusion or compound formation. Compositionally sharp and topologically smooth interfaces with good optical contrast are required for high reflectivity. Atomic-scale surface or interface roughness causes scattering losses and lowers the reflectivity but more importantly, it produces non-specular (diffuse) scattering. This scattering concentrated around the ML Bragg reflections reduces the imaging contrast when it falls into the field of view. From this point of view, interdiffusion and chemical reactions at the interfaces are less detrimental to the quality of a ML mirror than is the roughness itself, because diffuse but smooth interfaces only decrease the reflectivity [2, 3]. Moreover, these reflectivity losses may be compensated for if a large enough number of interfaces are penetrated. Furthermore, a ML with steep concentration gradients at the interfaces is inherently unstable and may experience interface degradation under high radiation loads and/or high temperatures to which ML mirrors are often exposed (e.g. synchrotron beam lines, intense plasma source diagnostics). Consequently, the specular reflectivity drops irreversibly. Therefore, interface quality must be preserved under both of these conditions.

Amorphous layers are generally preferred in ML mirrors since polycrystalline grains may lead to rough interfaces. Another adverse effect of polycrystallinity is grain boundary diffusion which is usually very fast and may result in discontinuous layers. Grain boundary diffusion can be avoided and atomically flat interfaces achieved in epitaxial superlattices which were tested for mirror applications. In particular, Al-based metallic superlattices with ultra-short periods (less than 1.5 nm) were prepared by atomic layer epitaxy (ALE) and molecular beam epitaxy (MBE) [4]. Such structures would make further progress in mirror performance possible when amorphous MLs have approached their inherent limits. The main problem is the severe restriction in selecting suitable material pairs which leads in turn to compromising the optical criteria. Compatibility of ALE and MBE techniques with pre-figured substrates must also be addressed. Interdiffusion is prevented in immiscible combinations of materials. If such materials do not satisfy the optical criteria at given wavelengths, interdiffusion of an optically suitable but miscible couple can be suppressed using a pair of materials in thermodynamic equilibrium or using compound layers instead of elemental ones [5–8]. Incorporation of thin diffusion barriers below 1 nm (e.g. C, B_4C, Ru) is an alternative approach [9].

Minimization of optical aberrations by maximizing the angle of incidence is general effort. While in the EUV region normal-incidence geometry does not put strong requirements on the ML period, for soft X-rays, shorter periods are needed which go down below 2 nm in extreme cases (ultra-short periods). Such mirrors are needed for the so-called water window (2.33–4.36 nm). The water window is located between the K absorption edges of oxygen and carbon where the absorption coefficient of water is very low while at the same time organic or biological materials, which contain carbon, are absorbing. Consequently, living cells can be observed in vivo in their natural environment [10]. Several material combinations optically convenient for the water window were studied, e.g. V/Al_2O_3 [11], Ni/C/Ti/C [12], W/Sc and Cr/C [13]. Another application example of ultra-short period MLs is coatings in hard X-ray telescopes working above 50 keV, e.g. W/Si and Pt/C [14]. Here, the angle of incidence is inherently grazing even for ultra-short periods.

The Debye–Waller attenuation factor for X-ray reflectivity (XRR) drops down rapidly with reduction of the ML period if the interface width is not reduced in the same way. Therefore, the interface quality becomes inherently crucial for the layer thicknesses below 1 nm. For example, the interface width of 0.1 nm (rms) reduces the reflectivity by $\approx$50% for the period of 1 nm (the value of the Debye–Waller factor). Actually, the ability of a material pair to yield MLs with minimum interface roughness rather than with high X-ray optical contrast is important for ultra-short period mirrors [15]. The materials must also provide continuous layers, the critical thickness being mostly 0.5–1 nm. No less important is the uniformity of the ML stack. It was shown [16] that period fluctuations should not exceed $\approx 1/N$ (N is the number of periods) which gives fluctuations less than 0.0025 nm for a period of 1 nm and $N = 450$. Thermal stability is also more critical here than in conventional MLs. Hence, the preparation of ultra-short period MLs is challenging for thin film technology and for materials science in general. The deposition process requires tight control of the parameters, stability and fast switching between the materials. A detailed interface characterization in situ provides a necessary feedback.

In this contribution, several types of ultra-short period MLs with different miscibilities of the constituents were fabricated by different techniques to study and to compare them in terms of the interface quality and thermal stability. Specular/non-specular X-ray reflectometry, transmission electron microscopy and X-ray and electron diffraction were employed for the analysis. An insight into interface phenomena is a prerequisite for a proper choice of material pairs which are compatible with ultra-short periods and subsequently for a knowledge-based and targeted optimization of fabrication conditions resulting in high-performance interference mirrors.

24.2 Sample Choice and Preparation

Because of lower absorption far above the K absorption edges, Cu and Ni are promising refractory metals to replace W or Pt in ultra-short period grazing-incidence X-ray mirrors working close to 100 keV. When combining them with a low-absorption spacer material, optical constants of such pairs fulfil well criteria worked out for material choice which guarantees high XRR of MLs [1].

Long-term room temperature (RT) stability and narrow interfaces below 0.3 nm were reported for dc-sputtered Cu/Si MLs with the period down to 2 nm [14, 17]. However, a rather large solubility and diffusivity of Cu in Si of 75% and $1.6 \times 10^{-10}\,\mathrm{cm}^2\,\mathrm{s}^{-1}$, respectively, may pose a problem at elevated temperatures regardless of the deposition technique, especially for ultra-short periods. As a part of our work, we studied this in detail in the MLs prepared by electron-beam evaporation in ultra-high vacuum (UHV).

Ni/C MLs were studied more frequently, e.g. in connection with the development of Göbel mirrors for laboratory X-ray sources or normal-incidence mirrors working at C–K edge. Carbon-based MLs are known to have a good thermal stability. Ni/C mirrors were found to be stable up to 300°C [18, 19] but the interface width in the as-deposited state proved to be larger than that in Cu/Si MLs. Therefore, Ni/C mirrors with periods down to 3 nm are rather exceptional. Common Ni/C mirrors were prepared mainly by rf/dc sputtering [20–23] or by pulsed laser deposition (PLD) [24–26]. Energetic adatoms in these techniques result in relatively smooth interfaces. On the other hand, PLD is known to produce mixing and interdiffusion which may broaden interfaces and deteriorate reflectivity. Electron-beam UHV evaporation is less prone to these effects but when used for Ni/C mirrors, an in situ ion-beam etching of Ni layers had to be applied to reduce the geometrical interface roughness [27, 28]. Energy dissipation from hot spots created by the ions impinging under a particular angle smoothens the deposited surface. As a cheaper and simpler alternative to ion-beam etching, we tested UHV electron-beam evaporation with in situ substrate heating to prepare ultra-short period Ni/C MLs. A larger surface mobility of adatoms is expected to heal the interface roughness and to yield thin layers, but the substrate temperature has to be optimized to avoid an enhanced constituent mixing. A low solubility of C in Ni (0.55% at 1,300°C) and practically zero solubility of Ni in C render the Ni/C couple very suitable for such a technique. Moreover, bulk Ni carbides are formed far above 1,000°C and are therefore no problem.

Cu/Si and Ni/C samples were deposited in a UMS 500 Balzers apparatus in a vacuum of 10^{-7} Pa onto silicon wafers with a native oxide layer. Typically, 10–15 periods of 1.5–3.2 nm were deposited starting with Si (Ni) which also served as a cover layer. Different deposition temperatures up to 160°C were tested for both couples to optimize the ML stack. Vacuum furnace annealing at 10^{-4} Pa was applied to test the thermal stability.

Thermal stability of MLs can be improved by using compound layers following a simple argument that compound decomposition is required before the formation of new mixed phases [29]. Recently, thin B_4C interlayers were applied as diffusion barriers for Mo/Si and Sc/Si systems [30, 31]. We tested the merit of spacer compound layers by application of B_4C instead of C in combination with Ni. For compound layers, sputtering is necessary. A sputter deposition technique based on distributed electron cyclotron resonance (DECR) was used to prepare Ni/B_4C MLs. The advantages of this method were discussed elsewhere [32]. A low argon gas pressure of 10^{-1} Pa results in smooth interfaces, so that a large number of periods was needed to utilize X-ray diffuse scattering for interface characterization. In particular, 400 periods of 1.59 nm were coated on a Si wafer starting with B_4C, the spacer material being used also for the top protective layer of the double period thickness. A rapid thermal annealing (RTA) in pure nitrogen in addition to the long-term vacuum annealing was employed to study the decay of the ML structure in detail.

For nanometre-scale period MLs, it is useful to utilize mutually immiscible materials to reduce the mixing at interfaces. The Sc/Cr couple with a positive heat of formation has been proposed several times for microscopy in the water window [33–36] but the growth process and interface properties are not well understood yet. A low solubility and mixing cannot be excluded since the heat of formation, although being positive, is rather small (1 kJ mol^{-1}). Moreover, ultra-short period immiscible MLs are prone to degradation to granular systems under a thermal load via crystallization and grain boundary diffusion. We inspected the interface quality and thermal stability of Sc/Cr mirrors prepared by ion-beam sputtering of the ECR type operating at the lowest Ar gas pressure of 10^{-3} Pa. These mirrors were prepared at Tohoku University, Sendai. The details of the deposition may be found elsewhere [37]. Up to 250 periods of 1.3–1.75 nm were deposited onto Si wafers starting with Sc. The thermal treatment was done in a vacuum of 10^{-4} Pa.

24.3 Sample Measurements and Characterization

The specular XRR and grazing-incidence X-ray diffuse scattering (GIXDS) coming from rough interfaces were measured on a high-resolution STOE diffractometer equipped with a GaAs double crystal monochromator located in the primary beam and providing $CuK_{\alpha 1}$ radiation at the output. A point scintillation detector was used to perform different types of scans: different cuts throughout the reciprocal space. The reciprocal space maps (RSMs) were measured on a home-built diffractometer with a linear position-sensitive detector, which provided a resolution element in the reciprocal space of $2.10^{-4} \times 2.10^{-4}\,nm^{-2}$, and on a Bruker D8 Discover Super Speed diffractometer equipped with a Göbel mirror. A rotating anode TXS generator provided enough intensity (10^9 photons s^{-1} in the primary beam) to measure RSM with a point scintillation detector within a reasonable time. Finally, atomic structure of the layers was examined by X-ray diffraction on a Bragg–Brentano powder diffractometer equipped with a focusing graphite monochromator in the diffracted beam. CuK_α radiation was used for all X-ray measurements.

For particular samples, transmission electron microscopy (TEM), high-resolution TEM (HR TEM) and electron diffraction were utilized to complete the X-ray studies. A JEOL 200 CX microscope (200 kV, point-to-point resolution of 0.34 nm) was used for TEM and JEOL 2010 (200 kV) and 3010 (300 kV) microscopes integrated with energy-dispersive X-ray analysis, both point-to-point resolution of 0.17 nm, were used for HR TEM and electron diffraction. The samples were observed in cross section. A standard sample thinning by a 5-keV Ar ion beam was followed by a 300-eV Ar ion-beam treatment to reduce the damaged surface layer for HR TEM observations.

The specular XRR provides the basic parameters of the ML stack in terms of the layer thicknesses t, layer densities ρ and effective interface roughness (total interface width) σ_{eff}. The last parameter includes both compositional grading and morphological imperfections. The GIXDS, being produced solely by genuine (geometrical) interface roughness, is fully controlled by the morphology of particular interfaces. The specular XRR was evaluated by the optical recursive approach [38] supposing error function interface profiles while the GIXDS was calculated by the distorted-wave Born approximation (DWBA) [39]. For amorphous and polycrystalline layers found in the MLs under study, randomly rough interfaces are typical. Therefore, we adopted a statistical interface description based on cross-correlation functions between particular interfaces which can be divided into lateral and vertical parts. The lateral correlation function is not known a priori, different types being proposed. A correlation function [40]

$$C(r) = \sigma^2 \exp\left[-\left(\frac{r}{\xi}\right)^{2H}\right] \tag{24.1}$$

was convenient for all interfaces in Cu/Si MLs and Ni-based MLs. Here, r is the length along an interface, σ is the root-mean-square (rms) geometrical interface roughness, ξ is its lateral correlation length and H is the Hurst parameter which is connected with the fractal interface dimension as $D = 3-H$. Generally, the Fourier transform of the correlation function provides the power spectral density (PSD). The Fourier integral for the correlation function given by (24.1) can be calculated only numerically providing no analytical expression for PSD. Another type of a so-called K correlation function was applicable for Sc/Cr ML. In this case, the GIXDS calculation could utilize the corresponding PSD which reads

$$\mathrm{PSD}(Q_x) = \frac{4\pi H \sigma^2 \xi^2}{\left(1 + Q_x^2 \xi^2\right)^{1+H}}. \tag{24.2}$$

Here, Q_x is the lateral wave vector transfer which takes place on a particular Fourier component of the interface roughness. An exponential decay of the vertical correlation (replication) of the interface profiles controlled by a single vertical correlation length L_{vert}

$$C(z) = \exp\left(-\frac{z}{L_{\mathrm{vert}}}\right) \tag{24.3}$$

was convenient for the interval of the roughness frequencies covered by the GIXDS measurements in all cases. Here, z means vertical distance between particular interfaces.

The following strategy was used for data evaluation. Layer thicknesses t and effective interface roughness σ_{eff} were first extracted from the XRR fit, being rather independent parameters. The precision of determination of both parameters is ± 0.1 nm or better. Alternatively, layer densities ρ were set as free parameters to adjust the XRR fit around the critical angle. All these parameters were further fixed in the GIXDS simulations to obtain σ, ξ and L_{vert} values which do not affect the XRR in the first-order DWBA. The precision of the correlation lengths is generally within $\pm 25\%$. The RSMs were measured to get a qualitative overview of different GIXDS features coming from the interface correlations while particular cuts provided quantitative parameters. As a rule, several particular cuts were measured and fitted for each sample to avoid fit ambiguity as much as possible. Finally, the H value, influencing mainly a small region around the specular reflection, was refined separately. This parameter may reach values between zero and unity but the simulation is rather insensitive to values larger than 0.5.

24.4 Results and Discussion

The Cu/Si couple provided very regular UHV-evaporated ML stacks with amorphous or microcrystalline layers at different deposition temperatures, the interface roughness σ and mixing quantified as $\sigma_{\mathrm{eff}} - \sigma$ being minimized at 80°C. Examples of the XRR and GIXDS simulations are shown in Fig. 24.1 with the simulation parameters gathered in Table 24.1.

An asymmetry between the roughness of Cu/Si and Si/Cu interfaces shows that Cu layers are created rougher while the mixing is comparable. The interface widths σ_{eff} are effectively independent of the ML period down to $\approx$2 nm. Below $\approx$2 nm, the ML Bragg peak in the XRR curve is poorly resolved above the background while Kiessig fringes are still visible. This indicates that mixing at the interfaces destroys the ML stack. The vertical correlation length L_{vert} is comparable to the total ML thickness, giving rise to a pronounced resonant diffuse scattering around the Bragg peaks. A quickly saturating reflectivity drop at RT, which reached $\approx$25% on the first Bragg peak 2 weeks after the deposition, was observed. A severe suppression of the ML Bragg peaks due to a strong mixing and interdiffusion was induced by an annealing at 100°C for 1 h while the ML collapsed upon annealing at

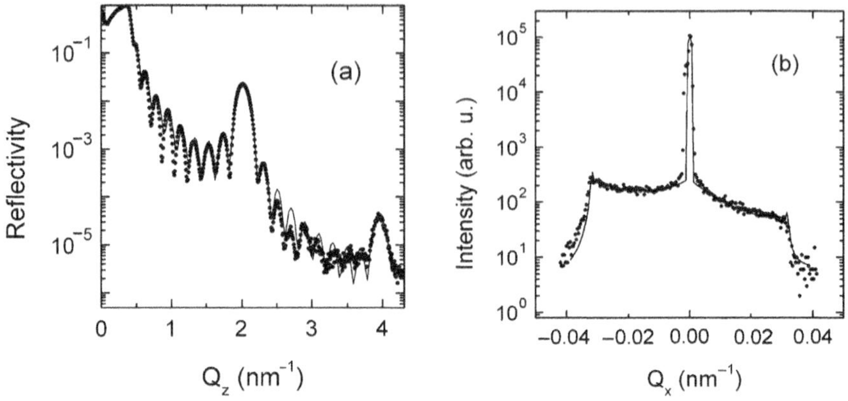

Fig. 24.1. The specular XRR (**a**) and rocking curve (**b**) simulations for the Cu/Si ML deposited at 80°C with the simulation parameters indicated in Table 24.1. The latter curve was measured by a sample scan with the detector fixed at the first Bragg maximum. Q_z and Q_x stand for the vertical and lateral (along the surface) components of the scattering vector (wave vector transfer), respectively (*dots* – measured points and *line* – simulation)

Table 24.1. The XRR and GIXDS simulation parameters of Cu/Si MLs (10 periods) for the deposition temperatures indicated

Simulation parameter	RT	80°C	80°C
t_{Cu} (nm)	1.22	1.27	0.65
t_{Si} (nm)	2.25	1.9	0.98
Λ (nm)	3.47	3.17	1.63
$\sigma_{\mathrm{eff\ Si\text{-}on\text{-}Cu}}$ (nm)	0.7	0.5	0.7
$\sigma_{\mathrm{eff\ Cu\text{-}on\text{-}Si}}$ (nm)	0.6	0.45	0.6
$\sigma_{\mathrm{Si\text{-}on\text{-}Cu}}$ (nm)	0.4	0.3	–
$\sigma_{\mathrm{Cu\text{-}on\text{-}Si}}$ (nm)	0.35	0.25	–
ξ (nm)	10	10	–
L_{vert} (nm)	30	30	–
H	0.2	0.3	–

For $\Lambda < 2$ nm, the GIXDS around the first Bragg peak was too low above the background to be simulated reliably. Bulk densities were used for the simulations t layer thickness, Λ ML period, σ_{eff} interface width, σ geometrical interface roughness, ξ lateral correlation length, L_{vert} vertical correlation length, H Hurst parameter

200°C/1 h. Here, an onset of η' (Cu, Si) tetragonal phase formation (JCPD–ICDD data set file no. 23-0224) was revealed by X-ray diffraction, starting presumably in the mixed regions. Fast grain boundary diffusion of Cu into Si layers mediated by this interface phase then accelerated the ML degradation. Nevertheless, interdiffusion with the substrate was not developed yet after the 280°C/1 h annealing as suggested by the presence of the Kiessig fringes in the XRR curve.

Table 24.2. The XRR and GIXDS simulation parameters of a Ni/C ML (10 periods) deposited at 80°C

Simulation parameter	As-deposited	100°C	150°C	200°C	250°C	300°C
t_{Ni} (nm)	1.26	1.26	1.28	1.32	1.33	1.34
t_C (nm)	1.89	1.89	1.92	1.98	1.99	2.01
Λ (nm)	3.15	3.15	3.2	3.3	3.32	3.35
ρ_{Ni} (g cm^{-3})	8.9	8.9	8.9	8.5	8.5	8.0
ρ_C (g cm^{-3})	2.2	2.2	2.1	2.0	2.0	1.8
$\sigma_{\text{eff C-on-Ni}}$ (nm)	0.65	0.65	0.7	0.75	1.1	1.6
$\sigma_{\text{eff Ni-on-C}}$ (nm)	0.45	0.45	0.5	0.55	0.7	1.2
$\sigma_{\text{C-on-Ni}}$ (nm)	0.35	–	–	–	0.5	0.8
$\sigma_{\text{Ni-on-C}}$ (nm)	0.25	–	–	–	0.3	0.4
ξ (nm)	5	–	–	–	8	10
L_{vert} (nm)	10	–	–	–	3	1
H	0.25	–	–	–	0.18	0.25

Four-hour annealings at the temperatures indicated were successively applied to the same sample. The GIXDS was not measured in the temperature region where the XRR did not change qualitatively

The quality of the ML stack proved to be more sensitive to the UHV deposition temperature for the Ni/C couple where a well-resolved ML Bragg peak in the XRR was observed only around 80°C. A comparison with Cu/Si MLs of similar periods shows slightly rougher and more mixed metalloid-on-metal interfaces and shorter correlation lengths after deposition (Table 24.2). HR TEM revealed that Ni layers contained grains with locally well-ordered regions inside (Fig. 24.2a) which control the interface morphology. The interface width obtained from the XRR reaches the value of 1 nm for the ML period below 3 nm suggesting agglomeration effects in Ni. This observation compares well with a percolation threshold around 2 nm found in sputtered samples [41] at which Ni forms coalescent layers. The interface correlation lengths indicated by the GIXDS are much shorter in Ni/C than in Cu/Si MLs while fractal behaviour ($H < 1$) is observed in all cases. Such a behaviour was reported for Ni/C MLs previously from post-deposition atomic force microscopy measurements [23].

A step-like vacuum annealing from 100°C up to 350°C with 50°C steps and with a duration of 4 h at each step was applied to the sample with the nominal ML period of 3 nm. After each annealing, the sample was cooled down in vacuum and measured at RT. The XRR and GIXDS simulation parameters are also shown in Table 24.2. An increase of the σ and $\sigma_{\text{eff}} - \sigma$ values indicates an advanced interface roughening and interdiffusion, respectively, before the ML breakdown at 350°C.

Cross-sectional HR TEM images of the collapsed ML showed well-developed grains embedded in a disordered matrix (Fig. 24.2b). The grains are composed of Ni of face-centred cubic (fcc) symmetry according to

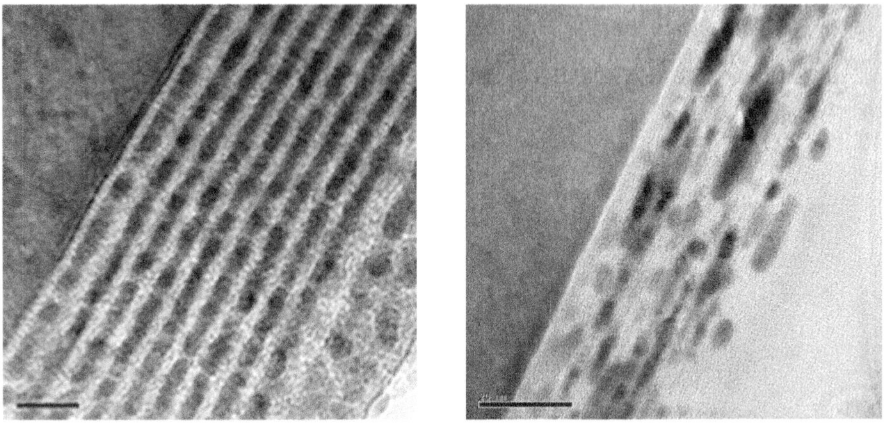

(a) (b)

Fig. 24.2. Cross-sectional HR TEM images of the as-deposited (**a**) and collapsed (**b**) Ni/C ML whose XRR and GIXDS simulation parameters are given in Table 24.2. *Black bars* at the bottom correspond to 10 nm (**a**) and 20 nm (**b**)

energy-dispersive X-ray and electron diffraction analyses while the matrix is formed by C and a small fraction of fine granular fcc Ni. Obviously, the annealing stimulates the growth and coalescence of the original Ni grains found in the as-deposited state, which governs also interface morphology as reflected in an increase of the lateral correlation length and a decrease of the vertical correlation of the interface roughness (decay of interface conformity). As soon as Ni grains are well developed and the ML becomes discontinuous, diffusion of C along Ni grain boundaries may also contribute to the ML breakdown. A complete diffusion of C into Ni layers was reported in the past [21]. The observed thermal stability is comparable to sputtered and PLD Ni/C mirrors [18, 19, 25].

It is worth noting that the ML period steadily increases and C layer densities decrease when increasing the temperature above 100°C. Similar effects were reported in sputtered Ni/C MLs with larger periods [21, 42] and were attributed to a transformation of the amorphous into the graphitic-like structure. Although we were not able to trace this effect directly by HR TEM or X-ray diffraction due to very thin C layers, the observed growth and coalescence of Ni grains connected with a long-distance collective diffusion of Ni atoms across C regions may induce graphitization. It was shown that such a metal-driven graphitization is preferred to carbide formation when C is in excess [43] which explains also the absence of carbide formation in the temperature range applied. Once initiated, the graphitization due to Ni diffusion proceeds even at RT as evidenced by ≈8% increase of the ML period which was observed on the sample annealed at 300°C after a 16-month RT storage.

Table 24.3. The XRR simulation parameters of the as-deposited Ni/B_4C ML (400 periods)

t_{Ni} (nm)	t_{B_4C} (nm)	Λ (nm)	$\sigma_{eff\ Ni\text{-}on\text{-}B_4C}$ (nm)	$\sigma_{eff\ B_4C\text{-}on\text{-}Ni}$ (nm)
0.78	0.81	1.59	0.28	0.26

Ni/B_4C MLs prepared by DECR do not suffer from agglomeration effects in Ni as a ML with a period below 2 nm and an extremely small interface width below 0.3 nm could be deposited (Table 24.3). Obviously, much higher adatom mobilities with DECR sputtering than with UHV deposition have a healing effect on the geometrical interface roughness and layer continuity at small thicknesses which goes along with the amorphous character of the layers as confirmed by X-ray and electron diffraction. On the other hand, the presence of compound layers does not favour mixing effects. Significantly, substitution of C by B_4C has no detrimental effect on the theoretical optical contrast. Due to very smooth interfaces, a large number of periods ($N = 400$) was necessary to visualize GIXDS effects in RSM. Only the GIXDS around the first ML Bragg maximum is visible (Fig. 24.3a) as the second maximum emerges directly from the instrumental background and the higher orders cannot be seen at all. This fact is a consequence of the ultra-short ML period when the measurements of RSM are especially instructive. A concentration of GIXDS in the form of a sheet around the Bragg peak (resonant diffuse scattering) is a clear sign of a partial vertical correlation of the interface roughness while its distinct asymmetry is exceptional. HR TEM revealed that the interfaces are wavy and partly copy each other.

The as-deposited Ni/B_4C ML was first exposed to several isothermal annealings at 300°C up to 8 h of total time with no significant changes of XRR but a slight improvement of the peak reflectivity on the first Bragg maximum. Contrarily, an additional annealing at 400°C for 2 h destroyed the ML completely and the Bragg peak disappeared. Therefore, an intermediate 350°C/2 h annealing, which brought about a decrease of the peak reflectivity by $\approx$1 order of magnitude, was done independently. Such a pre-annealed sample was then processed by a series of 5- or 10-min RTAs with a step-like increase of temperature up to 520°C which resulted in a further severe reduction of the peak reflectivity and a reduction of the ML period from 1.59 to 1.54 nm, the first Bragg maximum being still well resolved. This reduction may be attributed to the annealing-out of the excess free volume typical for amorphous structure. HR TEM inspection showed that the layered structure without mixed regions but with many topological defects was still preserved. The RSM (Fig. 24.3b) exhibits substantial changes in comparison with the as-deposited state. Though it was not possible to simulate RSM with common correlation functions presented in the previous section, a qualitative change due to the annealing could be simulated when doubling the lateral correlation

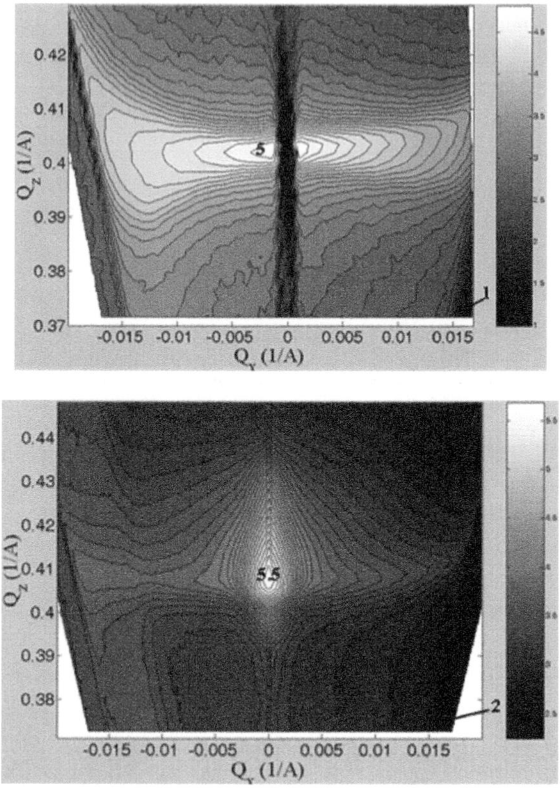

(a)

(b)

Fig. 24.3. Reciprocal space maps of the (**a**) as-deposited and (**b**) annealed up to 520°C Ni/B_4C ML around the first Bragg peak. In the first map, the shadow of a beam stopper protecting a position-sensitive detector hides the XRR whose simulation parameters are shown in Table 24.3. The second map was measured with a point detector. The *labels* denote the logarithms of the extreme values of the intensity

length from 10 to 20 nm. This increase cannot be attributed to the grain growth as in the case of Ni/C couple as no crystallization inside the ML was observed. However, large re-crystallized regions in the substrate at the interface with the ML, presumably Ni silicide grains, could be seen locally by TEM. After the ML breakdown, the original layered structure was transformed into an inhomogeneous amorphous-like structure with only one diffuse ring in electron diffraction pattern. This collapsed ML still keeps a sharp interface with the substrate. HR TEM revealed a rare occurrence of crystallographically ordered regions, probably (111) planes of fcc Ni. This fact suggests that Ni diffusion controls the ML breakdown as in Ni/C couple but the mechanism of decay of the compound layers is unclear in the absence of a

Table 24.4. The XRR and GIXDS simulation parameters of the as-deposited Sc/Cr ML (250 periods)

t_{Sc}(nm)	t_{Cr}(nm)	Λ(nm)	$\sigma_{Cr\text{-on-}Sc}$(nm)	$\sigma_{Sc\text{-on-}Cr}$(nm)	ξ(nm)	L_{vert}(nm)	H
0.83	0.93	1.76	0.28	0.25	7	35	1

crystalline phase formation. Nevertheless, any deviation from the compound stoichiometry, which is common in extremely thin layers, may affect thermal stability of Ni/B_4C mirror adversely.

Ultra-short periods could be achieved and many periods ($N = 250$) were deposited also for the Sc/Cr couple using ECR ion source. Similarly as for Ni/B_4C, only the region around the first Bragg maximum could be used for GIXDS analysis due to the ultra-short ML periods (Table 24.4). The presence of the resonant diffuse scattering sheet in RSM (Fig. 24.4a) suggests here as well a partial vertical correlation of the interface roughness. In contrast to previous cases, where the lateral correlation function according to (24.1) was used, PSD given by (24.2) proved to be convenient to simulate GIXDS for Sc/Cr MLs (Table 24.4; Fig. 24.4b). Different cuts throughout the calculated RSM gave a good quantitative agreement with particular scans measured separately. The vertical correlation length represents only ≈8% of the total ML thickness. The σ_{eff} and σ values are practically identical with a slight asymmetry between Sc/Cr and Cr/Sc interfaces, showing no mixing as expected because of the negligible miscibility of the elements. No mixing and a high regularity of the ML stack were confirmed by HR TEM and TEM, respectively. Electron diffraction showed only one ring typical for an amorphous structure but narrower than that for Ni/B_4C which implies a higher short-range ordering.

Thermal stability tests were performed on a ML with a period of 1.25 nm repeated 150 times. The sample was first exposed to several isothermal annealings at 280°C up to 33 h of total time which brought about a decrease of the peak reflectivity on the first Bragg maximum by ≈25%. This decrease was connected with an increase of the initial interface roughness of 0.28 nm to the final value of 0.30 nm for Cr-on-Sc interfaces while a similar increase from 0.26 to 0.29 nm was observed for Sc-on-Cr interfaces. Simultaneously, the ML period increased to 1.28 nm which may be ascribed to a structural ordering inside the layers. A further 450°C/4 h annealing decreased the peak reflectivity by one order of magnitude and a subsequent 650°C/4 h one destroyed the layered structure completely, as reflected in the disappearance of the ML Bragg maxima. No crystalline phase could be detected by X-ray diffraction. Presumably, a fine granular structure typical for immiscible components was formed.

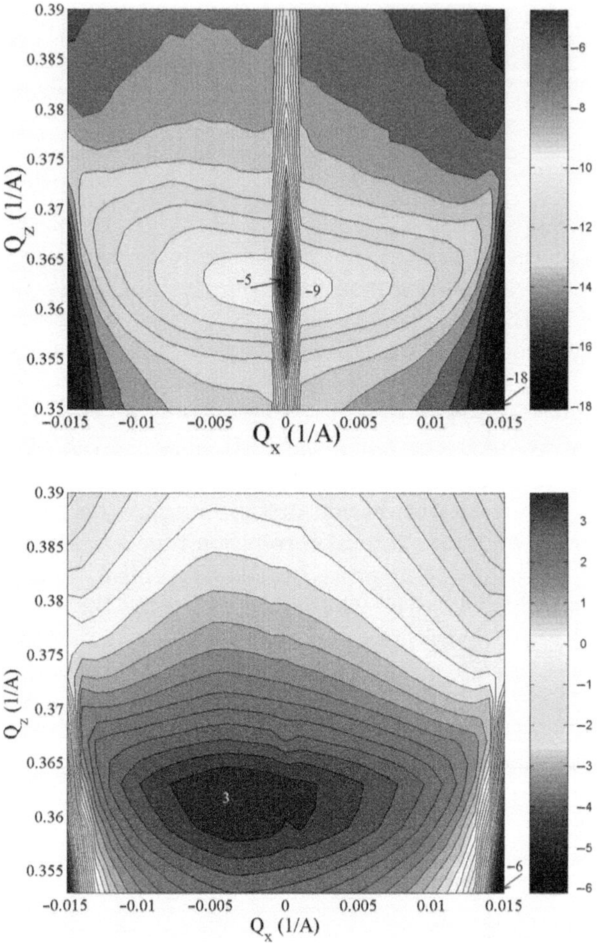

(a)

(b)

Fig. 24.4. Measured (**a**) and simulated (**b**) reciprocal space maps of the Sc/Cr ML around the first Bragg peak. The simulation parameters are shown in Table 24.4. The *labels* denote the logarithms of the extreme values of the intensity

24.5 Conclusions and Outlook

The ability to form thermally stable ultra-short period MLs was tested for material couples of different compositions and miscibilities. Such periods are inherently required for high-quality imaging in the water window or for very hard X-rays and put strong requirements on the interface quality.

The Cu/Si couple is attractive for very hard X-rays because it is low absorbing and forms very regular ML stacks easily down to small periods. On the other hand, it is a miscible pair of materials. We found a limit of $\approx$2 nm for the ML period which still yields a well-resolved ML Bragg peak in the XRR curve for UHV-deposited MLs and which compares well with dc-sputtered ones prepared previously. Thermal stability restricts their use to below 100°C and therefore does not qualify this material pair to be used for ML mirrors working in high heat load conditions.

Ni/C mirrors were studied mostly for applications below the C–K edge (284 eV) or for hard X-rays of several keV (e.g. Göbel mirrors). A shift to ultra-short periods would render them useful for applications close to 100 keV with a low-absorption coefficient. A better thermal stability than for the Cu/Si couple may also be expected due to a low mutual solubility of Ni and C. We found a minimum limit of ≈2.5 nm for the ML period imposed by agglomeration effects in polycrystalline Ni layers. Such a ML is stable up to 300°C, the breakdown being controlled by the growth and coalescence of Ni grains and Ni-induced graphitization of C. The thermal stability is comparable to that of Ni/C MLs with larger periods studied previously, which suggests that it is independent of the ML period when continuous layers are formed. Vertical correlation of the interface roughness is much weaker than for Cu/Si MLs with positive implications for their use when a good specular imaging contrast is required.

From the technological point of view, UHV deposition with an in situ substrate heating proved to be a cost-effective promising alternative to in situ ion-beam etching for Cu/Si and Ni/C couples. Nevertheless, ultra-short periods down to 2 nm or less are not accessible with them.

An ultra-short ML period far below 2 nm could be achieved by a substitution of C by B_4C in Ni/C MLs and application of DECR sputtering for deposition of Ni/B_4C MLs. A high adatom energy allows formation of extremely smooth amorphous layers which are continuous at very small thicknesses. The presence of compound layers enhances thermal stability to 350°C on a long-term annealing and above 500°C on RTA. The amorphous character of the layers plays also some role as it excludes fast grain boundary diffusion. The mirror collapses by Ni diffusion and decomposition of B_4C layers without formation of a well-developed crystalline phase.

The Sc/Cr couple is also able to provide MLs with an ultra-short period far below 2 nm using ECR sputtering. The MLs have excellent quality and good thermal stability up to 500°C as expected for the elements with negligible miscibility. The vertical correlation length of the interface roughness constitutes less than 10% of the total ML thickness, so that its detrimental effect on the specular imaging contrast is minimized. The ML breakdown is presumably controlled by the formation of fine granular phase typical for immiscible elements.

For the future, new material combinations suitable for ultra-short period ML mirrors aimed at specific applications have to be tested in detail in a way similar to that demonstrated in this work. In particular, material pairs yielding structures with minimum interface roughness have to be searched for. It has to be stressed that even traditional material pairs represent new challenges for technology when MLs with layer thicknesses below 1 nm and several hundred periods have to be deposited. At present, an optimization of the reflectivity in UHV electron-beam evaporated ML stacks is done in situ by ion milling and partial material removal after deposition of each layer. In this way the layers, which have been deposited thicker than nominal values, are thinned to optimum thicknesses with in situ control. For the mirrors deposited by sputtering, a high quality of the ML stack is achieved by employing stable sputtering parameters and a precise control of the motion of the sample with respect to the targets. ECR sputtering with the inert gas pressure far below the thermalization threshold proves to be superior to other techniques in terms of yielding extremely smooth interfaces. To fabricate regular ML stacks with ultra-short periods, further progress is expected by making a still tighter control of the deposition process including an in situ monitoring with a resolution below 10^{-1} nm to provide feedback for subsequent correction procedures.

Thermal stability requirements are also more critical with ultra-short periods below 2 nm, particularly in view of the advent of free electron laser (FEL) sources whose brilliance is several orders of magnitude higher than that at present synchrotron beam lines. To set and manipulate thermal stability limits of ultra-short period mirrors working with femtosecond intense pulses, ML response to short-time power loads far below a second and with the ramp edge of several 100 K s^{-1} must be studied. Here, advanced time-resolved in situ diagnostics of the processes at the interfaces, like in situ ellipsometry with several ms read-out, must be applied.

Though much knowledge has already been gained, the research and development of ultra-short period ML mirrors with the individual layer thicknesses on the sub-nanometre scale is still open to new creative ideas. The design and the production of mirrors with atomic level control will shift frontiers not only in the ML optics but also in nanotechnology as a whole.

Acknowledgements

The authors are grateful to M. Yamamoto from IMRAM Tohoku University, Sendai, for providing Sc/Cr ML samples and to V. Holý from Charles University, Prague, for providing DWBA software. Z. Bochníček from Masaryk University, Brno, is acknowledged for X-ray reciprocal space measurements. Supports by COST P7 Action, Scientific Grant Agency VEGA (contracts no. 2/4101/26 and 2/6030/26) and Center of Excellence SAS project CE-PI I/2/2006 are acknowledged.

References

1. M. Yamamoto, T. Namioka, Appl. Optics **31**, 1622 (1992)
2. E. Spiller, *Soft X-Ray Optics* (SPIE Optical Engineering Press, Bellingham, 1994)
3. D.G. Stearns, M.B. Stearns, Y. Cheng, J.H. Smith, N.M. Ceglio, J. Appl. Phys. **67**, 2415 (1990)

4. T. Kingetsu, M. Yamamoto, Surf. Sci. Rep. **45**, 79 (2002)
5. V.V. Kondratenko, Yu.P. Pershin, O.V. Poltseva, A.I. Fedorenko, E.N. Zubarev, S.A. Yuchin, I.V. Kozhevnikov, S.I. Sagitov, V.A. Chirkov, V.E. Levatsov, A. Vinogradov, Appl. Optics **32**, 1811 (1993)
6. M. Hansen, K. Anderko, *Constitution of Binary Alloys* (McGraw-Hill, New York, 1958)
7. H.J. Stock, U. Kleineberg, A. Kloidt, B. Schmiedeskamp, U. Heinzmann, M. Krumrey, P. Müller, F. Scholze, Appl. Phys. Lett. **63**, 2207 (1993)
8. R. Senderák, M. Jergel, Š Luby, E. Majková, V. Holý, G. Haindl, F. Hamelmann, U. Kleineberg, U. Heinzmann, J. Appl. Phys. **81**, 2229 (1997)
9. A. Patelli, J. Ravagnan, V. Rigato, G. Salmaso, D. Silvestrini, E. Bontempi, L.E. Depero, Appl. Surf. Sci. **238**, 262 (2004)
10. D. Morris, C.J. Buckley, G.R. Morrison, A.G. Michette, P.A.F. Anastasi, M.T. Browne, R.E. Burge, P.S. Charalambous, G.F. Foster, J.R. Palmer, P.J. Duke, Scanning **13**, 7 (1991)
11. A. Nefedov, H. Zabel, F. Schäfers, Nucl. Instrum. Meth. A **467–468**, 345 (2001)
12. H. Takenaka, H. Ito, K. Nagai, Y. Marumatsu, E. Gullikson, R.C. Perera, Nucl. Instrum. Meth. A **467–468**, 341 (2001)
13. S.S. Andreev, M.S. Bibishkin, N.I. Chkhalo, E.B. Kluenkov, K.A. Prokhorov, N.N. Salashchenko, M.V. Zorina, F. Schäfers, L.A. Shmaenok, J. Synchrotron Radiat. **10**, 358 (2003)
14. P.H. Mao, F.A. Harrison, D.L. Windt, F.E. Christensen, Appl. Optics **38**, 4766 (1999)
15. F. Schäfers, Physica B **283**, 119 (2000)
16. M.S. Bibishkin, A.A. Fraerman, A.E. Pestov, K.A. Prokhorov, N.N. Salashchenko, Yu.A. Vainer, Nucl. Instrum. Meth. A **543**, 333 (2005)
17. D.L. Windt, Appl. Phys. Lett. **74**, 2890 (1999)
18. G.S. Lodha, S. Pandita, A. Gupta, N.V. Nandedkar, K. Yamashita, J. Electron Spectrosc. Rel. Phenom. **80**, 453 (1996)
19. K. Nakajima, S. Aoki, S. Sudo, S. Fujiwara, Jpn. J. Appl. Phys. **31**, 2864 (1992)
20. J. Friedrich, I. Diel, C. Kunz, S. Di Fonzo, B.R. Müller, W. Jark, Appl. Optics **36**, 6329 (1997)
21. V. Dupuis, M.F. Ravet, C. Tete, M. Piecuch, Y. Lepetre, R. Rivoira, E. Ziegler, J. Appl. Phys. **58**, 5146 (1990)
22. C. Borchers, C. Michaelsen, Philos. Mag. A **82**, 1195 (2002)
23. M. Ulmeanu, A. Serghei, I.N. Mihailescu, P. Budau, M. Enachescu, Appl. Surf. Sci. **165**, 109 (2000)
24. N.V. Kovalenko, S.V. Mytnichenko, V.A. Chernov, JETP Lett. **77**, 80 (2003)
25. V.A. Chernov, E.D. Chkalo, N.V. Kovalenko, S.V. Mytnichenko, Nucl. Instrum. Meth. A **448**, 276 (2000)
26. R. Dietsch, T. Holz, T. Weissbach, R. Scholz, Appl. Surf. Sci. **197**, 169 (2002)
27. Y.P. Pershin, Y.N. Zubarev, V.V. Kondratenko, O.V. Poltseva, A.G. Ponomarenko, V.A. Servryukova, J. Verhoeven, Metallofiz. Nov. Tekhnol. **24**, 795 (2002)
28. E.I. Puik, M.J. Van der Wiel, H. Zeulenmaker, J. Verhoeven, Appl. Surf. Sci. **47**, 251 (1991)
29. A.F. Jankowski, L.R. Schrawyer, M.A. Wall, J. Appl. Phys. **68**, 5162 (1990)
30. T. Bottger, D.C. Meyer, P. Paufler, S. Braun, M. Moss, H. Mai, E. Beyer, Thin Solid Films **444**, 165 (2003)

31. A.F. Jankowski, C.K. Saw, C.C. Walton, J.P. Hayes, J. Nilsen, Thin Solid Films **469**, 372 (2004)
32. Ch. Morawe, J.-C. Peffen, O. Hignette, E. Ziegler, Proc. SPIE **3773**, 90 (1999)
33. N.N. Salashchenko, E.A. Shamov, Opt. Commun. **134**, 7 (1997)
34. F. Schäfers, H.C. Mertins, F. Schmolla, I. Packe, N.N. Salashchenko, E.A. Shamov, Appl. Optics **37**, 719 (1998)
35. J. Birch, F. Eriksson, G.A. Johansson, H.M. Hertz, Vacuum **68**, 275 (2002)
36. T. Kuhlmann, S. Yulin, T. Feigl, N. Kaiser, T. Gorelik, U. Kaiser, W. Richter, Appl. Optics **41**, 2048 (2002)
37. K. Sakano, M. Yamamoto, Proc. SPIE **3767**, 238 (1999)
38. J.H. Underwood, T.W. Barbee, AIP Conf. Proc. **75**, 170 (1981)
39. V. Holý, T. Baumbach, Phys. Rev. B **49**, 10668 (1994)
40. S.K. Sinha, E.B. Sirota, S. Garoff, H.B. Stanley, Phys. Rev. B **38**, 2297 (1998)
41. C. Borchers, P. Ricardo, C. Michaelsen, Philos. Mag. A **80**, 1669 (2000)
42. T. Djavanbakht, V. Carrier, J.M. Andre, R. Barchewitz, P. Troussel, J. Phys. IV **10**, 281 (2000)
43. R. Sinclair, T. Itoh, R. Chin, Microsc. Microanal. **8**, 288 (2002)

25

Specially Designed Multilayers

J.I. Larruquert, A.G. Michette, Ch. Morawe, Ch. Borel, and B. Vidal

Abstract. Periodic multilayers, utilising Bragg reflection at single angle or wavelength, are established as efficient reflectors from the hard X-ray down to the extreme ultraviolet (XUV) region of the electromagnetic spectrum. More recently, both laterally and depth-graded multilayers have been designed and fabricated; they allow either reflection of divergent beams or over a broad angular or wavelength range, or a combination of both. Recent developments in aperiodic structures, along with advances in ultra-short period and transmission mutilayers, are discussed in this chapter. Modelling methods to provide designs for specific purposes are described, as are advances in manufacturing techniques and quality control. In addition to peak reflectivity at a specific wavelength or angle, high integrated reflectivity over a given wavelength or angular range is considered, along with flat reflectivity profiles. Important potential applications of flat response mirrors are X-ray micro-spectroscopy, X-ray diffraction, XUV polarimetry, and any other technique requiring both high reflectivity and broad bandwidth.

25.1 Introduction

Periodic multilayers, utilizing Bragg reflection at single angles or wavelengths, are established as efficient reflectors from the hard X-ray to the extreme ultraviolet (XUV) region of the electromagnetic spectrum. Recently, laterally graded multilayers allowing reflection of divergent beams and depth-graded multilayers providing broad angular or wavelength ranges have been designed and made; combinations having both types grading have also been considered. Recent developments in such aperiodic structures, along with advances in ultra-short period and transmission multilayers, are discussed in this chapter. Modeling methods to provide designs for specific applications are described, but manufacturing techniques and quality control are not discussed explicitly, as these are similar to those used for conventional multilayers. In addition to peak reflectivity at a specific wavelength or angle, high integrated reflectivity over a given wavelength or angular range is considered, along with flat reflectivity profiles.

Important potential applications of flat response mirrors include X-ray microspectroscopy, X-ray diffraction, XUV polarimetry, and any other technique requiring both high reflectivity and broad bandwidth. Importantly, in the recent years, the use of multilayer-based X-ray optics at third-generation synchrotron sources has seen a considerable growth [1,2]. The principal applications are focusing or collimating optics and broadband monochromators.

25.1.1 Periodic Multilayers

For many years, the only effective way of manipulating X-ray beams was through the phenomenon of "total" external reflection. For this, the grazing angle of incidence must be less than the critical angle which, for most materials, is less than about 1° for X-rays. Such small grazing angles result in several major disadvantages including aberrations and small effective apertures. However, at low energies, single surface reflectors could be used, while at high energies crystal optics were (and still are) employed.

The problem was in the intermediate energy range; it was this gap that periodic multilayer mirrors were initially designed to fill. Such coatings can, in principle, be deposited on substrates of any form and can be used as reflectors for energies of ~0.01–10 keV at incidence angles ranging up to, in some cases, normal incidence. Periodic multilayers have a typical energy resolution of 1–10% which is about two orders of magnitude larger than possible with perfect crystals, leading to a significant gain in flux. At the same time, they allow for larger effective apertures than grazing incidence mirrors, making them considerably shorter and easier to handle. When properly prepared, they conserve the properties, such as coherence, of the X-ray beam, and they can be tailored to specific applications using the refraction-corrected Bragg equation

$$m\lambda = 2d\sqrt{n^2 - \cos^2\theta}. \tag{25.1}$$

Here, m is the diffraction order, λ is the (vacuum) wavelength, d is the multilayer period, n is the average refractive index in the multilayer stack, and θ is the glancing angle at the multilayer surface. However, periodic multilayers work only within specific wavelength or angular ranges, where (25.1) is approximately satisfied, which limits their applicability. Such restrictions can be alleviated by using optimized designs.

25.2 Optimized Multilayers

In any geometry where the X-ray beam is incident over a range of glancing angles, the multilayer structure has to be deposited with a lateral thickness gradient to fulfill the Bragg condition (25.1) along the full length of the mirror [3]. Similarly, whenever broader reflectivity profiles are required, depth-graded or nonperiodic layered structures have to be deposited [4]. Nonperiodic and laterally graded multilayers can be combined into single devices providing fixed focus optics over broad energy intervals [5]; such optics have wide applicability in, e.g., microspectroscopy and diffraction experiments.

25.2.1 Laterally Graded Multilayers

Most multilayer-based synchrotron optics, including focusing and collimating devices using curved multilayers and flat monochromators [6], require lateral thickness gradients to fulfill the Bragg condition for fixed photon energy along the full beam footprint. Depending on the geometry, different lateral gradients have to be applied; three commonly used geometries are elliptical, parabolic, and flat. For an elliptical surface,

$$\sin^2\theta = \frac{b^2}{pq}, \tag{25.2}$$

where θ is the glancing angle, b is the semiminor axis of the ellipse, and p and q are the object and image distances. For a parabolic profile,

$$\sin^2\theta = \frac{f}{2p}, \tag{25.3}$$

where $f/2$ is the distance from the focus of the parabola to its apex, and for a flat surface,

$$\sin^2\theta = \frac{s^2}{p^2}, \tag{25.4}$$

where s is the perpendicular distance from the surface to the source point.

To obtain the corresponding bilayer lateral thickness gradients, the angles for the appropriate surface must be used in (25.1). Strictly speaking, the local angle of incidence varies also with depth, but this effect is typically negligible compared to the widths of the multilayer Bragg peaks.

In practice, lateral thickness gradients are obtained by differential coating techniques. In a static setup, masks modify the particle fluxes while dynamic approaches use moving masks or the relative motion between source and substrate. In the sputter deposition system at the European Synchrotron Radiation Facility (ESRF), for example, the substrate remains fixed while the sources are scanned with variable speeds [1].

Figure 25.1 illustrates the case of a laterally graded parabolic W/B_4C multilayer collimator [1]. Here, both W and B_4C layer thicknesses were varied individually to optimize the peak reflectivity along the total length of the optic. The agreement between theory and experiment is remarkable. For the multilayer to be efficient, the required accuracy of the thickness profile must be considerably better than the width of the multilayer Bragg peak. With the present ESRF equipment, relative errors of $<0.5\%$ RMS can be achieved.

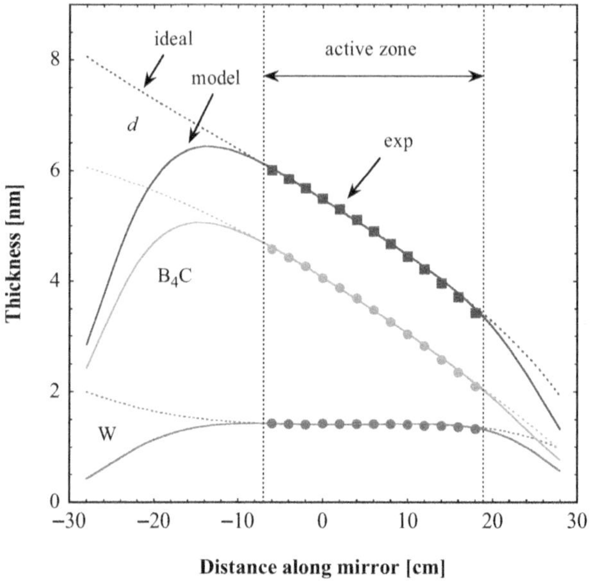

Fig. 25.1. Thickness profile of a laterally graded parabolic W/B_4C multilayer. *Broken lines* indicate the ideal case deduced from the given geometry. *Solid lines* show a simulation based on realistic parameters of the coating system. The *symbols* are experimental data obtained from X-ray reflectivity measurements

25.2.2 Depth-Graded Multilayers

One way of increasing the bandwidth of a multilayer is simply to reduce the number of reflecting bilayers. For hard X-rays, this approach is penalized by the rapidly decreasing reflectivity. A more efficient option is to vary the d-spacing in such a way that X-rays of different energies, or incident at different angles, are reflected from different depths within the total stack. This concept of depth grading was first introduced in the design of supermirrors for neutron optics [7], but for X-rays, the stronger absorption complicates the design. However, different mathematical models have been developed, leading to nonperiodic layered structures [8–10]. These typically involve starting with a periodic structure, or some other analytically derived layer thickness distribution, and then randomly changing the position of a randomly selected boundary between two layers in at attempt to improve some "merit function" (MF). If, for example, the requirement is for a flat reflectivity response over a given angular range at a fixed wavelength, then an appropriate MF would be

$$\mathrm{MF} = \sum_{i=1}^{N} [R(\theta_i) - R_0]^2, \tag{25.5}$$

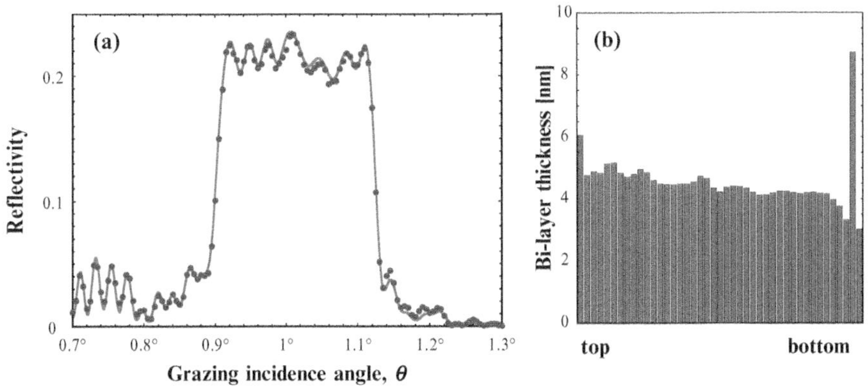

Fig. 25.2. (**a**) Specular reflectivity of an aperiodic Ni/B_4C multilayer as a function of the incidence angle at an X-ray energy of 8.048 keV. *Circles* indicate the experimental data points; the *line* is a simulation based on the bilayer structure of (**b**)

where $R(\theta_i)$ is the reflectivity at a selected angle from N sample points and R_0 is the target reflectivity. The reflectivities are calculated using the standard Parrat recursive formalism [11].

Figure 25.2a shows the reflectivity spectrum of such a multilayer, designed and made at the ESRF [4]. It was designed to provide a flat reflectivity profile over an angular bandwidth of about 20% at 8 keV. It consists of 43 Ni/B_4C bilayers of variable thickness as indicated in Fig. 25.2b. Numerical fits to the data indicate that the residual undulations on the flat plateau are mainly caused by oxide formation on top of the sample. It is mandatory that all key parameters of the sample structure (thickness, density, interdiffusion, oxidation) have to be known in advance, since they enter directly into the design algorithm.

Such mirrors can also be designed for lower energies at larger grazing incidence angles. An example of the performance of a W/Si mirror designed to act as a spectral filter on a titanium K_α (4.5 keV) X-ray source, for applications in radiobiology, is shown in Fig. 25.3. The mirror was designed and made in collaboration with Tongji University, Shanghai, and the measurements were done at the Gray Cancer Institute, UK.

Similar depth-graded multilayers can be designed to work over a range of energies at fixed incidence angle. A simple example of this would be to arrange a series of periodic mirrors, one on top of the other, such that each reflects a specific energy with the upper mirrors being relatively transparent to the radiation reflected from the lower mirrors. An example of the stack structure of such a mirror [12–16], consisting of 20 W/Si bilayers, and the corresponding reflectivities are shown in Fig. 25.4. The calculated reflectivity at a grazing incidence angle of 0.6° is over 40% in the energy range 6–9 keV. The first six bilayers (zone 1) define the shape of the calculated reflectivity

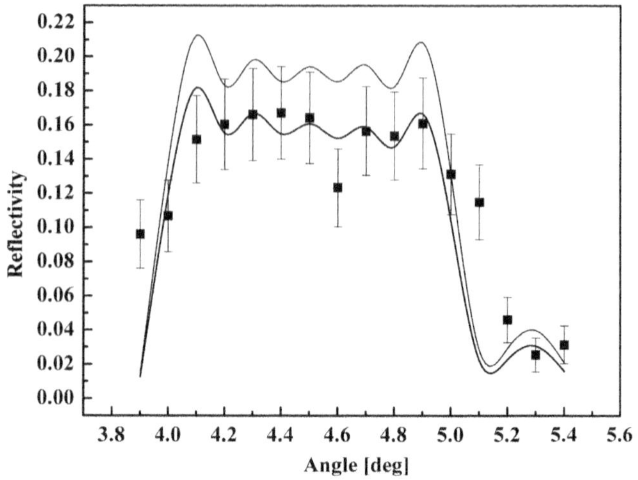

Fig. 25.3. Reflectivity of a broad angular range multilayer designed for an energy of 4.5 keV. The *thin curve* is the design reflectivity (with the aim of a flat response of 20%) and the *thick curve* is a fit to the measured points with an interlayer roughness of 0.4 nm

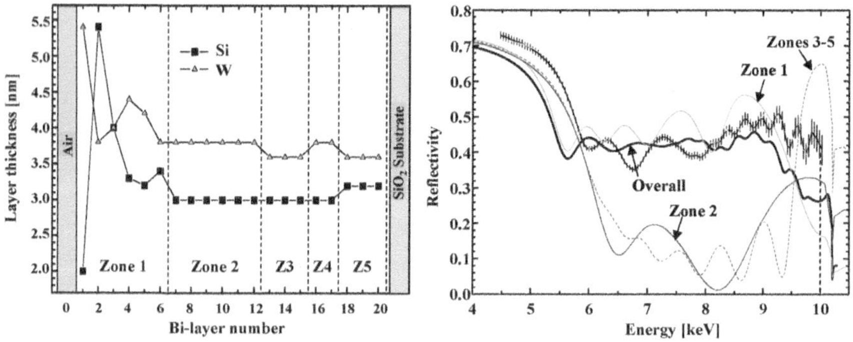

Fig. 25.4. Layer thickness distribution of a five-zone depth-graded multilayer (*left*) and the corresponding reflectivities (*right*); the measured reflectivities are indicated by the *line* with error bars. The overall reflectivity takes into account absorption in the upper layers, and the sudden drop at an energy of just over 10 keV is due to the tungsten L_β absorption edge

curve for energies up to 9.5 keV. The layers of zones 2–5 are primarily responsible for the reflectivity at 9–10 keV. In the 6–9 keV range, the reflectivity curve is flat with some small oscillations which can be attributed to the mixing of second-order reflectivity between zone 1 and the deeper layers of the mirror. The decrease in the total reflectivity at energies higher than 9 keV is mainly due to absorption in the tungsten layers and to the effects of roughness, which increase exponentially with the square of the photon energy. The experimental reflectivity is higher than the calculated values, especially for the

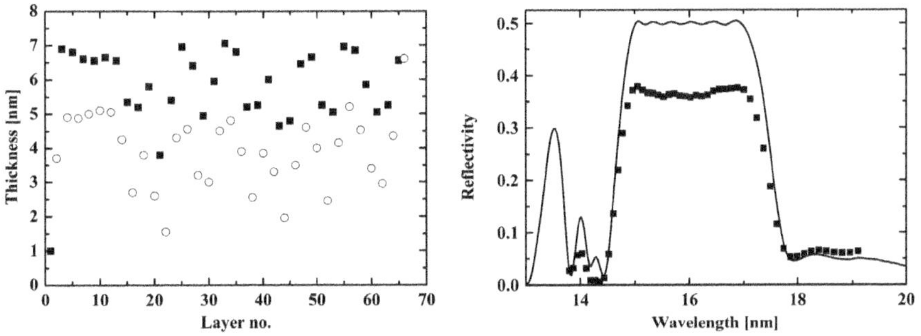

Fig. 25.5. A depth-graded Mo/Si multilayer designed for a flat reflectivity profile in the wavelength range 15–17 nm at the Brewster angle. The *left-hand plot* shows the layer thickness distribution and the *right-hand graph* shows the calculated (*line*) and measured reflectivities

"total" reflectivity part of the spectrum (below 5 keV) and at energies greater than 8 keV. These differences are partially due to a $\pm 0.01°$ systematic shift in the experimental angle corresponding to the goniometer angular precision.

A more flexible approach allows each layer thickness to be varied in the optimization. An appropriate MF is then

$$\mathrm{MF}_E = \sum_{i=1}^{N} [R(E_i) - R_0]^2 \quad \text{or} \quad \mathrm{MF}_\lambda = \sum_{i=1}^{N} [R(\lambda_i) - R_0]^2, \tag{25.6}$$

where $R(E_i)$ and $R(\lambda_i)$ are the reflectivities at selected energies and wavelengths, respectively, from N sample points, and R_0 is the target reflectivity. Here, examples of mirrors designed for extreme ultraviolet (EUV) radiation, where wavelength is the more usual variable, are considered [17–22].

Figure 25.5 shows the parameters and performance of a depth-graded Mo/Si multilayer designed for a reflectivity of 50%, in the wavelength range 15–17 nm, at the Brewster angle which, since the refractive indices are very close to unity, is very close to 45°. The lower measured reflectivities may be attributed to interlayer roughness or diffusion with a length scale of about 1 nm. The performance of a similar mirror, designed for a wider wavelength range, is shown for comparison in Fig. 25.6; because of the wider wavelength range, the measured performance is compromised compared to that of the mirror of Fig. 25.5. The mirrors were designed and made in collaboration with Tongji University, Shanghai, and the measurements were carried out at the BESSY synchrotron.

Since these mirrors were designed to work at the Brewster angle, they also act as polarizing elements. The plotted reflectivities are for the s-component of polarization, R_s. Figure 25.6 also shows the design and measured polarizing capabilities, P, of the two mirrors, where

$$P = \frac{R_\mathrm{s} - R_\mathrm{p}}{R_\mathrm{s} + R_\mathrm{p}}. \tag{25.7}$$

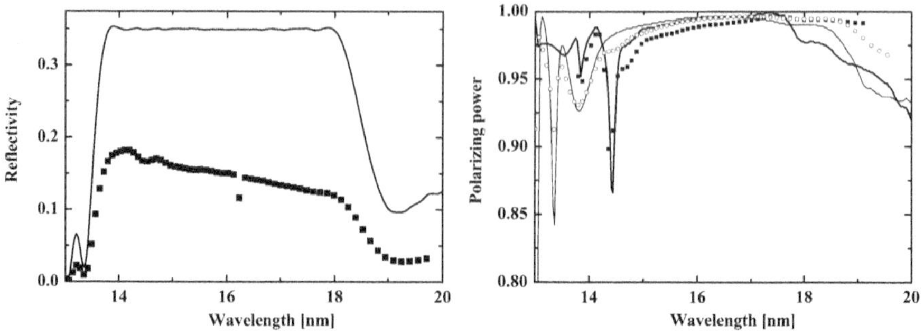

Fig. 25.6. Calculated and measured reflectivities (*left*) of a Mo/Si multilayer designed for a reflectivity of 35% in the wavelength range 14–18 nm. The *right-hand plot* shows the calculated (*light curve*) and measured (*open circles*) polarizing capabilities of this mirror as well as those of the mirror of Fig. 25.5

Similar multilayers can also be made on transmissive substrates. When used in conjunction with reflective multilayers, in a polarizer/analyzer arrangement, these allow polarization analysis to be carried out over an extended wavelength range without having to change the optics [22].

25.2.3 Doubly Graded Multilayers

The concepts of laterally and depth-graded multilayers may be combined into single structures; such a device would be capable of broadband focusing with fixed focal distance [23]. Rewriting the Bragg equation (25.1) leads to the energy dispersion relation for a multilayer

$$E = \frac{hc}{2d\sqrt{n^2 - \cos^2\theta}}, \tag{25.8}$$

where E is the X-ray energy. The angular dependence can be interpreted in terms of the focal distance at each point on the mirror and θ can be substituted by the appropriate expression from (25.2) to (25.4).

The concept is similar to that of DuMond diagrams [24] and can be illustrated by plotting the dispersion of the energy E of the reflected photons against the focal distance q along the optics. The example in Fig. 25.7 shows the case of two elliptically shaped Ru/B_4C multilayer mirrors. Periodic multilayers without lateral gradient are characterized by the curved thick solid lines. Such structures reflect X-rays only in a narrow "intersection volume" in (E, q) space, defined by the intrinsic multilayer line width, which is typically a few percent. With a lateral period thickness gradient, the energy response becomes constant (thick horizontal solid line in Figure 25.7). In practice, this has the advantage that the multilayer reflects at the same energy over the whole mirror length. An aperiodic multilayer without lateral gradient widens the energy bandpass and therefore opens the dispersion area, thereby increasing the intersection zone in (E, q) space. The most attractive solution,

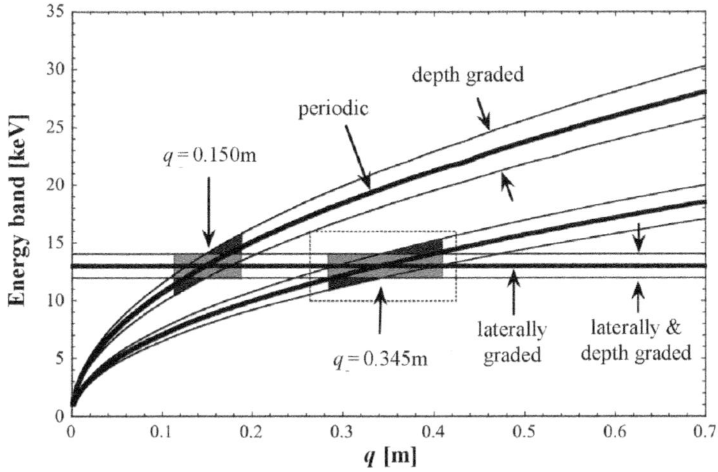

Fig. 25.7. Energy dispersion of different Ru/B_4C multilayers, with periods of 6.7 nm, showing the reflected energy bands against the distance q from the focal spot along the multilayer surface. *Thick solid lines* represent the periodic multilayers either with or without lateral thickness gradient. A purely depth-graded multilayer covers the area limited by the thin curves. A combination of both is indicated by the zone between the straight lines. The *lighter shaded areas* show the responses of two Ru/B_4C Kirkpatrick–Baez mirrors

however, is a combination of lateral and depth gradient. It provides a broad and constant energy bandwidth over its whole length.

As described in Sect. 25.2.2, the spectral response of a nonperiodic multilayer can be designed locally. The overall intensity spectrum I_q of a curved focusing device, however, depends on the additional lateral gradient and on the geometrical influence of the curved mirror. The local reflectivity profile along the mirror can be calculated based on the knowledge of both lateral and depth gradients. In addition, the local intensity $\mathrm{d}I_q/\mathrm{d}\alpha_q$ as a function of the angular acceptance $\mathrm{d}\alpha_q$ has to be taken into account (see inset in Fig. 25.8). In the given configuration, the X-ray source can be considered to be isotropic over the angular range used. In the case of an elliptical profile, and at a given incidence angle θ, the local intensity contribution $\mathrm{d}I_q$ in the focus coming from a given angular range, $\mathrm{d}\alpha_q$, is proportional to the angle, $\mathrm{d}\alpha_p$, seen from the source. Since, at the reflection point of the beam,

$$q\mathrm{d}\alpha_q = p\mathrm{d}\alpha_p, \tag{25.9}$$

the local intensity is

$$\frac{\mathrm{d}I_q(\theta)}{\mathrm{d}\alpha_q(\theta)} \approx \frac{\mathrm{d}\alpha_p(\theta)}{\mathrm{d}\alpha_q(\theta)} = \frac{q(\theta)}{p(\theta)} = M(\theta), \tag{25.10}$$

where M is the demagnification. Using (25.2), the local intensity can now be directly expressed as a function of θ. This approach is valid when the detector integrates over the whole focal spot. The results for the two Ru/B_4C mirrors of Fig. 25.7 are shown in Fig. 25.8. Only the relevant branch, with $p > q$, is shown. The minimum possible grazing angle occurs at the vertex of the ellipse where $p = q$ and where the local intensity is normalized to unity. Very similar results can be obtained for parabolic mirrors [5]. The total spectral response can be obtained by integrating the local reflectivity spectra multiplied by (25.10) over the full length of the optics.

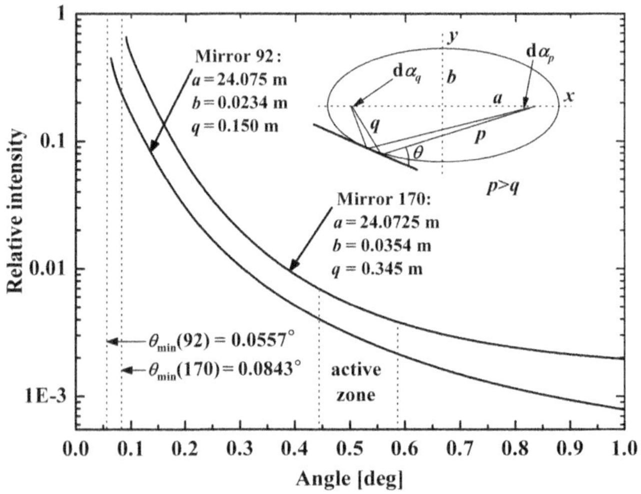

Fig. 25.8. Relative intensity after reflection from two elliptically shaped mirrors as a function of the incidence angle θ. The curves were calculated using (25.10) and correspond to the branch $p > q$ of the ellipse showing both multilayer mirrors. The illuminated (active) angular zone and the smallest angles of incidence (for $p = q$) are indicated by *dotted lines*

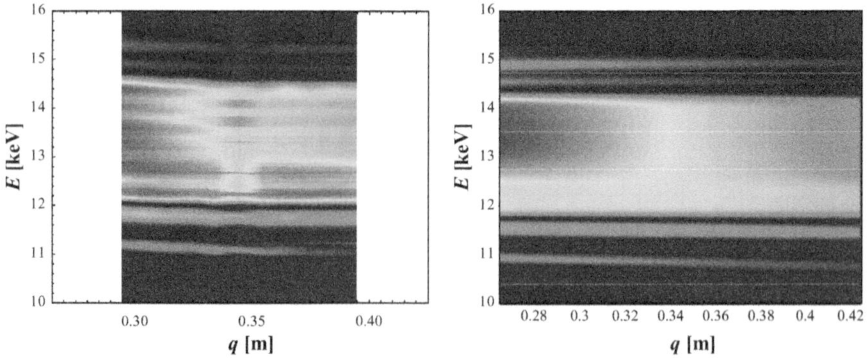

Fig. 25.9. Gray scale images of the experimental (*left*) and simulated (*right*) X-ray reflectivity data of multilayer 170, plotted as functions of focal distance q and photon energy E

Figure 25.9 (left image) shows a two-dimensional plot in (E, q) space based on the experimental data set of the multilayer 170 (see Fig. 25.8) and the corresponding simulation (right image). The area in (E, q) space selected here is indicated by the dotted rectangle in Fig. 25.7. The plots show broad horizontal bands of high reflectivity and confirm the more general theoretical approach described above. The intensity fluctuations of the experimental data of Fig. 25.7 are caused by subtle deviations from the ideal layer structure. In the main features, however, there is a good agreement.

25.3 Multilayers with Strongly Absorbing Materials

In the discussions of the previous sections, although absorption had to be taken into account in determining the multilayer properties, at the wavelengths considered, it is sufficiently low that reasonable performances can still be obtained. However, reflective multilayer coatings for the wavelength range 50–105 nm, in the EUV region of the electromagnetic spectrum, have not previously been developed because all conventional materials are strongly absorbing. Due to this strong absorption, radiation would be mostly absorbed in the few outermost layers of a multilayer; only single layer coatings are useable. A new concept, sub-quarter-wave multilayers (SQWMs), has been recently developed; with this, a reflectivity enhancement is obtained over that of a single layer even when the materials are strongly absorbing. For wavelengths shorter than ~50 nm, the intrinsic absorption decreases but remains moderately high. Thus, reflective coatings based on two-material multilayers are feasible, but even so SQWMs may provide valuable reflectivity increases. According to this new multilayer concept, enhancements in reflectivity can be obtained by the superposition of thin films of two or more materials when certain material selection rules are satisfied. The main results on SQWMs are summarized in the following sections. Section 25.3.1 highlights the main steps in the derivation to obtain the new multilayers. Section 25.3.2 exhibits a few examples of multilayers involving materials with a strong absorption. Section 25.3.3 demonstrates that the new multilayers can also be applicable in spectral regions of moderate absorption where standard two-material multilayers are already possible. Finally, Sect. 25.4 summarizes new algorithms for multilayer optimization.

25.3.1 Sub-Quarter-Wave Multilayers

A general multilayer coating can be considered to consist of m thin films of different absorbing materials on an opaque substrate. Figure 25.10 shows a diagram of such a general multilayer. The complex refractive indices of the films and substrate are referred to as $\tilde{n}_i$, $i = 1, 2, \ldots, m + 1$, starting with the outermost film; $\tilde{n}_0$ is the refractive index of the incidence medium. Since the materials absorb radiation, all the $\tilde{n}_i$ are complex numbers, and in the

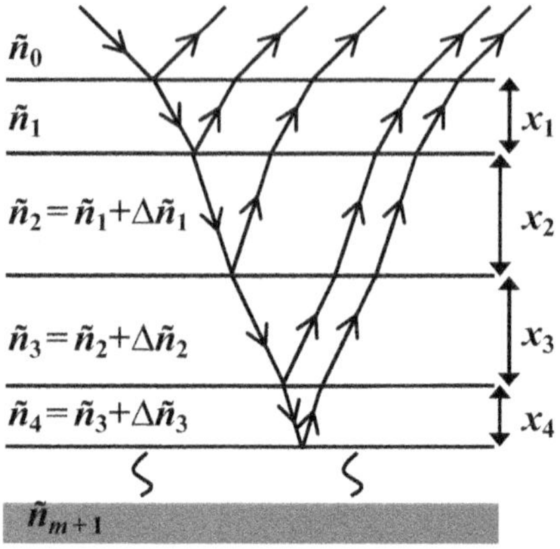

Fig. 25.10. Incident and reflected rays on a general multilayer

most general case, they may all be different. The reflectivity of the multilayer can be obtained through recurrent calculation [11] starting at the innermost interface. The aim is to obtain the material selection rules linking the $\tilde{n}_i$ to allow the design of multilayers that provide a reflectivity enhancement through every single layer. However, the exact equations that determine the general trends of the reflectivity in terms of the refractive indices are difficult to handle. An appropriate approximation, which provides manageable equations, is based on the assumption that all the materials in the multilayer have close refractive indices, although the applications of SQWMs, as will be shown in Sect. 25.3.2, are not restricted to such materials.

The refractive index differences of the materials across the i, $i+1$ interface are represented as $\Delta\tilde{n}_i = \tilde{n}_{i+1} - \tilde{n}_i$, $i = 1, 2, \ldots, m$, with $\Delta\tilde{n}_i = \Delta n_i + \mathrm{i}\Delta k_i$. The above assumption implies that $|\Delta\tilde{n}_i| \ll |\tilde{n}_i|$. The multilayer reflectivity is then developed to first order in $\Delta\tilde{n}_i$. The first-order expansion is a good description when the materials in the multilayer are assumed to absorb radiation strongly and/or when the refractive index changes $|\Delta\tilde{n}_i|$ are small, so that multiple reflections can be neglected.

Using this technique, the following conditions were obtained for the refractive indices of the materials to provide an m-dimensional reflectivity maximum with m sub-quarter-wave films on an opaque substrate [25,26]. For the outermost layer ($i = 1$), and for s-polarized radiation,

$$\mathrm{Im}\left[\frac{F_0^*\left(1 - F_0^2\right)\Delta\tilde{n}_1}{\cos\theta_1}\right] < 0, \qquad (25.11)$$

while for p-polarized radiation

$$\mathrm{Im}\left[F_0^*\left(1 - F_0^2\right)\Delta\tilde{n}_1\cos\theta_1\left(1 - \frac{\tilde{n}_0^2\sin^2\theta_0}{\tilde{n}_1^2\cos^2\theta_1}\right)\right] < 0, \qquad (25.12)$$

where F_0 is the Fresnel reflection coefficient of the outermost interface and θ_i is the propagation angle in the ith layer. For the inner layers ($i = 2, 3, \ldots, m$)

$$\mathrm{Im}\left[\frac{\Delta\tilde{n}_i}{\Delta\tilde{n}_{i-1}}\right] < 0, \tag{25.13}$$

the latter equation being valid for both polarizations. For normal incidence and for radiation incident in a vacuum, the material selection rules become, for the outermost layer,

$$\Delta n_1\left(1 + n_1^2 + k_1^2\right)k_1 + \Delta k_1\left(1 - n_1^2 - k_1^2\right)n_1 > 0, \tag{25.14}$$

and for the inner layers

$$\Delta n_{i-1}\Delta k_i < \Delta n_i \Delta k_{i-1}. \tag{25.15}$$

If a set of $m + 1$ materials satisfy these conditions, then an SQWM can be designed with a reflectivity enhanced by each added layer. Condition (25.13) can be easily interpreted through a graphical representation. If the complex refractive indices are plotted in the n–k plane, the condition is satisfied if the switch from material i to $i + 1$ is by clockwise rotation; an example of this thumb rule is given in Sect. 25.3.2. Even if (25.11)/(25.12) and (25.13) are not satisfied for some trial materials arranged in a certain sequence, the same refractive indices in a different sequence may provide a solution. In addition, a set of materials may not satisfy (25.13), but a subset may.

The highest possible reflectivity is obtained when using the most extreme available refractive indices, as will be shown in Sects. 25.3.2 and 25.3.3. From a theoretical point of view, more materials in the multilayer (satisfying (25.13)) will provide a higher calculated reflectivity. For L different materials, with refractive indices $\tilde{n}_1$ to $\tilde{n}_L$, after making use of all the available materials, completing a full period, the procedure can be repeated for more periods so long as the reflectivity enhancement continues. In this sense, these mirrors are similar to standard two-material multilayers:

- In the standard case, one full period is completed after two layers; whereas in SQWMs, it is completed after L layers.
- In the standard case, the refractive index oscillates between two values; in SQWMs, it rotates in the correct sequence among all the materials.
- The integrated optical path through one full period is $\lambda/2$ in both cases.

This shows that SQWMs represent a generalization of standard two-material multilayers to any number of materials. An SQWM with just one material and only one period can be modeled step by step by removing one material at a time (and correspondingly reducing the reflectivity) until a single layer coating remains.

The reflectivity enhancement obtained with SQWMs is a result of the more interfaces per constant optical path than in standard multilayers. These extra interfaces provide extra contributions to the reflectivity coming from radiation that has propagated a short distance and hence has been absorbed less. The phase shift in the reflection at any interface between two contiguous materials, that are selected as described above, supplements the otherwise insufficient phase shift after propagation through sub-quarter-wave layers; the superposition of the different contributions results in the higher reflectivity. Absorption is intrinsic to the benefit of SQWMs: if there was a material in nature with no absorption, then 100% reflectivity would be theoretically feasible with a two-material multilayer [27] and obviously no further enhancement would then be possible.

The optimum film thicknesses that provide an m-dimensional reflectivity maximum space for an m-layer multilayer are given when the first derivative of the reflectivity with respect to all film thicknesses is zero. The second derivative must satisfy the conditions for the extreme value to be a maximum, to discard minima or saddle points. If

$$x^0_{\max,L} = \left(x^0_{\max,1}, x^0_{\max,2}, \ldots, x^0_{\max,m}\right) \tag{25.16}$$

is the point at which the first-order expansion of the first derivative of reflectivity has an extreme value, then the optimum film thicknesses are given by

$$x^0_{\max,L} = -\frac{\lambda}{4\pi}\varphi_z, \quad i = 1, 2, \ldots, m, \tag{25.17}$$

where φ_z is the phase of complex number z

$$z = F_0^* \left(1 - F_0^2\right) \tilde{n}_i F_i \cos\theta_i, \quad i = 1, \tag{25.18}$$

$$z = \frac{\tilde{n}_i F_i \cos\theta_i}{\tilde{n}_{i-1} F_{i-1} \cos\theta_{i-1}}, \quad i = 2, 3, \ldots, m, \tag{25.19}$$

and F_i is the Fresnel reflection coefficient at the i, $i+1$ interface. In the first-order expansion, each optimum layer thickness depends only on the parameters of the specific interface and not on those of other interfaces. The phases in (25.17) are multivalued functions; phases differing from a certain solution by an integer multiple of 2π are also solutions. However, only those satisfying the condition on the second derivative will be maxima. To minimize absorption, the maximum obtained with the thinnest possible layer is needed, so that the smallest solution for φ_z is that required. When materials satisfy the material selection rule (25.13), the condition on the second derivative is automatically satisfied for the thinnest solution of (25.17). A trial material that gives a reflectivity maximum with a smallest phase value larger than π is not suitable for SQWM.

25.3.2 Applications of SQWM with Strongly Absorbing Materials

As an example, the design of an SQWM with the highest possible reflectivity at the OII 83.4 nm spectral line is considered. This line is of interest for atmospheric physics [28], but observations at this and nearby wavelengths are sometimes difficult due to the low intensity of the sources. Hence, enhanced reflectivities would be of great benefit, compared to thin films of SiC which is among the materials with the largest reflectivity at 83.4 nm.

Figure 25.11 shows the refractive indices of five materials selected, using the formalism described in Sect. 25.3.1, from those with the most extreme optical constants at 83.4 nm. The arrows indicate the sequence, according to (25.13), in which the materials have to be deposited from the outermost to the innermost layer to obtain reflectivity enhancements. The calculated reflectivities as functions of wavelength for multilayers consisting of one to six layers are shown in Fig. 25.12. It can be seen that a reflectivity increase is obtained every time a new material is added. However, due to the high absorption of all the materials at 83.4 nm, the reflectivity saturates after a few layers, and no further increase is obtained by adding more layers beyond (outermost) SiC/B_4C/C/Al_2O_3/MgF_2/SiC (innermost). Note that aluminum, even though it has adequate optical constants, was not used in this example, but it will be considered in others. It is noteworthy that the increase in reflectivity, although optimized at 83.4 nm, extends over a wide spectral range. Similarly, for a multilayer optimized at normal incidence, a reflectivity increase over that of a single layer is obtained for a wide range of incidence angles. The slow dependences of multilayer reflectivity with wavelength and angle mean

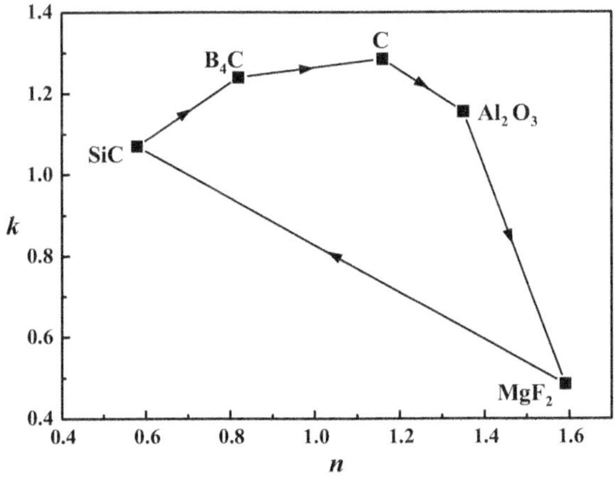

Fig. 25.11. Representation in the complex plane of the refractive indices of five materials at 83.4 nm. The *arrows* point to the material of the underlying film (after [26])

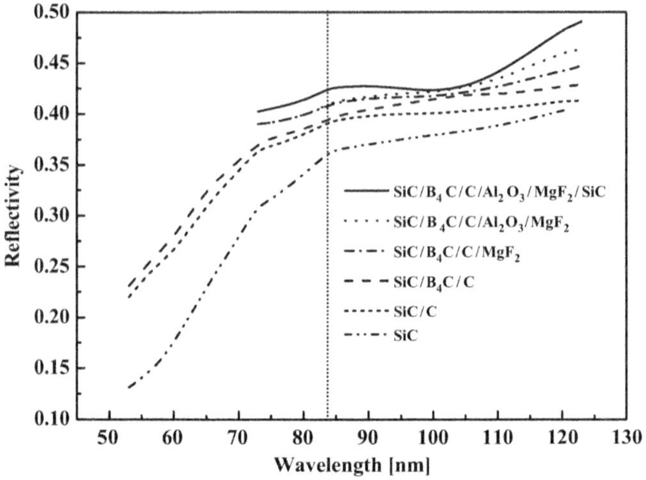

Fig. 25.12. Calculated reflectivities of an opaque layer of ion-beam deposited SiC, and of multilayers optimized at 83.4 nm (indicated by the *dotted vertical line*). In each case, the first named material is the outermost layer (after [25])

that there is a high tolerance for layer thicknesses and for errors in the optical constants of materials. An approximate value of the layer thicknesses is given by in the derivation of SQWM theory, it was assumed that the refractive indices all have close values. However, the current example shows that, even when the refractive index differences are large, the material selection rules are still a helpful tool for giving a reasonable film thickness estimates.

Experimental demonstrations of SQWMs optimized at a wavelength of 92 nm have been performed [29, 30]. Multilayers were designed and prepared between two and four layers of different materials. The multilayers show increased reflectivities in the 50–200 nm spectral range.

25.3.3 Extension of the Mechanism of Reflectivity Enhancement to Moderately Absorbing Materials

Multilayer coatings have been developed over the last decades with increasing reflectivities for radiation at wavelengths below ∼50 nm. However, the largest theoretical reflectivities of such multilayers are still limited because all materials absorb radiation in these spectral regions. Although there has been some work on using more than two materials [31–33], most XUV multilayers have been made so far by alternating layers of two materials with optimized thickness of each layer to provide the highest possible reflectivities at a desired wavelength (angle), or range of wavelengths (angles), as described in Sect. 25.1. Multilayer coatings with just two different materials will be referred to here as standard multilayers.

For moderately absorbing materials, with $k/n < \sim 0.2$ which is the case for most materials at wavelengths below ~50 nm, the first-order expansion of Sect. 25.3.1 that allows multiple reflections to be neglected is no longer valid. However, the ideas of SQWMs may still be applied with some modifications. For standard multilayers, the smallest contributions to the reflectivity are from radiation reflected at the innermost interfaces and mainly arise from the reflection at a single interface. The contributions of reflections from more than one inner interface provide only small corrections to the smallest contribution. Thus, a first-order expansion of the reflectivity accurately describes the contributions of the innermost interfaces. This means that a reflectivity enhancement may be obtained by replacing the innermost bilayers of a standard multilayer with an SQWM structure using the material selection rule determined in Sect. 25.3.1. In practice, further reflectivity enhancement is obtained by replacing all bilayers of a standard multilayer with SQWM structures.

As an example, consider SQWMs with high reflectivity at the HeII 30.4 nm line [34], compared to the several multilayer coatings which have been designed and made with various two-material combinations, such as C/Si [35], Ir/Si [36], and Al/Nb [37]. These have theoretical normal-incidence reflectivities of ~0.30 at 30.4 nm and measured values of ~0.25 (for C/Si). A higher normal reflectivity is obviously desirable for, e.g., astrophysics applications.

A reflectivity enhancement can be obtained with a multimaterial SQWM. The material selection is performed by plotting the refractive indices of all materials available at the desired wavelength, followed by selecting all the materials with the most extreme refractive indices which define a polygon that encloses all the other materials. The multilayer with the optimum film thicknesses including all the materials of the polygon boundary will provide a reflectivity that is higher than that which can be obtained with a multilayer designed with just some of these materials. Selecting a set of materials inside the polygon will result in a lower reflectivity, regardless of the number of materials. It should be noted that the optimum layer thicknesses of SQWMs with moderately absorbing materials are not given accurately by (25.16), and so they have to be obtained as part of the optimization process.

The optical constants of many materials at 30.4 nm are shown in Fig. 25.13. The candidate materials for the multilayer with the largest reflectivity are those with the most extreme values, i.e., on the boundary of the polygon, in a sequence given, from the outermost to the innermost layer, by clockwise rotation in the n–k plane. The calculated normal-incidence reflectivities of optimized SQWMs for various material combinations are given in Table 25.1. In the calculations, the interfaces were assumed to be smooth and abrupt, with no material diffusion. The number of periods was limited to 25, because the reflectivity enhancement with more periods was negligible.

A significant reflectivity enhancement, by a factor of 1.13, is obtained when the number of materials increases from two to three. Adding further materials, up to a total of six, results in further noticeable enhancement.

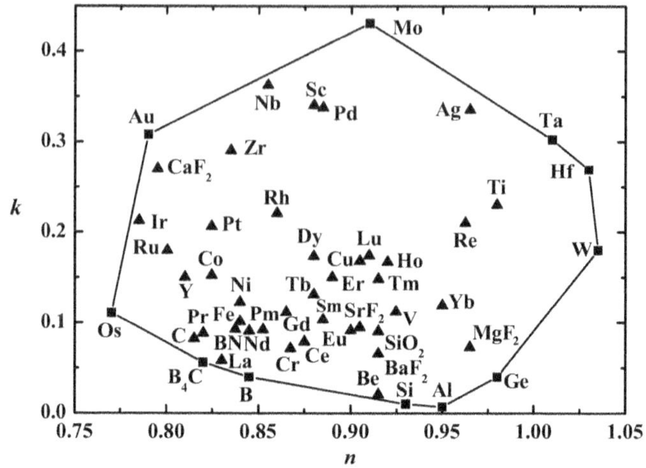

Fig. 25.13. The optical constants for many materials at 30.4 nm (after [34]). The optical constant data were obtained from the web site of the Center for X-Ray Optics (Lawrence Berkeley National Laboratory) [38]

Table 25.1. Calculated normal-incidence reflectivities of SQWMs optimized for 30.4 nm. The multilayers contain 25 periods. The material sequences are from the outermost layer (after [34])

Materials in the period	Reflectivity
Al/Os	0.446
Al/B_4C/Mo	0.504
Al/B/Os/Mo	0.517
Al/B/Os/Mo/W	0.526
Al/Si/B/Os/Mo/W	0.532
Al/Si/B/Os/Mo/W/Ge	0.533
Al/Si/B/Os/Au/Mo/W/Ge	0.534
Al/Si/B/B_4C/Os/Au/Mo/W/Ge	0.535
Al/Si/B/B_4C/Os/Au/Mo/Hf/W/Ge	0.536

In each case the combination of materials providing the largest

Thereafter, the increases become smaller; for ten materials in each period the total increase, over two materials, is 20%. The highest reflectivities are obtained when aluminum is the outermost material, and so the calculation should incorporate the effect of the thin oxide film that readily grows on aluminum in contact with the atmosphere. This was not taken into account in this example, since the purpose is to show the benefit of multimaterial SQWMs compared to standard two-material multilayers.

In this and other examples investigated, it was found that the increase in the number of materials does not necessarily affect the spectral bandwidth. The intrinsic bandwidth depends more on the specific optical constants of the materials in the multilayer than on the number of materials. A similar result was obtained for the angular bandwidths of SQWMs.

A similar analysis has been used to design multilayers with enhanced reflectivities at 13.4 nm [39], the wavelength of interest for the future EUV lithography. The polygon boundary in the n–k plane is now defined by the materials Ru, Rh, Pd, Ag, Ni, CsI, Ce, Si, Sr, Nb and Mo; multilayers have been designed using combinations of these materials with ruthenium or rhodium as the outermost layer. The calculated reflectivities of optimized normal-incidence SQWMs are shown in Table 25.2 for the best combinations of between two and 11 materials. In these calculations, the number of periods was limited to 50 since the enhancement with more periods was negligible. The reflectivity enhancement increase is significant up to about four materials; a further small increase is obtained for multilayers with 5–9 materials and a marginal increase with ten and 11 materials.

The enhancement, 5% over a standard Si/Mo multilayer, is smaller at 13.4 nm than at 30.4 nm, due to the higher transparency of the materials so that more layers contribute to the reflectivity. The extra interfaces of SQWMs close to the outer surface (compared to standard multilayers) provide contributions to the reflectivity from radiation that has propagated a shorter distance and has been less strongly absorbed. Thus, the benefit of having extra interfaces closer to the outer surface is less important when more periods contribute.

Due to the large number of reflectors in an EUV projection lithography system (~9), a small reflectivity increase of each individual mirror may still

Table 25.2. Calculated normal-incidence reflectivities R and R^9, and R^9 integrated over the bandwidth, of multilayers optimized for 13.4 nm. The multilayers contain 50 periods with an opaque substrate of the same material as that of the outermost (left) layer (after [39])

Materials in the period	R	R^9	$\int R^9$ ($\times 10^{-2}$ nm)
Si/Mo	0.7495	0.0746	2.412
Mo/Si	0.7628	0.0874	2.814
Rh/Si/Mo	0.7602	0.0848	2.585
Rh/Si/Sr/Mo	0.7795	0.1062	2.986
Ru/Si/Mo	0.7661	0.0909	3.014
Ru/Si/Sr/Mo	0.7846	0.1127	3.443
Ru/Rh/Si/Sr/Mo	0.7852	0.1135	3.358
Ru/Rh/Ce/Si/Sr/Mo	0.7858	0.1142	3.298
Ru/Rh/Ce/Si/Sr/Nb/Mo	0.7863	0.1149	3.317
Ru/Rh/Ag/Ce/Si/Sr/Nb/Mo	0.7867	0.1154	3.1486
Ru/Rh/Ag/CsI/Ce/Si/Sr/Nb/Mo	0.786912	0.11570	3.0692
Ru/Rh/Ag/Ni/CsI/Ce/Si/Sr/Nb/Mo	0.786921	0.11571	3.0639
Ru/Rh/Pd/Ag/Ni/CsI/Ce/Si/Sr/Nb/Mo	0.786929	0.11572	3.0632

provide a significant throughput enhancement. The reflectivity after nine reflections at the target wavelength and integrated within the reflectance band are also shown in Table 25.2. At the design wavelength, the overall enhancement is 55% (51%) for 11 (4) materials, but note that the integrated reflectivity decreases for more than four materials, due to decreased bandwidth.

There are, of course, important practical considerations that must be taken into account when manufacturing such multilayers; one, as mentioned previously, is the stability of the multilayer in the environment of use. The material selection rules discussed in Sect. 25.3.1 can also be used to specify suitable capping layers [39]. Increasing the number of materials results in more difficult preparation of multilayers; in practice, it may be difficult to implement multilayers with more than about four materials. Also, some selected materials may not be compatible with one another, and so before preparing SQWMs the stability and suitability of the materials to be deposited as thin and smooth films must be assessed.

25.4 New Layer-by-Layer Multilayer Design Methods

The design of a multilayer coating with the highest possible reflectivity at a given wavelength is simple when a pair of absorption-free materials with different refractive indices is available. In this case, the multilayer just has a sufficiently large number of quarter-wave layers of the two materials. However, when either or both materials absorb radiation, the largest reflectivity is not given by the quarter-wave layer design [27]. Now, an optimized multilayer will have decreasing (increasing) thicknesses for the more (less) absorbing material from the innermost to the outermost layer. This complicates the calculation of the layer thicknesses, more so when optimizing SQWMs with three or more materials. The recipe of (25.17) still provides an estimate of the optimum thicknesses; in this section, algorithms that provide the exact values are described.

For s- or p-polarized radiation, or for normal incidence with any polarization, the layer thicknesses of the multilayer with the largest reflectivity at a given wavelength can be optimized sequentially from the innermost to the outermost layer. Thus, it is not necessary to perform a new optimization of the entire stack when new layers are added on top of an optimized system [40]; a mathematical description of this is given in [41]. For standard two-material multilayers, a common optimization algorithm starts with the three innermost layers and optimizes their thicknesses for maximum reflectivity. In the following step, the next two layers are added and the three outer layers are optimized. The process is continued by adding more periods until the multilayer is completed [27]. The three outer layers are used at every step since the outermost interface after the addition of the jth layer pair becomes an internal interface when the layer pair $j+1$ is added. The multilayer calculation is then transformed from one optimization problem with $2N$ parameters, where N is the number of periods, into N optimizations with three parameters. When $M \geq 2$ different materials are used in each period, the calculation results in N optimizations with $M+1$ parameters each.

Another layer-by-layer multilayer design method was developed by Yamamoto and Namioka [42]. This is based on the smooth connection of the reflectivity as a function of the layer thickness at the switching point from one material to the other. With this method, the multilayer design process is transformed into $2N$ optimizations with one parameter, resulting in a much faster algorithm. However, some derivations of the method, which is based on a graphical representation, were qualitative, and no explicit function to be optimized was provided.

To improve these algorithms, a new mathematically rigorous layer-by-layer multilayer design method has been developed [41, 43]. This provides the optimum layer thicknesses, regardless of the number of different materials used. The following summarizes the main details of the design of a multilayer coating consisting of m thin films of various absorbing materials deposited on an opaque substrate, assuming that the interfaces are abrupt and smooth. Let $\tilde{n}_i$ be the refractive indices of the films and substrate ($i = m+1$), the outermost film corresponding to $i = 1$. In the most general case, all the $\tilde{n}_i$ may be different. The algorithm searches for an m-dimensional reflectance maximum in thickness space. At this maximum, the first derivative of the multilayer reflectivity with respect to every film thickness must be zero; this condition provides the layer thicknesses x_i and results in [43], for $i = 1$,

$$\mathrm{Re}\left(r_0^* \frac{\partial r_0}{\partial x_1}\right) = R_0 \cdot \mathrm{Re}\left(\frac{1}{r_0}\frac{\partial r_0}{\partial x_1}\right) = 0 \tag{25.20}$$

and, for $i = 2$ to m,

$$\mathrm{Im}\left(\frac{\partial r_{i-1}/\partial x_i}{\frac{\partial r_{i-2}}{\partial x_{i-1}}\Big/\frac{\partial r_{i-2}}{\partial r_{i-1}}}\right) = -\frac{\lambda}{4\pi}\mathrm{Re}\left(\frac{\partial r_{i-1}/\partial x_i}{\tilde{n}_{i-1} r_{i-1}\cos\theta_{i-1}}\right) = 0. \tag{25.21}$$

In these equations, r_i is the (amplitude) reflectivity of the ith layer, θ_i is the propagation angle through the layer, and R_0 is the multilayer reflectivity.

25.4.1 Two Algorithms for Multilayer Optimization

Two different approaches starting with (25.20) and (25.21) have been used to develop two equivalent algorithms. Here, the main details of the two algorithms will be described; a rigorous development of the algorithms can be found elsewhere [41, 43].

In the first approach, the conditions on the first derivative given by (25.20) and (25.21) result in

$$\mathrm{Im}(u_i) = 0, \tag{25.22}$$

for all i. In (25.22),

$$u_i = \frac{\tilde{n}_i \cos\theta_i}{\tilde{n}_{i-1}\cos\theta_{i-1}} \frac{\left(1 - F_{i-1}^2\right) r_i \exp\beta_i}{\left(F_{i-1} + r_i \exp\beta_i\right)\left(1 + F_{i-1} r_i \exp\beta_i\right)}, \tag{25.23}$$

where

$$\beta_i = \frac{4\pi}{\lambda} i\tilde{n}_i x_i \cos\theta_i \tag{25.24}$$

and F_i is the Fresnel reflection coefficient at the i, $i+1$ interface. Here, r_i is the amplitude reflectivity of the submultilayer including interfaces from i to $m+1$. Equation (25.22) is important to the algorithm because it is the basis of the calculation of every layer thickness. Since $\mathrm{Im}(u_i)$ only depends on the layer thickness of the ith to mth layers, the optimization can be performed iteratively starting with the innermost layer. To select the solutions of (25.22) that correspond to reflectivity maxima (rather than minima or saddle points), the conditions on the second reflectivity derivative are used, expressed by

$$\frac{\partial}{\partial x_i} \mathrm{Im}(u_i) \prod_{j=1}^{i-1} \mathrm{Re}(u_j) > 0, \tag{25.25}$$

for all i. A drawback of these conditions is that they cannot be expressed in a recurrent form, such as (25.24), because of the presence of the terms $\mathrm{Re}(u_j)$. All conditions given by (25.25) depend on these terms which, however, will not be known until the multilayer design is completed.

A search on the sign of $\mathrm{Re}(u_i)$, through several examples of multilayers with two or more materials in different wavelength ranges, shows that the conditions

$$\mathrm{Im}(u_i) = 0, \quad \frac{\partial \mathrm{Im}(u_i)}{\partial x_i} > 0 \Rightarrow \mathrm{Re}(u_i) > 0 \tag{25.26}$$

are always satisfied for all m. This is important since it means that every solution selected by the positive sign of $\partial\, \mathrm{Im}(u_i)/\partial x_i$ ultimately results in a maximum reflectivity, enabling a recurrent design starting with the innermost layer. The multilayer design is performed in the following steps:

- Search for the thinnest x_m satisfying $\mathrm{Im}(u_m) = 0$ and $\partial\, \mathrm{Im}(u_i)/\partial x_i > 0$.
- Proceed iteratively from the innermost to the outermost layer.
- Check *a posteriori* that $\mathrm{Re}(u_i) > 0$ for all i.

An infinity of thicknesses satisfying $\mathrm{Im}(u_i) = 0$ can be found, with alternating signs of $\partial\, \mathrm{Im}(u_i)/\partial x_i$. The thinnest solutions satisfying $\partial\, \mathrm{Im}(u_i)/\partial x_i > 0$ should be selected to maximize the reflectivity.

In the second algorithm, a new function *inreflectance*, $\Im$, whose maximum provides the layer thicknesses of the optimum multilayer, is defined from (25.20) and (25.21):

$$\Im_{i-1} = \left| r_{i-1}^{(\tilde{n}_{i-1}\cos\theta_{i-1})^{-1}} \right|. \tag{25.27}$$

This function involves, instead of the simple reflectivity modulus, the modulus raised to the power of inverse of the complex refractive index of the next (outer) layer, including an inclination term. Thus inreflectance is not defined for the outermost layer. Maximizing the inreflectance at every internal layer, proceeding sequentially outward, results in a multilayer with the largest reflectivity. At a given layer in the multilayer, inreflectance and reflectivity differ when the next (outer) material absorbs radiation, i.e., when the exponent in (25.27) is complex, but are equivalent, for optimization purposes, when the next outer layer is transparent. The outermost layer of the multilayer is intrinsically different to the internal layers, and is optimized through reflectivity, not inreflectance. All inreflectance maxima are found at larger layer thicknesses than that for the reflectivity maxima when the next outer material absorbs radiation.

The reflectivity and inreflectance for the normal-incidence multilayer of Fig. 25.12, optimized for 83.4 nm, are compared in Fig. 25.14 as functions of thickness of the growing multilayer [26]. The discontinuities at layer boundaries are due to the change of the top medium from which radiation penetrates to the optimizing layer. In the case of (amplitude) reflectivity, there is a sudden change from r_{i+1} as the ith layer is completed to $(F_{i+2} + r_{i+1})/(1 + F_{i+2}r_{i+1})$ at the start of layer $i + 1$. The shapes of the two functions are clearly different; for a growing layer, inreflectance increases to a maximum at the optimum thickness, whereas reflectance is not only decreasing at the inreflectance maximum, but also it is already decreasing at the start of the growth of Al_2O_3, C, and B_4C layers. SiC is an exception as it is the outermost layer and

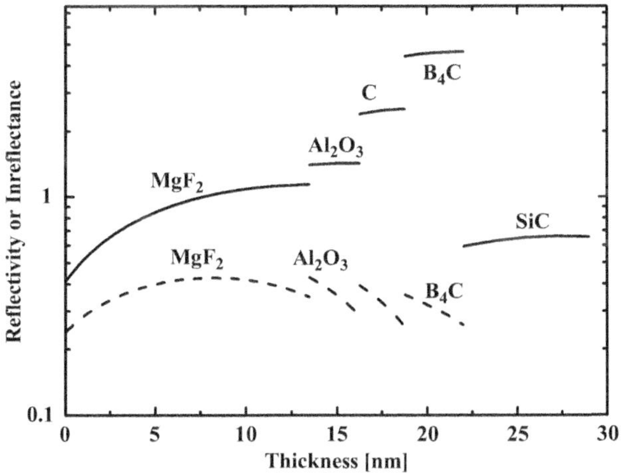

Fig. 25.14. Comparison ofMultilayer systems!reflectivity and inreflectance for comparison of inreflectance (*full lines*) and reflectivity (*dashed lines*) as a function of thickness for a $SiC/MgF_2/Al_2O_3/C/B_4C/SiC$ multilayer optimized for the largest normal-incidence reflectivity at 83.4 nm

hence the optimum thickness is given by the maximum reflectivity, which in this case is coincident with the maximum inreflectance because the incidence medium (vacuum) does not absorb radiation. This discussion clearly shows that inreflectance is a better measure of when the optimum thickness has been reached.

The first derivative of inreflectance is connected to the u_i functions of the first algorithm that were defined in (25.22)–(25.24) through

$$\frac{\partial}{\partial x_i}\left|r_{i-1}^{(\tilde{n}_{i-1}\cos\theta_{i-1})^{-1}}\right| = -\frac{4\pi}{\lambda}\left|r_{i-1}^{(\tilde{n}_{i-1}\cos\theta_{i-1})^{-1}}\right|\mathrm{Im}(u_i), \tag{25.28}$$

for $i = 1$ to $m-1$, and

$$\frac{\partial|r_{m-1}|}{\partial x_m} = -\frac{4\pi}{\lambda}|r_{m-1}|\mathrm{Im}(u_m), \tag{25.29}$$

for $i = m$. This demonstrates that the two algorithms are equivalent. They result in extremely fast multilayer optimization, with computing time roughly proportional to the number of layers, regardless of the number of different materials.

25.4.2 Layer-by-Layer Design of Multilayers with Barrier Layers

The layer-by-layer multilayer design methods described in Sect. 25.4.1 can be supplemented by adding the possibility of alternating fixed thickness layers along with layers whose thicknesses are optimized for high reflectivity at a given wavelength [43, 44]. This allows the design of multilayers with barrier layers to prevent diffusion and/or reaction between the multilayer constituents, or when intermixed zones develop at multilayer interfaces, or when capping layers are used. Modifications of the algorithms presented in Sect. 25.4.1 have been developed to enable the design of such multilayers with any number of layers of any optical constants. The optimization can be performed for either normal or nonnormal incidence and for either s- or p-polarized radiation. Here, only a few details of the extension of the first algorithm are summarized.

In the description below, it is assumed that the layers from the outermost (1) to $j_1 - 1$ have fixed thicknesses and that layer j_1 is the outermost layer to be optimized. Then, layers $j_1 + 1$ to $j_2 - 1$ have fixed thicknesses and layer j_2 ($j_2 > j_1$) is the second layer to be optimized; and so on. If h is the function that provides the subset of layers whose thicknesses have to be optimized, then $h(1) = j_1$, $h(2) = j_2$, etc. If m is the number of layers that have to be optimized out of the M total number of layers, then $h(m) < M$ when there are fixed layers between the innermost layer to be optimized and the substrate; otherwise, $h(m) = M$.

A similar mathematical derivation to that of Sect. 25.4.1 in which the derivative of reflectivity with respect to every nonfixed (free) film thickness, layers $h(1)$, $h(2), \ldots, h(\mathrm{m})$, is set to zero results in a generalization of the u_i functions that were defined in (25.22)–(25.24):

$$U_i = \prod_{j=h(i-1)+1}^{h(i)} u_j. \tag{25.30}$$

There is one U function per free layer. Every U function involves not only the free layer, but also the upper fixed layers that are below the next outermost layer to be optimized. Equation (25.22) now becomes, for $i = m$,

$$\mathrm{Im}(U_i) = 0 \tag{25.31}$$

and the two conditions generalized from those of (25.26) are

$$\mathrm{Im}(U_i) = 0, \quad \frac{\partial \mathrm{Im}(U_i)}{\partial x_i} > 0. \tag{25.32}$$

Equation (25.32) summarizes the optimization method for multilayers with fixed thickness layers along with those to be optimized. The design is performed in the following steps:

- Calculate the reflectivity of the fixed innermost layers, if any.
- Search for the thinnest x_m (the thickness of the innermost layer to be optimized) satisfying (25.32), using (25.30). This involves calculations with $h(m)$, the layer to be optimized, along with the block of fixed layers from $h(m-1)+1$ to $h(m)-1$.
- Proceed iteratively from the innermost ($i = m$) to the outermost ($i = 1$) layers to be optimized. Every layer optimization involves all the fixed layers on top of it that are below the next outer layer to be optimized. The outermost layer to be optimized involves all the upper (capping) fixed layers (1 to $h(1)-1$).
- Check *a posteriori* that $\mathrm{Re}(U_i) > 0$ for $i = 1$ to m.

As for multilayers without fixed layers, an infinity of thicknesses satisfying $\mathrm{Im}(U_i) = 0$ can be found, with alternating signs of $\partial \mathrm{Im}(U_i)/\partial x_i$. The thinnest solution satisfying $\partial \mathrm{Im}(U_i)/\partial x_i > 0$ must be selected to maximize the reflectivity.

This algorithm has been applied to the optimization of multilayers for EUV lithography at ~13.4 nm [44]. The performance of the algorithm was demonstrated by optimizing Si/Mo multilayers with:

- Intermixed zones of $MoSi_2$ that were taken into account in the optimization, resulting in a slight reflectivity increase compared to designs ignoring the intermixing zones
- B_4C barrier layers; when the barrier layers are taken into account, reflectivities are slightly higher than for simpler designs
- A range of barrier layer materials, including the use of different materials at the Si-on-Mo and Mo-on-Si interfaces, which could result in a reflectivity increase.

25.4.3 Multilayers with Continuous Refractive Index Variation

The algorithms described in Sects. 25.4.1 and 25.4.2 may readily be extended to the design of a multilayer coating in which the refractive index varies continuously in depth. In this case, an inhomogeneous coating with a refractive index profile can be optimized, in principle, at every thickness element to obtain the largest possible reflectivity at a given wavelength. In practice, several difficulties may arise. First, materials with suitable continuously varying refractive indices must be available. Second, the long series of ultrathin layers, each with the correct refractive index, may be complicated to deposit; the stability of the structure may also be an issue. Further, there is a physical limit to ultrathin layers determined by the atomic/molecular scale of the materials. On the plus side, inhomogeneous coatings may sometimes be produced naturally when materials react or diffuse across interfaces, and a good understanding of the optimization of such processes may allow advantage to be taken of these usually undesirable effects.

In the design of such multilayers, the complex refractive index is assumed to take values within a given continuous domain and in a given sequence. The coating design is generated through a series of layer elements with small refractive index contrast across interfaces; the thickness of the element is calculated in terms of the refractive index increment at the interface. The coating is optimized element by element starting from the substrate. When the refractive index varies both continuously and smoothly, the thickness element is first order in the refractive index increment. A convenient way to calculate the refractive index profile in depth is to express the available domain of the refractive index $\tilde{n}(t)$ as a function of a real parameter t defined within a certain range. With this notation, the elemental thickness corresponding to the range t to $t + \Delta t$ is [45]

$$\Delta x(t) = \frac{\lambda}{4\pi} \frac{\mathrm{Im}\left\{ \frac{\dot{\tilde{n}}^2}{2\tilde{n}^2}[2v(r) - v^2(r)] + \frac{\ddot{\tilde{n}}}{\dot{\tilde{n}}} v(r) \right\}}{\mathrm{Re}\left[\dot{\tilde{n}}\left(\frac{1}{r} - r\right)\right]} \Delta t, \tag{25.33}$$

where $\dot{\tilde{n}}(t)$ and $\ddot{\tilde{n}}(t)$ are, respectively, the first and second derivatives of the refractive index with respect to t, and

$$v(r) = \frac{1}{r} + 2 + r. \tag{25.34}$$

To determine whether the reflectivity has a maximum at each thickness element, rather than a minimum or saddle point,

$$\mathrm{Re}\left[\Delta\tilde{n}_i\left(\frac{1}{r_i} - r_i\right)\right] > 0 \tag{25.35}$$

must be satisfied; this is valid when the refractive index varies continuously and smoothly. However, (25.35) has a larger field of application since it was derived only under the condition that the refractive index increments must be small. It can thus be used to select the materials and their sequence in a coating to obtain the largest possible reflectivity. The simpler material selection rule defined in Sect. 25.3.1 has, in a few cases, failed to discriminate whether a given set of materials deposited in a given sequence will provide a reflectivity enhancement [34,46]; these failures always occurred when the expression given by (25.13) was positive but close to zero, i.e., when the three refractive indices involved lie close to a straight line. Condition (25.35) may be more accurate than (25.13) since fewer assumptions were made in deriving it, but it depends on the reflectivity of the inner part of the coating and on the thickness of the specific layer, which reduces its applicability.

As an example, consider an arbitrary refractive index function for $\lambda = 13.4\,\text{nm}$ incident in a vacuum given by [45]

$$\tilde{n}(t) = \tilde{n}_c + \rho \exp(it) \tag{25.36}$$

with $\tilde{n}_c = 0.9756 + 0.0304i$, $\rho = 0.0281$, the innermost element has $t = 0$, and t takes increasingly positive values for each added element. The substrate is assumed to be opaque with a refractive index equal to that of the innermost element, $\tilde{n}_0 = \tilde{n}(t = 0) = \tilde{n}_c + \rho$. The refractive index has values within a circumference of radius ρ centered at $\tilde{n}_c$ and t represents the angle in radians, covering the range 0–$2\pi \times 40$ in the coating. There is a correspondence between a 2π increment in t and the period of a multilayer: for a few discrete refractive indices (usually 2), a period has one layer of each material. Hence, the given range of t covers an optical path equivalent to that of a multilayer with 40 periods. Figure 25.15 shows $\Delta x/\Delta t$ for the optimized coating as a function of t, calculated using. In this calculation, the parameter increment was $\Delta t = 0.00436\,\text{rad}$ (0.25°).

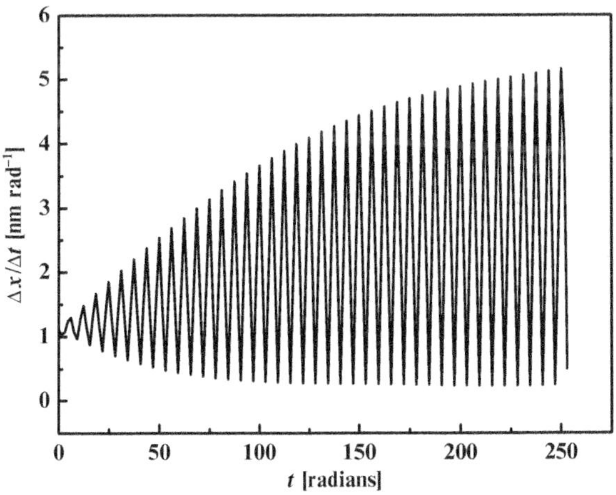

Fig. 25.15. The ratio of element thickness to parameter increment as a function of the parameter t, using (25.36) to define the refractive index in terms of t

25.4.4 Multilayer Design for Nonnormal Incidence and Partially Polarized Radiation

The most general case of multilayer optimization for the largest reflectivity at a given wavelength when radiation is partially polarized (or even linearly polarized in an arbitrary direction) has been specifically addressed only recently [47]. The equations that provide the condition for maximum reflectivity in this general case are not amenable to the simplifications that may be performed when the radiation is s- or p-polarized. Such simplification is essential to demonstrate sequential optimization, which implies that multilayer optimization for partially polarized radiation may not be sequential.

The importance of nonsequential optimization for a specific combination of materials has been demonstrated [47]. It has also been shown that more than one maximum may be found within optical path differences smaller than $\lambda/2$, contrary to what is observed for multilayers optimized for s- or p-polarization.

However, most complex refractive index combinations result in negligible differences between the exact nonsequential optimization and an approximate solution obtained through sequential optimization. Empirically, it appears that the failure of the optimization is a more likely when the materials have large differences in both real and imaginary parts of the refractive index, and the larger real refractive index material has a low extinction coefficient.

25.5 Conclusions

This chapter has demonstrated a range of techniques for designing optimized multilayers, in terms of both lateral and depth grading, to provide enhanced performances compared to standard periodic or quasiperiodic systems. Advantages include the flexibility of such optics which makes them very attractive for various applications on third-generation synchrotron sources. The ability to adapt layered structures to a given beam divergence or energy spectrum allows for the design of very efficient devices. One fundamental problem in the design is how to impose a lateral thickness gradient on a nonperiodic structure without deteriorating the reflectivity spectra away from the point of reference. A more complex formulation of the merit function including the total length of the mirror will be required. A further advantage is the possibility to perform broadband polarization analysis without having to change optical components, important in studies of magnetic and structural properties of materials [48]. Additionally, the ability to make multilayers that are efficient at longer wavelengths where materials are absorbing will widen the range of applications.

Acknowledgments

In addition to the European Science Foundation support, various parts of the research described in this chapter have been supported by national funding. The work described in Sect. 25.2 was based in part on collaboration with L. Claustre, I.V. Kozhevnikov, J.Ch. Peffen, L. Peverini, and E. Ziegler, supported by the ESRF management. Other work in that section was supported by the Royal Society, London (contract no. NC/China/16660), the National Natural Science Foundation of China (contract no. 10435050), the National 863 Sustentation Fund of China (contract no. 2006AA12Z139), the Program for New Century Excellent Talents in University, China (contract no. NCET-04-037), and the European Union through the BESSY-EC-IA-SFS contract (BESSY-ID.05.2.165). The principal Chinese collaborators were Z. Wang and H. Wang of Tongji University, Shanghai. This work is also part of that being carried out by the UK Smart X-Ray Optics (SXO) consortium funded by the Basic Technology Programme of Research Councils UK (grant code EP/D04880X/1). The members of the SXO consortium are University College London (including the Mullard Space Science Laboratory), King's College London, the Gray Cancer Institute, the Scottish Microelectronics Centre at the University of Edinburgh, the University of Birmingham, the University of Leicester, and STFC Daresbury Laboratory. Silson Ltd is an associate member. The work described in Sects. 25.3 and 25.4 research was performed under the following financial support: ESP88-0450, ESP93-0324, ESP96-0508, ESP2001-4517-PE, ESP2002-01391, and ESP2005-02650 from the Spanish National Programme for Space Research, and AYA-2001-0826, from the Spanish National Programme for Astronomy and Astrophysics.

References

1. Ch. Morawe, J.-Ch. Peffen, O. Hignette, E. Ziegler, Proc. SPIE **3773**, 90 (1999)
2. R.L. Headrick, K.W. Smolenski, A. Kazimirov, C. Liu, A.T. Macrander, Rev. Sci. Instrum. **73**, 1476 (2002)
3. Ch. Morawe, P. Pecci, J.Ch. Peffen, E. Ziegler, Rev. Sci. Instrum. **70**, 3227 (1999)
4. Ch. Morawe, E. Ziegler, J.Ch. Peffen, I.V. Kozhevnikov, Nucl. Instrum. Methods A **493**, 189 (2002)
5. Ch. Morawe, J.-Ch. Peffen, E.M. Dufresne, Y.S. Chu, A.T. Macrander, Proc. SPIE **5195**, 1 (2003)
6. T. Bigault, E. Ziegler, Ch. Morawe, R. Hustache, J.Y. Massonnat, G. Rostaing, Proc. SPIE **5195**, 12 (2003)
7. F. Mezei, Commun. Phys. **1**, 81 (1976)
8. I.V. Kozhevnikov, I.N. Bukreeva, E. Ziegler, Proc. SPIE **3448**, 322 (1998)
9. A.V. Vinogradov, R.M. Fechtchenko, Proc. SPIE **3773**, 128 (1999)
10. Z. Wang, J. Cao, A.G. Michette, Opt. Commun. **177**, 25 (2000) and references therein
11. L.G. Parrat, Phys. Rev. **95**, 359 (1954)
12. F. Mezei, P.A. Dagleish, Commun. Phys. **2**, 41 (1977)
13. I. Anderson, Proc. SPIE **983**, 84 (1988)
14. B. Vidal, Z. Jiang, F. Samuel, Proc. SPIE **1738**, 30 (1992)

15. M. Schuster, H. Göbel, L. Brügemann, D. Bahr, F. Burgäzy, C. Michaelsen, M. Störmer, P. Ricardo, R. Dietsch, T. Holz, H. Mai, Proc. SPIE **3767**, 183 (1999)
16. Y. Koujun, Nucl. Instrum. Methods A **529**, 59 (2004)
17. Z. Wang, H. Wang, J. Zhu, F. Wang, Z. Gu, L. Chen, A.G. Michette, A.K. Powell, S.J. Pfauntsch, F. Schäfers, J. Appl. Phys. **99**, 056108 (2006)
18. Z.S. Wang, H.C. Wang, J.T. Zhu, F.L. Wang, Z.X. Gu, L.Y. Chen, A.G. Michette, A.K. Powell, S.J. Pfauntsch, F. Schäfers, Opt. Express **14**, 2533 (2006)
19. H. Wang, J. Zhu, Z. Wang, Z. Zhong, S. Zhang, W. Wu, L. Chen, A.G. Michette, A.K. Powell, S.J. Pfauntsch, F. Schäfers, A. Gaup, Thin Solid Films **515**, 2523 (2006)
20. Z. Wang, H. Wang, J. Zhu, Y. Xu, C. Li, F. Wang, Z. Zhang, Y. Wu, X. Cheng, L. Chen, A.G. Michette, S.J. Pfauntsch, A.K. Powell, F. Schäfers, A. Gaupp, M. MacDonald, Appl. Phys. Lett. **89**, 241120 (2006)
21. Z. Wang, H. Wang, J. Zhu, Z. Zhang, S. Zhang, Y. Xu, S. Zang, W. Wu, F. Wang, B. Wang, L. Chen, A.G. Michette, S.J. Pfauntsch, A.K. Powell, F. Schäfers, A. Gaupp, M. MacDonald, Appl. Phys. Lett. **90**, 031901 (2007)
22. Z. Wang, H. Wang, J. Zhu, Z. Zhang, S. Zhang, F. Wang, W. Wu, L. Chen, A.G. Michette, S.J. Pfauntsch, A.K. Powell, F. Schäfers, A. Gaupp, M. Cui, L. Sun, J. Zhu, M. MacDonald, Appl. Phys. Lett. **90**, 081910 (2007)
23. C. Morawe, C. Borel, E. Ziegler, J.-Ch. Peffen, Proc. SPIE **5537**, 115 (2004)
24. J.W.M. DuMond, Phys. Rev. **52**, 872 (1937)
25. J.I. Larruquert, J. Opt. Soc. Am. A **18**, 1406 (2001)
26. J.I. Larruquert, J. Opt. Soc. Am. A **18**, 2617 (2001)
27. E. Spiller, *Soft X-Ray Optics* (SPIE, Bellingham, 1994), p. 143
28. M. Gruntman, H.J. Fahr, J. Geophys. Res. **105**, 5189 (2000)
29. J.I. Larruquert, R.A.M. Keski-Kuha, Appl. Opt. **38**, 1231 (1999)
30. J.I. Larruquert, R.A.M. Keski-Kuha, Appl. Opt. **41**, 5398 (2002)
31. T. Graf, A.G. Michette, A.K. Powell, J. Phys. IV France **104**, 243 (2003)
32. T. Graf, A.G. Michette, A.K. Powell, Proc. SPIE **5193**, 134 (2003)
33. T. Graf, Characterisation of Laser Plasma XUV Sources for Spectromicroscopy, Ph.D. Thesis, London University, 2004
34. J.I. Larruquert, J. Opt. Soc. Am. A **19**, 391 (2002)
35. M. Grigonis, É.J. Knystautas, Appl. Opt. **36**, 2839 (1997)
36. R.A.M. Keski-Kuha, Appl. Opt. **23**, 3534 (1984)
37. J.B. Kortright, Nucl. Instrum. Methods Phys. Res. A **246**, 344 (1986)
38. http://henke.lbl.gov/optical_constants/ (last accessed 7 May 2007)
39. J.I. Larruquert, Opt. Commun. **206**, 259 (2002)
40. C.K. Carniglia, J.H. Apfel, J. Opt. Soc. Am. **70**, 523 (1980)
41. J.I. Larruquert, J. Opt. Soc. Am. A **19**, 385 (2002)
42. M. Yamamoto, T. Namioka, Appl. Opt. **31**, 1622 (1992)
43. J.I. Larruquert, J. Opt. Soc. Am. A **22**, 1607 (2005)
44. J.I. Larruquert, J. Opt. Soc. Am. A **21**, 1750 (2004)
45. J.I. Larruquert, J. Opt. Soc. Am. A **23**, 99 (2006)
46. J. Gautier, F. Delmotte, M. Roulliay, F. Bridou, M.F. Ravet, A. Jéreome, Appl. Opt. **44**, 384 (2005)
47. J.I. Larruquert, J. Opt. Soc. Am. A **23**, 1967 (2006)
48. M. Hecker, S. Valencia, P.M. Oppeneer, H.-Ch. Mertins, C.M. Schneider, Phys. Rev. B **72**, 054437 (2005)

Part V

Diffraction Optics

26

Diffractive-Refractive Optics: X-ray Crystal Monochromators with Profiled Diffracting Surfaces

J. Hrdý and J. Hrdá

Abstract. X-ray refractive lenses are now well known and have been used successfully for focusing synchrotron radiation. However, refraction also occurs during Bragg and Laue diffraction from perfect crystals. This phenomenon is studied in detail in this chapter, and a possible application for X-ray monochromators with profiled diffracting surfaces is described. It is shown that by machining the diffracting crystal surface into a suitable shape it is possible to create a sagittally or meridionally focusing monochromator, as well as monochromators for filtering higher harmonics or the p polarization component over a broad range of wavelengths.

26.1 Introduction

Perfect single crystals are important hard X-ray optical elements – X-ray monochromators. These elements are based on Bragg or Laue diffraction. If a polychromatic parallel beam impinges on a single crystal under a so-called Bragg angle θ_B (Fig. 26.1), then the diffracted beam has a wavelength, λ, given by the well-known Bragg law (see, e.g., [1])

$$2d \sin \theta_B = k\lambda. \tag{26.1}$$

Here d is the relevant lattice plane spacing and k is an integer ($k = 1, 2, \ldots$).

It means that besides the fundamental harmonic ($k = 1$), higher harmonics ($k = 2, 3, \ldots$) may be diffracted under the same angle θ. The diffraction of X-rays on perfect crystals is described by the dynamical theory of diffraction (see, e.g., [2] or [3]). From this theory it follows that a monochromatic beam is diffracted in a finite angular region, ω, and the center of this angular region is larger than the Bragg angle by $\Delta\theta$ as calculated from (26.1). This means that the Bragg law (26.1) does not hold exactly. The values ω and $\Delta\theta$ are of the order of arc seconds. This small deviation from the Bragg law is a manifestation of refraction. For the *symmetrical* Bragg diffraction (the surface of crystal is parallel to diffracting crystallographic planes) the angle between the incident beam and the diffracting planes, θ_0, and the angle between the diffracted beam and diffracting planes, θ_h, are equal and the crystal behaves

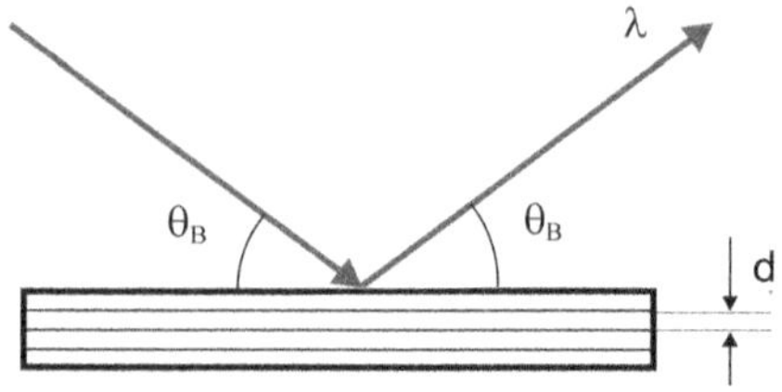

Fig. 26.1. Symmetric Bragg diffraction on a crystal with a lattice plane spacing d

like a mirror. Thus, from the X-ray optics point of view the refraction is not too interesting.

The situation is different if the crystal surface is not parallel with the diffracting crystallographic planes. In this case the crystal does not behave exactly as a mirror any more. The incident and diffracted beams are not symmetrical with respect to the surface normal (which is trivial) but they are also not symmetrical with respect to the normal to the diffracting crystallographic planes. There are two limiting cases. The *asymmetric diffraction* corresponds to the situation when the surface normal lies in the plane of diffraction, i.e., the plane determined by the incident beam and the normal to the diffracting planes. In this case we will call the crystal *asymmetric crystal.* The diffracted beam lies in the plane of diffraction; thus the diffraction is still *coplanar.* The *inclined diffraction* occurs when the plane determined by the surface normal and the normal to the diffracting planes is perpendicular to the plane of diffraction. Here we will call the crystal *inclined crystal.* As will be shown later, the diffraction is noncoplanar. The *general asymmetric diffraction* is the combination of the asymmetric and the inclined diffraction.

26.1.1 Asymmetric Diffraction

For the asymmetric diffraction the values ω_0 and $\Delta\theta_0$ for an incident beam and the values ω_h and $\Delta\theta_h$ for a diffracted beam are not identical and are different from the values ω_s and $\Delta\theta_s$ for the symmetric diffraction. The following set of relations holds [4]:

$$\begin{aligned}
\omega_0 &= \omega_s b^{-1/2}, \\
\omega_s &= (2r_e\lambda^2 P|F_{hr}|e^{-M})/\pi V \sin 2\theta_B, \\
\Delta\theta_0 &= (1/2)(1+1/b)\Delta\theta_s, \\
\Delta\theta_s &= r_e\lambda^2 F_{0r}/\pi V \sin 2\theta_B, \\
\omega_h &= \omega_s(b)^{1/2}, \\
\Delta\theta_h &= (1/2)(1+b)\Delta\theta_s, \\
\theta_0 &= \theta_B + \Delta\theta_0, \\
\theta_h &= \theta_B + \Delta\theta_h, \\
b &= \sin(\theta_B - \alpha)/\sin(\theta_B + \alpha).
\end{aligned} \tag{26.2}$$

Here V is the unit-cell volume, $r_e = e^2/mc^2$ (classical electron radius), F_{hr} is the real part of the structure factor F_h (h stands here for Müller indices hkl), P is the polarization factor, and e^{-M} is the temperature factor. The index s stands for the symmetrical diffraction. The angle α is the angle between the diffracting planes and the surface and is taken as positive for grazing incidence. The asymmetry index, b, is defined according to Matsushita and Hashizume [4]. The typical values of $\Delta\theta_s$ and ω_s are from fractions to tens of angular seconds. The angle θ_B is the Bragg angle calculated from the Bragg law (1). For the cross sections $\boldsymbol{CS}_\mathbf{0}$ And $\boldsymbol{CS}_\mathbf{h}$ of the incident and the diffracted beams the following holds:

$$\boldsymbol{CS}_\mathbf{h} = \boldsymbol{CS}_\mathbf{0}/b \tag{26.3}$$

and together with (26.2),

$$\omega_h \boldsymbol{CS}_h = \omega_0 \boldsymbol{CS}_0. \tag{26.4}$$

The consequence of the above relations may be demonstrated with the help of the DuMond graph (Fig. 26.2a). The real situation is shown in Fig. 26.2b. Let us suppose that a parallel and polychromatic beam is impinging on an asymmetrically cut (= asymmetric) crystal with some angle, θ, between the beam and the lattice planes. The deviation from a mirror-like behavior may be described by the quantities δ and $\Delta\delta$. It obviously holds (for one harmonic):

$$\begin{aligned} \delta(\alpha) &= \Delta\theta_0 - \Delta\theta_h, \\ \Delta\delta(\alpha) &= |\omega_0 - \omega_h|. \end{aligned} \tag{26.5}$$

The deviation δ and the spread $\Delta\delta$, which may be changed by changing α resembles the refraction of light on a prism. The only difference is that the

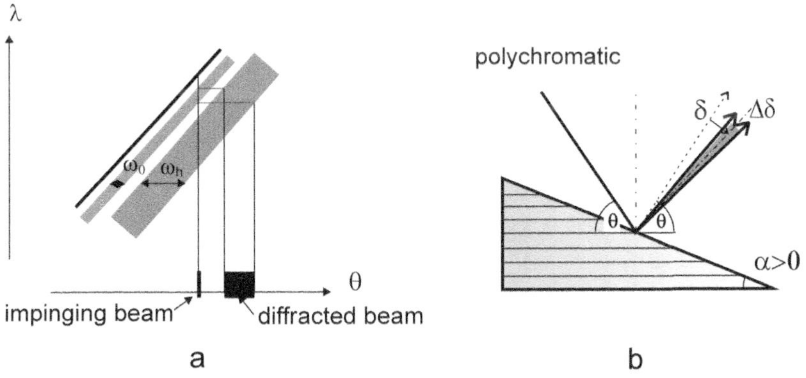

Fig. 26.2. (**a**) DuMond diagram of an asymmetric Bragg diffraction showing that as compared with a symmetric diffraction the diffracted beam is deviated and spread. Figure 26.2b shows the asymmetric Bragg diffraction of a polychromatic pencil beam in real space

prism accepts a broad range of wavelengths, whereas the crystal accepts only narrow wavelength interval $\omega_0(\mathrm{d}\lambda/\mathrm{d}\theta)$ and thus the interval $\Delta\delta$ is narrow. Nevertheless, it exhibits a wavelength dispersion, as in the case of a prism. One may deduce that this refraction effect (δ, $\Delta\delta$) may play an important role when the surface of crystal is curved (not bent), as in the case of refraction lenses in classical optics. The asymmetric diffraction with flat crystals is used to either compress or to extend the diffracted beam. This will be treated elsewhere in this book. (see the Chap. 29)

26.1.2 Inclined Diffraction

The inclined Bragg diffraction is a noncoplanar diffraction. The behavior of the diffracted beam may be seen in Fig. 26.3a, which shows the wave vectors and the dispersion surfaces in reciprocal space [5,6]. The points $P_{0,1}$ and $P_{0,2}$ are the origins of the impinging vectors, directed into the origin, O, of the reciprocal space, points $P_{h,1}$ and $P_{h,2}$ are the origins of diffracted vectors for symmetric Bragg diffraction ($\beta = 0$). The points $P_{h,1,\beta}$ and $P_{h,2,\beta}$ are the origins of diffracted vectors for the inclined diffraction. The indices 1 and 2 represent the limiting beams within the diffraction region, ω. It is seen that if the impinging monochromatic and parallel beam is scanned through the diffraction region, $\omega_0(\omega_0 = \omega_s)$, then the diffracted beam is deviated from the plane of diffraction, and this deviation grows during the scan. The consequence of this is demonstrated in Fig. 26.3b. It shows that if a parallel and polychromatic beam impinges on an inclined crystal with an inclination angle β, then the diffracted beam is deviated from the plane of diffraction in a sagittal direction (perpendicular to the plane of diffraction) and the beam

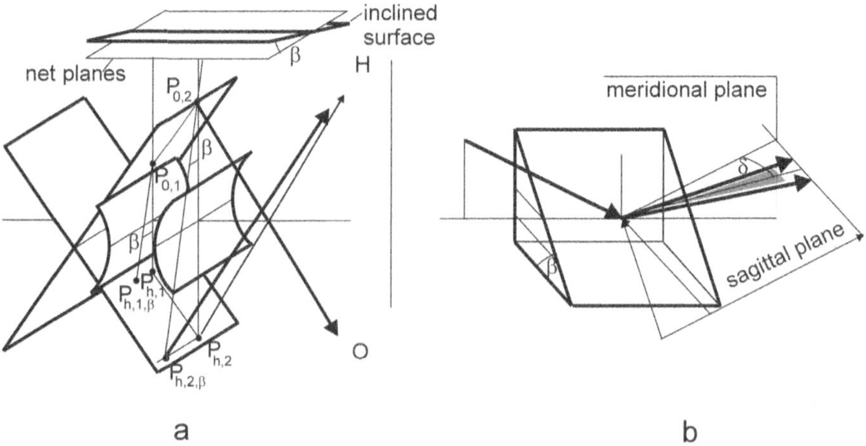

Fig. 26.3. Wave vectors in reciprocal space for an inclined diffraction (**a**). Inclined diffraction in a real space (**b**)

is sagittally spread. The deviation, δ, of the central beam from the plane of diffraction is

$$\delta = K \tan \beta \tag{26.6}$$

where

$$K = (2r_e F_0 / \pi V) d_{hkl} \lambda. \tag{26.7}$$

For Si crystals $K = 1.256 \times 10^{-3} d_{hkl}$ (nm) λ (nm). The inclined crystal monochromators based on the inclined diffraction are used to decrease the impinging radiation power density of synchrotron radiation. Here the deviation and the spread of the diffracted beam is the manifestation of refraction. As in the asymmetric diffraction, here the beam spread is also limited by the wavelength acceptance of a crystal for a given incidence angle. From the above it is clear that, as in the asymmetric case, interesting applications may be expected if the diffracting surface is machined into a suitable shape.

The refraction effect exists also in general asymmetric diffraction (the combination of asymmetric and inclined diffraction) and in Laue diffraction. These will be discussed later.

X-ray refractive lenses are now commonly known and are successfully used for focusing synchrotron radiation [7]. The aim of our work was to study the diffraction on crystals with curved diffracting surface and to investigate the possible applications of effects based on refraction described earlier.

The idea of a crystal monochromator with a curved (not bent) diffracting surface is not new. The Johansson spectrometer [8, 9] is a Bragg crystal with a circular profile machined into the working surface of the crystal. The crystal is then bent. Such a crystal focuses the monochromatic radiation on the Rowland circle. Spieker [10], designed a channel-cut crystal monochromator with profiled working surfaces such that the position of the exit beam remains fixed when tuning the wavelength. These two methods are based only on geometry; the refraction effect is completely neglected.

26.2 Bragg Diffraction on a Transverse Groove (Meridional Focusing)

From what was explained in section 26.1.1 and from Fig. 26.2b it may be deduced that the radiation diffracted on properly designed transverse groove may be meridionally focused. This is demonstrated in Fig. 26.4a. The diffracted beam 1 is deviated to the right due to the asymmetric diffraction and it is spread. The beam 3 is also spread and is deviated in the opposite direction, i.e., to the left. The beam 2, which is diffracted from the bottom of the groove where the diffraction is symmetrical, is neither deviated nor spread. The problem is to find the function $g(x)$ describing the shape of the groove, such that the centers of all diffracted beams (i.e., centers of the fans) will be concentrated into one point, the focus. Substituting (26.2) into (26.5) we obtain

$$\delta = 2\Delta\theta_s \tan\theta_B \tan\alpha / (\tan^2\theta_B - \tan^2\alpha). \tag{26.8}$$

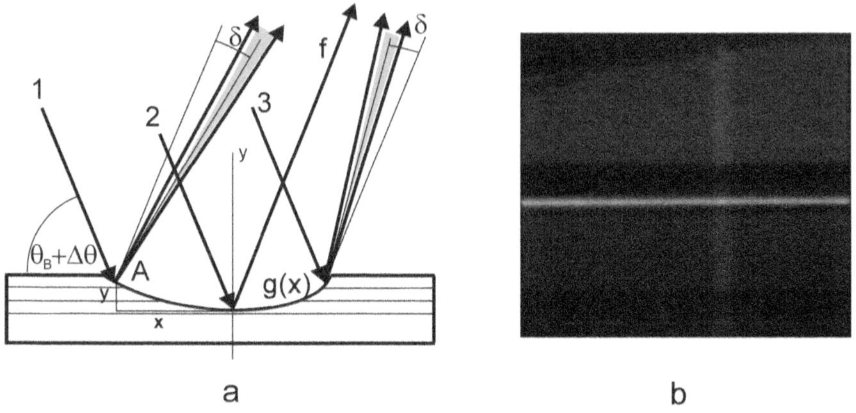

Fig. 26.4. Bragg diffraction on a transverse groove machined into a symmetric crystal (**a**). The diffracted beam is convergent. Figure 26.4b shows image of an X-ray beam diffracted on a crystal with a transverse groove

Let the profile of the groove be described by a function $y = g(x)$ (see Fig. 26.4a). Let us suppose that the impinging radiation is parallel. In order that the beam impinging on the surface of the groove at a certain point, $A(x, y)$, be diffracted to the focus, the deviation, δ, must be [11]

$$\delta = [-x \sin(\theta_B + \Delta\theta_0) + y \cos(\theta_B + \Delta\theta_0)]/f, \tag{26.9}$$

where f is the focal distance.

Taking into account that $\tan\alpha = -g'(x)$ ($g' = \mathrm{d}g/\mathrm{d}x$) and neglecting $\Delta\theta_0$ in (26.9), then (26.8) and (26.9) gives the differential equation [11]

$$[x \sin\theta_B - g(x)\cos\theta_B]/f = 2\Delta\theta_s \tan\theta_B g'(x)/\{\tan^2\theta_B - [g'(x)]^2\}, \tag{26.10}$$

which describes approximately the shape of the transverse groove. In [12] this equation was further modified to include the finite divergence of the impinging radiation.

The shape of the groove obviously depends on the wavelength, λ, the focusing distance, f, and the source–crystal distance, S. In order that such a focusing monochromator could be used for a broad wavelength region, it is necessary to produce either several parallel grooves for various λs or only one groove whose shape changes along the groove axis. The focusing conditions could be then adjusted by a translation of the crystal.

We have demonstrated this kind of focusing by an experiment performed in ESRF at the BM5 beamline [12]. The transverse groove, machined into a Si(111) crystal, was calculated for $\lambda = 0.15\,\mathrm{nm}$, $S = 40\,\mathrm{m}$, and $f = 2\,\mathrm{m}$. Figure 26.4b shows the image of the diffracted radiation at the distance of 2 m from the crystal. The figure shows, in the upper and lower parts, the image of the radiation diffracted from the flat part of the crystal and between them there is the image of the radiation which is diffracted on the groove and is concentrated into a narrow bright line. The width of the groove was about 2.5 mm.

Because of the spread, $\Delta\delta$, of the diffracted beam the focus cannot be sharp. Even without any refraction effect the diffracted radiation would be concentrated at the right side of the groove because of the asymmetric diffraction. It makes sense to compare the peak intensity in the focal plane after diffraction from the groove and from a flat asymmetric crystal with the asymmetry corresponding to the right side of the groove. The ray-tracing simulation of the experiment showed that the groove would give about 3.3 times higher intensity in the peak at the focal plane than an asymmetric concentrator.

26.3 Harmonics Free Channel-Cut Crystal Monochromator with Profiled Surface

Another application of diffraction on the meridionally profiled (curved) surface may be a channel-cut crystal monochromator which suppresses higher harmonics in the broad region of the Bragg angles. Let us suppose that the first diffracting surface (the first wall of the channel) is flat and symmetrically cut (its surface is parallel with diffracting crystallographic planes). If the second wall is also flat but asymmetrically cut, then $\Delta\theta_h$ and ω_h for the first wall ($\Delta\theta_h = \Delta\theta_s$ and $\omega_h = \omega_s$) is not equal to $\Delta\theta_0$ and ω_0 for the second wall, and only a part (or none) of the radiation diffracted from the first wall is diffracted from the second wall: i.e., diffraction is detuned. This depends on the degree of overlap of the corresponding Darwin–Prins (DP) curves. The values $\Delta\omega$ and ω decrease with the order of diffraction k. This means that for a certain asymmetry of the second wall, the DP curves for higher harmonics do not overlap any more but the overlapping of DP curves for the fundamental harmonic is still sufficient. The radiation diffracted from the channel-cut crystal is then practically free of higher harmonics. This way of obtaining harmonics rejection, which is valid for one λ and its close neighborhood was suggested by Matsushita and Hashizume [4]. The mathematical description of this situation is following:

$$(P|F_{hr}^{(k)}|\mathrm{e}^{-M(k)})/|F_{0\mathrm{r}}| = (1/2)|1/b^{1/2} - 1| \qquad (26.11)$$

where b corresponds to the second surface and k (>1) stands for the order of diffraction.

As the left part of (26.11) is independent of θ, b must also be independent of θ. This implies that α must change with θ, which means that the second wall must be curved in order that the channel-cut crystal monochromator rejects higher harmonics in the whole region of θ or λ [13].

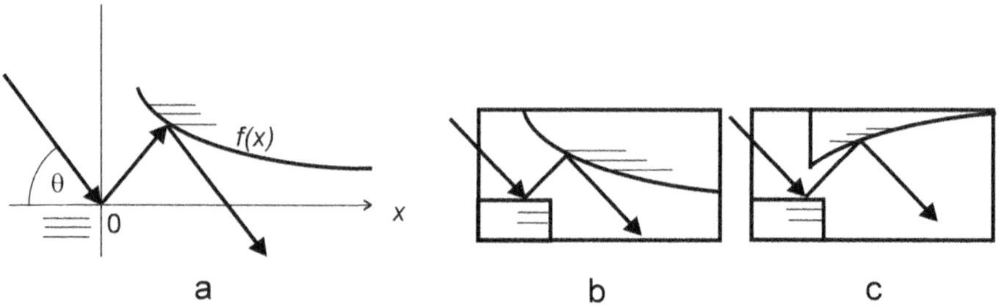

Fig. 26.5. Harmonics-free channel-cut crystal monochromator: the derivation of the shape of the second diffracting surface (**a**). The second diffracting surface may be convex (**b**) or concave (**c**)

The equation for b (see (26.2)) may by rewritten as follows:

$$\tan\alpha = [(1-b)/(1+b)]\tan\theta = B\tan\theta. \tag{26.12}$$

Let us introduce the axes of the coordinates with the origin on the first wall, such that the X-ray beam is impinging at the origin on the first wall (Fig. 26.5a). The axis of rotation of the monochromator also passes through the origin. Let the profile of the second wall be described by the function $f(x)$. Then (26.12) may be rewritten in the following form:

$$\mathrm{d}f(x)/\mathrm{d}x = Bf(x)/x. \tag{26.13}$$

The angle α is here taken as negative. Thus $b > 1$ and $B < 0$. The solution of this differential equation is

$$f(x) = Cx^B \tag{26.14}$$

where B is negative and the second wall of the crystal is convex.

It is obvious, that by a similar consideration as above, the second wall of the channel-cut crystal may be cut so that the angle, α, is positive [14]. In this case the DP acceptance curve for the diffraction on the second wall is shifted toward higher angles, θ. The condition when the DP curves just touch each other is the same as in the previous case, however, $|1/b^{1/2} - 1| = 1/b^{1/2} - 1$, because $b < 1$. This leads to (26.14) where B is positive and the second wall is concave. Both kinds of channel-cut crystal monochromators are schematically shown in Fig. 26.5b, c.

It is obvious that the monochromator with a convex wall concentrates the diffracted beam but slightly increases its divergence, as follows from paragraph 26.2 (or section 26.2 or 26.2). The monochromator with concave beam creates a broad beam and slightly decreases its divergence or may even create a slightly convergent beam if the impinging beam is almost parallel, as it is in the case of synchrotron radiation. This has been discussed in detail in [14]. Obviously, the harmonics rejection here is the consequence of the dependence of refraction on α and the order of diffraction. The harmonics-free channel-cut crystal monochromator discussed above has not yet been tested experimentally.

The width of Darwin–Prins function for the π polarization is $\cos 2\theta$ times smaller than for the σ polarization component. As was shown by Hart and Rodrigues [15], a double crystal monochromator in a nondispersive $(+,-)$ setting which is detuned may reject the π polarization component similarly as it rejects higher harmonics [16, 17]. Only the degree of detuning is different. It is obvious that there should exist a channel-cut crystal monochromator with a suitably curved diffracting surface such that it rejects π polarization components for a broad region of θ. This will be treated in detail elsewhere [18].

26.4 Bragg Diffraction on a Longitudinal Groove (Sagittal Focusing)

In section 26.1.2 it was shown that in the case of an inclined diffraction the diffracted beam is deviated sagittally (perpendicularly to the plane of diffraction). Let us suppose that a longitudinal groove is produced in the diffracting surface of a crystal, as shown in Fig. 26.6a. The opposite walls of the groove deviate the beam in opposite directions. It is clear, that a properly designed shape of the groove may sagittally concentrate the diffracted beam at certain distance, f, from the crystal. The geometry of the diffraction is shown in Fig. 26.6b. For the determination of the shape of the groove we will suppose

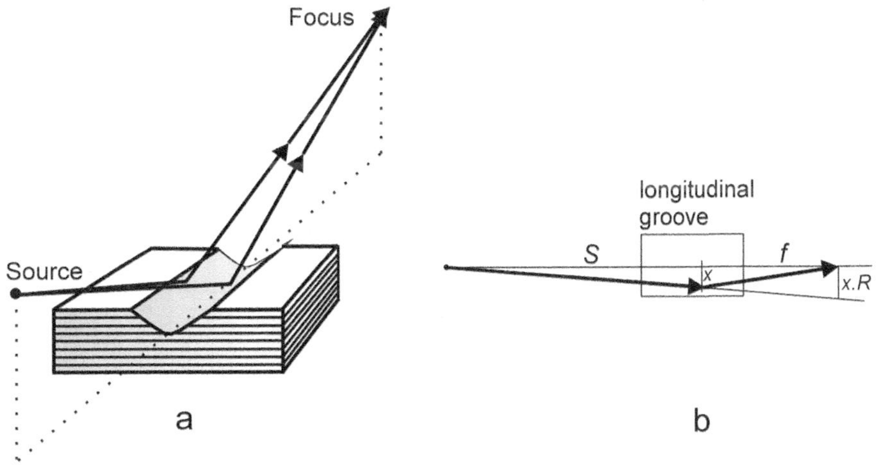

Fig. 26.6. Bragg diffraction on a crystal with a longitudinal groove (**a**). The diffracted beam is convergent. Geometry of the sagittal focusing due to the longitudinal groove (*top view*) (**b**)

that the distance of the grooved crystal monochromator from a point source is S and the focal length is f. Let the shape of the groove be described by a function $y(x)$. For the groove to act as a lens, it is necessary that the beam impinging on the crystal (groove) at a distance x from the longitudinal axis of the groove, be deviated by an angle

$$\delta \cong \tan\delta \cong [x(S+f)/S]/f = xR/f. \tag{26.15}$$

Equation (26.6) may be rewritten in the following way:

$$\tan\delta = K(\mathrm{d}y/\mathrm{d}x). \tag{26.16}$$

Equations (26.15) and (26.16) give a differential equation with the solution [6]

$$y = (R/2Kf)x^2 + \text{constant}. \tag{26.17}$$

The meaning of the above result is that the longitudinal parabolic groove focuses the radiation and thus acts as a sagittally focusing lens. For synchrotron radiation two crystals in a parallel, nondispersive (+, −) orientation are commonly used. The parabolic longitudinal groove may be then produced in both crystals. If only one crystal or more crystals in a nondispersive position are used, then the advantage of the sagittal focusing is deteriorated by two effects. The first one is shown in Fig. 26.7a. The vertical size of the beam increases after each diffraction. This depends on the depth of the groove. The second effect is the sagittal spread of the deviated beam which prevents the focus from being sharp.

Both effects mentioned above (aberrations) may be canceled by using a dispersive arrangement of crystals. From Fig. 26.7b it is clearly seen that the vertical broadening which appears after diffraction from first two crystals is completely canceled after diffraction on the following two crystals. The dispersion arrangement also cancels the sagittal spread seen in Fig. 26.3b.

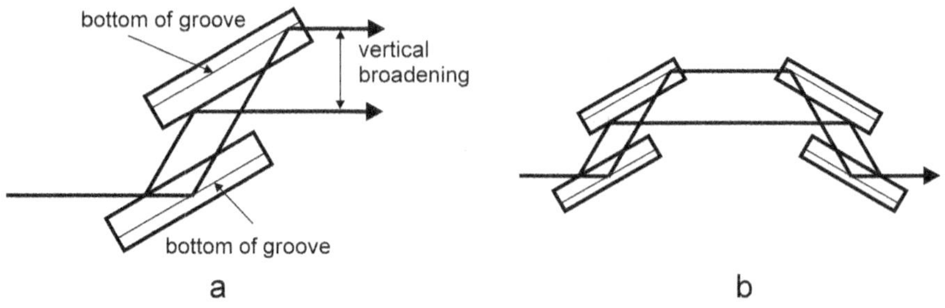

Fig. 26.7. A longitudinal groove broadens the diffracted beam vertically (**a**). Dispersive four crystal arrangement with longitudinal grooves cancels the vertical broadening of a diffracted beam originating from the first two crystals (**b**)

The nature of the sagittal spread is shown in Fig. 26.3a. When the impinging (monochromatic) beam spans the diffraction region, ω_0, from smaller to higher θ, then the sagittal deviation grows. For example the beam corresponding to a smaller θ at the beginning of the diffraction region leaves the crystal with minimal sagittal deviation. Let us suppose that there is another crystal adjusted in dispersion position with respect to the first crystal. This beam impinges on the second crystal at the end of the diffraction region corresponding to a higher θ and the sagittal deviation is maximal. The resulting deviation after diffraction on both crystals is 2δ for any beam impinging on the crystals within the region ω_0. The angle δ is the average deviation as shown in Fig. 26.3b. This holds for any θ within the region ω. This very important result shows that the $(-,+,+,-)$ arrangement, shown in Fig. 26.7b, is ideal [19]. The second and the third crystals cancel the aberrations discussed above and the first and fourth crystals keep the direction of impinging and exit beams the same, which is important for synchrotron radiation. The position of the exit beam remains independent of θ. Moreover, the dispersion arrangement is the high resolution one. This arrangement should provide practically point-to-point focusing, which means that we may expect a sharp focus. The practical expressions important for the design of the parabolic groove are following

$$y(\text{mm}) = a(\text{mm}^{-1})(x(\text{mm}))^2, \tag{26.18}$$

$$a = (S + f)/2NKfS, \tag{26.19}$$

where f (mm) is the focusing distance, S (mm) is the monochromator-source distance and N is the number of diffraction events on the grooves. For the four crystal arrangement shown in Fig. 26.7b, $N = 4$ provided that the beam is diffracted only once on each crystal. The focusing distance f may be determined from

$$f = S/(2aNKS - 1). \tag{26.20}$$

The parabolic groove may also be cut into an asymmetrically cut crystal. This is treated in detail by Hrdý [20]. It was shown there that for this case all the above formulae may be used. Only K must be replaced by

$$K' = K\ [(2 + b + 1/b)/4\cos\alpha]. \tag{26.21}$$

The difference between the function of the symmetrically and asymmetrically cut grooved crystals may be seen in Fig. 26.8. It shows a dependence of the focusing distance, f, on the Bragg angle θ for a four-crystal $(+,-,-,+)$ monochromator with the same crystals and grooves. Crystals cut symmetrically is compared with asymmetric crystals with $\alpha = 12.38°$. The monochromator with the asymmetric crystals gives a shorter focusing distance. In the angular region around $\theta = 22°$ the focusing distance is almost constant and for the highly asymmetric case, close to $\theta = \alpha$, the focusing distance is very small. The above expressions enable one to be able to design the crystals, the grooves and the asymmetry angle to meet the experimental requirements.

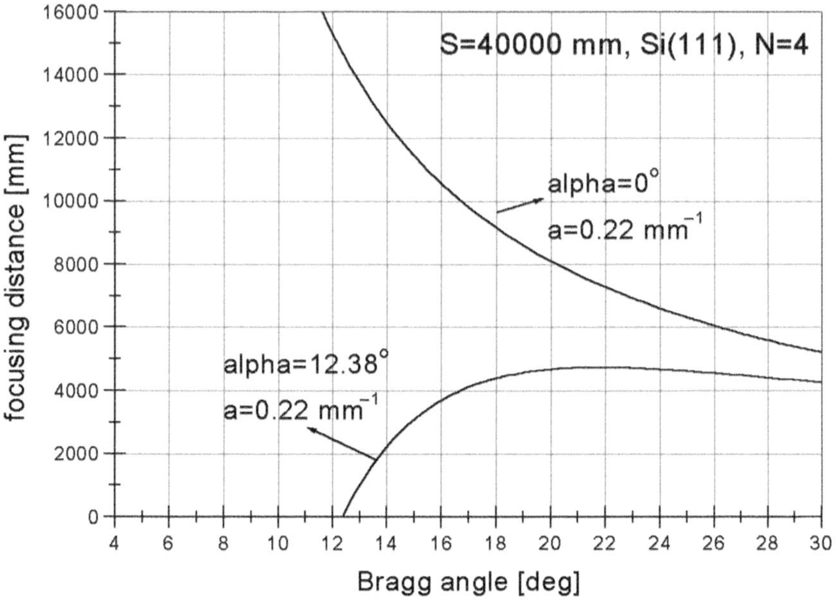

Fig. 26.8. The dependence of the sagittal focusing distance, f, on a Bragg angle, θ, for a symmetric ($\alpha = 0$) and an asymmetric Bragg diffraction

As was shown earlier, the ideal shape of the longitudinal groove is a parabolic groove. Our first experiment [19], which successfully demonstrated the sagittal focusing, used two channel-cut crystals with the parabolic grooves cut into each diffracting surface (Fig. 26.9a). To produce the grooves a precise diamond tool with a parabolic profile was ordered. As the tool was expensive, we later chose another approach [21]. We drilled a circular hole into a Si single crystal either parallel to diffracting planes (to simulate the symmetrical diffraction) or under some angle, α, to simulate an asymmetric diffraction. The X-ray beam is then diffracted twice inside the hole. The diameter D of the hole is given by

$$D = 1/a, \tag{26.22}$$

where a is the parameter of the parabola (26.18). The circular hole is then a good approximation of the parabolic groove if the sagittal beam size is small. One such crystal with three holes of different diameters, for three different λs, drilled in a Si crystal and used for an experiment at APS is shown in Fig. 26.9b. The more sophisticated design of the asymmetric channel-cut crystal with circular grooves is shown in Fig. 26.9c. This crystal monochromator was designed such that it accepts a relatively large sagittal extent of a synchrotron radiation beam (BM5 beamline at ESRF) and creates a focus at the distance of 20 m. Finally we produced asymmetric crystals with parabolic holes (Fig. 26.9d) but the experiment has not been done so far. The crystals with a parabolic hole seem to be the ideal solution because it is compact and

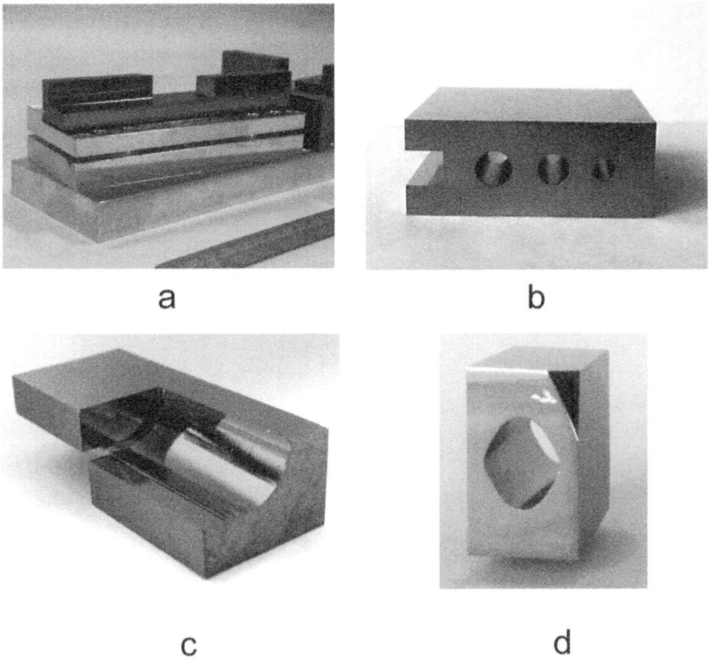

Fig. 26.9. Various realizations of sagittally focusing monochromator crystals. Two such crystals in a dispersive arrangement were used. A channel-cut crystal with parabolic grooves is shown in Fig. 26.9a. Figure 26.9b shows a symmetric crystal with three cylindrical holes for three different wavelengths. Diffraction occurs twice inside the hole. The more sophisticated asymmetric crystal with cylindrical holes of large diameter, D, is shown in Fig. 26.9c. An asymmetric crystal with a parabolic hole is shown in Fig. 26.9d

practically aberration free even though the crystal production is rather laborious. A disadvantage of the crystal with holes is that they cannot be detuned to reject harmonics. In fact, however, this is not a problem. Higher harmonics are practically not focused because of low refraction. It means that in the focus the fundamental harmonic is concentrated whereas higher harmonics are located mostly out of the focus.

To simulate the focusing described above, a ray-tracing program had to be developed which included the refraction effect during Bragg diffraction [22]. This program, although being based on some simplifications like rectangular shape of DP curves, proved to be very useful when discussing the influence of misalignments and shape imperfections of crystals. Later on, another program based on precise dynamical theory was developed by another group [23] and was used to study the aberration of the $(-,+,+,-)$ arrangement with four longitudinal parabolic grooves [24]. This showed that the system is practically aberration free. In spite of this, the size of the focus in our experiments was always somewhat larger than the theoretical one. This is due to the quality of the groove. First, the circular shape introduces some aberration. Second, it is difficult to produce an exact shape into the Si crystal, because after machining the surface must be etched and mechano-chemically polished. This is done by hand, and to keep the exact groove profile with high precision is very difficult.

The focusing distance in the $(-,+,+,-)$ arrangement depends on the parameter a of the parabola and the wavelength, λ. For example in the first experiment at NSLS [19] the width of the groove in the Si channel-cut crystals was 2 mm, the depth was slightly more than 1 mm and the energy $E = 15$ keV. The focusing distance was 4.5 m and the beam, whose size at the focusing distance would be 2.7 mm without focusing, was squeezed to a spot of 0.29 mm despite the relatively bad quality of the groove surface. The experiment at the 5ID beamline at APS [25] with a hole diameters of 7.2 mm for 8.048 keV and 4.4 mm for 13 keV, the focusing distance of 20 m, the crystals–source distance of 55 m, and Si(111) symmetrical diffraction gave the focus size of 417 μm, which is 110 μm larger than the size of the demagnified image of the source. Finally, in the experiment performed at the BM5 beamline at ESRF [26], we used an asymmetric Si(111) channel-cut crystal of a very special design with hole diameters of 22 mm such that the size of the crystals could be reasonably small (Fig. 26.9c). The source-to-crystals distance was 40 m, the focusing distance was about 20 m, and the energy about 8 keV. The crystals accepted 6 mm of the horizontal size of the beam and created the focus of the size of 0.4 mm. Figure 26.10 shows the focused beam. The focal spot is clearly seen against the broad background of higher harmonics which are practically not focused because of low refraction. The width of the unfocused radiation at the focusing distance was 8.8 mm. The size of the image of the demagnified source at the focal distance was 0.12 mm. In the last two experiments mentioned, the diffracting surfaces were mechano-chemically polished.

It is seen that for the long focusing distances the size of the demagnified image of the source is also large. There are three ways by which one can decrease the focusing distances. The first one is to use strongly asymmetric diffraction as is shown in Fig. 26.8. This is possible only for a narrow wavelength region. The second way is to use a multiple diffraction arrangement,

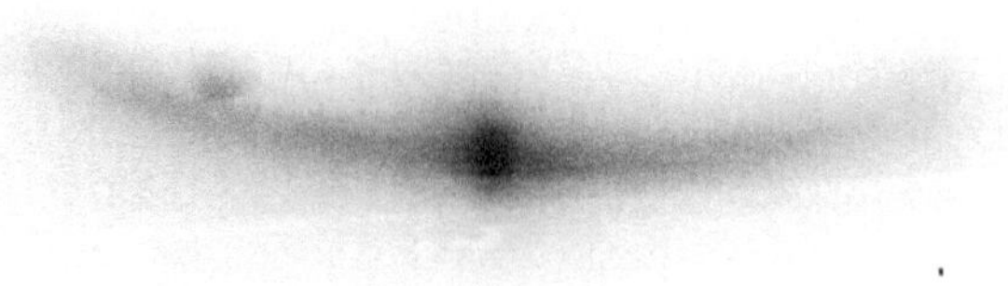

Fig. 26.10. Sagittally focused first harmonics on the background of higher harmonics which are practically not focused. The image was taken at 20 m from the crystal shown in Fig. 26.9c. The width of the image of the higher harmonics is about 9 mm

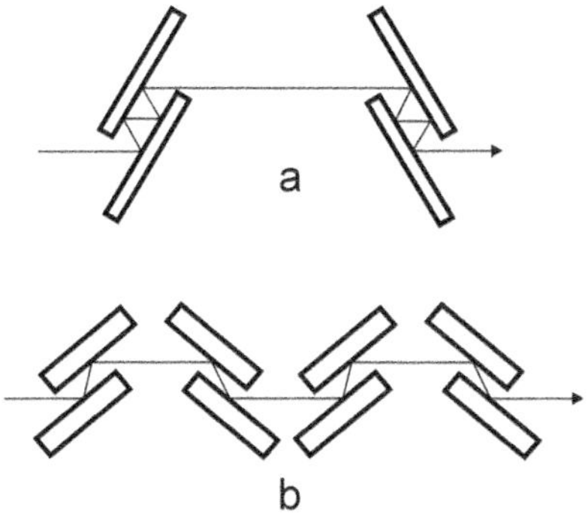

Fig. 26.11. Multiple crystal arrangements to decrease the focusing distance

Fig. 26.12. "Toothed" profile of a longitudinal groove to reduce the size of a crystal and the vertical broadening of a diffracted beam

for example $(-,+,-,+,+,-,+,-)$ (Fig. 26.11a). When using channel-cut crystals with four reflections on each crystal, because the vertical size of the beam grows after each reflection, the required size of the crystal might be larger than is technically feasible. To avoid this, several (even number) of channel cut crystals may be arranged in a dispersive position to create the arrangement $(-,+,+,-,-,+,+,-,\ldots)$ (Fig. 26.11b). The alignment of such a crystal arrangement would obviously be complicated. The third possibility is to use a very narrow groove (large a) which leads to a very low angular acceptance. From the above it is seen that the microfocusing by this method is principally possible but with the present technology of growing Si crystals and preparing their perfect curved surfaces it is difficult. By this method, however, it is very easy to concentrate the diffracted radiation sagittally at long focusing distances, because a large change of β results in a small change of the sagittal deviation, δ. It means that the precision of the parabolic groove (or the hole) does not have to be extremely high. As compared with the classical method, i.e., two crystals with the second crystal sagittally bent [27], the diffractive–refractive optics is compact, first two reflecting surfaces may be easily cooled as a whole when using a channel-cut crystals or crystals with holes. However, the tunability range is smaller and the acceptance is also smaller with presently available Si crystal sizes. On the other hand, the focus should be sharper because of the negligible aberration. Theoretically, the horizontal (sagittal) acceptance may be large but if the parabola describing the shape of the groove is narrow (large a) then the height of the crystal needed may be too large. This problem may be solved by a more complicated profile of the groove, as shown in Fig. 26.12. The tunability range may be increased by producing several holes or grooves with different sizes and different asymmetry angles in each crystal. Switching between different wavelength regions may be accompanied by switching between different grooves or holes. Another possibility may be to produce grooves or holes with variable parameters along their axes (e.g., a conical groove or hole). The change of wavelength may then be accomplished simply by shifting the crystals along the axis of the groove.

So far focusing only in one direction has been described. As suggested in [11] it should in principle be possible in a $(-, +, +, -)$ arrangement to produce longitudinal grooves, for example, in the second and the third crystals and transverse grooves in the fourth crystal. This should concentrate the beam in both directions. Another possibility may be to produce a properly designed depression in the surface of one crystal. Neither of the methods has been tested experimentally.

26.5 Laue Diffraction on a Profiled Surface (Sagittal Focusing)

A logical continuation of the work described above is the study of the possible application of the refraction effect occurring during Laue diffraction. Recently, the sagittal deviation of a beam diffracted from a sagittally inclined surface was studied for Laue asymmetric diffraction [28,29]. In this work the sagittal deviation of the beam diffracted from a flat asymmetric Laue crystal with a sagittally inclined wedge was observed experimentally. The simple theory presented in [28] gives the formula for the sagittal deviation, δ, of the beam (Fig. 26.13):

$$\delta = (|\mathrm{PN}|\lambda)\tan\beta = (|\mathrm{LP}|\ [\cos\theta/\cos(\theta+\alpha)]\lambda)\tan\beta, \tag{26.23}$$

where

$$|\mathrm{LP}| = [r_\mathrm{e}\lambda/(2\pi V\cos\theta_\mathrm{B})][F_{0\mathrm{r}} - \rho|F_{h\mathrm{r}}|\exp(-M)]. \tag{26.24}$$

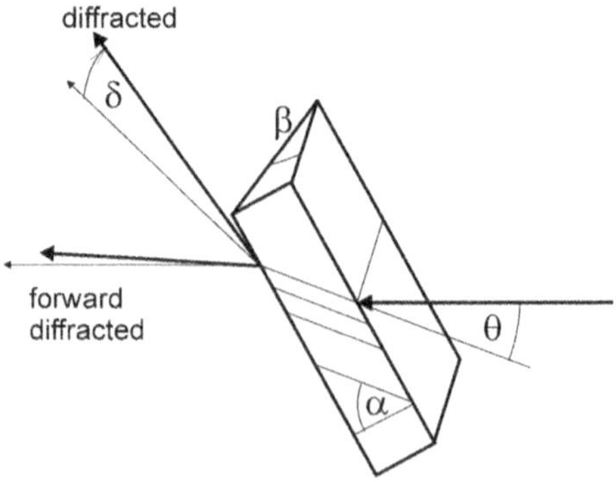

Fig. 26.13. Laue diffraction: diffracted and forward diffracted beam; both are sagittaly deviated

Here r_e is again the classical electron radius, V is the volume of the unit cell, θ_B is the Bragg angle, ρ is the polarization factor, and F_{0r} and F_{hr} are the real parts of the structure factors of the corresponding reflections (see, for example, [2]). The angle α is the deviation of the entrance surface from that in the symmetrical Laue case and β is the inclination angle. Here, it is assumed that in the vicinity of the Laue point the Ewald spheres may be replaced by planes.

As was suggested in [28] the effect of sagittal deviation discussed above may be utilized to sagittally focus the synchrotron radiation by a Laue crystal with a parabolic profile of one or both diffracting surfaces. Equation (26.23) is analogous to the formula (26.6). As in the Bragg case, this leads directly to a parabolic shape $y = ax^2$ of the diffracting surface. The parabolic surface may be approximated by a circular hole with the diameter D, where $D = 1/a$. Obviously, (26.23) could be applied to the design of a sagittally focusing Laue crystal with the profiled diffracting surface (parabolic or circular) if K from (26.19) is replaced by $(|\mathrm{LP}|[\cos\theta/\cos(\theta+\alpha)]\lambda)$. This is valid only if the exit surface is profiled and for the diffracted, not a forward diffracted, beam (see Fig. 26.13). The sagittal deviation of the forward diffracted beam is much smaller. The equally profiled entrance surface influences the sagittal deviation of the diffracted beam much less if the geometry is as shown in Fig. 26.13. This is discussed in detail in [30]. To understand the meaning of |LP| and |PN| in reciprocal space see [28].

For the experiment we used an asymmetric Si Laue crystal shown in Fig. 26.14. The diffracting part is the space between two cylindrical holes with the diameter $D = 1/a = 8\,\mathrm{mm}$. (The circle with the diameter $D = 1/a$ is a good approximation of parabola $y = ax^2$ for small x). The walls of the holes represent a sagittal tilt. The (111) diffracting crystallographic planes are deviated from the holes axes by 7.95°, so that $\alpha = 82.05°$. Both the

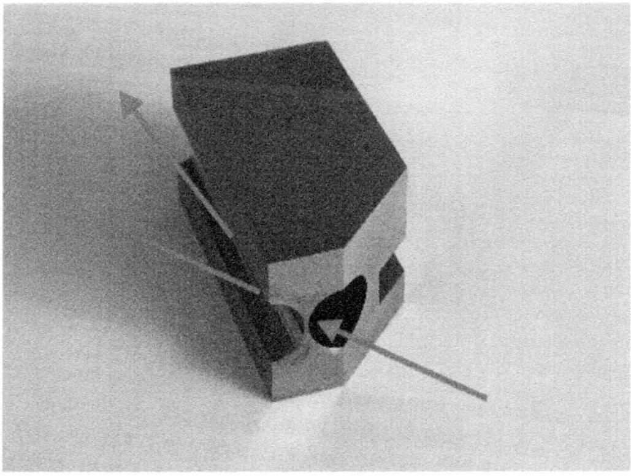

Fig. 26.14. Asymmetric Laue crystal with profiled entrance and exit surface. The working area is the space between the two circular holes

entrance and the exit diffraction surfaces were mechano-chemically polished. The experiment was performed at the BM05 beamline at ESRF (MI751).

After the Laue crystal we used a Bragg Si(111) symmetric crystal to redirect the Laue diffracted beam into a horizontal direction. To limit the presence of higher harmonics in the beam and to avoid the problem with energy determination, we used a primary monochromator which was set to 15.35 KeV ($\theta = 7.4°$) and was detuned. There were two beams which were diffracted from the crystal. The forward diffracted beam was not used, because the refraction effect is small (the refraction effect is stronger for beams forming a smaller angle with a surface). The diffracted beam which was used formed the angle of 0.55° with the crystal exit surface and deviated from the horizontal plane by the angle $2\theta = 14.8°$. Even if we had used a white beam, this combination of the Laue and Bragg crystals would have allowed us to reject higher harmonics, the angular distributions of harmonics for Bragg and Laue diffraction being different. The beam size was delimited by a slit 3 × 3 mm located before the Laue crystal. The crystals–source distance was 35 m.

Figure 26.15a shows the image of the diffracted beam just after the Bragg crystal. The horizontal (sagittal) dimension of the spot, i.e., the distance between the border beams A and B is 3.22 mm. The shape of the spot is a narrow "smile" because of the circular profile of the diffracting surfaces and the asymmetric diffraction. The image taken at 20 m from the crystal (Fig. 26.15b) is more complicated: it has the shape of a horseshoe. The border beams A and B are sagitally deviated such that their distance is 1.720 mm. Without focusing, the distance between both beams would be $3.22 \times (55/35) = 5$ mm (the crystals–source distance is 35 m and the detector–source distance is 55 m). Together with the sagittal deviation, the beams are also sagittaly spread (see the dimensions a and b). This situation is analogous

a

b

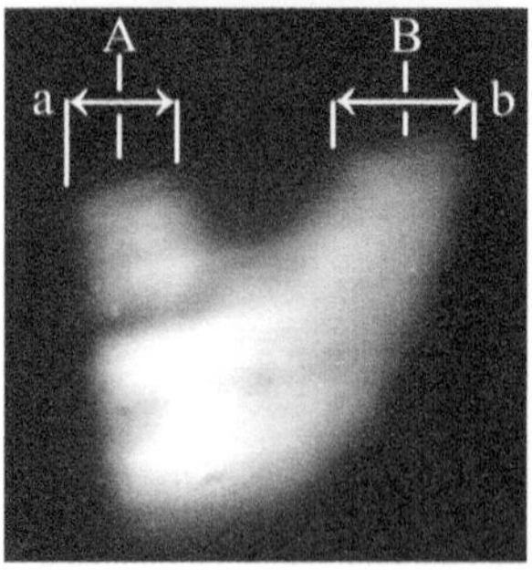

Fig. 26.15. The image of the Laue diffracted beam from the crystal shown in Fig. 26.14. Figure 26.15a shows the image taken close to the crystal, Fig. 26.15b is the image taken 20 m from the crystal

to the Bragg diffractive–refractive optics if only one crystal is used. The higher the deviation is, the higher is the spread. The consequence of this is that one crystal, or more crystals in a nondispersive arrangement, cannot create a sharp focus. In the Bragg case only two or more crystals in the dispersive arrangement completely cancel the sagittal spread so that the focus may be sharp [19]. We suppose that the same holds for the Laue diffractive–refractive optics.

From the distances between the beams A and B near the crystals and at 20 m from the crystals we can deduce that the diffracted radiation is convergent and the focusing distance is about +43 m, which is, however, different from the theoretical prediction. This indicates that further theoretical and experimental work is necessary.

The experiment described above shows that the idea of diffractive–refractive optics developed for Bragg diffraction in the past can be extended also to Laue diffraction. It was shown that the asymmetric Laue crystal with profiled diffracting surfaces may concentrate the diffracted beam and thus increase the intensity in the diffraction spot.

To improve the quality of the focus, i.e., to cancel the sagittal and vertical spreads, two Laue crystals with profiled surfaces in dispersive setting should be used (see the analogy with the Bragg diffractive–refractive optics). The experimental test of this idea is under preparation.

Such sagittal focusing might in some cases serve as an alternative to the sagittal focusing by a sagittally bent asymmetric Laue monochromator described by Zhong et al. [31, 32].

26.6 Conclusion

Practically all existing X-ray crystal monochromators are produced with a flat diffracting surface. Such crystals may be bent to focus diffracted radiation. In this chapter it is shown that if the diffracting surface is machined into a suitable shape, the diffracted monochromatic beam may gain some additional useful properties. It may be sagittally or meridionally focused or either higher harmonics or the π polarization component may be filtered over a broad range of Bragg angles. It is shown that these properties are the result of refraction.

Acknowledgments

The project was supported by the Institutional Research Plan AV CR No: AVOZ 10100522, by MSMT of the Czech Republic (Contract No: OC P7.001), and the GAAV of the Czech Republic No: IAA 100100716.

References

1. L.V. Azároff, R. Kaplow, N. Kato, R.J. Weiss, A.J.C. Wilson, R.A. Young, *X-Ray Diffraction* (McGraw-Hill, New York, 1974)
2. B.W. Batterman, H. Cole, Rev. Mod. Phys. **36**, 681 (1964)

3. A. Authier, *Dynamical Theory of X-Ray Diffraction* (Oxford University Press, Oxford, 2001)
4. T. Matsushita, H. Hashizume, in *X-Ray Monochromators*, ed. by E.E. Koch. Handbook on Synchrotron Radiation, vol. 1A (North-Holland Publishing, Amsterdam, 1983) pp. 261–314
5. J. Hrdý, O. Pacherová, Nucl. Instrum. Methods in Phys. Res. A **327**, 605 (1993)
6. J. Hrdý, J. Synchrotron Radiat. **5**, 1206 (1998)
7. A. Snigirev, V. Kohn, I. Snigireva, B. Lengeler, Nature **384**, 49 (1996)
8. T. Johansson, Z. Phys. **82**, 507 (1933)
9. E.P. Bertin, *Principles and Practice of X-Ray Spectrometric Analysis* (Plenum, New York, 1970)
10. P. Spieker, M. Ando, N. Kamile, Nucl. Instrum. Methods in Phys. Res. **222**, 196 (1984)
11. J. Hrdý, J. Hrdá, J. Synchrotron Radiat. **7**, 78 (2000)
12. J. Hrdý, E. Ziegler, N. Artemiev, F. Franc, J. Hrdá, Th. Bigault, A. Freund, J. Synchrotron Radiat. **8**, 1203 (2001)
13. J. Hrdý, Czech. J. Phys. **40**, 361 (1990)
14. J. Hrdá, J. Hrdý, Czech. J. Phys. **56**, 1421 (2006)
15. M. Hart, A.R.D. Rodrigues, Philos. Mag. B **40**, 149 (1979)
16. M. Hart, A.R.D. Rodrigues, J. Appl. Crystallogr **11**, 248 (1978)
17. M. Hart, A.R.D. Rodrigues, D.P. Siddons, Acta. Cyst. A **40**, 502 (1984)
18. J. Hrdá, J. Hrdý, To be published
19. J. Hrdý, D.P. Siddons, J. Synchrotron Radiat. **6**, 973 (1999)
20. J. Hrdý, J. Synchrotron Radiat. **8**, 1200 (2001)
21. N. Artemiev, J. Hrdý, S. Peredkov, A. Artemev, A. Freund, R. Tucoulou, J. Synchrotron Radiat. **8**, 1207 (2001)
22. N. Artemiev, J. Hrdý, S. Peredkov, A. Artemev, J. Synchrotron Radiat. **11**, 157 (2004)
23. P. Mikulík, A. Kuběna, Unpublished
24. J. Hrdý, A. Kuběna, P. Mikulík, J. Phys. D **38**, 4325 (2005)
25. J. Hrdý, N. Artemiev, A.K. Freund, J. Quintana, Proc. SPIE **4501**, 88 (2001)
26. N. Artemiev, J. Hrdý, T. Bigault, J. Hoszowska, S. Peredkov, Proc. SPIE **5195**, 104 (2003)
27. C.J. Sparks, G.E. Ice, B.W. Batterman, Nucl. Instrum. Methods in Phys. Res. **195**, 73 (1982)
28. J. Hrdý, J. Hoszowska, C. Mocuta, N. Artemiev, A. Freund, J. Synchrotron Radiat. **10**, 233 (2003)
29. J. Hrdý, J. Hoszowska, V. Mocella, Unpublished
30. J. Hrdý, V. Mocella, P. Oberta, L. Peverini, K. Potlovskiy, J. Synchrotron Radiat. **13**, 392 (2006)
31. Z. Zhong, C.C. Kao, D.P. Siddons, J.B. Hastings, J. Appl. Cryst. **34**, 504 (2001)
32. Z. Zhong, C.C. Kao, D.P. Siddons, J.B. Hastings, J. Appl. Cryst. **34**, 646 (2001)

27

Neutron Multiple Reflections Excited in Cylindrically Bent Perfect Crystals and Their Possible use for High-Resolution Neutron Scattering

P. Mikula, M. Vrána, and V. Wagner

Abstract. Bent prefect crystals can give many strong multiple-reflection processes through utilising two sets of lattice planes which are mutually in the dispersive setting. They provide doubly reflected beams with a very narrow bandwidths and high collimation. Such strong multiple-reflection processes realized in cylindrically bent perfect crystals have been proven to be suitable for ultra high-resolution monochromating. Calculations indicate that the multiple-reflection effects could easily be identified. Some examples of powder diffraction and high resolution radiography are also presented.

27.1 Introduction

In many cases new samples and necessity of measurement of finer effects require a substantial increase of angular and/or energy resolution of conventional diffractometers operating in a conventional performance mode. In such a case, a convenient monochromator plays a key role. Bent perfect crystal (BPC) slabs as neutron monochromators have been proved as an excellent alternative of conventional mosaic crystals. They provide a way how to increase luminosity and angular/energy resolution of some scattering devices installed usually at steady state sources [1, 2]. An increase of the luminosity is carried out by focusing in real space, while a higher resolution can be achieved by focusing in momentum space and rather small effective mosaicity of the BPCs. However, in the case of TOF scattering devices, the BPC elements practically have not been used and with respect to the TOF techniques the Bragg diffraction optics is far from being fully explored. We have already demonstrated that Si BPCs in fully asymmetric diffraction geometry and in combination with a linear position sensitive detector could be very efficient analyzers for high-resolution TOF spectrometry [3, 4]. New possibilities of more effective use of neutron scattering devices have recently opened an employment of sandwich type BPC monochromators/analyzers. They provide, e.g., multiple wavelength monochromatized beams [5] and a larger range of curvatures permitting an easier luminosity and resolution optimization of some dedicated instruments [6, 7]. Together with construction of new powerful neutron sources, new scattering instruments with improved resolution properties are designed. One of the candidates of monochromators for very high resolution neutron diffractometers and spectrometers appear so-called dispersive monochromators based on a dispersive double diffraction process. It can be realized by means of two independent crystals [8, 9] or by exciting a strong multiple reflection effect inside one elastically deformed perfect crystals [10, 11].

27.2 Multiple Bragg Reflections in Elastically Bent Perfect Crystals

The effects of multiple Bragg reflections (MBR) in a single crystal can be observed when more than one set of planes are simultaneously operative for a given wavelength, i.e., when more than two reciprocal lattice points are at the Ewald sphere. MBR-effects can result in an increase of the intensity of a weak primary reflection (*Umweganregung*). The extreme case of this is the simulation of forbidden primary reflection. For X-rays, the positive MBR-peaks were first observed by Renninger [10] and Moon and Shull for neutrons in mosaic crystals [11]. However, it has been found that strong MBR-effects can be excited also in BPC elements [12–16]. In the simplest case of the MBR-effect, a particular weak or forbidden primary reflection is strengthened by a successive cooperation of the two allowed reflections (secondary and tertiary). All these reflections are defined by the scattering vectors $\boldsymbol{g}_1$, $\boldsymbol{g}_2$, and $\boldsymbol{g}_3$, respectively. Then, the doubly reflected beam has the same direction as the one that could have been reflected by the particular primary set of planes (see Fig. 27.1). Scattering vectors $\boldsymbol{g}_2$ and $\boldsymbol{g}_3$ are in relation to $\boldsymbol{g}_1$ as $\boldsymbol{g}_1 = \boldsymbol{g}_2 + \boldsymbol{g}_3$.

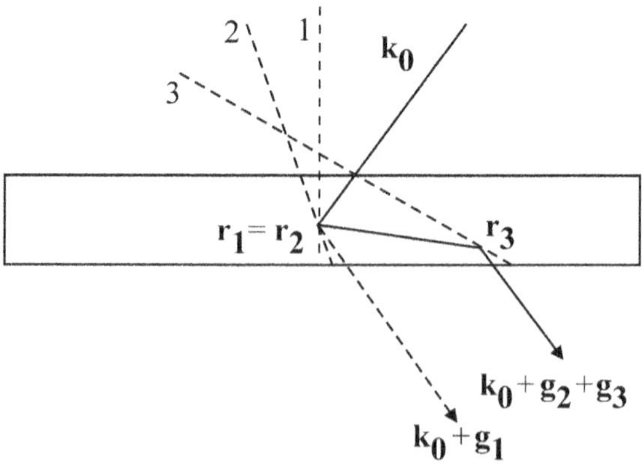

Fig. 27.1. Schematic diagram of a two-step multiple Bragg reflection simulating a weak or forbidden reflection. The numbers 1, 2, and 3 represent the primary, secondary and tertiary reflection planes, respectively

It follows from the crystal symmetry that when a secondary reflection fulfils the Bragg condition simultaneously with the primary one, there exists automatically a tertiary reflection defined by $\boldsymbol{g}_3 = \boldsymbol{g}_1 - \boldsymbol{g}_2$. In cases when an MBR – monochromator would be envisaged for a practical employment the choice of a suitable elastic deformation and the chosen diffraction geometry pay very important role in order to minimize the primary reflection as well as the higher order contributions which accompany the MBR-effect. It is known that the integrated reflectivity of a deformed crystal is a function of the scalar product $(\boldsymbol{g} \cdot \boldsymbol{u})$ [17,18] which may be zero for the primary and its higher order reflections, i.e., $(\boldsymbol{g}_1 \cdot \boldsymbol{u}) = 0$ and the corresponding integrated reflectivity is independent of the deformation represented by the displacement $\boldsymbol{u}$ [18]. This is also valid in our case of cylindrical bending and symmetric transmission geometry. On the other hand $(\boldsymbol{g}_2 \cdot \boldsymbol{u}) = -(\boldsymbol{g}_3 \cdot \boldsymbol{u})$ need not be zero and the deformation can bring about a large increase of the MBR-effect keeping the integrated reflectivity related to the higher order reflections constant.

It was already experimentally proved that due to the fact that the second and third system of reflection planes are mutually in the dispersive setting, the doubly reflected beam has a narrow bandwidth $\Delta\lambda/\lambda$ in the range 10^{-4}–10^{-3} and collimation of the orders of minutes of arc [15]. Furthermore, it was also pointed out that the MBR-effect could possibly be used for an ultrahigh resolution monochromatization [16]. Such MBR-effect is in fact a result of the dispersive double-reflection process realized in one crystal.

For investigation of MBR-effects usually two methods are usually used:

(a) Method of azimuthal rotation of the crystal lattice around the scattering vector $\boldsymbol{g}_1$ of the primary reflection for a fixed wavelength [12–14].
(b) Method of $\theta - 2\theta_{\mathrm{D}}$ scan in the white beam for a fixed azimuthal angle.

In the present case we preferred just the latter method of $\theta - 2\theta_{\mathrm{D}}$ scan when setting the cylindrically bent perfect Si-crystal slabs for diffraction in symmetric transmission geometry. Three Si-slabs of different cuts were used (see Fig. 27.2). All three slabs then permit simulation of forbidden primary reflections 222 or 002.

Fig. 27.2. Schematic diagram displaying the cuts of the used crystal slabs

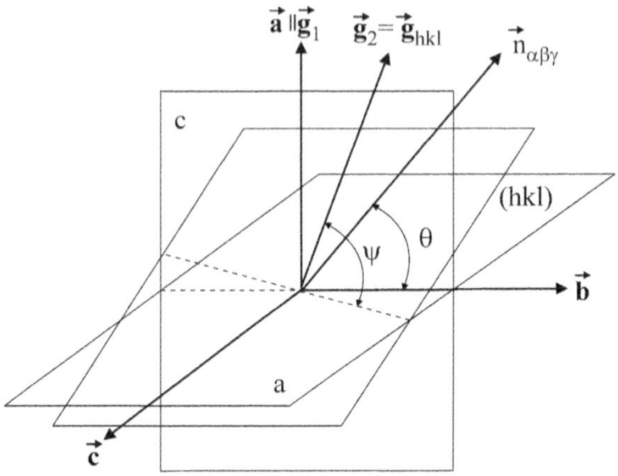

Fig. 27.3. Coordinate system describing the MBR reflection occurring with the primary reflection

27.3 Calculation

From the geometry shown in Fig. 27.3 it is possible to derive the following relations

$$\begin{aligned} \boldsymbol{n}_{\alpha\beta\gamma}\, \boldsymbol{g}_1 &= |\boldsymbol{g}_1| \sin\theta \\ \boldsymbol{n}_{\alpha\beta\gamma}\, \boldsymbol{g}_{hkl} &= |\boldsymbol{g}_{hkl}| \sin\psi \\ \boldsymbol{n}_{\alpha\beta\gamma}\, \boldsymbol{b} &= |\boldsymbol{b}| \cos\theta , \end{aligned} \tag{27.1}$$

where $|\boldsymbol{n}_{\alpha\beta\gamma}| = 1$ from the definition and

$$\sin\psi = (d_1/d_{hkl}) \sin\theta \tag{27.2}$$

is given from the Bragg condition (d is the lattice spacing). Then, for crystal slabs used in our experiment we could derive the relations providing Bragg angles θ of the primary reflection where secondary reflection represented by the Miller indices h, k, l can participate in the MBR-process.

1. For the crystal slab with the largest surface parallel to $(\underline{1}10)$ and the longest edge parallel to [111]

$$\tan\theta = (3/2)^{1/2}(-h+k)/[(h^2+k^2+l^2)/m - h - k - l]. \tag{27.3}$$

2. For the crystal slab with the largest surface parallel to $(\underline{11}2)$ and the longest edge parallel to [111]

$$\tan\theta = (1/2)^{1/2}(-h-k+2l)/[(h^2+k^2+l^2)/m - h - k - l]. \tag{27.4}$$

3. For the crystal slab with the largest surface parallel to $(\underline{1}10)$ and the longest edge parallel to [001]

$$\tan\theta = (1/2)^{1/2}(-h+k)/[(h^2+k^2+l^2)/m - l]. \tag{27.5}$$

The parameter m in the relations (27.3)–(27.5) means the order of the primary reflection.

Using the formulae (27.3)–(27.5), in the range of $\theta - 2\theta_D$ scans one can easily identify a lot of secondary and tertiary reflections participating in MBR-process with respect to a chosen primary reflection. However, when setting the crystal in the polychromatic incident beam, similarly to a single reflection case, the presence of higher (or lower) orders can be found. For example, if in our case we are interested in MBR studies with respect to a forbidden primary Si(222) reflection, in many cases one can also find an accompanied MBR-effect with respect to the primary reflections Si(111), Si(333), Si(444) etc., simultaneously. In practice, all undesirable higher (or lower) order contributions can be eliminated, e.g., by a neutron wavelength selector. Sometimes, the wavelength distribution of neutrons passing through the neutron guide can be favorable, when undesirable higher order contributions could be automatically eliminated, because their corresponding shorter wavelength neutrons are not present in the spectrum. Therefore, in our case we omitted all MBR-contributions corresponding to the neutron wavelengths smaller than 0.08 nm. Furthermore, one should consider that the reflection probability related to individual secondary and tertiary reflections is rather small for shorter neutron wavelengths [19, 20]. Contrary to the mosaic crystals, in our case of bent crystals for a detailed estimation of the individual MBR-contributions the value of $|\boldsymbol{g}_{2,3} \cdot \boldsymbol{u}|/(|\boldsymbol{u}| \cdot |\boldsymbol{g}_{2,3}|)$ plays also an important role [18].

27.4 Search for Strong Multiple Bragg Reflection Effects

The experimental search of the MBR-effects appearing at different wavelengths was done on the two axis POLDI diffractometer installed at the end of the thermal neutron guide in GKSS Geesthacht. The Si-slab was situated in the white beam in the place of the monochromator. Having at a disposal three cylindrically bent Si-crystal slabs of different cut, after setting them for symmetric transmission, we carried out $\theta - 2\theta_D$ scans in the θ-range from 7° to 60° (for one azimuthal position) with the largest surface of the crystal perpendicular to the primary scattering plane. In such a case the deformation brought about by bending had no influence on the reflectivity of the primary reflection but a strong effect on the secondary and tertiary reflections participating in MBR-process. Figures 27.4–27.6 display parts of the individual scans where the MBR-effects were strongly excited. For the sake of comparison, the scans with a flat nonbent crystal as well as the background are also introduced in Fig. 27.4. It should be pointed out that the individual crystal slabs are of different thickness whose parameter is also very important for a final estimation of the reflectivity power of the MBR-monochromator. Simply, it can be said that the intensity of the MBR-effect is linearly proportional to the crystal thickness. Then, of course, the divergence $\Delta\theta$ and the $\Delta\lambda$ spread of the diffracted beam correspondingly increase. It should be pointed out that before some practical use of the MBR- monochromator, an optimization procedure of the curvature with respect to the individual MBR-effects should be carried out. On the basis of the relations (27.3)–(27.5), we could easily determine the secondary and tertiary planes contributing to the observed MBR-effects as well as the corresponding Bragg angles θ. Table 27.1 shows the results corresponding to the first four strongest peaks from the Fig. 27.4 which are related

Table 27.1. Calculated reflections and the Bragg angles of the strongest peaks

Peak N.	Primary, secondary, tertiary reflections	θ/deg	Peak N.	Primary, secondary, tertiary reflections	θ/deg
1	$111/\underline{5}13/60\underline{2}$	11.536	2	$111/\underline{1}3\underline{3}/2\underline{2}4$	13.763
1	$111/06\underline{2}/1\underline{5}3$	11.536	2	$111/151/0\underline{4}0$	13.763
1	$111/\underline{5}3\underline{3}/6\underline{2}4$	11.536	2	$111/\underline{4}00/511$	13.763
1	$111/\underline{2}64/3\underline{5}\underline{3}$	11.536	2	$111/\underline{2}24/3\underline{1}\underline{3}$	13.763
1	$111/\underline{3}1\underline{3}/404$	11.536	3	$111/\underline{1}53/2\underline{4}\underline{2}$	14.705
1	$111/044/1\underline{3}\underline{3}$	11.536	3	$111/\underline{4}2\underline{2}/5\underline{1}3$	14.705
1	$111/\underline{3}73/4\underline{6}\underline{2}$	11.536	4	$222/\underline{3}1\underline{1}/513$	29.956
1	$111/\underline{6}4\underline{2}/7\underline{3}3$	11.536	4	$222/\underline{3}13/51\underline{1}$	29.956
1	$111/133/0\underline{2}\underline{2}$	11.536	4	$222/15\underline{1}/1\underline{3}3$	29.956
1	$111/\underline{2}0\underline{2}/313$	11.536	4	$222/153/1\underline{3}\underline{1}$	29.956

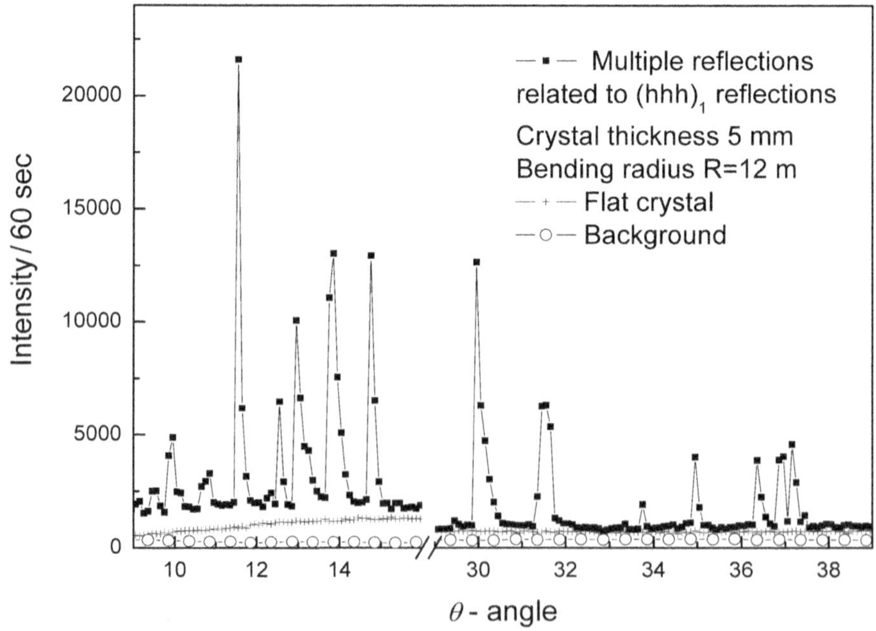

Fig. 27.4. Part of $\theta - 2\theta_D$ scan with the crystal slab (the largest face parallel to (110)) set for $(hhh)_1$ reflections in the symmetric transmission geometry

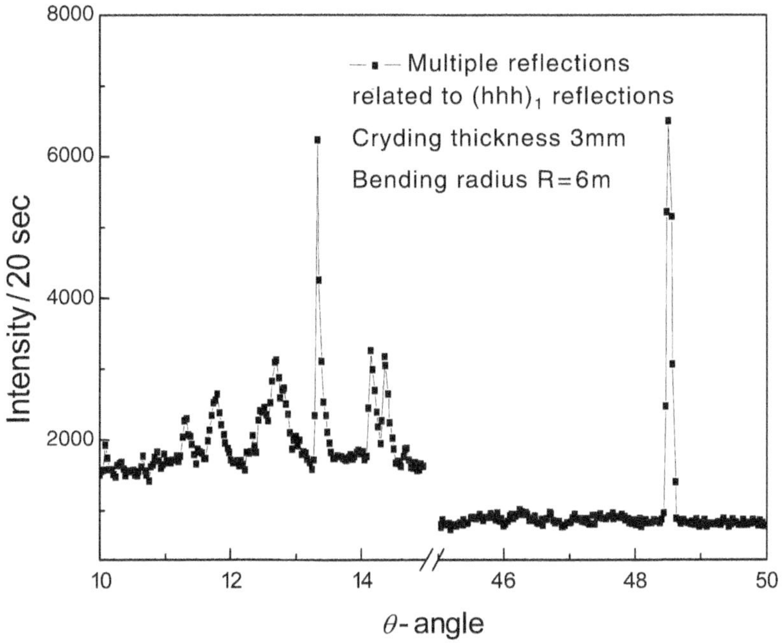

Fig. 27.5. $\theta - 2\theta_D$ scan with the crystal slab (the largest face parallel to (112)) set for $(hhh)_1$ reflections in the symmetric transmission geometry

to the allowed Si(111) or forbidden Si(222) reflections. The wavelength λ is then determined simply by Bragg equation applied on the primary reflection. It can be seen from Table 27.1 that the MBR-peak observed at $\theta = 11.536°$ for $\lambda = 0.125$ nm is in fact an additional "strengthening" of the intensity of Si(111) primary reflection by ten simultaneous multiple reflections. Similarly, the calculation can be carried out for other peaks observed at different cuts of crystal slabs. It can be seen from Figs. 27.4–27.6 that the density of the MBR-peaks is higher for smaller Bragg angles, i.e., for shorter neutron wavelengths because the density of reciprocal lattice points on the Ewald sphere increases with $1/\lambda$. It is clear that the calculations can provide only angular positions of possible MBR-effects on the $\theta - 2\theta$ scan. However, the height of the MBR-signal depends on several parameters. First of all, it is the value of crystal curvature. Then, taking into account the importance of the value of $|\boldsymbol{g}_{2,3} \cdot \boldsymbol{u}|/(|\boldsymbol{u}| \cdot |\boldsymbol{g}_{2,3}|)$, the diffraction geometry (symmetric transmission, symmetric reflection, or asymmetric alternatives) plays also an important role. Furthermore, one can expect a linear dependence of the signal on the thickness of the crystal slab. Due to these facts, for finding the optimum parameters of the MBR-monochromator and the performance of the scattering device, Monte Carlo simulations would be desirable [21].

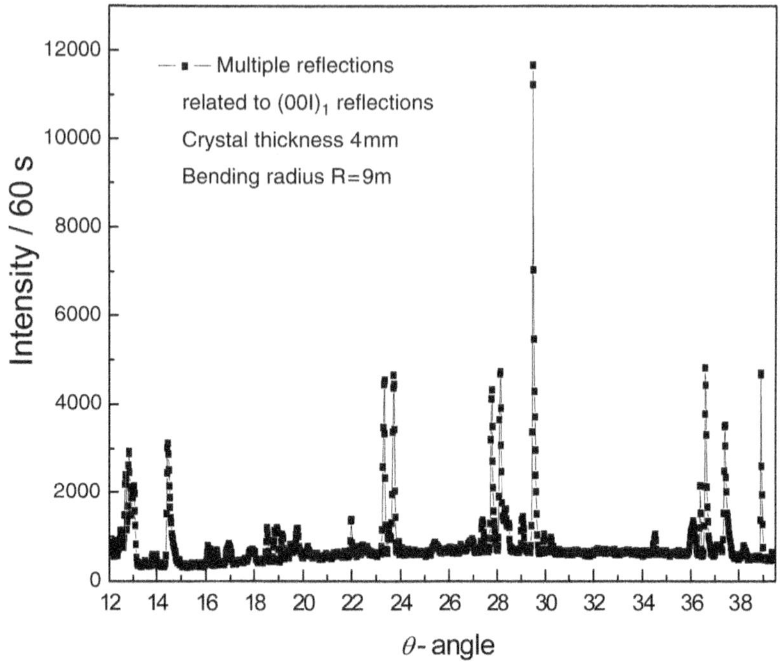

Fig. 27.6. $\theta - 2\theta_{\mathrm{D}}$ scan with the crystal slab (the largest face parallel to (110)) set for $(001)_1$ reflections in the symmetric transmission geometry

27.5 Powder Diffraction Experimental Test

For the first experimental powder test we chose the effect observed at $\theta = 29.956°$ which is related to the simulation of the forbidden 222 reflection by a cooperative action of $153/1\underline{3}\underline{1}$ and $\underline{3}1\underline{1}/513$ (secondary/tertiary) reflections at $\lambda = 0.156$ nm (see Fig. 27.4). The powder diffraction test carried out with a solid α-Fe polycrystalline standard sample of the diameter of $\Phi = 2$ mm (see Fig. 27.7) displays several diffraction profiles that clearly prove the applicability of the MBR-monochromator for high-resolution diffraction studies. *FWHM* of the diffraction profiles are, however, determined by the spatial resolution of the used PSD (1.5 mm) and the width of the sample ($\Phi = 2$ mm). Therefore, it can be considered as an upper limit. Then, this performance we used for investigation of Fe-reflections in an induction hardened S45C steel rod ($\Phi = 20$ mm) having different phase composition at different distances from the rod axis [22]. The gauge volume was determined by 2 mm wide slits in the incident as well as diffracted beam. Figure 27.8 displays the diffraction profile obtained at the distance 8 mm from the axis. Similarly, Fig. 27.9 displays the diffraction profile obtained at the distance 6 mm from the axis. Thanks to the used high-resolution monochromatic beam, after a fitting procedure we could reliably determined contributions of the individual phases.

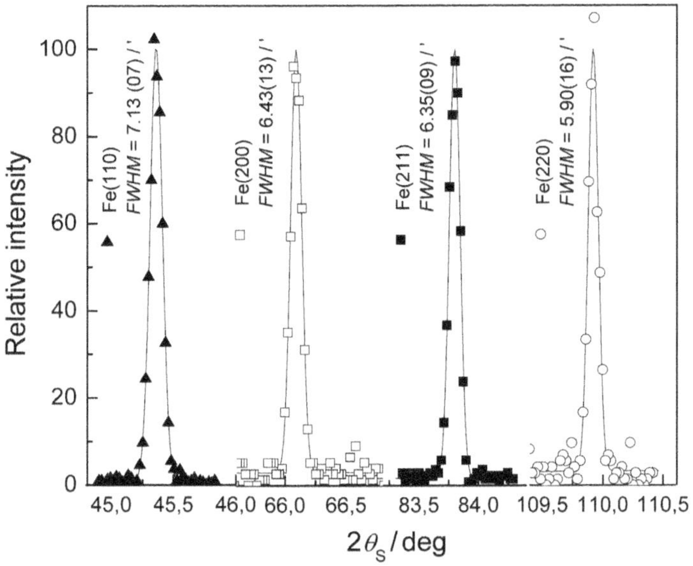

Fig. 27.7. Examples of the α-Fe profiles taken with the MBR-monochromator

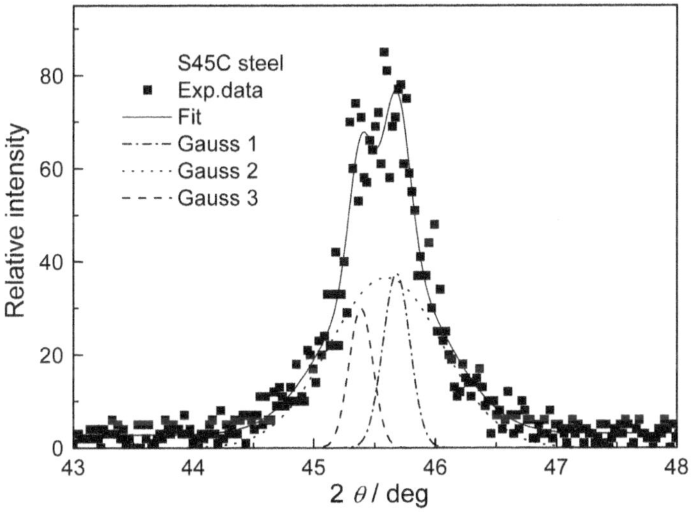

Fig. 27.8. Induction-hardened S45C steel diffraction profile taken at 2 mm under the surface with the fitted profiles related to the perlitic, ferritic and martensitic phases

27.6 Neutron Radiography Experimental Test

Using MBR-effect observed with the crystal slab having the main face parallel to the planes (112) (see Fig. 27.5) at $\theta = 48.526°$ and $\lambda = 0.235\,\mathrm{nm}$ we have carried another test of a possible application in the field of neutron

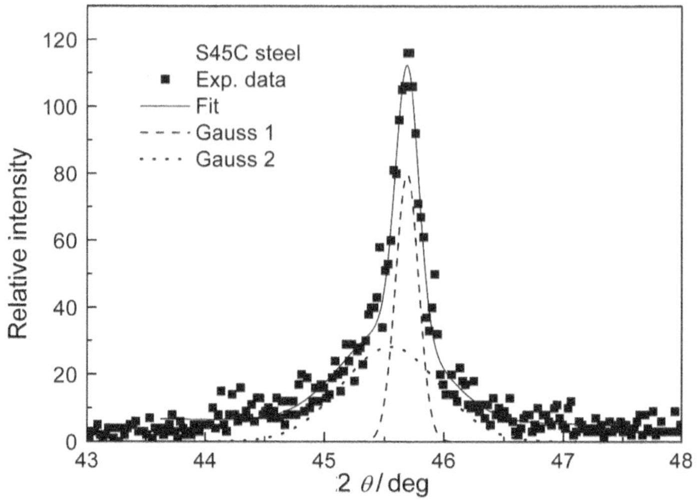

Fig. 27.9. Induction-hardened S45C steel diffraction profile taken at 4 mm under the surface with the fitted profiles related to the ferritic and martensitic phases

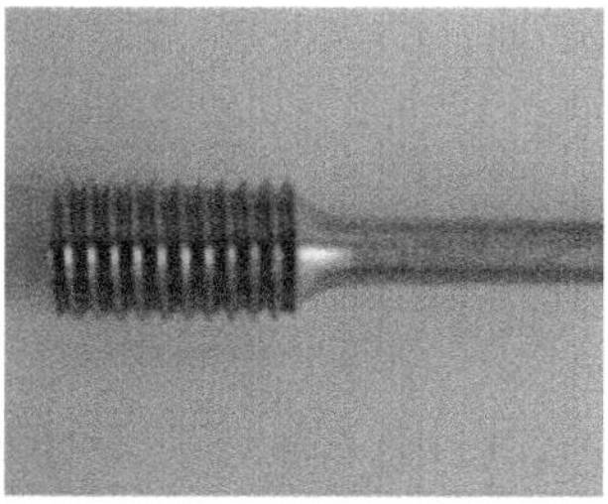

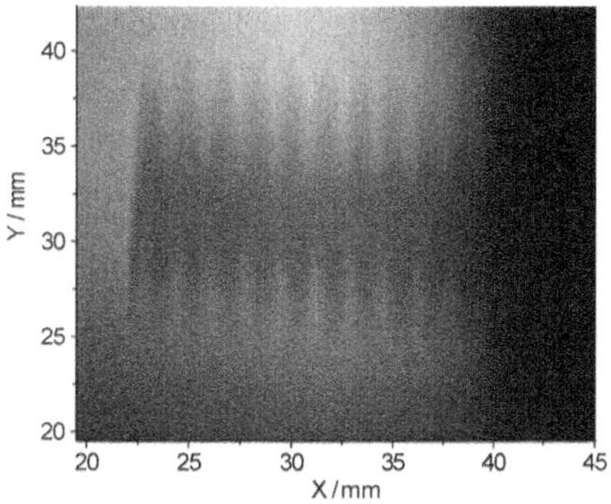

Fig. 27.10. The radiography image (*right*) of the screw of a steel sample (*left*) which is usually used in a tension/compression rig for tensile tests in mechanical engineering

radiography. The idea consists in a fact that thanks to the excellent property of the MBR-beam (high collimation and high monochromaticity) the imaging plate does not need to be necessarily in a close contact with the irradiated object but at some distance up to 1 m. Moreover, in this case, the cross section of the MBR-beam is several cm^2 which is useful for 1:1 image. Figure 27.10 displays the radiography image of a screw taken by the image plate situated at the distance of 70 cm from the sample. As can be seen from Fig. 27.10, the obtained image really confirms very good collimation and spatial resolution of the beam; however, in this case it is only in the horizontal plane. In the vertical plane the collimation is still rather high, because all planes participated in the MBR-process are perpendicular to the horizontal (scattering)

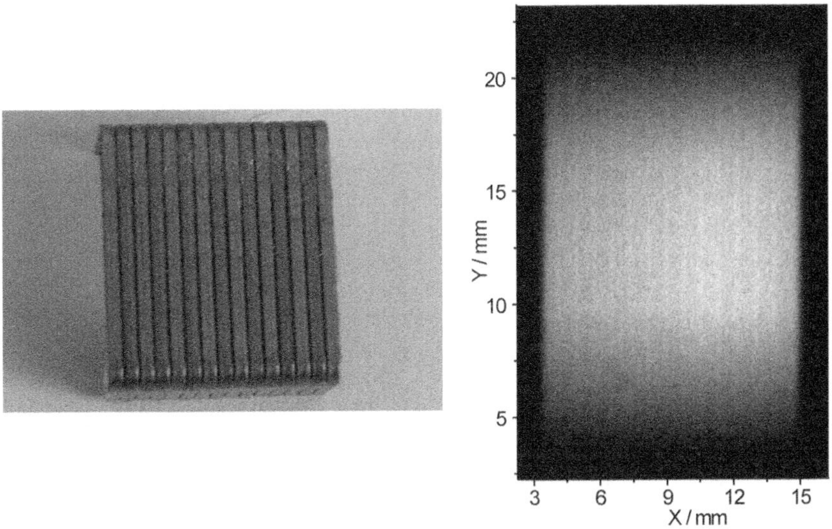

Fig. 27.11. The radiography image (*right*) and the photo of the steel office staples (*left*)

plane. Then the dispersive arrangement between the secondary and tertiary lattice planes has much weaker influence on the collimation and monochromaticity in the vertical plane. Similarly Fig. 27.11 displays the radiography image of the office staples also taken at the distance of 70 cm. In this case a new remarkable effect was observed. If we take into account that the thickness of the office staples is only a few tenths of millimeter, the obtained image is not based on absorption, but thanks to a high-beam monochromaticity, on the refraction principle occurring at the edges between the individual needles. This new radiography method can be considered as a complementary one to the so-called phase contrast radiography [23–25]. It is clear that much attention should be paid in developing this new radiography technique. First of all, other MBR-planes should be tested in order to achieve high collimation and monochromaticity of the neutron beam in both horizontal and vertical directions. Then, the MBR-effects should be studied on the cold neutron source at longer neutron wavelengths where much stronger refraction effects are expected.

Acknowledgments

We appreciate very much GENF in GKSS for providing us the beam time and all necessary support by means of the ACCESS program of EU-FP6-NMI3, contract no: RII3-CT-2003-505925. In the Czech Republic this research has been supported by the projects GACR N. 202/06/0601, COST – OC P7.003, AV0Z10480505, and MSM2672244501.

References

1. M. Popovici, W.B. Yelon, J. Neutron Res. **3**, 1 (1995)
2. P. Mikula, V. Wagner, M. Vrána, Physica B **283**, 289 (2000)
3. P. Mikula, M. Furusaka, S. Haryo, Y. Yashushige, Appl. Phys. A, **74**, 207 (2002)
4. P. Mikula, M. Vrána, M. Furusaka, Y. Yasushige, in *Proceedings of International Conference ICANS XVI*, Düsseldorf-Neuss, Germany, May 12–15, 2003, ed. by G. Mank and H. Conrad, pp. 401–405
5. M. Vrána, P. Mikula, P. Lukáš, V. Wagner, Physica B, **241–243**, 231 (1998)
6. J. Kulda, J. Šaroun, Nucl. Instrum. Methods A **379**, 155 (1995)
7. P. Staron, H.U. Ruhnau, M. Mamotti, P. Mikula, R. Kampmann, Physica B, **276–278**, 158 (2000)
8. P. Mikula, M. Vrána, M. Furusaka, V. Wagner, Y.N. Choi, M.K. Moon, V.T. Em, C.H. Lee, Nucl. Instrum. Methods in Phys. Res. A **529**, 138 (2004)
9. P. Mikula, M. Vrána, V. Wagner, Y.M. Choi, S.A. Kim, H.S. Oh, K.H. Sung, C.H. Lee, in *Proceedings of International Conference. on Advanced. Technology in Experimental Mechanics, ATEM'03*, CD ROM Published by the Japan Society of Mechanical Engineers No. 03-207, Sep. 10–12, 2003, Nagoya, paper OS04W0288
10. R. Renninger, Z. Kristallographie **113**, 99 (1937)
11. R. Moon, C.G. Shull, Acta Cryst. **17**, 805 (1964)
12. P. Mikula, R.T. Michalec, M. Vrána, J. Vávra, Acta Cryst. A **35**, 962 (1979)
13. P. Mikula, M. Vrána, R.T. Michalec, J. Kulda, J. Vávra, Phys. Stat. Sol. A **460**, 549 (1980)
14. M. Vrána, P. Mikula, R.T. Michalec, J. Kulda, J. Vávra, Acta Cryst. A **37**, 459 (1981)
15. P. Mikula, J. Kulda, M. Vrána, R.T. Michalec, J. Vávra, Nucl. Instrum. Methods, **197**, 563 (1982)
16. P. Mikula, B. Chalupa, J. Kulda, M. Vrána, R.T. Michalec, L. Sedláková, in *Proceedings of the Symposium Neutron Scattering*, 6.-8.8.1984, West Berlin, HMI-B 411, ed. by M.S. Lehmann, W. Saenger, G. Hildebrandt, H. Dachs, pp. 233–235
17. J. Kulda, P. Mikula, Phys. Stat. Sol. A **92**, 95 (1985)
18. S. Takagi, Acta Cryst. **15**, 1311 (1962)
19. J. Kulda, Acta Cryst. A **40**, 120 (1984)
20. P. Mikula, J. Kulda, M. Vrána, B. Chalupa, J. Appl. Cryst. **17**, 189 (1984)
21. J. Šaroun, J. Kulda, Physica B **234–236**, 1102 (1997)
22. P. Mikula, M. Vrána, V. Wagner, K. Inoue, Z. Kristallographie, Suppl. **23**, 205 (2006)
23. N. Kardjilov, Further developments and applications of radiography and tomography with thermal and cold neutrons. Ph.D. Thesis, Technische Universität München, 2003
24. N. Kardjilov, S.W. Lee, E. Lehmann, I.C. Lim, C.M. Sim, P. Vontobel, Nucl. Instrum. Methods Phys. Res. A **542**, 100 (2005)
25. M. Bastürk, N. Kardjilov, H. Rauch, P. Vontobel, Nucl. Instrum. Methods Phys. Res. A **542**, 106 (2005)

28

Volume Modulated Diffraction X-Ray Optics

A. Erko, A. Firsov, D.V. Roshchoupkin, and I. Schelokov

Abstract. Systematic experimental and theoretical investigations of different types of Bragg-Fresnel gratings, both static and dynamic, are discussed. Static gratings are produced by etching in a multilayer or by evaporating gold or nickel masks on the surfaces of symmetric or asymmetric Si [111] crystals. These have been used to obtain X-ray diffraction at different energies. The properties of both sagittal and meridional diffraction gratings are discussed. Dynamic diffraction gratings are produced by propagating a surface acoustic wave along a piezoelectric crystal. Experimental data are compared with the theoretical calculations.

28.1 Introduction

Experimental and theoretical data point to a new application of nanometer radiation in diagnostics, transmission and processing of information: X-ray electronics as a branch of the general science of electromagnetic radiation control. This application is based on the development of powerful radiation sources in the nanometer range, synchrotrons, storage rings and X-ray lasers in the future, and on the methods of microelectronics technology that enable fabrication of structures of X-Ray electronic devices with submicron or nanometer element sizes, i.e. micro-photonic devices. As in other fields of engineering associated with receiving, transmission, processing and storage of information, the evolution of micro-photonics is primarily due to the development of elements and methods to control X-Ray beams, their focusing, modulation, etc.

Conventionally the elements of X-ray optics can be divided into the following groups:

- Passive elements designed for deviation or focusing of beams, analogous to optical lenses, mirrors, static beam splitters
- Dispersive elements for spectral devices, monochromators
- Active elements for beam scanning, controlled adaptive optics
- Elements for transforming information of electric, acoustic, optical signals into X-Ray beam modulation

Since the wavelength of nanometer X-Ray radiation is several orders of magnitude less than that of optical radiation, it is possible to investigate radiation that is close to fundamental absorption edges of substances (K, L, M, etc. atom electron shells). The spatial resolution of X-ray nano-photonics systems reaches 0.1 nm [1], which makes possible active elements with a capacity up to 10 Tbit cm^{-2}, exceeding by four orders of magnitude potential of optical recording. A signal modulation frequency may amount to 10^{10} Hz. X-Ray channel energy losses by diffractive beam divergence are six orders less than those of optical systems, which is promising for long-range space communication.

For the fabrication of planar submicron structures with sizes on the order of X-Ray wavelengths, the deposition and growth of thin films of different materials have enabled fabrication of diffraction optical elements in the nanometer range. Optics for X-ray beams means creating effective focusing elements with the structure of three-dimensional Fresnel zones: combined microstructure X-Ray optics [2] or Bragg–Fresnel optics [3]. Multilayer mirrors and crystals provide the basis for such elements and are the primary elements of microphotonics, since they enable transformation of electrical, optical or acoustic signals into X-ray beam modulation [4].

In this chapter we report on systematic theoretical and experimental investigations of volume (Bragg–Fresnel) gratings: (a) static, made with etching technology including a metallic structure on the surface of multilayers and crystals, and (b) dynamic, produced by surface and volume acoustic waves. A Bragg–Fresnel grating is a basic optical element for the construction of a variety of X-ray optical devices including fast X-ray modulators. Understanding its properties is essential for effectively designing high resolution, focusing dispersive X-ray optics.

28.2 Static Volume Grating Properties

One can consider two types of volume gratings: gratings etched into a multilayer/crystal mirror, called *etched gratings*, and those made by evaporation (sputtering) metals on the surface of a crystal or multilayer mirror, called *surface gratings*. Such gratings are the basic type of Bragg–Fresnel optics [5]. Once the properties of these elements are known, it will be possible to predict the major properties of more complicated structures, such as Bragg–Fresnel lenses and X-Ray holograms [6, 7].

This chapter describes the important points in the computer simulation and the testing of the particular case of lamellar gratings with a rectangular groove profile.

The gratings are investigated in two different experimental geometries: sagittal diffraction and meridional diffraction. The definitions of a sagittal and of a meridional grating are given in Figs. 28.1 and 28.3.

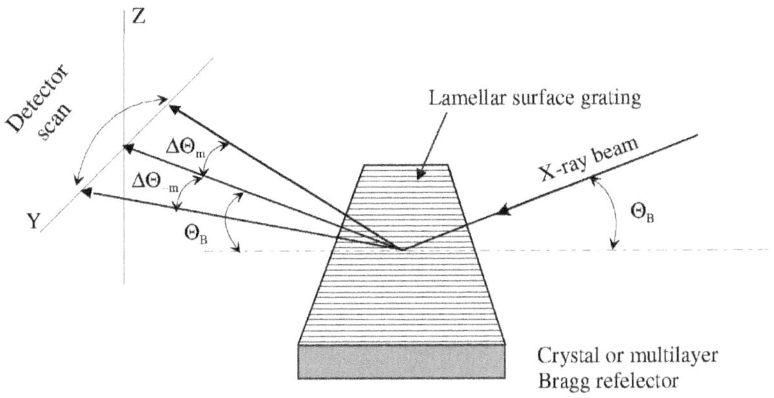

Fig. 28.1. Definitions for a sagittal grating

28.2.1 Sagittal Bragg–Fresnel Gratings

In sagittal geometry, the grating grooves are aligned parallel to the optical plane, as defined by the ingoing and outgoing X-ray beams. In this case diffraction takes place perpendicular to the optical plane. In meridional geometry, the grating grooves are aligned perpendicular to the optical plane and diffraction takes place in the optical plane.

Surface Sagittal Grating

The simplest type of a Bragg–Fresnel grating consists of a surface of a Bragg reflector – crystal or multilayer – covered with a phase shifting metal mask, made, for example, of gold stripes. The definition of a sagittal grating geometry is given in Fig. 28.1. For these gratings the groove direction is parallel to the beam direction. Θ_B corresponds to the Bragg angle on the crystal substrate. $\Delta\Theta_m$ and $\Delta\Theta_{-m}$ correspond to the directions of positive and negative diffraction orders in the sagittal (horizontal) plane.

The properties of sagittal Bragg–Fresnel surface gratings are very similar to those of transmission diffraction gratings. To produce an optimum phase-shift, Φ_{opt}, between adjacent grooves for a transmission grating at normal incidence, i.e. to ensure the maximum diffraction efficiency, a structure with optimal thickness, t_{opt}, is required.

The maximum diffraction efficiency of a sagittal grating is limited to the value of 0.4 for completely transparent phase material, i.e. with an absorption constant $\beta = 0$. The Bragg reflector, in the case of a sagittal grating, serves only as an energy-selective mirror reflecting a monochromatic X-ray beam.

One example would be a gold mask grating on the surface of a Si(111) single crystal reflector. The optical constants for a gold layer at the energy of 8,500 eV are refraction constant $\delta_{\text{au}} = 4.2 \times 10^{-5}$, absorption constant $\beta_{\text{au}} = 4 \times 10^{-6}$. In this particular case for a grating with a groove width equal to the free space between grooves, the relative integral diffraction efficiency of a corresponding transmission grating should be of the order of 31% with an optimal groove thickness of 1,650 nm. Taking into account the Bragg angle of a Si(111) reflection at 8,500 eV ($\lambda = 0.146$ nm, $\Theta_B = 13.43°$), one can determine that the optimal groove thickness for the sagittal Bragg–Fresnel coated grating is 192 nm. For the multilayer reflector with the period of 3.0 nm, the value of Bragg angle will be reduced to $\Theta_B = 1.396°$ at the same wavelength. The optimal groove thickness in this case will be as small as 20.1 nm. Therefore, with a grazing incidence beam it is possible to achieve the theoretical maximum diffraction efficiency for a grating using a considerably reduced mask thickness. For a reflection sagittal grating (RGS) the optimum thickness of material is defined by the equation

$$t_{\mathrm{opt}}(\mathrm{RGS}) = \frac{\Phi_{\mathrm{opt}} \lambda \sin \Theta_{\mathrm{B}}}{4\pi\delta}. \tag{28.1}$$

The X-ray beam is transmitted twice through the thickness t_{opt} at the angle Θ_B. In comparison with a transmission grating at normal incidence, t_{opt} is reduced by a factor of $0.5 \sin \Theta_B$.

The difference in the optimal thickness for transmission gratings and reflection sagittal grating is largest for high energies. For the calculations, a Si(111) crystal with a lattice period of 0.31 nm is used and a multilayer mirror with a period of 3 nm as the reflector. According to (28.1), for an aspect ratio of about 1:1 (layer thickness equal to a groove width) the minimum zone width of the Bragg–Fresnel grating at 30 keV can be as small as 0.2 μm. The grating can be easily fabricated for photon energies as high as 100–200 keV, which is a great problem for conventional transmission gratings due to the very high material thickness (aspect ratio) required.

The angular dispersion of the diffraction orders relative to zero reflection can be written in the small angle approximation:

$$\Delta\Theta_m = \arcsin\left(\frac{\lambda m}{d}\right), \tag{28.2}$$

where $\Delta\Theta_m$ is the diffraction angle for the mth diffraction order.

The experimental measurements were done using synchrotron radiation [8] and the results are shown in Fig. 28.2. A gold mask on the surface of a Si(111) crystal was fabricated by electron-beam lithography and lift-off technology. The mask period was 2 μm ($1 \times 1\,\mu\mathrm{m}^2$) with a thickness of 192 nm. A detector scan was done in the horizontal plane (Fig. 28.1), which corresponds to the sagittal diffraction plane. The theoretical position and efficiencies of the mth diffraction orders (*) are calculated using (28.2).

The results of the measurements demonstrate the validity of the theoretical estimate of the optimal thickness of the phase-shifting material.

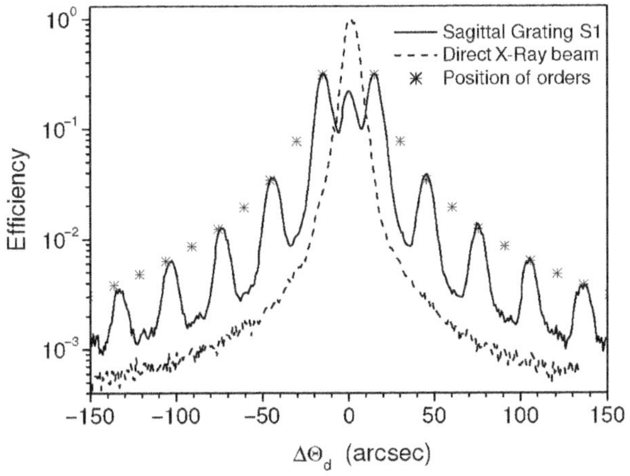

Fig. 28.2. The diffraction spectra of a surface sagittal grating made with a gold thickness of 192 nm and period of 2 µm

Etched Multilayer Sagittal Grating

This type of a grating is made by etching the groove profile into a multilayer or crystal substrate. For the calculations of diffraction efficiency of this grating an average refraction index, $\bar{\delta}$, of the diffraction structure can be used. This index is defined by the material and structural properties of a Bragg reflector, and the depth of profile must be equal to the X-ray penetration depth in the multilayer, the so-called extinction depth.

In this case ordinary diffraction on a thin grating takes place, yielding many diffraction orders simultaneously in the detector (2θ) scan mode similar to Fig. 28.2. Combining (28.1) for the optimal depth of profile with the average value of the multilayer refractive index

$$\bar{\delta} = \frac{\delta_1 d_1 + \delta_2 d_2^2}{d} \tag{28.3}$$

leads to the optimal depth of grating profile for a multilayer sagittal grating (MLSG) as [9]

$$t_{\mathrm{opt}}(\mathrm{MLSG}) = \frac{\Phi_{\mathrm{opt}}\lambda^2}{8\pi(\delta_1 d_1 + \delta_2 d_2)}, \tag{28.4}$$

where δ_i are the optical refraction indices of the multilayer materials, d_i are the corresponding thicknesses of the materials ($d = d_1 + d_2$), λ is the wavelength. For the particular case of a W/Si mirror with a 3.16 nm multilayer period ($d_{\mathrm{W}} = 1.0$ nm and $d_{\mathrm{Si}} = 2.16$ nm) one will have $t_{\mathrm{opt}} = 47.3$ nm or ~16 bilayers. The optical constants at the wavelength 0.154 nm are the following: for tungsten, $\delta_{\mathrm{W}} = 4.57 \times 10^{-5}$ and $\beta_{\mathrm{W}} = 4 \times 10^{-6}$ and for silicon, $\delta_{\mathrm{Si}} = 7.56 \times 10^{-6}$

and $\beta_{\text{Si}} = 1.7 \times 10^{-7}$. The efficiency of the grating can be calculated by the Kirz formula corresponding to an amplitude phase grating with

$$\chi = \frac{\beta_1 d_1 + \beta_2 d_2}{\delta_1 d_1 + \delta_2 d_2} \cdots . \tag{28.5}$$

For the given materials of a Si/W multilayer mirror, $\chi = 0.22$ and $e_{\text{eff}} = 0.8$, one will have $E_1 \sim 0.26$, which is in good agreement with the numerical computation and experimental measurements [10]. Therefore, a sagittal multilayer etched grating has the properties of a classical amplitude–phase transmission grating with corresponding optical parameters.

The same value of an optimal etching profile for a maximum absolute order efficiency has been calculated using the differential method, developed in [11]. For a sagittal grating the measured peak efficiency is in agreement with calculations using (28.4). The value of the 'resonance phase' depends on the mean value of the refractive index of a multilayer structure and corresponds roughly to the extinction depth.

Etched Crystal Sagittal Grating (ECSG)

The optimal depth of the profile for a crystal sagittal Bragg–Fresnel grating also corresponds to the extinction depth of a crystal material and is equal to [12]

$$t_{\text{opt}}(\text{ECSG}) = \frac{\lambda \sin \Theta_{\text{B}}}{2|\chi_0|}, \tag{28.6}$$

where χ_0 is the 0th order Fourier coefficient of crystal polarization. In the case of a crystal one can obtain diffraction from different crystallographic orientations with the corresponding grating period, d_{hkl},

$$d_{hkl} = \frac{a}{\sqrt{h^2 + k^2 + l^2}}, \tag{28.7}$$

where a is the lattice parameter of the crystal, equal to 0.543 nm for Si, for example.

Finally for a particular crystal orientation one can calculate the depth of the profile using the simple equation

$$t_{\text{opt}}(\text{ECSG}) = \frac{\lambda^2 \sqrt{h^2 + k^2 + l^2}}{4\, a|\chi_0|} \cdots . \tag{28.8}$$

As an example, the value of χ_0 for a Si crystal is equal to 1.5×10^{-5} at the energy of 8 keV. Therefore, at this energy an optimum depth of profile is equal to 1.26 μm for the Si(111) reflection. A more detailed theory of the Bragg–Fresnel sagittal diffraction and focusing has been developed by Kohn [13]. Sagittal etched crystal lenses have been used in the works of Snigirev et al. for diffraction focusing of synchrotron radiation [14–16].

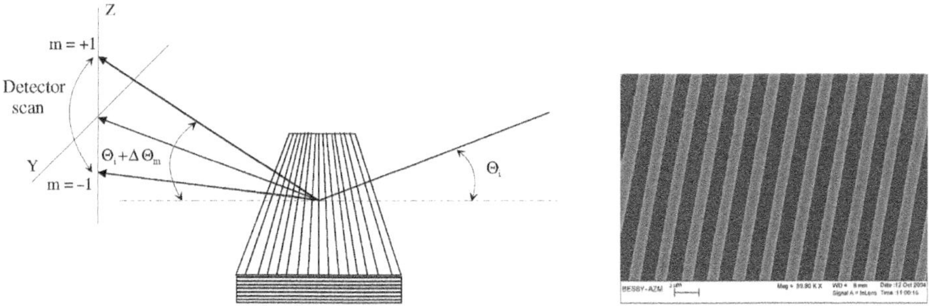

Fig. 28.3. Definitions for a meridional grating (*left*) and micro-photograph of the grating in a scanning electron microscope (*right*)

28.2.2 Meridional Bragg–Fresnel Gratings

Surface Meridional Grating

The definitions for the meridional grating geometry are given in Fig. 28.3. The grating grooves are perpendicular to the beam direction. The Θ_B corresponds to the Bragg angle on the crystal substrate. $\Delta\Theta_m$ and $\Delta\Theta_{-m}$ correspond to the directions of positive and negative diffraction orders in the meridional (here, vertical) plane.

The diffraction properties of a meridional grating on a multilayer mirror or crystal surface can be described to a first approximation using the conventional grating equation. The basic grating equation may be written as

$$\lambda m = d[\cos\Theta_\mathrm{i} - \cos(\Theta_\mathrm{i} + \Delta\Theta_m)], \tag{28.9}$$

with the grazing incidence angle Θ_i, which, in general, is not equal to the Bragg angle, and $\Delta\Theta_m$ is the angular dispersion.

A detector scan is made in the vertical optical plane, which corresponds to the meridional diffraction plane (Fig. 28.3).

In the case of a meridional grating, the angular dispersion in the mth order can be defined using the formula

$$\Delta\Theta_m = \arcsin\left(\cos\Theta_\mathrm{i} + \frac{\lambda m}{d}\right) - 90^\circ + \Theta_\mathrm{i} \cdots . \tag{28.10}$$

The difference between the values of positive and negative orders is important for small grazing angles, Θ_i. The value of the angle of negative diffraction orders is limited by the grazing angle of incidence of the incoming beam. The efficiency of the different diffraction orders for a meridional grating with a crystal (multilayer) substrate cannot, in general, be calculated by the same equations as applicable for a sagittal (transmission) grating. These formulas are valid only for meridional gratings on a non-dispersive mirror substrate, for example, for a total external reflection mirror. A crystalline or multilayer substrate has an angular selectivity according to the Bragg law. Such a mirror

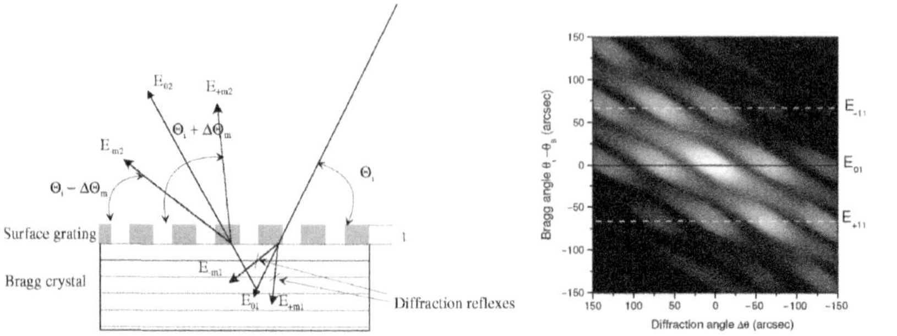

Fig. 28.4. X-ray diffraction on a meridional surface grating with the thickness t. Bragg diffraction inside the crystal rejects all diffraction orders that do not fulfill the Bragg condition (*left*). Two-dimensional intensity distribution for the meridional Bragg–Fresnel grating on the Si(111) crystal with a grating period of 2 µm (*right*)

can reflect a monochromatic beam only in the limited angular interval inside the so-called *rocking curve*.

The diffraction on a meridional Bragg–Fresnel grating can be divided into three steps (Fig. 28.4). First, the incoming beam is diffracted on the surface transmission grating and penetrates into the crystal. If the angular dispersion of the grating, $\Delta\Theta_m$, is larger than the intrinsic rocking curve of the crystal, only the zero order, corresponding to the Bragg condition, is reflected by the crystal. The intensities of the diffraction orders outside the Bragg peak will be rejected and absorbed in the substrate. Second, the reflected zero order beam is diffracted a second time on the surface transmission grating. Finally, the crystal substrate does not limit the angular spectrum of the outgoing beam but limits the angular acceptance of the Bragg–Fresnel grating.

The definitions for meridional grating measurements are given in Fig. 28.4. The Θ_i corresponds to the grazing incidence angle on the crystal substrate, Θ_d corresponds to the direction on the detector and $\Delta\Theta_m$ and $\Delta\Theta_{-m}$ correspond to the directions of positive and negative diffraction orders in the meridional (here, vertical) plane.

The first diffraction process on a surface grating leads to a loss of output intensity. The accepted angular spectrum of the input beam is limited by the crystal rocking curve. The second diffraction on the surface grating produces the higher diffraction orders measured in the experiment.

If the grazing incidence angle corresponds to the Bragg condition of the crystal, then only the zero order of the primary diffracted beam will be diffracted a second time on a surface grating after reflection on the crystal substrate. The total efficiency of the diffraction orders on a Bragg–Fresnel grating can be written as

$$E_m = \frac{1}{4m^2\pi^2}\left\{\begin{matrix}(1+\exp(-2\chi\Phi_{\mathrm{opt}}))^2- \\ -4\cos^2(\Phi_{\mathrm{opt}})\exp(-2\chi\Phi_{\mathrm{opt}})\end{matrix}\right\}\cdots . \tag{28.11}$$

The maximum diffraction efficiency of a meridional grating is limited according to (28.11) to the value of 0.1 as compared to 0.4 for a sagittal Bragg–Fresnel grating. Consequently, the diffraction efficiency of an amplitude (absorbing) meridional grating is limited to 0.025 as compared to 0.1 for a sagittal Bragg–Fresnel grating. The value of the optimal phase-shift for a meridional grating is less than that for a sagittal grating (28.11). The optimal phase-shift for a gold mask at 8,500 eV is equal to 2.89 rad for a sagittal grating and 1.48 rad for a meridional one. The optimal mask thickness has in both cases almost the same value because of the factor of 2 in the equation for the optimal thickness of a meridional Bragg–Fresnel grating (MBFG):

$$t_{\mathrm{opt}}(\mathrm{MBFG}) = \frac{\Phi_{\mathrm{opt}} \lambda \sin \Theta_{\mathrm{B}}}{2\pi\delta}. \tag{28.12}$$

The optimal value of a phase-shift in a gold mask of 1.48 rad yields the theoretical maximum grating efficiency of 7.7%. Taking into account the corresponding Bragg angle of 13.43° for the Si(111) reflection together with the optical parameters of gold, one obtains the value of optimal thickness of 190 nm.

The validity of the concept of *the three step diffraction process* on a meridional surface grating was verified by the experimental measurements of a meridional grating with a period of 2 µm. The same gold mask on the surface of a Si(111) crystal, as used in the case of the sagittal grating, was used for the meridional grating tests. A two-dimensional angular scan by the sample and detector was done in the vertical (optical) plane at each point around the Bragg angle. The result is represented in Fig. 28.4. The crystal analyzer was used instead of a slit to improve the angular resolution [17].

The diffraction angles were measured by a detector scan for each value of grazing incidence angle around the Bragg maxima of the crystal. In accordance with Fig. 28.4, the crystal does not reflect the X-ray beam at angles that do not correspond to the Bragg condition. The secondary resonances, appearing when −1st and +1st diffraction orders fulfill the Bragg condition, are shown in Fig. 28.4 (E_{-11} and E_{+11}). At these angles in the detector plane one can measure the same angular spectrum as at the exact Bragg condition for the primary beam.

According to (28.10) and (28.12), it is possible to increase the angular dispersion of a diffraction grating and decrease the optimal grating thickness by using a small grazing angle for the primary beam. Conventional crystals such as Si or Ge provide Bragg angles on the order of 10°–15° in the energy range around 10 keV. Smaller reflection grazing angles can be realized using asymmetrical crystals with a high asymmetry parameter or multilayer mirrors with nanometer-scale periods [17].

28.2.3 Etched Meridional Gratings

The properties of etched meridional Bragg–Fresnel gratings are more complicated than those for the sagittal case. The diffraction process on a meridional grating can, in general, no longer be described as the superposition of two independent diffraction structures but instead by a complex volume interference phenomena.

Several methods have been developed for the meridional grating calculations. A description in the kinematic approximation, which neglects interactions between incident and diffracted beams, was used by several authors [18]. In the calculations [19], rigorous theory based on Maxwell's equations which take into account this interaction (dynamical theory) was used. Computations can be done using a differential method and modal theory [20]. In the modal theory, the solution is sought for the whole structure. Therefore, structures with any number of layers can be calculated at the same time. In addition, using this method, one can quickly compute the dispersive curves in reciprocal space and the Bragg angle scanning of the diffracted orders. However, this method is valid only for a lamellar grating. It is not applicable, for example, for a sawtooth profile grating. In the present chapter, the differential technique, developed in [21], is used for multilayer meridional Bragg–Fresnel grating calculations. These calculations are carried out layer-by-layer, making the method suitable for any kind of profile. Lamellar gratings have been studied by several authors both theoretically and experimentally [22–25].

The nature of diffraction on a three-dimensional grating/multilayer structure strongly depends on the optical properties of the materials and the characteristic size of a grating period with the 'lattice parameter' a_1 and a multilayer period with the 'lattice parameter' a_2. As a consequence, two different approximations can be used for different limiting conditions.

Multilayer Etched Meridional Grating

In the first approximation, the 'double dispersion' phenomena of multilayer gratings can be described as a combination of Bragg diffraction on reflecting layers and surface diffraction on a planar grating. In this simplest case, Bragg diffraction limits the output energy and the angular spectra of the reflected beam, and the planar grating produces an additional angular dispersion. The measurements of meridional gratings with a 'large' period, which exceeds criteria described later, show a simple combination of grating and multilayer, as if they were used separately, one after other, similar to a surface grating. The detector scan spectrum at the fixed Bragg angle, Θ_B, shows several diffraction peaks from the surface grating inside the broad Bragg peak of the multilayer mirror. The property of the 'short' period, etched volume grating is not the simple 'overlapping' of the two independent structures. Instead, one must refer to the theory of crystal diffraction. The characteristics of a volume grating can be demonstrated by the dependence of the absolute efficiency of the +1 order on the depth of the grating profile with a lamellar grating period being taken as the variable parameter. These curves are shown in Fig. 28.5. For

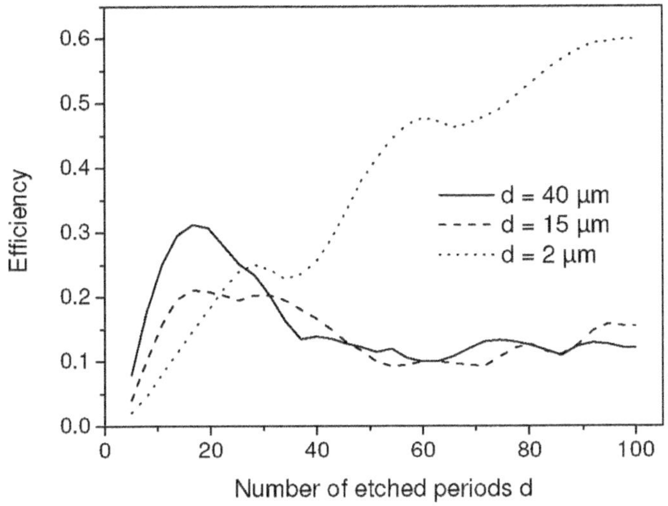

Fig. 28.5. Maximum efficiency of the +1 diffraction order vs. the number of etched periods. The W/Si multilayer mirror is performed with 100 bilayers with a period of 3 nm

the calculations, the same multilayer parameters as used for calculations of a surface grating were taken by using (28.8).

According to the differential method for relatively large grating periods (in the range of 200–20 μm), one can observe a peak of efficiency that corresponds to an appropriate phase condition for both sagittal and meridional gratings. The value for the optimal depth, obtained using the differential method, agrees very well with the optimal depth calculated by the analytical formula (28.4) for a sagittal grating.

The value of the 'resonance phase' depends on the mean value of the refractive index of a multilayer structure, $\overline{\delta}$, and corresponds to the extinction depth in a multilayer. Properties of such gratings are the same as for conventional reflection phase gratings, except for the Bragg selectivity. For short period gratings, i.e. with lateral periods of less than 10 μm, the behavior is different. The aforementioned W/Si multilayer with the spacing of 3 nm is an example. According to the differential model the 'phase peak' of efficiency is shifted into the depth of the multilayer, and the +1 order intensity continuously increases with the increasing depth of the grating profile (Fig. 28.5). This phenomenon cannot be explained without involving volume diffraction effects.

Let us describe a multilayer grating as a two-dimensional crystal with two different translation vectors, a_1 in the direction along a surface (X) and a_2 in the depth of a multilayer (crystal) (Z). Such a macro-crystal has two main crystallographic directions along the z and x axes. The multilayer crystal structure is shown schematically in Fig. 28.6. As in a natural crystal with two different lattice parameters, one can define crystallographic directions

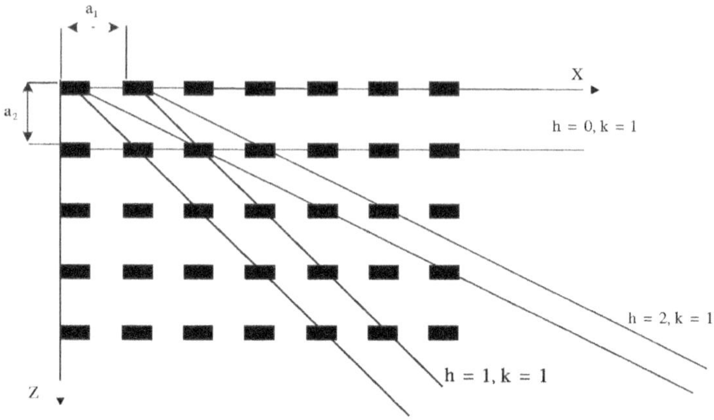

Fig. 28.6. Schematic representation of a 'multilayer crystal'

corresponding to different translation vectors of the lattice with an absolute value of ha_1 and ka_2.

Using crystallographic indexes one can define diffraction parameters $d_{h,k}$ for a short period grating ($a_2 \cong a_1$) as

$$d_{h,k} = \frac{a_1 a_2}{\sqrt{(ha_1)^2 + (ka_2)^2}}. \tag{28.13}$$

For a long period grating ($a_2 << a_1$) an effective diffraction occurs only between waves diffracted from the top and the bottom of the grooves. A phase shift between these waves depends only on the different optical paths in the multilayer groove and vacuum. Multilayer mirrors act like a monochromatic reflector with the phase reflecting grating on the top. The properties are the same as for a sagittal grating with a period of $d_{h,k} \approx a_1$. Looking at the efficiency dependence vs. the depth of the etched profile (Fig. 28.5), one can see an increase in the absolute reflectivity up to 0.3, which corresponds to the phase maximum (π phase shift) between diffracted waves. Diffraction orders are located inside of a multilayer Bragg peak and cannot be observed without zero order diffraction.

One can define these two limiting cases even more precisely taking into account the extinction depth of a multilayer or crystal structure (28.11). As already mentioned, for a sagittal grating the depth of profile is optimal if it is equal to the value of extinction depth, t^z_{ext}. Extending the definitions, one can introduce an extinction depth value for the grating along the X direction, which could be defined as

$$t^x_{\text{ext}} \approx \frac{t^z_{\text{ext}}}{\sin(\Theta_\text{B})} \cdots . \tag{28.14}$$

The volume properties of a meridional etched grating become essential if the period of a lateral grating is less than t^x_{ext}. For example, a W/Si multilayer with a period of 3 nm and $\gamma = 0.3$ has an extinction depth on the order of 50 nm at 0.154 nm wavelength. The corresponding 'volume effect' parameter has a value of $t_{\text{ext}}^{x} \sim 1.95\,\mu\text{m}$. The same parameter for a Si(111) single crystal reflection is equal to $t_{\text{ext}}^{x} \sim 5.3\,\mu\text{m}$.

The meridional multilayer grating properties have been experimentally measured on lamellar multilayer grating samples having a variable grating period and etching depth. All experimental measurements were performed in the two-dimensional ($\Theta_i - \Theta_d$) scan mode at the energy of 8 keV (Fig. 28.3). In this mode, the diffracted field was scanned with the detector slit at the angle Θ_d for each incident angle, Θ_i, in order to record the intensity of all 'n' orders.

The diffracted efficiency distribution in three dimensions vs. incident Θ_i as well as the diffracted $2\Theta_i$ angle was measured and plotted. As an example, Fig. 28.7 represents the results of the grating measurements in the detector scan mode for the 230 nm profile depth and 4 µm grating period. The angle of incidence was in the range of 1.5°–1.65°. For each incident angle, Θ_i, the diffracted field was scanned with the detector slit in the same range. Using such a method, diffraction orders −2, −1, 0, +1, +2 can easily be resolved (see Fig. 28.7).

As can be seen from this plot, the maximum intensity for the minus first diffraction order corresponds to the minimum of the zero order. A similar result was described by Neviere [26] for another type of multilayer grating, one coated on large period blaze echelette grating. In that paper, the structure

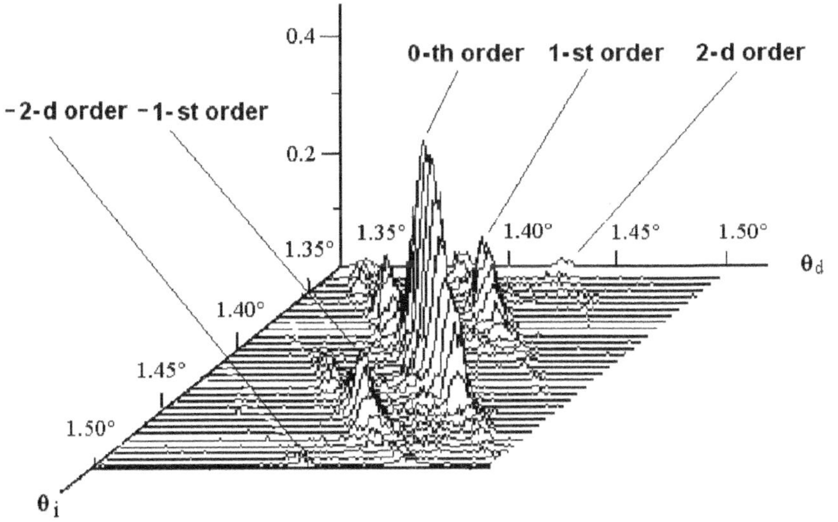

Fig. 28.7. $\Theta_i - \Theta_d$ plot in the detector scan mode for the meridional multilayer grating with a profile depth of 230 nm

of the multilayer grating was totally different from our structure, which is a short period lamellar grating etched in a multilayer.

To measure the depth dependence of the diffraction efficiency, seven points on the grating with a variable profile have been tested experimentally.

28.3 Dynamic Diffraction Gratings based on Surface Acoustic Waves

This chapter presents the application of a surface acoustic wave (SAW) of the Rayleigh type as a diffraction grating for X-ray radiation. Propagation of SAWs in the crystal leads to the sinusoidal modulation of a crystal lattice and sinusoidal modulation of a crystal surface. Since the phase velocity of the SAW (2,000–4,000 m s^{-1}) is much lower than the speed of the X-rays, the acoustic deformation can be considered as quasi-static and characterized by its wavelength and amplitude. However, the use of an ultrasonic super-lattice in the X-ray wavelength range has some limitations. First, it is necessary to apply acoustic waves with a very short wavelength ($\Lambda \sim 1$–$10\,\mu m$) in order to produce a large angular dispersion between diffraction satellites [27–32]. This requirement is related to the large Bragg angles, Θ_B, for the real piezoelectric crystals such as quartz, $LiNbO_3$, $LiTaO_3$, $La_3Ga_5SiO_{14}$, $La_3Ga_{5.5}Ta_{0.5}O_{14}$, which lie between 3° and 40°. Therefore, it is attractive to use a multilayer X-ray mirror under the Bragg angle on the order of 1° [33–36] or in a total external reflection mode ($\alpha_i \sim 0.1°$–$0.3°$) [37, 38], where the SAW with a wavelength of $\Lambda = 10$–$40\,\mu m$ produces considerable angular dispersion with X-rays. Total external reflection is interesting for two other reasons. First, a high reflectivity, typically 90%, is possible. Second is the high efficiency of scattering by the surface acoustic waves, the amplitude of which is nearly comparable to the depth of the penetration of the evanescent X-ray wave of the order of 10 nm.

It is also possible to control both the wavelength and the amplitude of a dynamic SAW grating by changing the amplitude of the input high-frequency electric signal and the excitation frequency. These possibilities can be used to optimize the space–time modulation based on X-ray diffraction by surface acoustic waves [39, 40].

28.3.1 The SAW Device

Figure 28.8a, b show the SAW device based on a piezoelectric crystal. To excite a Rayleigh SAW, an interdigital transducer (IDT) is deposited on the crystal surface by photolithography or e-beam lithography. An IDT transforms the high-frequency signal into acoustic oscillations of the crystal lattice, which propagate along the crystal surface. The SAW amplitude on the crystal surface can be changed linearly from zero to several angstroms by varying the amplitude of the high-frequency electrical voltage supplied by a high-frequency

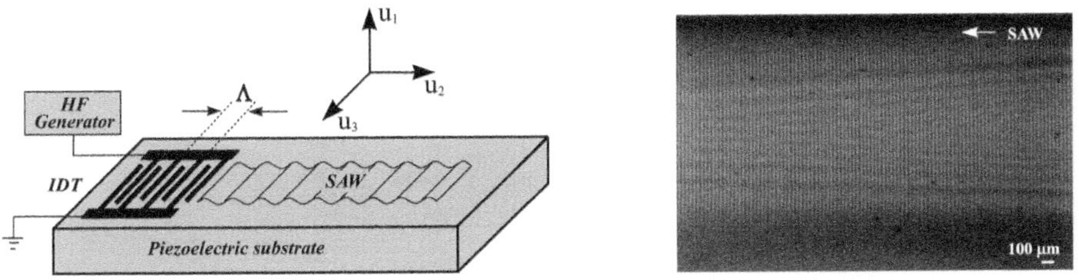

(a) (b)

Fig. 28.8. (**a**) SAW device. (**b**) SAW propagation in the *YZ*-cut of a $LiNbO_3$ crystal. $\Lambda = 30\,\mu m$

generator to the IDT. Figure 28.8b presents the scanning electron microscopy image of the SAW propagation in the *YZ*-cut of a $LiNbO_3$ crystal with the velocity of $V = 3{,}488\,m\,s^{-1}$. The SAW with wavelength $\Lambda = 30\,\mu m$ was excited at the resonance excitation frequency $f = 116.3\,MHz$. It is seen that SAW behaves like a strongly periodic sinusoidal diffraction grating.

SAW propagation causes a sinusoidal deformation of the crystal lattice and crystal surface in the first approximation. A Rayleigh SAW is actually elliptically polarized, but in the case of a symmetric reflection geometry, in-plane displacements of the crystal lattice do not influence diffraction. The deformation involved in the diffraction process can be written as

$$h = h_0 u_1 \sin(Kx), \tag{28.15}$$

where $K = 2\pi/\Lambda$ is the SAW wave vector and h_0 is the SAW amplitude on the crystal surface, which can be controlled by varying the input signal on the IDT.

28.3.2 Total External Reflection Mirror Modulated by SAW

The diffraction of light by ultrasound has been investigated theoretically [41,42] and experimentally [43–45]. Theoretical curves (see Figs. 28.10–28.12) show excellent agreement with experimental results. A detailed description of the diffraction theory on a surface grating can be found in [38].

Figure 28.9 depicts a double-crystal X-ray diffractometer used to study X-ray diffraction on the surface of the *YZ*-cut of a $LiNbO_3$ crystal modulated by surface acoustic waves under total external reflection. An X-ray tube with a rotating copper anode (Cu Kα radiation, $\lambda = 0.154\,nm$, running at 40 kV and 60 mA) was used as the source of X-ray radiation. A plane X-ray wave behind a double Si(111) crystal-monochromator was collimated by a $10\,\mu m$ slit. For diffraction studies under total external reflection, the crystal surface was treated by chemical dynamic polishing so that the roughness does not exceed 1 nm. This treatment is very important because the roughness decreases the value of the critical angle. An IDT with an $8\,\mu m$ finger width that corresponds to a $\Lambda = 32\,\mu m$ SAW was deposited on the surface

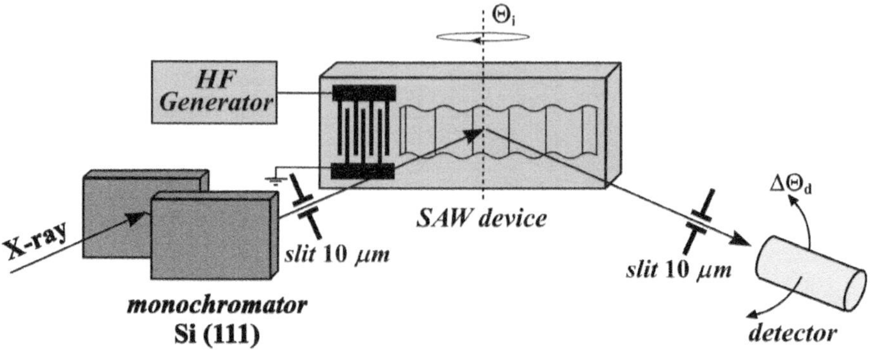

Fig. 28.9. Diagram of the double-crystal X-ray diffractometer

of the sample so that the SAW propagates along the Z axis with a velocity $V = 3.488\,\mathrm{km\,s^{-1}}$. The resonance frequency of the IDT was $f_0 = 109\,\mathrm{MHz}$. For the experiment described here the collimated plane X-ray wave falls on the crystal surface modulated by the surface acoustic wave at the incident angle $\Theta_\mathrm{i} = 0.22°$, slightly below the experimentally measured critical angle of the YZ-cut of a $LiNbO_3$ crystal, $\alpha_\mathrm{c} = 0.30°$. The X-ray plane wave diffracts on the ultrasonic superlattice so that the angular position of diffraction satellites can be determined from the grating equation:

$$k \cos \Theta_m = k \cos \Theta_\mathrm{i} + mK, \tag{28.16}$$

where $k = 2\pi/\lambda$, $K = 2\pi/\Lambda$ and m is the diffraction order.

According to (28.16), the X-ray radiation is expected to diffract on the crystal surface modulated by the SAW so that the angular divergence should be 0.090° and 0.063° for $m = 0$ and $m = +1$ (-1), respectively. The diffracted X-ray radiation is recorded by a scintillation detector behind a 10 μm slit. In all results, the diffracted X-ray intensity was normalized to the intensity of the incident beam.

Figure 28.10 shows the experimental (a) and calculated (b) curves of the diffracted X-ray radiation intensity, I, as a function of the detector scanning angle, $\Delta\Theta_\mathrm{d}$, obtained at the X-ray incident angle $\Delta\Theta_\mathrm{i} = 0.22°$. The resonance excitation frequency of the SAW was $f_0 = 109\,\mathrm{MHz}$ and values of the amplitude of the input sinusoidal signal on the IDT ranged from $U = 2$–$17\,\mathrm{V}$. The sinusoidal amplitude of the SAW, h, is a linear function of the amplitude of the input signal on the IDT. In the calculated curves, h is assumed to be between 0.2 and 1.7 nm. In Fig. 28.10, diffraction satellites are observed at the angles $\Delta\Theta_1 = 0.090°$ and $\Delta\Theta_{-1} = 0.063°$ from the intense reflected beam. These values are in a good agreement with those calculated from expression (28.15) for the -1 and $+1$ diffraction orders. The maximum intensity of the $m = -1$ diffraction order makes up 10.5% of the intensity of the incident X-ray beam for an amplitude of the input signal on the IDT $U = 17\,\mathrm{V}$. The great difference in the diffraction order intensities $(E_{-1} > E_1)$ and angular

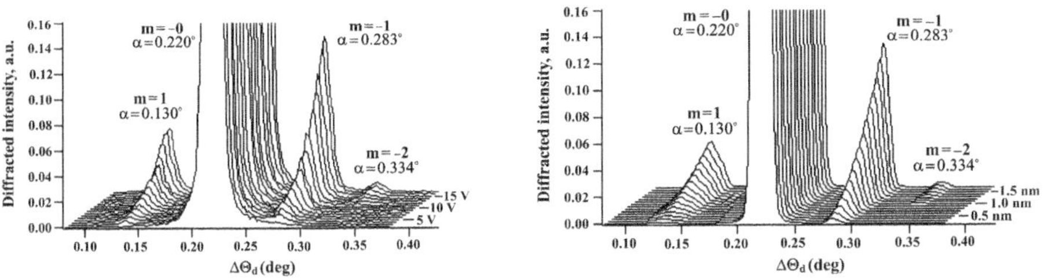

(a) (b)

Fig. 28.10. Experimental (**a**) and calculated (**b**) diffracted X-ray intensity I as a function of the detector scanning angle $\Delta\Theta_{\mathrm{d}}$, obtained at the X-ray incident angle $\Theta_{\mathrm{i}} = 0.22°$, resonance excitation frequency of the SAW, $f_0 = 109$ MHz and at different amplitudes of the SAW: $U = 2$–17 V

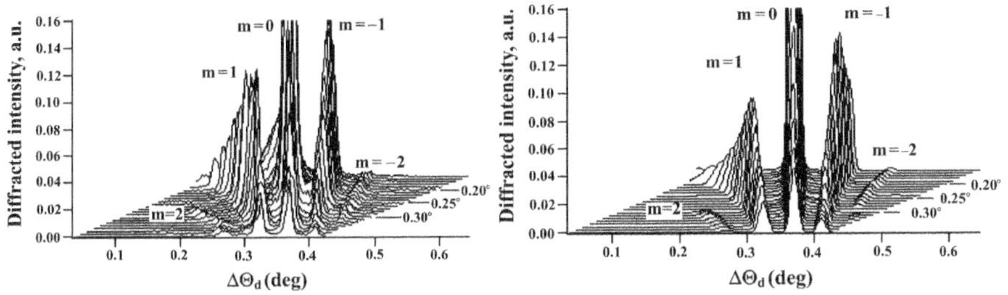

(a) (b)

Fig. 28.11. Experimental (**a**) and calculated (**b**) diffracted X-ray radiation intensity, I, as a function of the detector scanning angle, $\Delta\Theta_{\mathrm{d}}$, obtained at the resonance excitation frequency of the SAW $f_0 = 109$ MHz, amplitude of the input signal on the IDT $U = 17$ V and at different values of the incident angle $\Theta_{\mathrm{i}} = 0.15°$–$0.37°$

divergences between the diffraction orders ($\Delta\Theta_{-1} < \Delta\Theta_1$) is a consequence of the small X-ray incident angle $\Delta\Theta_{\mathrm{i}} = 0.22°$. It is observed that the linewidth is larger for the $m = +1$ peak, which is closer to the surface. This is an effect of the divergence of the incident beam and can be understood by calculating $\mathrm{d}\Theta_m/\mathrm{d}\Theta_{\mathrm{i}}$ from (28.16).

Figure 28.11 shows the experimental (a) and calculated (b) diffracted X-ray radiation intensity, I, as a function of the detector scanning angle, $\Delta\Theta_{\mathrm{d}}$, obtained at the resonance excitation frequency of the SAW, $f_0 = 109$ MHz, with an amplitude of the input signal on the IDT, $U = 17$ V, and at different values of the X-ray incident angle, $\Delta\Theta_{\mathrm{i}} = 0.15°$–$0.37°$. In the calculated dependence, the sinusoidal amplitude, h, is assumed to be 1.7 nm. Figure 28.12 represents the experimental dependence and theoretical curves (full lines) of the diffracted X-ray intensity, I, as a function of the incident angle, $\Delta\Theta_{\mathrm{i}}$, obtained at the resonance excitation frequency of the SAW, $f_0 = 109$ Hz, and at an amplitude of the input signal on the IDT, $U = 17$ V. These dependencies (Figs. 28.11 and 28.12) demonstrate that the $m = +1$ (-1) diffraction order

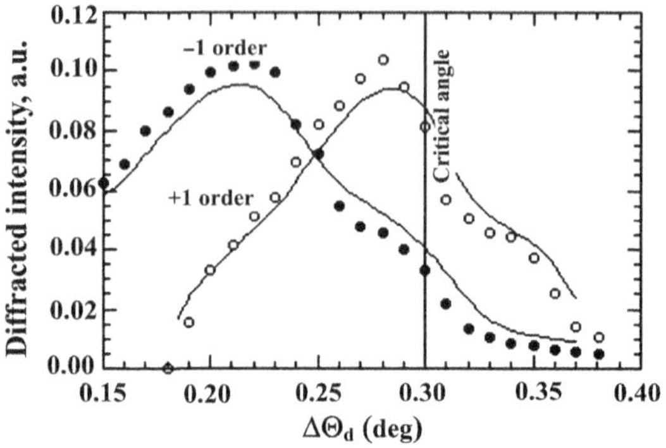

Fig. 28.12. Diffracted X-ray intensity I as a function of the incident angle obtained at the resonance excitation frequency of the SAW, $f_0 = 109\,\mathrm{MHz}$, and at an amplitude of the input signal on the IDT $U = 17\,\mathrm{V}$. The full line shows the calculated values

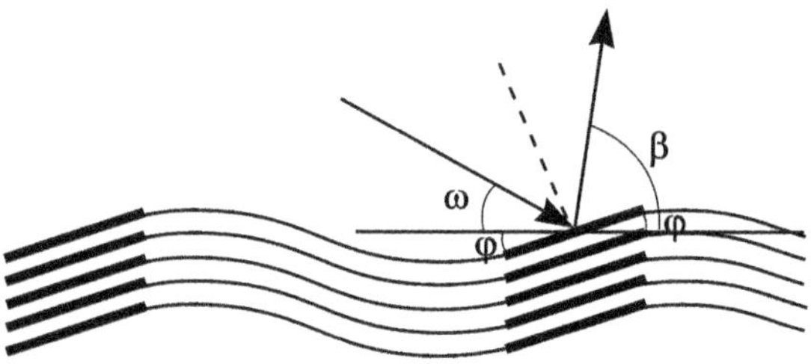

Fig. 28.13. Formation of the reflecting pseudo-lattices

has a maximum intensity at an incident angle of the X-ray beam of $\Theta_\mathrm{i} = 0.22°$ and $0.28°$, respectively.

28.3.3 Multilayer Mirror Modulated by SAW

The main restriction of the total external reflection technique is the generally low efficiency of the diffraction satellites (around 20%) [38]. By using an X-ray mirror this efficiency can be increased.

The next considerations help to predict which incident angle is likely to favor a given diffraction order. Because of the presence of the acoustic wave, the incident angle on the surface varies between $\omega - \varphi$ and $\omega + \varphi$ (Fig. 28.13). Therefore, a strong Bragg reflection occurs if the incident angle, Θ, fulfills the inequality

$$\Theta_\mathrm{B} - \varphi < \omega < \Theta_\mathrm{B} + \varphi \cdots . \qquad (28.17)$$

This situation is indeed possible, since at an incident angle corresponding to (28.17), some parts of the acoustic wave form a new family of reflecting pseudo-planes, for which the Bragg condition is fulfilled (Fig. 28.13)

$$\omega + \varphi = \Theta_\mathrm{B}. \qquad (28.18)$$

The multilayer interference X-ray mirror modulated by SAW thus acts as a diffraction grating, reflecting the maximum intensity in the direction determined by the angle β (Fig. 28.13):

$$\beta = \omega + 2\varphi \cdots . \tag{28.19}$$

The incident angle, ω, giving a maximum intensity in the mth diffraction order, can be determined from (28.15).

Solving (28.18) and (28.15), we obtain

$$\cos\omega - \cos(2\Theta_\mathrm{B} - \omega) = m\lambda/\Lambda \tag{28.20}$$

or, for small incident angles,

$$\omega \approx \Theta_\mathrm{B} - m\lambda/2\Lambda\Theta_\mathrm{B}, \tag{28.21}$$

which is in a good agreement with experimental results. Note that (28.20) is similar to the equation that gives the position of the peaks of maximum intensity on the rocking curve.

The propagation of an X-ray wave in a multilayer interference X-ray mirror modulated by surface acoustic waves can be investigated using the dynamic diffraction theory in distorted crystals presented in [46–48]. In this case, the multilayer acts as an artificial crystal. The deformation field in the crystal (as in the theory of elasticity) is described by the vector $\vec{u}$, representing the displacement of the atoms from the equilibrium position in the perfect crystal. This displacement must satisfy some limitations, the same as in the case of elastic wave propagation in a crystal. The next expression can be used to describe the polarizability of the distorted crystal [49–51]

$$\chi(\vec{r}) = \chi^*(\vec{r} - \vec{u}(\vec{r})), \tag{28.22}$$

where χ^* is the polarizability of the perfect crystal and $\vec{r}$ is the radius-vector [52–54].

The X-ray wave field in the crystal can be written as a sum of modulated waves:

$$\vec{E}(\vec{r}) = \exp\left\{-\mathrm{i}\vec{k}_0\vec{r}\right\} \sum_h \vec{E}_h(\vec{r}) \exp\left\{-\mathrm{i}\vec{h}\vec{r}\right\}, \tag{28.23}$$

where $\vec{k}_0$ is the wave vector of the incident wave and $\vec{h}$ is the vector of the reciprocal lattice.

The distribution of the X-ray wave field in the crystal in the case of two strong waves is described by the following fundamental equations:

$$\begin{cases} -\dfrac{2\mathrm{i}}{k_0}\dfrac{\partial E_0}{\partial s_0} = \chi_{00}E_0 + \chi_{0h}\exp\left(\mathrm{i}\vec{h}\vec{u}\right)E_h, \\ -\dfrac{2\mathrm{i}}{k_0}\dfrac{\partial E_h}{\partial s_h} = (\chi_{00} - \alpha)E_h + \chi_{h0}\exp\left(\mathrm{i}\vec{h}\vec{u}\right)E_0, \end{cases} \tag{28.24}$$

where $\chi^*_{hh'} = \dfrac{1}{V}\displaystyle\int_V \chi^*(\vec{r})\exp\left\{\mathrm{i}\left(\vec{h}' - \vec{h}\right)\vec{r}\right\}\mathrm{d}\vec{r}$

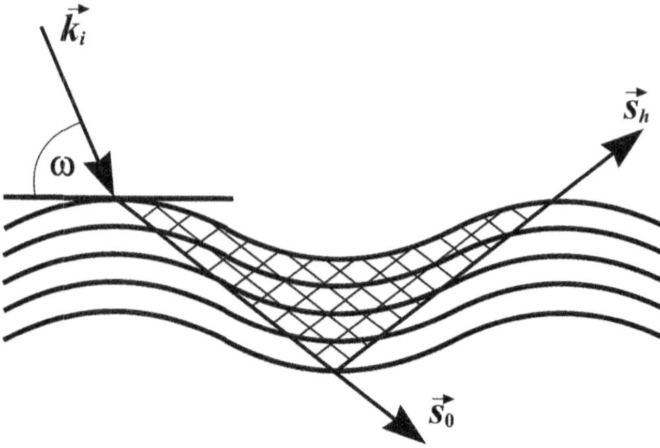

Fig. 28.14. Coordinate system for the calculation of the wave field in the multilayer

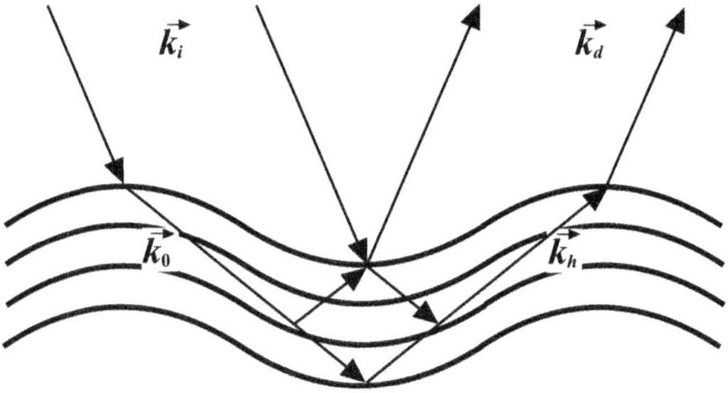

Fig. 28.15. Coordinate systems for the determination of the boundary conditions

are the Fourier coefficients of the polarizability; $\vec{s}_0$ and $\vec{s}_h$ are the unit vectors along the refracted and diffracted waves (Fig. 28.14), $\alpha = \left(k_{\mathrm{h}}^2 - k_0^2\right)/k_0^2$; and V is the volume of the unit cell. It is necessary to take into account some boundary conditions: the continuity of the wave on the surface gives (Fig. 28.15):

$$\begin{aligned}&\vec{E}_{\mathrm{i}}\left(\vec{r}_{\mathrm{e}}\right)\exp\left(-\mathrm{i}\vec{k}_i\vec{r}_{\mathrm{e}}\right)+\vec{E}_{\mathrm{d}}\left(\vec{r}_{\mathrm{e}}\right)\exp\left(-\mathrm{i}\vec{k}_{\mathrm{d}}\vec{r}_{\mathrm{e}}\right)\\&=\vec{E}_0\left(\vec{r}_{\mathrm{e}}\right)\exp\left(-\mathrm{i}\vec{k}_0\vec{r}_{\mathrm{e}}\right)+\vec{E}\left(\vec{r}_{\mathrm{e}}\right)\exp\left(-\mathrm{i}\vec{k}_h\vec{r}_{\mathrm{e}}\right),\end{aligned} \tag{28.25}$$

where indices i,d and 0,h correspond to the incident and the diffracted waves in the vacuum and in the crystal, respectively and $\vec{r}_{\mathrm{e}}$ is the radius-vector of the input surface. In (28.25) it is also assumed that the incident angle is large enough to neglect the reflection. The following two equations are obtained from (28.25):

$$\begin{cases} \vec{E}_0\left(\vec{r}_e\right) = \vec{E}_i\left(\vec{r}_e\right)\exp\left(-i\left(\vec{k}_i - \vec{k}_0\right)\vec{r}_e\right) \\ \vec{E}_d\left(\vec{r}_e\right) = \vec{E}_h\left(\vec{r}_e\right)\exp\left(i\left(\vec{k}_d - \vec{k}_h\right)\vec{r}_e\right). \end{cases} \tag{28.26}$$

These relations describe the phase shifts that occur on the two interfaces: vacuum-crystal and crystal-vacuum. In the case of a non-plane interface, this phase shift changes from one surface element to another and cannot be neglected in the case of a multilayer mirror modulated by SAW.

The calculations of the real diffraction pattern are carried out using the method described in [38]. Thus, we suppose that all the surface elements on which the plane wave, $\vec{E}_d$, falls act as secondary sources of spherical waves:

$$E(p) = -\frac{i}{2\lambda}\int_S E_d \exp\left(-\vec{k}_d\vec{r}_e\right)\frac{e^{ikr}}{r}\begin{Bmatrix} \cos\left(\vec{n}\,\hat{,}\,\vec{k}_d\right) - \\ -\cos\left(\vec{n}\,\hat{,}\,\vec{r}\right) \end{Bmatrix} dS. \tag{28.27}$$

Using (28.22) and taking into account that we investigate the diffracted wave in the far field region ($1/r \approx \text{const}$)

$$\begin{aligned} E(p) &= -\tfrac{i\cos\alpha}{\lambda}\int_S E_h \exp\left(-i\vec{k}_h\vec{r}_e\right)\tfrac{e^{ikr}}{r}dS \\ &= \text{const}\int_S E_h \exp\left\{-i\left[\left(k_{hx} - k\cos\alpha\right)x + \left(k_{hz} - k\sin\alpha\right)z\right]\right\}dS, \end{aligned} \tag{28.28}$$

where β is the diffraction angle, $k = 2\pi/\lambda$, $z = h\sin Kx$ the amplitude of the modulated surface, h the SAW amplitude, $K = 2\pi/\Lambda$, S the modulated surface. The exponential term describes the phase shift due to the refraction on the crystal–vacuum interface.

The following equation is used to calculate the diffracted X-ray intensity

$$I = E(p)E^*(p)\cdots. \tag{28.29}$$

In the calculated intensities, the sinusoidal amplitude, h, of the SAW is assumed to be 1.4 nm. Theoretical curves (see Figs. 28.18 and 28.19) show good agreement with experiments.

The multilayer mirror was deposited on the *YZ*-cut of a $LiNbO_3$ crystal treated initially by chemical dynamic polishing to decrease the roughness to 0.5 nm. The W/C multilayer was produced by magnetron sputtering. It is made of 60 bilayers of 5.3 nm each. The Bragg angle of this multilayer is $\Theta_B = 0.83°$.

An IDT for the SAW excitation with a 4 μm finger width corresponding to $\Lambda = 16$ μm SAW wavelength was also deposited on the free surface of the $LiNbO_3$ crystal. The SAW propagates along the Z axis with a velocity $V = 3.488$ km s^{-1}. The resonance excitation frequency of the IDT is $f_0 = 218$ MHz.

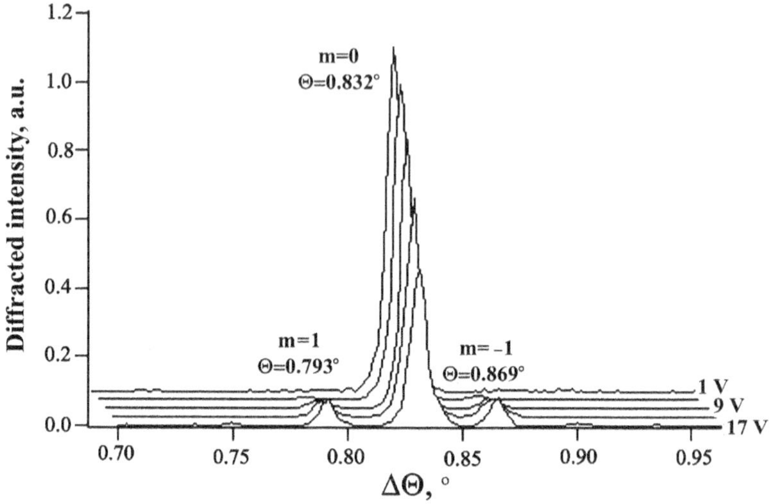

Fig. 28.16. Diffracted intensity as a function of the detector scanning angle $\Delta\Theta_{\mathrm{d}}$ obtained at the Bragg incident angle for different amplitudes of the input signal on the IDT

The diffracted X-ray intensity is recorded by a scintillation detector with a 10 μm input slit.

Figure 28.16 shows the experimental curves of the diffracted X-ray radiation intensity, I, as a function of the detector scanning angle, $\Delta\Theta$, where $\Delta\Theta_{\mathrm{B}} = 0.823°$ and for different values of the amplitude of the input sinusoidal signal on the IDT in the range of 1–17 V. The diffraction satellites are observed at the angular deviations $\Delta\Theta_{+1} = 0.039°$ and $\Delta\Theta_{-1} = 0.038°$ from the intense reflected beam. These values can be precisely predicted from the grating equation.

The intensities of the $m = +1, \ -1$ satellites increase and the zero order satellite decrease with the SAW amplitude. The maximum intensities of these diffraction orders make up 7% of the intensity of the diffracted beam without SAW excitation. The difference in the angular deviations between the diffraction orders $\Delta\Theta_{+1} > \Delta\Theta_{-1}$ is in agreement with (28.10).

Figure 28.17 presents rocking curves obtained with the detector placed in the Bragg position $\Delta\Theta_{\mathrm{B}} = 0.823°$, and for different amplitudes of the input signal on the IDT. The satellite intensities increase rapidly with the SAW amplitude while the zero order peak decreases. In the case of $U = 17\,\mathrm{V}$, the intensities of the $+1, \ -1$ satellites become higher than the zero order intensity. At this voltage, the intensity in the $+1$ and -1 satellites reaches 68% of the Bragg peak without SAW excitation.

Figure 28.18 shows the (a) experimental measurements and (b) calculations based on the model developed below for the diffracted X-ray intensity, E, as a function of the detector scanning angle, $\Delta\Theta_{\mathrm{d}}$, for different incident

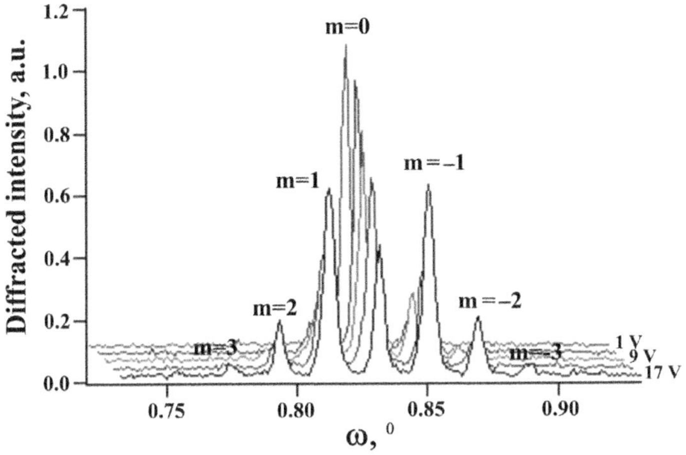

Fig. 28.17. Rocking curves with the detector in the Bragg position for different amplitudes of the input signal on the IDT

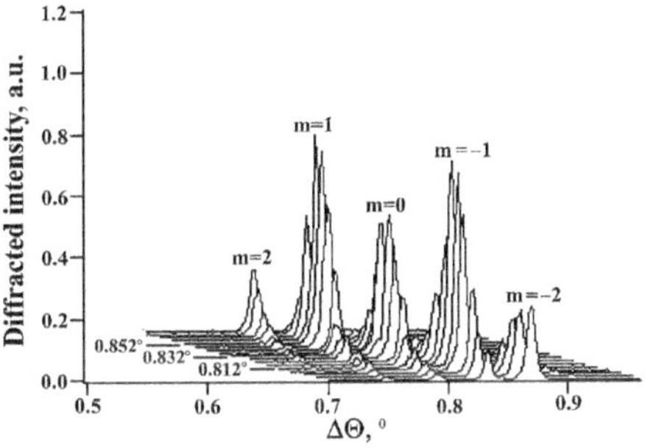

(a)

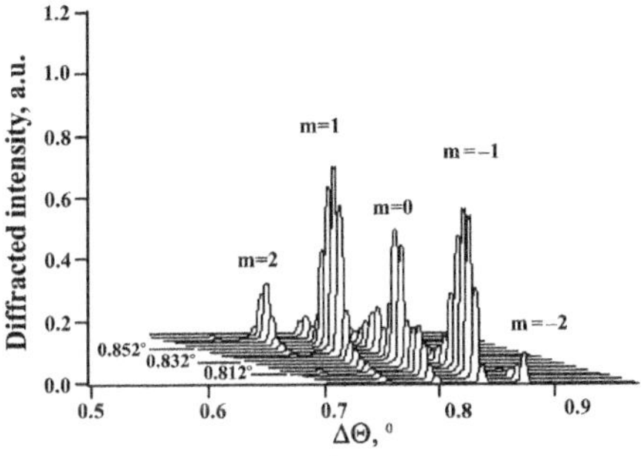

(b)

Fig. 28.18. Experimental (**a**) and calculated (**b**) diffracted intensity as a function of the detector scanning angle $\Delta\Theta$ obtained for $U = 17\,\mathrm{V}$ ($h = 1.4\,\mathrm{nm}$) and for various incident angles Θ between $0.792°$ and $0.872°$

angles. The $m = \pm 1$ diffraction orders reach a maximum intensity ~58% for an incident angle of $\Theta_\mathrm{i} = 0.808°$ and $0.848°$, respectively.

The same phenomena take place for the ± 2 order satellites: the maximum intensity ~22% was obtained for an incident angles $\Theta_\mathrm{i} = 0.792°$ and $0.872°$. This means that, to obtain the maximum energy diffracted towards a precise satellite, it is necessary to lightly shift the incident angle away from the exact Bragg angle of the multilayer.

In Fig. 28.19 are shown the experimental and calculated maximal intensities of the ± 1 and 0 order peaks for various incident angles, Θ_i. The maximum intensity in the diffraction orders corresponds to the minimum intensity of the zero order, as was demonstrated previously for the volume gratings, etched in a multilayer mirror (Fig. 28.7).

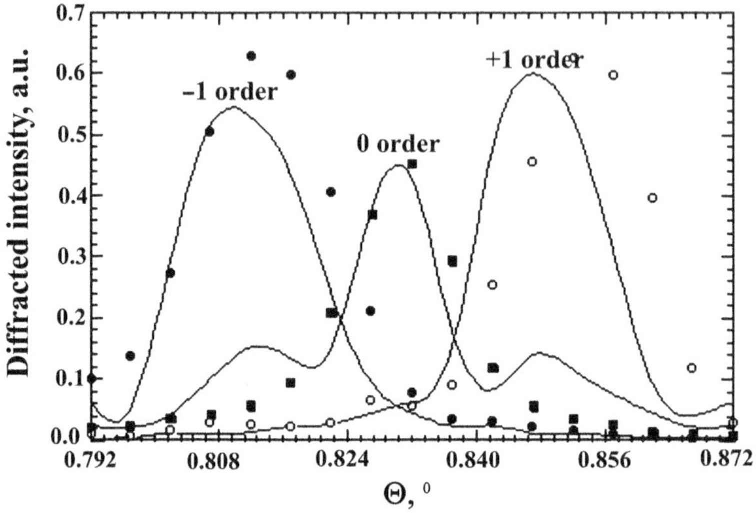

Fig. 28.19. Maximum diffracted intensity of the +1, 0, −1 peaks as a function of incident angle Θ_i, $f_0 = 218\,\mathrm{MHz}$, $U = 17\,\mathrm{V}$. The solid line shows calculated values

28.3.4 Crystals Modulated by SAW

In contrast to acoustically modulated multilayer mirrors, a crystal, modulated by SAW, acts much more effectively because, for the crystal, the value of the SAW amplitude can exceed the interplanar spacing. The same diffraction efficiency can be obtained with much lower acoustic amplitude than required for an X-ray mirror. In this section the X-ray diffraction by langasite (LGS) crystal, ($La_3Ga_5SiO_{14}$), excited by SAW is presented.

Figure 28.20 shows the calculated amplitude of the crystal lattice displacements in LGS caused by SAW propagation vs. crystal depth. The calculation [55] is based on the elastic and piezoelectric properties of the LGS [56]. It is seen that the SAW penetration depth inside the crystal is approximately one SAW wavelength (see component u_1 normal to the crystal surface). The longitudinal component, u_2, is parallel to the direction of the SAW propagation. The presence of the transverse displacement component, u_3, suggests that the propagation direction of the acoustic energy flow does not coincide with the SAW wave vector direction.

X-ray diffraction on acoustically modulated atomic planes gives rise to diffraction satellites on both sides of the Bragg peak. In case of symmetric Bragg reflections, the angular position of diffraction satellites can be determined from the grating (28.15).

The angle between adjacent satellites measured on a rocking curve can be deduced from (28.15)

$$\delta\Theta_{m\mathrm{RC}} = m\lambda/2\Lambda \sin\Theta_{\mathrm{B}} = md/\Lambda, \tag{28.30}$$

where d is the interplanar spacing.

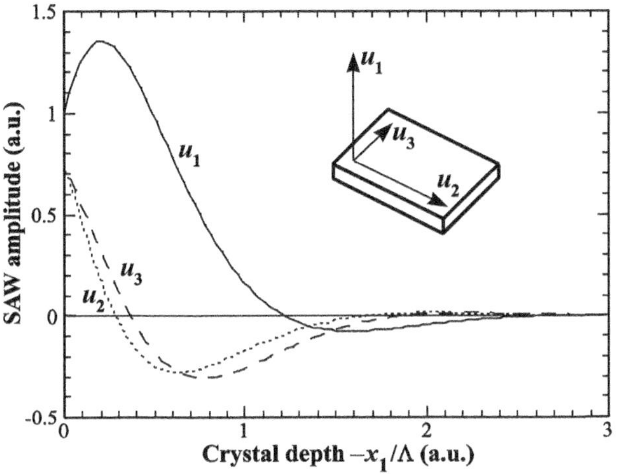

Fig. 28.20. Calculated SAW amplitudes vs. crystal depth: normal (u_1), longitudinal (u_2) and transverse components (u_3)

LGS is a piezoelectric crystal of space group symmetry 32. The crystal lattice is similar to that of quartz with the parameters $a = 0.817\,\mathrm{nm}$ and $c = 0.5095\,\mathrm{nm}$ [57].

An X-cut, (110) atomic planes parallel to the crystal surface, of LGS was used for this experiment. To excite a Rayleigh SAW, IDT was deposited on the crystal surface by photolithography. At the resonance excitation frequency, $f_0 = 192.5\,\mathrm{MHz}$, the SAW wavelength was $\lambda = 12\,\mu\mathrm{m}$ and the propagation velocity was $V = 2{,}310\,\mathrm{ms}^{-1}$.

Rocking curves were measured at various SAW amplitudes. The X-ray energy was 11 keV. The interplanar spacing for the (110) reflection in LGS is $d = 0.4087\,\mathrm{nm}$. In the kinematic approximation the X-ray penetration depth inside the crystal depends on the absorption in LGS as a function of energy given by

$$\mu_z^{-1}(E) = \sin(\Theta_\mathrm{B}(E))/2\mu_\mathrm{l}(E), \tag{28.31}$$

where μ_l is the linear absorption coefficient and Θ_B is the Bragg incident angle. This dependence is shown in Fig. 28.21. The K-edge of Ga at 10.47 keV causes a drastic change in the absorption coefficient. At the energy of 11 keV, the X-ray penetration depth reaches only $\mu_z^{-1} = 0.48\,\mu\mathrm{m}$, which is much less than the SAW penetration depth inside the crystal ($\mu_z^{-1}/\mu_\mathrm{SAW}^{-1} < 1$).

Figure 28.22 shows selected rocking curves for LGS (reflection (110)) excited by a $\Lambda = 12\,\mu\mathrm{m}$ SAW at 11 keV measured at various input voltages (U) supplied to the IDT. The Bragg incident angle is $\Theta_\mathrm{B} = 7.92°$. The FWHM of the Bragg peak without SAW excitation is 3.2 arcsec (Fig. 28.22a).

Figure 28.22 shows that the number of diffraction satellites observed on the rocking curve increases with the amplitude of the input signal supplied to the IDT, i.e. with the SAW amplitude.

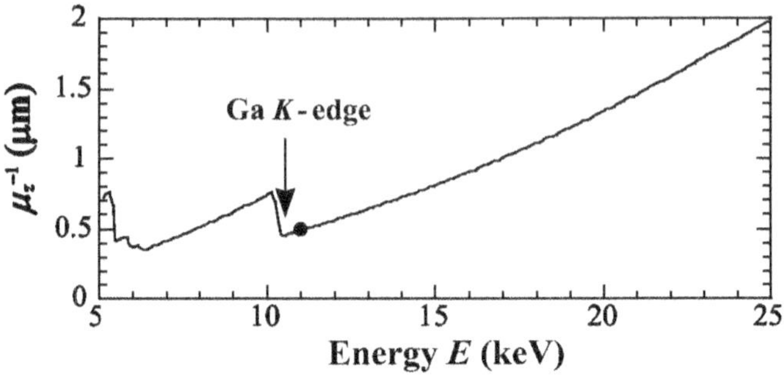

Fig. 28.21. X-ray penetration depth in an LGS crystal for (110) reflection vs. energy. The black circle shows the energy 11 keV

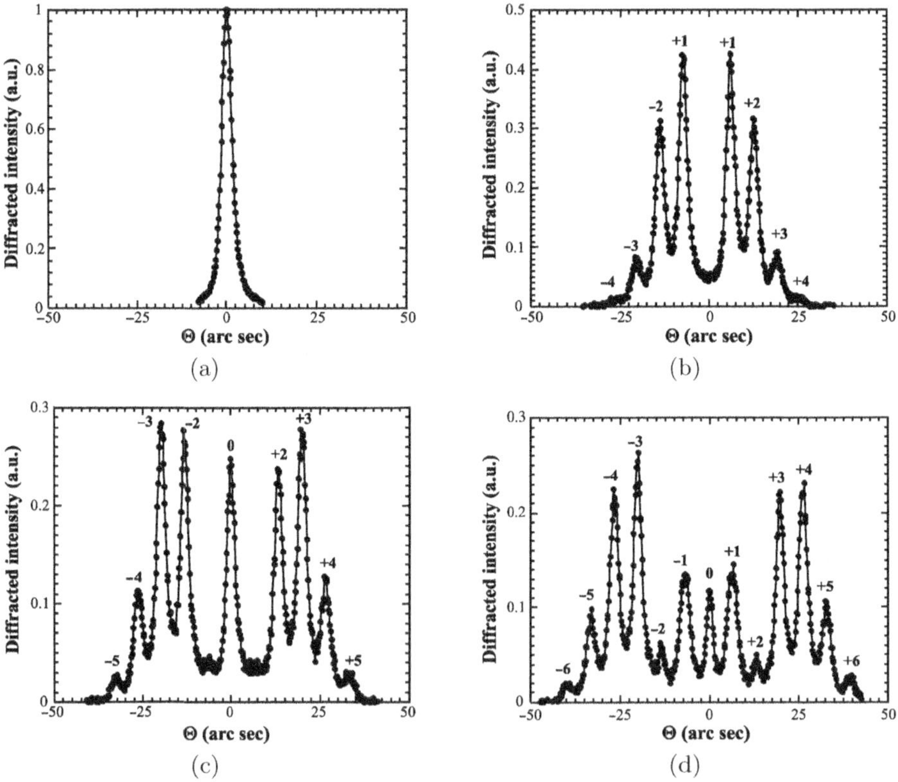

Fig. 28.22. Rocking curves measured for different amplitudes of the input signal supplied to the IDT: (**a**) $U = 0\,\text{V}$, (**b**) $U = 8.5\,\text{V}$, (**c**) $U = 14\,\text{V}$, (**d**) $U = 18\,\text{V}$

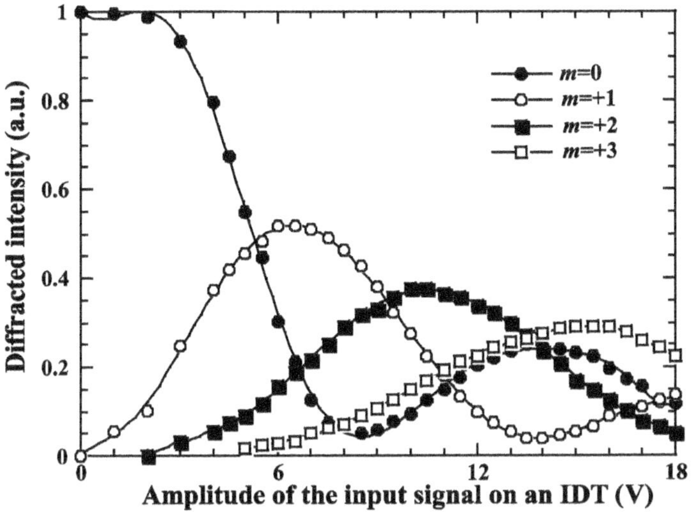

Fig. 28.23. Intensities of the diffraction satellites ($m = 0, 1, 2, 3$) vs. amplitude of the input signal supplied to the IDT. Black circles, squares, triangles and diamonds: experimental data. Solid lines: calculated data. $E = 11\,\text{keV}$; $\Lambda = 12\,\mu\text{m}$; (110) reflection

The angular divergence between two neighboring diffraction satellites, $\delta\Theta_{m\text{RC}}$, is 6.8 arcsec, which agrees quite well with the value 6.9 arcsec calculated from (28.30).

Intensities of selected diffraction satellites ($m = 0, 1, 2, 3$) as a function of the input voltage on the IDT are shown in Fig. 28.23. The intensity of the diffraction satellites, except for satellite $m = 0$, develops as soon as the acoustic amplitude reaches a threshold value, which increases with the diffraction order. After rapidly reaching a maximum, the satellite intensity decreases smoothly and oscillates.

It can be seen (Fig. 28.22b) that the intensity of the $m = 0$ diffraction satellite is equal to zero for $U = 8.5\,\text{V}$. For this specific SAW amplitude, the phase shift of the X-ray radiation diffracting into the zero satellite from the SAW minima and maxima regions (where the atomic planes are still parallel to the surface) is equal to π. This phenomenon can be observed only if the acoustic wave field probed by X-rays is very homogenous in amplitude. This is, therefore, only possible if the X-ray absorption is strong enough to avoid any interaction with deep regions of the crystal where the acoustic amplitude is strongly damped. In the case of $LiNbO_3$, the absorption is never high enough to achieve the complete extinction of a satellite except in the case of an asymmetric reflection for which the incident angle can be very small [29].

For $U = 14\,\text{V}$, the extinction of the $m = +1(-1)$ diffraction satellites is observed and can also be explained by the π-phase shift between crystal regions diffracting towards this satellite (Fig. 28.22c). The maximum value of the input signal amplitude is $U = 18\,\mathrm{V}$. At this amplitude, the intensity of the $m = +2(-2)$ diffraction satellite decreases, although no complete extinction of these satellites occurs (Fig. 28.22d).

As calculated in [29], the intensity of the mth diffraction satellite is proportional to

$$I_m \propto \left| \int_0^\infty \exp(-\mu_z z)\, J_m\left(h_0 q_z u_1(z)\right) \mathrm{d}z \right|^2 , \tag{28.32}$$

where J_m is the mth order Bessel function. Figure 28.23 shows diffraction satellite intensities (solid lines) vs. SAW amplitude calculated by (28.32). There is good agreement between experimental results and calculations showing that numerical calculations of u_1 based on the elastic and piezoelectric tensors of the langasite crystal and on the Rayleigh wave characteristics are correct. These results are much better than those for the case of SAW propagation in a $LiNbO_3$ crystal where kinematic simulations were useless especially at low acoustic amplitudes and for 0 and 1 order satellites [29]. This difference can be explained by the fact that for the X-ray penetration depth, being so small in a langasite crystal at 11 keV, the X-rays interact only with strongly distorted regions of the crystal. If this was not the case, dynamical theory should be necessary to take into account the contribution to the diffracted intensity coming from deep non-distorted (i.e. perfect) regions of the crystal [31].

References

1. A. Gupta, C. Meneghini, A. Saraiya, G. Principi, D.K. Avasthi, Nucl. Instrum. Methods Phys. Res. **B212**, 458 (2003)
2. T.W. Barbee Jr., SPIE Proc. **911**, 169 (1988)
3. V.V. Aristov, S.V. Gaponov, V.M. Genkin, Y.A. Gorbatov, A. Erko V.V. Martynov, L.A. Matveeva, N.N. Salaschenko, JETPh Lett. **44**(4), 265 (1986)
4. A. Erko, in *Materials Science Forum*, vol. 321–324 (Trans Tech Publications, Switzerland, 2000), pp. 174–178
5. A. Erko, V.V. Aristov, B. Vidal, *Diffraction X-Ray Optics* (IOP Publishing, Bristol, 1996)
6. A. Firsov, A. Svintsov, A. Erko, W. Gudat, S. Kuznetsov, M. Grigoriev, A. Asryan, M. Ferstl, S. Shapoval, V.V. Aristov, Nucl. Instrum. Methods Phys. Res. **A467–A468**, 366 (2001)
7. A. Firsov, A. Svintsov, S.I. Zaitsev, A. Erko, V.V. Aristov, Opt. Commun. **202**, 55 (2002)
8. A. Erko, A. Firsov, Proc. SPIE **5539**, 148 (2004)
9. A.I. Erko, B. Vidal, P. Vincent, Y.A. Agafonov, V.V. Martynov, D.V. Roschupkin, M. Brunel Nucl. Instrum. Methods Phys. Res. **A333**, 599 (1993)
10. A. Erko, Y.A. Agafonov, L.A. Panchenko, A. Yuakshin, P. Chevallier, P. Dhez, F. Legrand, Opt. Commun. **106**, 146 (1994)

11. P. Vincent, in *Electromagnetic Theory of Gratings*, ed. by R. Petit. Topics in Current Physics, vol. 22, (Springer, Berlin Heidelberg New York, 1980) pp. 101–121
12. E. Aristova, Ch. David, A. Freund, Ya. Hartman, B. Kaulich, A. Snigirev, V. Yunkin, in *X-Ray Microscopy IV*, ed by V.V. Aristov, A.I. Erko, (Bogorodskii Pechatnik, 1994) pp. 617–620
13. V.G. Kohn, The Theory of X-Ray Bragg-Fresnel Focusing by Flat and Elastically Bent Lens, Preprint IAE-5878/9. M., (1995)
14. V.V. Aristov, A.I. Erko, A.Y. Niculin, A.A. Snigirev, Opt. Commun. **58**, 300 (1986)
15. Ya. Hartman, A. Freund, A. Snigireva, A. Snigirev, A. Souvorov, Nucl. Instrum. Methods Phys. Res. **A385**, 371 (1997)
16. Ya. Hartman, V.A. Yunkin, A. Snigirev, J. X-Ray Sci. Technol. **6**, 249 (1996)
17. A. Erko, A. Firsov, Proc. SPIE **5539**, 148 (2004)
18. A. Pardo, J.-M. Andre, A. Sammar, J. Opt. (Paris), **22**(3), 141 (1991)
19. A. Sammar, J.-M. Andre, J. Opt. Commun. **86**, 245 (1991)
20. M. Brunel, A.I. Erko, V.V. Martynov, B. Vidal, P. Vincent, A. Yuakshin, D.V. Roshchoupkin, Nucl. Instrum. Methods Phys. Res. **A339**, 617 (1994)
21. B. Vidal, P. Vincent, P. Dhez, M. Neviere, Proc. SPIE, **556**, 142 (1985)
22. S. Bat, G. Soullie, A. Mirone, M. Idir, P. Gukrin, F.-R. Ladan, P. Troussel, R. Barchewitz, B. Vidal, Opt. Commun. **144**, 281 (1997)
23. V.A. Chernov, V.I. Erofeev, N.I. Chkhalo, N.V. Kovalenko, S.Y. Mytnichenko, Nucl. Instrum. Methods Phys. Res. **A405**, 310 (1998)
24. V.V. Martynov, Yu. Platonov, Rev. Sci. Instrum. **73**, 1551 (2002)
25. V.A. Chernov, V.I. Erofeev, N.I. Chkhalo, N.V. Kovalenko, S.Y. Mytnichenko, Nucl. Instrum. Methods Phys. Res. **A359**, 138 (1995)
26. M. Neviere, J. Opt. Soc. Am. **A8**, 1468 (1991)
27. R. Tucoulou, R. Pascal, M. Brunel, O. Mathon, D.V. Roshchupkin, I.A. Schelokov, E. Cattan, D. Remiens, J. Appl. Crystallogr. **33**, 1019 (2000)
28. W. Sauer, M. Streibl, T.H. Metzger, A.G.C. Haubrich, S. Manus, A. Wixforth, J. Peisl, Appl. Phys. Lett. **75**, 1709 (1999)
29. R. Tucoulou, F. de Bergevin, O. Mathon, D. Roshchupkin, Phys. Rev. **B64**, 134108(9) (2001)
30. D.V. Roshchupkin, D.V. Irzhak, R. Tucoulou, O.A. Buzanov, J. Appl. Phys. **94**, 6692 (2003)
31. I.A. Schelokov, D.V. Roshchupkin, D.V. Irzhak, R. Tucoulou, J. Appl. Crystallogr. **37**, 52 (2004)
32. R. Tucoulou, O. Mathon, C. Ferrero, V. Mocella, D.V. Roshchupkin, R.E. Kumon, J. Appl. Phys. **97**, 113505(5) (2005)
33. A.I. Erko, D.V. Roshchupkin, A.A. Snigirev, A.M. Smolovich, A. Yu Nikulin, G.V. Vereshchagin, Nucl. Instrum. Methods Phys. Res. **A282**, 634 (1989)
34. D.V. Roshchupkin, I.A. Schelokov, R. Tucoulou, M. Brunel, Nucl. Instrum. Methods Phys. Res. **B129**, 414 (1997)
35. R. Tucoulou, D.V. Roshchupkin, I.A. Schelokov, M. Brunel, L. Ortega, E. Ziegler, M. Lingham, C. Mouget, S. Douillet, Nucl. Instrum. Methods Phys. Res. **B132**, 207 (1997)
36. R. Tucoulou, D.V. Roshchupkin, O. Mathon, I.A. Schelokov, M. Brunel, E. Ziegler, C. Morawe, J. Synch. Radiat. **5**, 1357 (1998)
37. D.V. Roshchupkin, M. Brunel, F. de Bergevin, A.I. Erko, Nucl. Instrum. Methods Phys. Res. **B72**, 471 (1992)

38. D.V. Roshchupkin, I.A. Schelokov, R. Tucoulou, M. Brunel, IEEE Trans. Ultrason. Ferroelectrics Freq. Contr. **42**(1), 127 (1995)
39. D.V. Roshchupkin, M. Brunel, Rev. Sci. Instrum. **64**(2), 379 (1993)
40. R. Tucoulou, M. Brunel, D.V. Roshchupkin, I.A. Schelokov, Rev. Sci. Instrum. **69**(7), 2704 (1998)
41. J.W.S. Rayleigh, *Theory of Sound*, vol. 2 (Dover, New York, 1945)
42. A. Yariv, P. Yeh, *Optical Waves in Crystals* (Wiley, New York, 1984)
43. E. Salzmann, D. Weismann, J. Appl. Phys. **40**, 3408 (1969)
44. E.P. Ippen, Proc. IEEE **55**, 245 (1967)
45. A. Yariv, *Optical Electronics* (Holt, Renehart and Winston, New York, 1985)
46. S. Takagi, Acta Cryst. **15**, 1131 (1962)
47. S. Takagi, J. Phys. Soc. Jpn. **26**, 1239 (1969)
48. A.M. Afanas'ev, V.G. Kohn, Acta Cryst. **A27**, 421 (1971
49. N. Kato, J. Phys. Soc. Jpn. **18**, 1785 (1963)
50. N. Kato, J. Phys. Soc. Jpn. **19**, 67 (1963)
51. N. Kato, J. Phys. Soc. Jpn. **19**, 971 (1963)
52. N. Kato, J. Phys. Soc. Jpn. **18**, 1785 (1963)
53. N. Kato, J. Phys. Soc. Jpn. **19**, 67 (1963)
54. N. Kato, J. Phys. Soc. Jpn. **19**, 971 (1963)
55. E. Dieulesaint, D. Royer, *Ondes Elastique dans les Solids* (Masson, Paris, 1974)
56. J. Bohm, E. Chilla, C. Flannery, H.-J. Fröhlich, T. Hauke, R.B. Heimann, M. Hengst, U. Straube, J. Cryst. Growth **216**, 293 (2000)
57. E. Chilla, C.M. Flannery, H.-J. Fröhlich, U. Straube, J. Appl. Phys. **90**, 6084 (2001)

29

High Resolution 1D and 2D Crystal Optics Based on Asymmetric Diffractors

D. Korytár, C. Ferrari, P. Mikulík, F. Germini, P. Vagovič, and T. Baumbach

Abstract. The development of high resolution X-ray measurements and imaging in real and reciprocal space is related to the improvement of the optical elements available for use. Crystal diffractive optics still give the highest resolution in reciprocal space and in energy, and progress has also been made in improving resolution in real space. In this chapter a short introduction to the dynamical theory behind crystal diffractors and their coupling is given and modern one- and two-dimensional elements based on symmetric, asymmetric and inclined diffractions are introduced. The design, the modeling of the output parameters and the experimental results are presented for a special 2-bounce V-shaped monochromator, for a monolithic 4-bounce monochromator and for a monolithic 2D beam de/magnifier.

29.1 Introduction

The optical scheme of a high-resolution X-ray diffractometer (HRXRD) includes an X-ray source (laboratory or synchrotron), beam conditioning optics, sample, analyzer, and detector. A compromise must be found between high intensity on one side and high resolution in real and reciprocal space or high energy resolution on the other side. The best resolution in reciprocal space is achieved with crystal X-ray optics based mainly on Bragg diffraction from perfect crystals such as silicon and germanium. Si(111) double crystal monochromators for synchrotron and Ge (220/440) Bartels type monochromators for laboratory sources are typical examples. High resolution in reciprocal space is given by the small widths of Bragg diffraction peaks of these crystals, approaching theoretical values in the arcsec range as given by dynamical theory of X-ray diffraction. The full width at half maximum of these peaks can be further decreased using asymmetric diffraction in a grazing incidence setting. This setting is also known as a one-dimensional (1D) beam expander in real space. Combining two asymmetric diffractors with mutually perpendicular scattering planes a two-dimensional (2D) image magnification can be obtained.

This chapter gives a short introduction to modern 1D and 2D crystal optics for high-resolution diffractometry and imaging which are based on asymmetric diffractors. It describes channel-cut monochromators and monolithic 2D beam de/magnifiers, the theoretical background used in their design, the modeling of their parameters using the DuMond diagrams and the spectral functions, and beam tracing, as well as the experimental results obtained mainly with monolithic devices.

29.2 Scattering Geometries and Crystal Diffractors

There are various scattering geometries used for the X-ray diffraction characterization of crystals and thin layers. In the elastic scattering process the incident monochromatic plane wave with wavevector $\mathbf{K_0}$ is elastically scattered into a wave with the wavevector $\mathbf{K}$, where $|\mathbf{K_0}| = |\mathbf{K}| = \mathbf{K} = 2\pi/\lambda$, and λ is the wavelength. This process is characterized by the *scattering vector* (or *wavevector transfer*) $\mathbf{Q} = \mathbf{K} - \mathbf{K_0}$.

In a crystal or in a *crystal diffractor*, which is the physical part of a crystal characterized by the reciprocal lattice vector $\mathbf{H_{hkl}}$ and the in-crystal surface normal $\mathbf{n}$, *which is used to diffract the incident X-ray beam with wavevector* $\mathbf{K_0}$, (see Fig. 29.1), most of the scattered intensity is located in a small region surrounding the reciprocal lattice point, H. Such lattice points are characterized by corresponding lattice vectors $\mathbf{H} = \mathbf{H_{hkl}} = \mathbf{OH}$, where $|\mathbf{H_{hkl}}| = 2\pi/d_{hkl}$ and d_{hkl} is the interplanar distance of the lattice planes (*hkl*). The structure factor, F_{hkl}, determines the amplitude of the scattered wave. The purpose of *reciprocal space mapping* is to determine the scattered

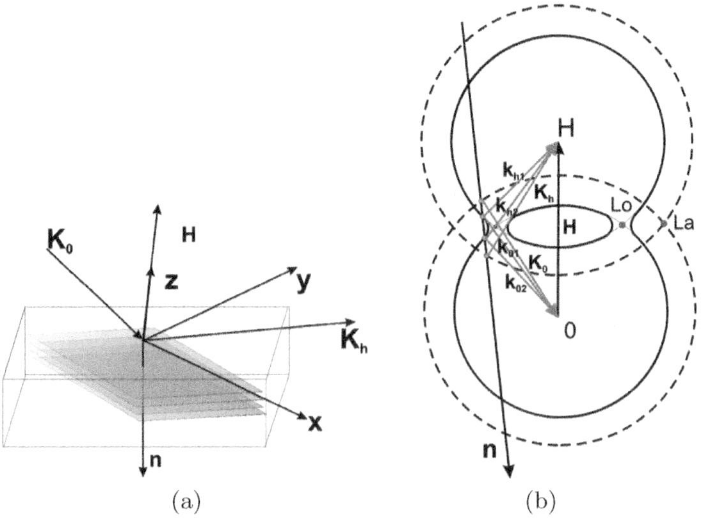

Fig. 29.1. A flat crystal X-ray diffractor in real (**a**) and reciprocal (**b**) space

intensity in reciprocal space. The kinematical theory of X-ray diffraction in an infinite crystal leads to the Bragg law in the vectorial form

$$\mathbf{K_H} - \mathbf{K_0} = \mathbf{H} \tag{29.1}$$

and implies the concept of Ewald sphere [1]. The scalar Bragg law has the form

$$2d_{hkl} \sin\theta_\mathrm{B} = \lambda, \tag{29.2}$$

where θ_B is the Bragg angle between the incident or the diffracted beam and diffracting planes.

If $\mathbf{n}$ is the in-crystal surface normal, the *plane of incidence* $(\mathbf{K_0}, \mathbf{n})$ is generally not parallel to the *dispersion plane* $(\mathbf{n}, \mathbf{H})$, and we have a *noncoplanar geometry*. In the *coplanar geometry*, where the *plane of incidence* is parallel to the *dispersion plane*, the reciprocal space maps of scattered intensity are functions of $\mathbf{Q}$ while in the *noncoplanar case* this intensity depends on both vectors $\mathbf{K_0}$ and $\mathbf{K_H}$ independently [7]. In the semiinfinite crystal, due to the presence of the vacuum-crystal interface, there is additional scattering from this interface which modifies the vacuum waves $\mathbf{K_0}$ and $\mathbf{K_H}$ into the *in-crystal* (*refracted*) waves $\mathbf{k_0}$ and $\mathbf{k_H}$. The requirement of momentum conservation leads to the continuity of the tangential component of the wavevectors (the in-crystal and vacuum wavevectors differ only by a component along the surface normal $\mathbf{n}$)

$$\mathbf{k_0} = \mathbf{K_0} + \kappa\mathbf{n} \tag{29.3}$$

$$\mathbf{k_H} = \mathbf{K_H} + \kappa_H\mathbf{n} = \mathbf{K_0} + \mathbf{H} + \kappa_H\mathbf{n}. \tag{29.4}$$

The quantities κ and κ_H have a magnitude of about $\sim K|\chi_0|/2$, where χ_0 is expressed generally by (29.10) (see below). Using (29.3) and (29.4) the wavevector of the diffracted wave in vacuum can be expressed as

$\mathbf{K_H} = \mathbf{K_0} + \mathbf{H}'$, where $\mathbf{H}' = \mathbf{H} + \mathbf{\Delta_H}$, $\mathbf{\Delta_H} = \Delta_H\ \mathbf{n}$, where $\Delta_H = \kappa - \kappa_H$. The additional *momentum transfer*, Δ_H, at the vacuum-crystal interface can be expressed as

$$\Delta_H = K\left(-\gamma_H \pm \sqrt{\gamma_H^2 - \alpha_H}\right), \tag{29.5}$$

$$\text{where } \gamma_H = \frac{(\mathbf{K_0} + \mathbf{H}) \cdot \mathbf{n}}{K} = \gamma_0 + \frac{(\mathbf{H} \cdot \mathbf{n})}{K}, \gamma_0 = \frac{\mathbf{K_0} \cdot \mathbf{n}}{K} \tag{29.6}$$

are the directional cosines of the vector $\mathbf{K_0} + \mathbf{H}$ and $\mathbf{K_0}$, respectively, with respect to the inward crystal surface normal, $\mathbf{n}$, and

$$\alpha_H = \frac{2\mathbf{K_0} \cdot \mathbf{H} + \mathbf{H}^2}{K^2}. \tag{29.7}$$

This dimensionless parameter is a function of the magnitude and the direction of $\mathbf{K_0}$ relative to $\mathbf{H}$. The direction cosine of the diffracted beam $\mathbf{K_H}$ is

$$\gamma_H' = \frac{\mathbf{K_H} \cdot \mathbf{n}}{K} = \pm\sqrt{\gamma_H^2 - \alpha_H}. \tag{29.8}$$

The general expressions above can be simplified for special cases such as symmetrical, asymmetrical, symmetrical inclined, and asymmetrical inclined cases. Two additional important cases, which are not covered by the usual approaches, are for a Bragg angle close to 90° and for an extreme asymmetry with the angle of incidence close to the critical angle. These are described in [18].

29.3 Basic Results of Dynamical Theory

Historically, several approaches have been taken to solve the problem of the scattering of an incident X-ray wave by a crystal. One of them is to solve Maxwell's equations for a plane monochromatic electromagnetic radiation wave $\mathbf{E}(\mathbf{r}, t) = \mathbf{E}_i \exp[\mathrm{i}(\mathbf{K_0 r} - \omega t)]$ scattered by a medium with an electric susceptibility $\chi(\mathbf{r})$. The susceptibility is a continuous function with the periodicity of the crystal lattice

$$\chi(\mathbf{r}) = \Sigma \chi_H \exp(\mathrm{i}\mathbf{Hr}), \tag{29.9}$$

where

$$\chi_H = -\frac{r_\mathrm{e} F_H}{\pi V}\lambda^2 \tag{29.10}$$

are its Fourier components (generally complex), r_e is the classical radius of electron, F_H the structure factor of the crystal unit cell and V is the volume of the unit cell. The solution gives a Bloch wave

$$D(r) = \sum_H D_H \mathrm{e}^{\mathrm{i}k_H r} \tag{29.11}$$

inside the crystal, composed of an infinite number of plane waves $\mathbf{k_H} = \mathbf{k_0} + \mathbf{H}$. The amplitudes of the electric induction vectors D_H obey the system of fundamental equations of the dynamical theory

$$\frac{k_H^2 - K^2}{K^2} D_H = \sum_{H'} \chi_{H-H'} D_{H'} \tag{29.12}$$

from which it comes that the only significantly excited component waves are those for which $k_H \sim K|1 + \chi_0|/2$, which is called the *excitation condition*. Single-, two-, three- and multiple- beam diffraction can occur. We will concentrate on two-beam cases.

For the two-beam case, the condition for the existence of a nonzero solution (determinant of the system of linear equations (29.12) equal to zero) leads to the dispersion equation in the form [9]

$$\left(k_0^2 \chi_\mathrm{p} - K^2\right)\left(k_H^2 \chi_\mathrm{p} - K^2\right) = C^2 k_0^2 k_H^2 \chi_H \chi_{\bar{\mathrm{H}}}, \tag{29.13}$$

where $\chi_{\mathrm{p}} = 1 - \chi_0$, and C depends on polarization ($C_\sigma = 1$ and $C_\pi = (\mathbf{k_0} \cdot \mathbf{k_H})/(k_0 k_H)$. This is generally an equation of the fourth degree in Δ_H with four solutions for in-crystal wavevectors as depicted in Fig. 29.1b).

A number of approximations were used for various simplified cases [1]. For the classical two-beam case there are two waves, $\mathbf{k_0}$ and $\mathbf{k_H}$, inside the crystal for each polarization and two fundamental equations (above) for the unknowns D_0 and D_H.

For the Bragg case for a semiinfinite crystal where the diffracted beam is directed back into the vacuum semispace, the amplitude, and wavevector boundary conditions of continuity at the interface [1, 9] lead to the full determination of the radiation fields, $\mathbf{D_H}$, and of their intensities $|\boldsymbol{D}_H|^2$ and reflectivities

$$R = |b_H|^{-1} \left|D_H(z_0)\right|^2 \Big/ |D_i|^2, \tag{29.14}$$

$$\text{where } b_H = \frac{\gamma_0}{\gamma_H} \tag{29.15}$$

is the asymmetry factor. The reflectivity, R, corresponding to a measurable dependence of the scattered intensity on the angular setting, the *rocking curve*, can be written in the form

$$R = \left|-y \pm \sqrt{y^2 - 1}\right|^2, \quad \text{where} \quad y = \frac{\alpha b + \chi_0(1-b)}{2\,|C\chi_H|\,\sqrt{|b|}}, \tag{29.16}$$

$$\text{where } \alpha = \frac{2\lambda}{d_H(T)} \left[\frac{\lambda}{2d_H(T)} - \sin\theta\right] \tag{29.17}$$

is a temperature dependent deviation parameter [18]. The full width at half maximum, w, of the rocking curve is

$$w = 2C\,|\chi_H|\,\frac{\sqrt{|b|}}{\sin(2\theta_{\mathrm{B}})}. \tag{29.18}$$

An example of the simultaneous measurement of the reflected and the diffracted beams at grazing incidence is shown in Fig. 29.4.

29.4 Penetration and Information Depths

The effective absorption coefficient, μ_{e}, is given by the imaginary parts of the wavevectors, $\mathrm{K}_0^{\mathrm{i}} = \mathrm{K}_h^{\mathrm{i}}$ as

$$\mu_{\mathrm{e}} = -4\pi\gamma_0 K_0^{\mathrm{i}} = \mu_0 + 2\pi\gamma_0 \frac{\mathrm{Im}(\eta + \sqrt{\eta^2 - 1})}{\Lambda_0}, \tag{29.19}$$

where $\mu_0 = 2\pi K\,|\chi_0|$, η is the reduced deviation parameter from the exact Bragg angle, and $1/\Lambda_0$ is the extinction distance [1]. *The penetration depth* is

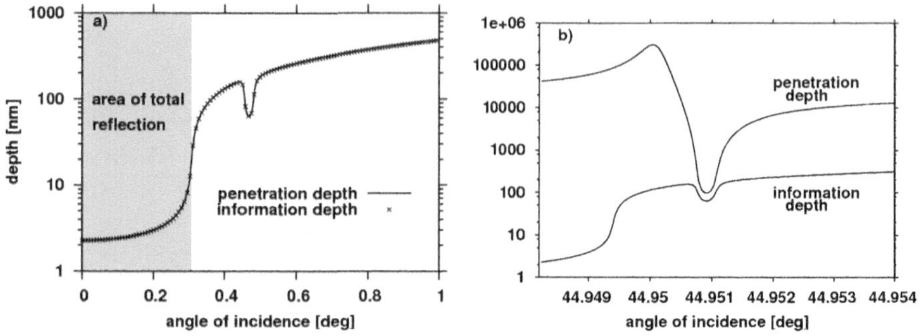

(a) (b)

Fig. 29.2. Penetration and information depths vs. the angle of incidence relative to the surface. $CuK_{\alpha 1}$ radiation, Ge(220) asymmetric coplanar diffractor with the angle of asymmetry 22.3°. (**a**) grazing incidence setting, (**b**) grazing emergence setting

the distance from the surface of the crystal for which the attenuation factor is $1/e$, from which one obtains $\exp(-\frac{\mu_e \ z_e}{\gamma_0}) = e^{-1}$, where $z_e = \gamma_0/\mu_e$. *The information depth* is given by the attenuation coefficients of both the incident and diffracted beams and is given by $1/z_{id} = 1/z_{e0} + 1/z_{eH}$. Figure 29.2 shows an example of the angular dependence of the penetration and information depths.

The penetration and information depths are important not only for the depth structural studies but also for the elements of crystal optics, namely because of their influence on the spatial resolution of crystal imaging optics. In addition, the temporal evolution of the X-ray beam pulses are affected by the dwell time of the beam in the crystal, which is important for the optics for the forthcoming generation of X-ray sources – free electron lasers.

29.5 Multiple Successive Diffractors in Coplanar and Noncoplanar Arrangements

By combining multiple diffractors successively, the incident beam can be diffracted successively several times and its size and its angular and spectral parameters can be modified accordingly. To have the maximum intensity throughput through the system, *the wavevector corresponding to the middle of the reflection range* should be considered for every successive diffractor using the real part of the wavevectors as done in [11]. If we denote the incident beam at the first diffractor by $(\mathbf{H}^{\mathbf{I}}, \mathbf{n}^{\mathbf{I}})$ as $\mathbf{K}_0^{\mathrm{I}}$ and the diffracted beam by $\mathbf{K}_{\mathbf{H}}^{\mathrm{I}}$, the last one being incident on the second diffractor $(\mathbf{H}^{\mathbf{II}}, \mathbf{n}^{\mathbf{II}})$ and diffracted by $\mathbf{K}_{\mathbf{H}}^{\mathrm{II}}$, etc., we can write a recurrent system of equations for the successively diffracted beams.

$$
\begin{aligned}
\mathbf{K}_H^{\mathrm{I}} &= \mathbf{K}_0^{\mathrm{I}} + \mathbf{H}^{\mathrm{I}} + \left(\kappa_0^{\mathrm{I}} - \kappa_H^{\mathrm{I}}\right)\mathbf{n}^{\mathrm{I}} \\
\mathbf{K}_H^{\mathrm{II}} &= \mathbf{K}_H^{\mathrm{I}} + \mathbf{H}^{\mathrm{II}} + \left(\kappa_0^{\mathrm{II}} - \kappa_H^{\mathrm{II}}\right)\mathbf{n}^{\mathrm{II}} \\
\mathbf{K}_H^{\mathrm{III}} &= \mathbf{K}_H^{\mathrm{II}} + \mathbf{H}^{\mathrm{III}} + \left(\kappa_0^{\mathrm{III}} - \kappa_H^{\mathrm{III}}\right)\mathbf{n}^{\mathrm{III}} \\
&\vdots \\
\mathbf{K}_H^{f} &= \mathbf{K}_H^{f-1} + \mathbf{H}^{f} + \left(\kappa_0^{f} - \kappa_H^{f}\right)\mathbf{n}^{f}
\end{aligned}
\tag{29.20}
$$

Multiplying the above equations for the middle of the reflection range by the corresponding lattice vectors, $\mathbf{H^i}$, we can obtain a system of linear equations for the coordinates of the unknown incident wavevectors, $\mathbf{K_0} = \mathbf{K_0^I}$ being able to diffract successively through the number of f spatially distributed diffractors:

$$
\begin{aligned}
\mathbf{K}_0 \cdot \mathbf{H}^{\mathrm{I}} =& -2K^2 \sin^2\theta_{\mathrm{B}}^{\mathrm{I}} - \kappa_0^{\mathrm{I}}(\mathbf{n}^{\mathrm{I}} \cdot \mathbf{H}^{\mathrm{I}}) \\
\mathbf{K}_0 \cdot \mathbf{H}^{\mathrm{II}} =& -2K^2 \sin^2\theta_{\mathrm{B}}^{\mathrm{II}} - \mathbf{H}^{\mathrm{I}} \cdot \mathbf{H}^{\mathrm{II}} - \left(\kappa_0^{\mathrm{I}} - \kappa_h^{\mathrm{I}}\right)(\mathbf{n}^{\mathrm{I}} \cdot \mathbf{H}^{\mathrm{II}}) - \kappa_0^{\mathrm{II}}(\mathbf{n}^{\mathrm{II}} \cdot \mathbf{H}^{\mathrm{II}}) \\
\mathbf{K}_0 \cdot \mathbf{H}^{\mathrm{III}} =& -2K^2 \sin^2\theta_{\mathrm{B}}^{\mathrm{III}} - \mathbf{H}^{\mathrm{I}} \cdot \mathbf{H}^{\mathrm{III}} - \mathbf{H}^{\mathrm{II}} \cdot \mathbf{H}^{\mathrm{III}} - \left(\kappa_0^{\mathrm{I}} - \kappa_h^{\mathrm{I}}\right)(\mathbf{n}^{\mathrm{I}} \cdot \mathbf{H}^{\mathrm{III}}) - \\
& - \left(\kappa_0^{\mathrm{II}} - \kappa_h^{\mathrm{II}}\right)(\mathbf{n}^{\mathrm{II}} \cdot \mathbf{H}^{\mathrm{III}}) - \kappa_0^{\mathrm{III}}(\mathbf{n}^{\mathrm{III}} \cdot \mathbf{H}^{\mathrm{III}}) \text{ etc.}
\end{aligned}
\tag{29.21}
$$

According to the value of the determinant of the system of linear equations (29.21) corresponding to the spatial arrangement of the diffractors we can obtain 0, 1, 2, or an infinite number of solutions. For example, a *coplanar arrangement* of diffractors implies an additional condition causing a degenerate system to give an infinite number of solutions.

Having the incident beam pass through the system of several (f) diffractors it is important to know the *spectral and angular composition* of the outgoing beam. This depends not only on the halfwidths of the individual diffractors but also on their spatial arrangement. Basically, a coplanar arrangement of diffractors can be in a *dispersive* (n, m) *setting* or in a *nondispersive* $(n, -n)$ *setting*. Simple *DuMond diagrams* (wavelength vs. angle dependence of the rocking curves widths) or the *spectral functions* (wavelength vs. angle dependence of the intensity transmitted through the system) are very useful for the calculation of the *degree of monochromatization and collimation.*

In noncoplanar arrangements of diffractors we have one special case when successive diffractors have mutually perpendicular scattering planes ($\sigma - \pi$ geometry). In the kinematical approximation the condition for this geometry is

$$\cos\phi_{\mathrm{I,II}} + \sin\theta_{\mathrm{B}}^{\mathrm{I}} \sin\theta_{\mathrm{B}}^{\mathrm{II}} = 0, \tag{29.22}$$

where $\phi_{\mathrm{I,II}}$ is the angle between diffracting planes [10].

29.6 Coupling of Multiple Successive Diffractors

Multiple successive diffractors can be prepared in one crystal block – a *monolithic system* of diffractors with the tightest possible coupling. They are especially advantageous because of their compact and monolithic design,

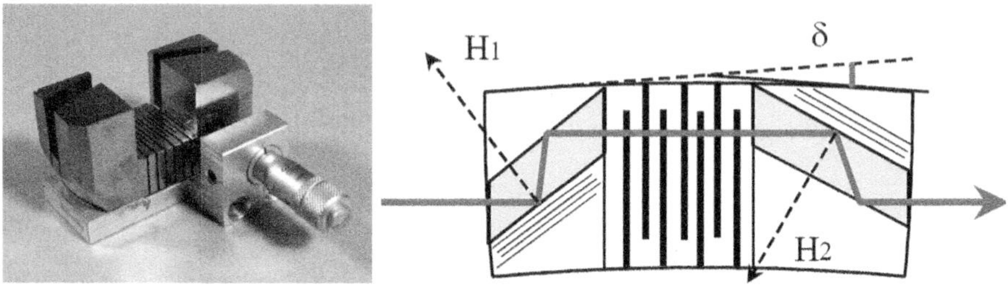

Fig. 29.3. Monolithic tiltable four-bounce monochromator, with a lamella-based hinge system for the adjustment of the diffraction vector $\mathbf{H}_2$–$\mathbf{H}_1$ by the rotation angle, δ

offering an extremely precise and stable wavelength, small dimensions, and a high thermomechanical stability. In a *polylithic system* when the diffractors are adjusted by independent goniometers we have a weak elastic link. In terms of the total reflectivity, this parameter is a product of the reflectivities of the individual diffractors (e.g., $R_{12} = R_2(\mathbf{H}^{\mathrm{II}}, \mathbf{K}_H^1)$. $R_1(\mathbf{H}^{\mathrm{I}}, \mathbf{K}_0)$ for a two-crystal arrangement).

There are another ways to couple successive diffractors, with the possibility of adjustment, e.g., by using elastic springs (Fig. 29.3) [4, 5].

Mathematically, the coupling and the adjustment are represented by processing the diffraction vectors $\mathbf{H}^{\mathrm{i}}$ with matrix transformations without changing the length of $\mathbf{H}^{\mathrm{i}}$ (pure rotation, mirror reflection, addition of a lattice vector) or changing it (adding a lattice vector, change of the lattice parameter by means of temperature or composition). A crystal diffractive–refractive 1D optics of this kind (multiple bounce) is represented by symmetric or asymmetric channel-cut monochromators and their combinations for which $\Sigma\mathbf{H}^{\mathrm{i}} = \mathbf{0}$. This condition has an important consequence, namely that the incident and outgoing beams are parallel. Such optics, either 1D or 2D, are called *in-line X-ray optics.*

Depending on the geometry, the outgoing beam can be linearly shifted relative to the incident beam. Due to refraction a small angular shift in the arcsec range is also possible. The in-line optical configuration with the beams arranged in one plane is important in synchrotron experiments because of the simpler instrumentation and in laboratory diffractometers because of the easier change of the optical elements without tedious readjustments.

Several types of in-line as well as non-in-line monochromators are presented in the following section. Namely, V-channel monochromators, monolithic Bartels-like monochromators, and a 2D beam compressor/expander.

29.7 Coplanar 1D Crystal Optics

Single flat crystal diffractors are sometimes used as analyzers or as reference crystals for testing purposes. Single and two-bounce diffractors with specially shaped surfaces have been used to focus synchrotron radiation [8]. In the next part we will concentrate on multiply bounced diffractors.

29.7.1 V-Shape 2-Bounce Channel-Cut Monochromators

A standard two-bounce V-channel Ge (220) monochromator with a final asymmetry factor $b = b^{\mathrm{I}} \cdot b^{\mathrm{II}} = 0.2$ (compression/expansion factor $M = 1/b = 5$) is used in combination with graded multilayer optics either to compress the beam (beam passing under Bragg condition from left to right in Fig. 29.4a, b) or to increase its intensity or cross section in the opposite direction in high-resolution diffractometers. Starting from the theoretical work of Servidori [17] we have designed and tested a new monochromator [6] in the beam expanding mode. Compared to the standard symmetrical channel-cut monochromator an intensity gain of 8.5 has been obtained at the expense of an output beam expansion by a factor of 4.5 in the scattering plane.

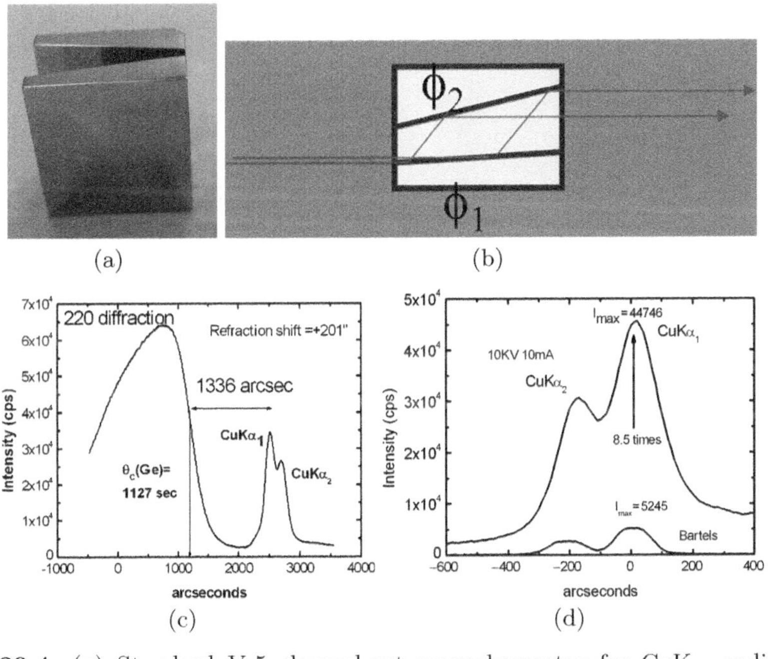

Fig. 29.4. (**a**) Standard V-5 channel-cut monochromator for CuK$_{\alpha1}$ radiation. (**b**) New, tested V-shaped monochromator with unequal asymmetry factors. (**c**) The asymmetry angle could be obtained precisely by recording simultaneously the reflected and doubly diffracted X-ray beam at grazing incidence. (**d**) Intensity gain compared to symmetrical channel cut monochromator

29.7.2 Monolithic 4-Bounce Monochromator for $CoK_{\alpha 1}$ Radiation

For the first time this Ge (220), (−2–20), (−440), (4–40) monochromator for the $CoK_{\alpha 1}$ radiation was considered in [3] and theoretically described in [15], where its suitability for a wavelength standard was also proposed. Its main advantage is the extremely stable wavelength and small dimensions. In the design presented in Fig.29.5 the energy bandpass is $\frac{\Delta E}{E} = 6.2 \times 10^{-5}$ and the divergence in the scattering plane $\Delta\Theta = 16.9$ arcsec. The limited precision of the wavelength calibration (∼1 meV) at a synchrotron source and the spread of the literature values of the germanium lattice parameter used in the design of the geometry of the monochromator lead to an energy shift of monochromator bandpass relative to the Lorentzian peak of the $CoK_{\alpha 1}$ emission line in Fig. 29.5c. To minimize this shift a monochromator temperature control is suggested.

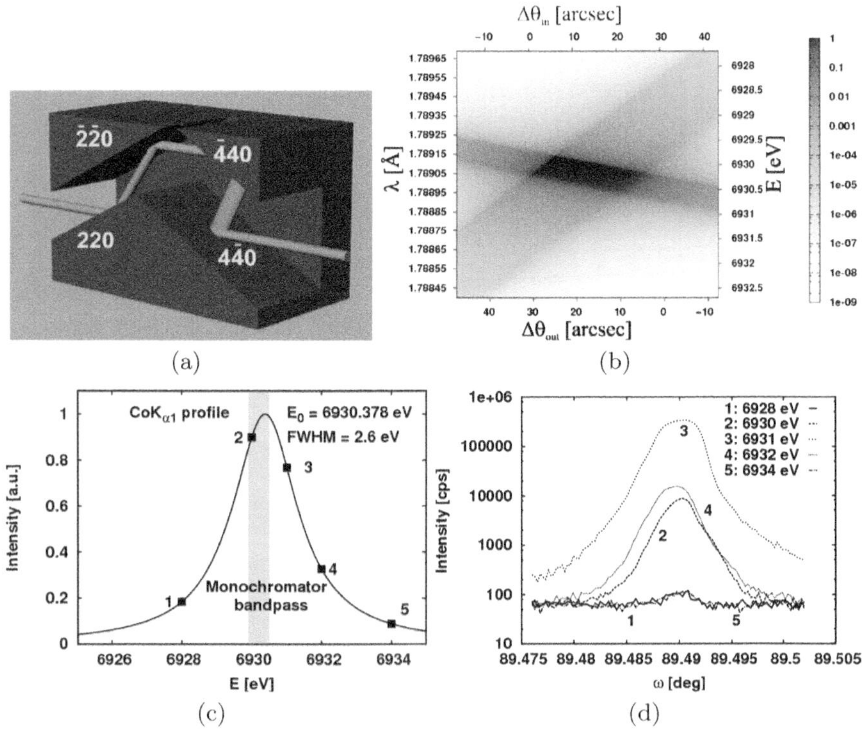

Fig. 29.5. (**a**) Monolithic Bartels-like germanium (220), (−2–20), (−440), (4–40) four-bounce monochromator for $CoK_{\alpha 1}$ radiation, (**b**) its spectral function, (**c**) Lorentz profile of natural emission line with the calculated monochromator bandpass, (**d**) synchrotron radiation intensity transmitted through the system for various photon energies around $E_0 = 6{,}930.32$ eV corresponding to the maximum of $CoK_{\alpha 1}$ emission line

29.8 Noncoplanar 2D Crystal Optics

The possibility of obtaining 2D image magnification using asymmetric diffractors was first demonstrated by Boettinger et al. [2] by their polylithic system. Spatial resolution and phase contrast effects have been treated in [19] and [16]. In a tomographic arrangement a volume resolution of 350 nm was achieved using a similar polylithic magnifier [20]. On the other hand, using a polylithic 2D beam compressor based on two independent $(n, -n)$ couples of asymmetric Bragg diffractors, adjusted to have perpendicular planes of diffraction, an X-ray microbeam of the size less than $7.6 \times 5.7\,\mu m^2$ was obtained and used to measure the strain distribution around the silicon dioxide edge on a silicon substrate [14, 21].

A monolithic 2D magnifier/demagnifier, designed according to the procedure presented in Sect. 29.5 and based on Si {311} diffractors, was recently successfully tested in its beam expanding and in beam compressing modes at ESRF [12, 13], see Fig. 29.6. In comparison with the X-ray magnifier, the exposition time on the same film and with nearly the same incident beam intensity decreased from 90 min to 30–120 s when testing the device in the beam compressing mode. This demonstrates a real 2D beam compression and concentration in this kind of X-ray optical element, which can open the way to a broader utilization of the device in beam conditioning optics.

29.9 Conclusions

The theoretical basis for 1D and 2D crystal, flat Bragg X-ray optics has been outlined. While 1D optics are important for reciprocal space mapping, the 2D optics are suitable for real space imaging. Special attention was given

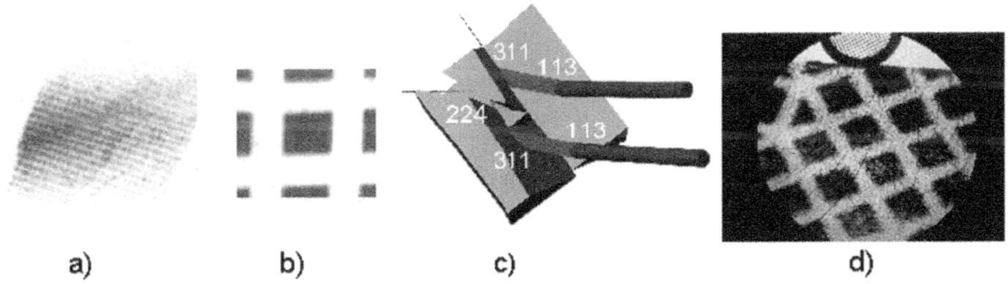

Fig. 29.6. Monolithic 2D X-ray beam compressor (for a beam passing from *right to left* in (**c**) and magnifier (*from left to right*) based on (113, −311, and 224) noncoplanar diffractors. (**a**) More than 400 parallel X-ray microbeams formed by the X-ray compressor from a wire mesh (**b**) put into the incident beam. The microbeam size is $7.6 \times 9.9\,\mu m$ ($H \times V$), their separation 3.3 µm horizontally and 4.3 µm vertically. (**d**) Microscopic Cu #300 mesh grid (*upper grid*) and its X-ray magnified image (*lower grid*). X-ray magnification 15 times at 9.6 keV. Grid diameter 2.3 mm, 32 µm wide stripes and 64 µm square windows (*with courtesy of Institute of Physics Publishing*)

to asymmetrical diffractors and to their coplanar and noncoplanar spatial arrangements. Some results of tests on selected V-channel monochromators and monolithic 2D beam de/magnifiers have been presented and ways to couple the individual elements have been proposed.

Acknowledgments

This work was supported by the Slovak Research and Development Agency under the contract No. APVV-0459-06 and by the Ministery of Education of the Czech Republic (contracts MSM 0021622410 and 1P04OCP07,004).

References

1. A. Authier, *Dynamical Theory of X-Ray Diffraction* (Oxford University Press, New York, 2001)
2. W.J. Boettinger, H.E. Burdette, M. Kuriyama, Rev. Sci. Instr. **50**, 26 (1979)
3. R.D. Deslattes, Appl. Phys. Lett. **12**, 133 (1968)
4. C. Ferrari and D. Korytár, Italian Patent 01299281, 2000
5. C. Ferrari and D. Korytár, J Appl. Cryst. **34**, 608 (2001)
6. C. Ferrari and D. Korytár, presented at COST P7 meeting Madrid (2005)
7. V. Holý, U. Pietch, T. Baumbach, *X-Ray Scattering from Thin Films and Multilayers* (Springer, Berlin Heidelberg New York, 1999)
8. J. Hrdý, E. Ziegler, N. Artemiev, F. Franc, J. Hrdá, T. Bigault, A. Freund, J. Synchrotron Rad. **8**, 1203 (2001)
9. X.R. Huang and M. Dudley, Acta Cryst. A **59**, 163 (2003)
10. D. Korytár, Czechoslovak J. Phys. **40**, 495 (1990)
11. D. Korytár, I. Bešše, P. Boháček, Czechoslovak J. Phys. **46** 1011 (1996)
12. D. Korytár, P. Mikulík, C. Ferrari, J. Hrdý, T. Baumbach, A. Freund, A. Kuběna, J. Phys. D: Appl. Phys. **36**, A65 (2003)
13. D. Korytár, T. Baumbach, C. Ferrari, L. Helfen, N. Verdi, P. Mikulík, A. Kubena, P. Vagovič, J. Phys. D: Appl. Phys. **38**, A208 (2005)
14. J. Matsui, Y. Tsusaka, K. Yokoyama, S. Takeda, M. Katou, H. Kurihara, K. Watanabe, Y. Kagoshima, S. Kimura, Nucl. Instr. Methods Phys. Res. B **199**, 15 (2003)
15. O. Pacherová, R. Bubáková, J. Appl. Cryst. **17**, 375 (1984)
16. P. Schäfer, R. Köhler, J. Phys. D: Appl. Phys. **36**, A113 (2003)
17. M. Servidori, J. Appl. Cryst. **35**, 41 (2002)
18. Y. Shvyd'ko, *X-Ray Optics* (Springer, Berlin Heidelberg New York, 2004)
19. R.D. Spal, Phys. Rev. Lett. **86**, 3044 (2001)
20. M. Stampanoni et al., Appl. Phys. Lett. **82**, 2922 (2003)
21. Y. Tsusaka, K. Yokoyama, S. Takeda, M. Urakawa, Y. Kagoshima, J. Matsui, S. Kimura, H. Kimura, K. Kobayashi, K. Izumi, Jpn. J. Appl. Phys. **39**, L635 (2000)

30

Thermal Effects under Synchrotron Radiation Power Absorption

V. Áč, P. Perichta, D. Korytár, and P. Mikulík

Abstract. Analyses of the effects of silicon crystal beam heating under static and dynamic synchrotron radiation (power loads are presented. This research is related to monochromator design and crystal optics in general. The aim of work is to analyze the conditions for insertion of the crystal X-ray optics into the high flux primary beam of the synchrotron. Suggestions for the optimization of the target geometry and cooling system arrangement are given.

30.1 Introduction

This chapter is devoted to the analysis of the effects of silicon crystal beam heating under static and dynamic synchrotron radiation (SR) power loads. Finite element (FE) simulations of the monocrystalline silicon target for surface and bulk temperature profiles and mechanical deformations are presented. The limitation of heat loads and the cooling geometry from the standpoint of diffraction properties are analyzed. The work is related to monochromator design and crystal optics in general [1] (see also Chap. 29). The aim of this work is to analyze the conditions for inserting the crystal X-ray optics into the high flux synchrotron primary beam. An optimal cooling system is needed. Hints for the optimization of the target geometry are given. Results presented here are a continuation of previous work [2, 3].

For the simulations, a regular 3D-mesh for the target geometry was implemented in the finite element model. The material is parameterized by anisotropic silicon mechanical properties. Our computation tool, in contrast to other known and often used tools, is optimized especially for SR crystal optics. The computation algorithm is adapted for 3D analyses of the temperature profile and an estimation of mechanical deformation for the resulting temperature field. The FE analyze of mechanical deformations is based on the balance of external mechanical forces and internal tensions and employs an interactive algorithm for fast computation convergence.

30.2 A Heat Transfer and Material Stress FE Model

The FE model used here consists of the integration of all thermal effects resulting from the absorption of radiation for real materials. These are the bulk heat absorption profile and effects related to the temperature field and the resulting mechanical deformations, which consequently affect the intensity distribution in the scattered beam.

30.2.1 Radiation Heat Absorption in the Matter

In spite of the different material dependencies for the description of the absorption of visible light and X-rays, the mathematical description for both follows an exponential decay law. The intensity, I, of X-ray photons transmitted through a thin foil of material for an incident intensity, I_0, follows the exponential attenuation

$$I = I_0 \exp(-\mu z), \tag{30.1}$$

where μ is the attenuation coefficient and z is the depth in the material. X-ray absorption depends on the energy of the X-ray photons and diminishes with increasing X-ray photon energy, E_{p} [4]. The heat absorption decreases nearly proportional to the cube of the energy $\left(\mu \approx 1/E_{\mathrm{p}}^3\right)$ except at absorption edges. The absorbed power density, q, is given by the formula

$$q = -E_{\mathrm{P}}\frac{\mathrm{d}I}{\mathrm{d}z} = E_{\mathrm{P}}\mu \exp(-\mu z). \tag{30.2}$$

The real absorbed heat profile can be different due to real conditions such as beam divergence and a variation of target material properties, for example.

30.2.2 Heat Transfer and Temperature Field

The heat transfer in anisotropic body is described by the universal heat transfer equation

$$\nabla \cdot \mathbf{q} + Q = \rho c \frac{\partial T}{\partial t} \quad \text{and} \quad \mathbf{q} = -\boldsymbol{\Lambda} \cdot \nabla T, \tag{30.3}$$

where $\mathbf{q}$ is the vector of the heat flow density, ρ is the material density, Q is the inner heat generation rate per unit volume, c is the heat capacity, $\boldsymbol{\Lambda}$ is the thermal conductivity tensor, T is temperature, and t is the time. Various boundary conditions can be specified to approximate real conditions:

- Constant temperature $T_{\mathrm{s}} = T(x, y, z, t)$ on defined target surface
- Heat convection $\mathbf{q}_{\mathrm{C}} = -h_{\mathrm{C}}\nabla T$ (h_{C} is the convection coefficient)
- Energy surface radiation $q_{\mathrm{sr}} = |\mathbf{q}_{\mathrm{sr}}| = \mu_{\mathrm{a}}\, q_{\mathrm{r}} - \sigma\mu_{\mathrm{e}}\, T^4$ (σ is the Stefan–Boltzmann constant, μ_{e} is the surface emission coefficient, μ_{a} is the surface absorption coefficient, and q_{r} is surface density of incoming heat flow) through the specified surface.

Generally, all parameters (ρ, c, $\mathbf{\Lambda}$, h, μ_e, μ_a) are dependent upon the local temperature, T. For computations of static mechanical deformations, it is necessary to specify the initial scalar temperature field $T_0(x, y, z, t) = T(x, y, z, 0)$. At computation of transient effects (for $t > 0$), the most important parameter is the thermal diffusivity α_T with dimension (m^2 s^{-1})

$$\alpha_\mathrm{T} = \frac{\Lambda}{\rho c}. \tag{30.4}$$

The α_T determines the dynamics of the heat transfer in the matter and Λ is the local scalar thermal conductivity. The static distribution of the temperature field is found by solution of (30.3) for $\frac{\partial T}{\partial t} = 0$ and $\frac{\mathrm{d}Q}{\mathrm{d}t} = 0$.

30.2.3 Mechanical Deformations

The local internal material stress is caused by thermal expansion (strain), ε_T. It depends on local temperature difference $\Delta T = T - T_0$, where T_0 is the initial temperature [5]. Then

$$\varepsilon_\mathrm{T} = \alpha \Delta T, \tag{30.5}$$

where α is a linear thermal expansion coefficient. The mechanical depression ε_m caused only by internal stresses can be expressed as

$$\varepsilon_\mathrm{m} = \sum_{k=1}^{3} \sum_{l=1}^{3} \mathbf{S}_{ijkl} \sigma_{kl}, \tag{30.6}$$

where $\mathbf{S}_{ijkl}$ is the compliance tensor (inverse matrix of the second order stiffness tensor $\mathbf{C}_{ijkl}$) and σ_{kl} is an internal stress. By superposing these two mechanisms, we obtain an expression for the total mechanical strain ε. Then

$$\varepsilon = \varepsilon_\mathrm{T} + \varepsilon_\mathrm{m} = \alpha(T - T_\mathrm{C}) + \sum_{k=1}^{3} \sum_{l=1}^{3} \mathbf{S}_{ijkl} \sigma_{kl}. \tag{30.7}$$

The balance criterion for FE analysis is formulated by the equivalence of internal thermal stresses for each neighbouring material element position

$$\sigma_{kl+} = -\sigma_{kl-}. \tag{30.8}$$

The finite element method yields the displacement field, which minimizes the total potential energy Π,

$$\Pi = \int_V \frac{1}{2} \varepsilon_\mathrm{T} \sigma \, \mathrm{d}V - \int_V \varepsilon_\mathrm{m} p_V \, \mathrm{d}V - \int_S \varepsilon_\mathrm{m} p_S \, \mathrm{d}S, \tag{30.9}$$

where p_V is the local body stress and p_S is the local surface stress.

30.2.4 Material Parameters

The most frequently used material for crystal X-ray optics is monocrystalline silicon. The surface orientation of monochromator crystals is typically (100), (110), and (111). A possible solution for improving the monochromator (or target) cooling efficiency is the use of isotopically pure ^{28}Si, which has a higher thermal conductivity than natural Si. The basic thermomechanical parameters of monocrystalline silicon are described in Table 30.1. The thermal conductivity temperature dependence $\lambda(T)$ of natural silicon and of ^{28}Si [6] and the temperature dependence of linear thermal expansion coefficients $\alpha(T)$ for monocrystalline silicon and CVD (Chemical Vapour Deposited) diamond are shown in Fig. 30.1. The anisotropic thermomechanical properties of silicon (Young's modulus, Poisson's ratio, surface and bulk shear modulus) are published in [7] and [8].

Despite better thermomechanical properties of diamond or isotopic pure silicon, the most interesting material for X-ray optics is still natural monocrystalline silicon because of its availability and cost. Diamond crystals are needed mainly for splitting white beam synchrotron radiation for subsequent monochromatization at several beamlines. Thus, at present great effort is being devoted to improve the homogeneity of large artificial diamond crystals.

30.3 Simulation of Monochromator Designs

In this section, we present simulations of perfect crystalline silicon targets with different shapes and under different heat conditions (incident radiation). The optimal target cooling system is investigated.

30.3.1 Silicon Target and Simulation Conditions

Let us study a typical cooled monochromator: a monocrystalline silicon target with block geometry as shown in Fig. 30.2 and with variable target body thickness h. The cooled back side (with cooling temperature $T_c = 293\,\mathrm{K}$) can have various geometries: a flat surface or a channeled surface (cooling channels). The X-ray beam is centered on the target with an incident angle $\theta = 5.14°$ and the beam spot is elliptical, $30 \times 10\,\mathrm{mm}^2$. The beam power density ($0.23\,\mathrm{W\,mm}^{-2}$) is defined to be homogenous in the whole irradiated surface area. The photon energy for surface absorption is $E_\mathrm{p} = 100\,\mathrm{eV}$, with an attenuation coefficient $\mu \approx 10^5\,\mathrm{m}^2\,\mathrm{kg}^{-1}$ and penetration depth in micrometer range. The photon energy for bulk absorption is $E_\mathrm{p} = 10\,\mathrm{keV}$ with $\mu \approx 3 \times 10^2\,\mathrm{m}^2\,\mathrm{kg}^{-1}$ and penetration depth in millimeter range.

Table 30.1. The thermomechanical parameters of natural monocrystalline silicon

Parameter	Symbol	Value	Range (K)	Notes
Thermal conductivity	Λ	$\lambda(T) = (49100/T) - 20$	$T \in (200, 1000)$	(J $s^{-1}m^{-1}$ K^{-1}) [6]
		$\lambda(T) = 786921\ T^{-1.485}$	$T \in (50, 1000)$	
Heat capacity	C	$c(T) = 5.10^{-6}T^3 - 0.008\,T^2 + 4.5575\,T - 72.888$	$T \in (200, 1000)$	(J kg^{-1} K^{-1})
		705	$T = 300$	
Density	ρ	2328.3	$T = 300$	(kg m^{-3})
Linear expansion coefficient	α	$\alpha(T) = 10^{-14}T^3 - 3 \cdot 10^{-11}T^2 + 2 \cdot 10^{-8}T - 2 \cdot 10^{-6}$	$T \in (100, 1000)$	(K^{-1}) [7]
Young modulus	Y	130.2–187.5		(GPa) [7]
Poisson ratio	V	0.064–0.361		[7]
Shear modulus	G	50.92–79.4		(GPa) [7]

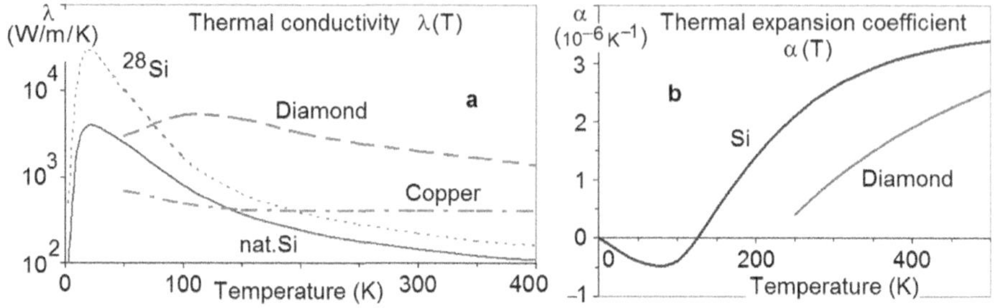

Fig. 30.1. The thermal conductivity (**a**) of natural and isotopically pure silicon ^{28}Si [6], CVD diamond and copper [9], and linear thermal expansion coefficient (**b**) for natural monocrystaline silicon [8] and CVD diamond [9]

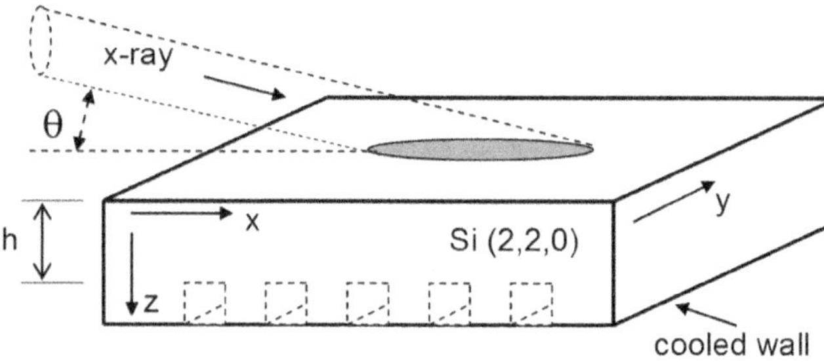

Fig. 30.2. The geometry of a silicon target and beam power parameters; the cooled back side wall is with or without cooling channels. Channel width is 1 mm and channel depth (not included in h) is 2 mm

30.3.2 Temperature Field and Surface Mechanical Deformations

The surface temperature, bulk temperature profile, surface deformation, and surface angle gradient for the channeled target with a body thickness $h =$ 2 mm and bulk absorption of the beam energy (photon energy, $E_{\mathrm{p}} = 10\,\mathrm{keV}$) are shown in Fig. 30.3. The reproduction of the channeled cooled surface (the back side of the target) is shown on the surface temperature and the mechanical deformations profile, respectively. This effect is smaller for higher target body thickness. The bulk heat absorption causes the asymmetry of temperature profile in the x-axis direction.

30.3.3 Dependence of Surface Mechanical Deformations on the Target Cooling Geometry

The mechanical deformations are investigated in terms of the surface slope angle, which is the most important parameter for diffraction and reflection properties of the crystal. The results for three target body thicknesses and flat and channeled cooling surface geometry are shown in Fig. 30.4. These results confirm the increase of the cooling effectivity with decreasing target body thickness.

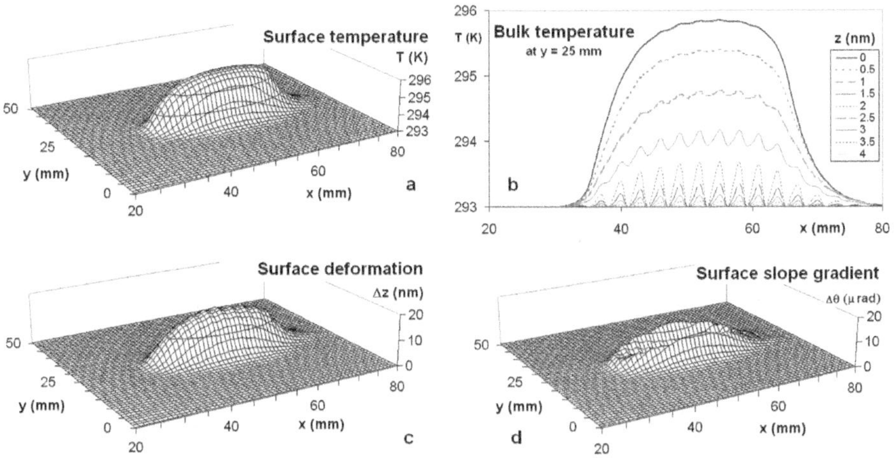

Fig. 30.3. Calculated surface temperature profile (**a**), bulk temperature profile (**b**), surface deformation (**c**), and surface slope gradient (**d**) for a slotted target with thickness $h = 2\,\text{mm}$ and bulk absorption of beam energy at $E_{\text{p}} = 10\,\text{keV}$

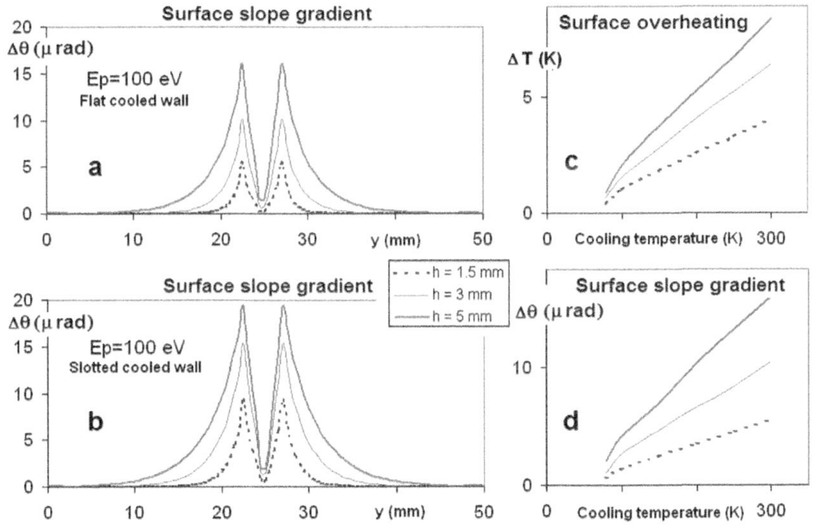

Fig. 30.4. The cross section of surface slope gradient in the middle of the target for target body thicknesses $h = 1.5$, 3, and 5 mm; surface heat absorption for flat cooling surface (**a**) and for channeled cooling surface (**b**); surface overheating (**c**) and surface slope gradient (**d**) vs. cooling temperature for flat cooling surface and various target body thicknesses h

30.3.4 Cooling Temperature

Decreasing the cooling temperature is the best way to reduce mechanical deformations caused by absorption of radiation heat. The increase of the thermal conductivity of silicon (Fig. 30.1) together with the decreasing thermal expansion coefficient as the cooling temperature is decreased makes it possible to significantly reduce the surface deformations. These results are documented in Fig. 30.4c, d. The cooling by liquid nitrogen (LN2) (78 K) can reduce the surface overheating and surface deformation by more than one order of magnitude in comparison to these effects using room temperature cooling.

30.3.5 Cooling Channels Variations

In this section, we will show how it is possible to reduce variations of the surface temperature and deformation due to the reproduction of the channel arrangement by optimization of cooling channels geometry. Two possible locations of rectangular cooling channels are seen in Fig. 30.5. The cooling channels are built in the silicon target (a) or in a cooler block (b), respectively. Usually, the cooler block is made from a high thermally conductive material (copper). Simulations in Fig. 30.5 show higher effectivity of the cooling channels made in copper block.

This fact is caused by the higher thermal conductivity in the bulk of the copper cooler together with its larger surface of the cooling channels in comparison to that in the silicon target. In addition, "print-through" of the channel geometry to the optical surface is reduced.

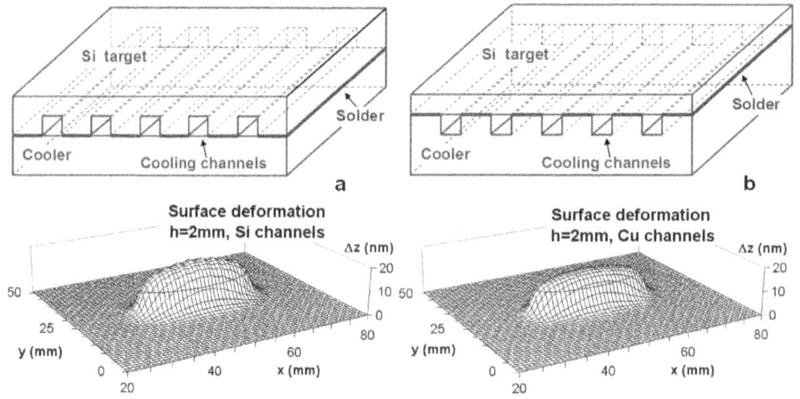

Fig. 30.5. Cooling channels variations and corresponding simulations of surface deformation (at photon energy, $E_p = 100\,\mathrm{eV}$) for cooling channels in the silicon target (**a**) and cooling channels in the copper cooler block (**b**)

30.3.6 Cooling Block Arrangement

The next example is devoted to the revision of cooling block efficiency. Two variations of the cooling block arrangement shown in Fig. 30.6 have been investigated. From mechanical point of view, the lateral side coolers are optimal for a higher target body thickness. Back side cooling is optimal for a smaller target body thickness. The simulation results for block Si target ($220\times120\times40\,\mathrm{mm}^3$) and rectangular beam spot ($5\times35\,\mathrm{mm}^2$, $E_\mathrm{p}=100\,\mathrm{eV}$, $0.43\,\mathrm{W\ mm}^{-2}$) shown in Fig. 30.6c, d confirm the comparable effectiveness of lateral side coolers compared to back side coolers if the target body thickness is comparable with lateral target size. The benefit of back side cooling is seen at smaller target

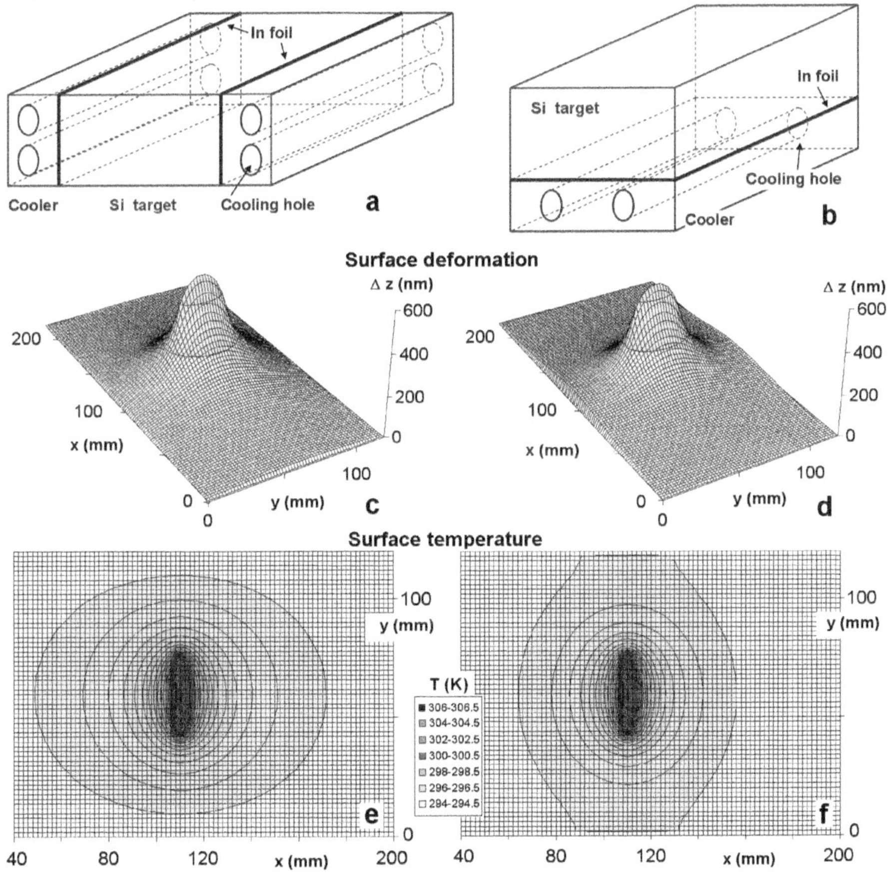

Fig. 30.6. The cooling block arrangement with (**a**) lateral side coolers, (**b**) back side cooler and simulations of surface deformations (**c**), and surface temperature (**e**) for lateral side cooling and (**d**, **f**) for back side cooling. The silicon target body thickness $h=40\,\mathrm{mm}$. The indium foil thermally connects the silicon target to the copper cooling block

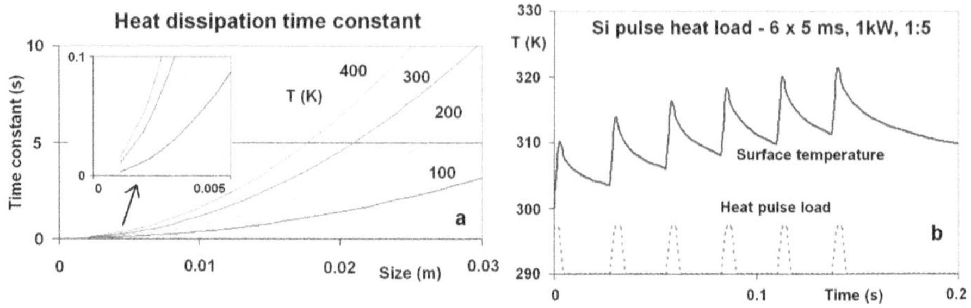

Fig. 30.7. Temperature and size dependence of the heat dissipation time constant $\tau_{\rm d}$ for monocrystalline silicon (**a**), and (**b**) shows the time dependence of the Si surface temperature under the load of a sequence of six pulses for a target body thickness $h = 3\,{\rm mm}$

body thickness. The significant effect is seen in a reduction of the surface deformation, which is the aim of the cooling system optimization. The minimum size of the target body thickness is given by the penetration depth of the X-ray beam, which depends upon its photon energy and the surface incident angle.

30.3.7 Dynamic Thermal Properties of Silicon

The dynamic thermal properties are significant in case of pulsed irradiation of the surface. The transient heat properties of the material are characterized by the heat dissipation time constant, $\tau_{\rm d}$, defined by the thermal diffusivity, $\alpha_{\rm T}$, and the target size. The typical temperature and target size dependences of $\tau_{\rm d}$ for monocrystalline silicon are shown in Fig. 30.7a. The time constant resulting from the solution of (30.3) with respect to (30.4) characterizes the dynamics of heat dissipation in the material after a pulsed heat load. The typical time dependence of the temperature at the start of the pulse load process (six pulses) is shown in Fig. 30.7b. The swing of surface temperature is dependent on the heat pulse frequency. The maximum of temperature under a long time periodic load goes to saturation. It is possible to minimize the temperature swing for Si target with the thickness in the millimeter range for pulse frequencies higher than 100 Hz.

30.4 X-Ray Diffraction Spot Deformation

In synchrotron beamlines, a crystal used as the first monochromator is subject to a white radiation. Monochromatization by Bragg reflection leads to a high heat load on the surface of the crystal. Heat conduction in the bulk and cooling at the bottom leads to a stationary distribution of crystal parameters. This is mainly the case for a Bragg reflecting crystal. The heated crystal exhibits deformations as shown in previous sections. Limiting ourselves to cooled monochromators, we will further focus only on diffraction in Bragg (reflection) geometry for semi-infinite crystals.

Let us first make a qualitative discussion about the expected diffraction image. It will be similar to X-ray topography images. This leads to the following:

1. Surface waviness follows the height profile. Surface roughness does not play a role in X-ray diffraction. Ray-tracing the scattered rays are incoherent. A surface elevation would shift the diffracted spot on the detector, which is negligible for micrometer pixel sizes and sub-micrometer elevations.
2. Crystal lattice waviness is equivalent to lattice misorientation. The angle of incidence of the Bragg maximum is shifted by the surface slope angle projection, α_x, which is given by the tangent to the crystal lattice plane (or the diffraction vector angle) as projected into the scattering plane. The direction of the diffracted beam (angle of α_f) will slightly change as well.
3. Crystal lattice deformation. This locally changes the Bragg angle θ_B. From the differential form of the Bragg law

$$\Delta\theta_\mathrm{B} = -(\Delta a/a)\tan\theta_\mathrm{B}, \tag{30.10}$$

where a is the crystal lattice constant.

In summary, the angle of incidence (α_i) of the Bragg peak maximum changes locally by

$$\Delta\alpha_\mathrm{i} = \alpha_x - (\Delta a/a)\tan\theta_\mathrm{B}. \tag{30.11}$$

Qualitatively, this value influences the diffracted intensity via rocking curve shift.

As an example, let us take a cooled Si (111) monochromator at 8 keV. The Bragg angle is 14.2°, the Bragg extinction length is 1.5 μm, and the Darwin curve fwhm is 7 arcsec (σ-polarization). The latter corresponds to 34 μrad or $\Delta a/a = 10^{-4}$, which are smaller than values calculated in the previous section. Thus, for a temperature change of a few degrees Kelvin, the diffraction image is kept almost unchanged. For lower energy and higher Bragg angles, the requirements are less strict.

Furthermore, real beam homogeneity is driven by its (a) divergence and (b) its wavelength spread. These are the effects that smooth the diffracted image. For example, a divergence of 10 arcsec is higher than the Bragg curve fwhm.

The diffraction image from a monochromator is simulated by usual ray-tracing methods. In brief, for perfect crystals, the usual dynamical diffraction is used, while for deformed crystals the Takagi-Taupin or (semi) kinematic approximation are adequate. For heated monochromators, there is a deformation gradient from the surface to the bulk. Qualitatively, the limit between dynamical theory for perfect and deformed crystals is the angular shift of the Bragg peak on the surface within a Borrmann fan. For a perfect crystal, it should be smaller than the Bragg peak width. Otherwise, for a large deformation on a micrometer scale, the use of the Takagi-Taupin approach for reflected wave intensity would be required.

Diffraction spot simulations by ray-tracing verify these qualitative conclusions. For a temperature field with a variation of several degrees, the diffracted image is homogeneous. The acceptable surface deformation limit in terms of the diffraction spot damage follows from (30.11) and depends not only on target surface deformation but also on beam divergence.

Acknowledgement

The authors acknowledge the support by the Slovak Ministry of Education (grants VEGA 1/4134/07 and APVV -0459-06) and by the Czech Ministry of Education (grants MSM 0021622410 and 1P04OCP07.004).

References

1. D. Korytár, P. Mikulík, C. Ferrari, J. Hrdý, T. Baumbach, A. Freund, A. Kuběna, Phys. D Appl. Phys. **36**, A65 (2003)
2. L. Zhang, J. Hoszowska, J.S. Migliore, V. Mocella, C. Ferrero, A.K. Freund, Nucl. Instrum. Methods Phys. Res. A **467–468**, 409 (2001)
3. V. Mocella, W.K. Lee, G. Tajiri, D. Mills, C. Ferrero, Y. Epelboin, J. Appl. Crystallogr. **36**, 129 (2003)
4. A. Thompson, D. Vaughan et al., *X-ray Data Booklet* (Lawrence Berkeley National Laboratory, Berkeley, CA 2001), http://xdb.lbl.gov/xdb.pdf
5. D.W. Nicholson, *Finite Element Analysis: Thermomechanics of Solids*, (CRC Press, West Palm Beach, FL, 2003) ISBN 0- 8493-0749-X
6. T. Ruf, R.W. Henn, M. Asen-Palmer, E. Gmelin, M. Cardona, H.-J. Pohl, G.G. Devyatych, P.G. Sennikov, Solid State Commun. **115**(5), 243 (2000)
7. J. Kim, D. Cho, R.S. Muller, in *Proceedings of the 11th International Conference on Solid State Sensors and Actuators*, Munich, 10–14 June 2001, pp. 662–665
8. J.J. Wortman, R.A. Evans, J. Appl. Phys. **36**(1), 153 (1965)
9. J.E. Graebner, J. Thermophys. **19**(2), 511 (1998)

Index

Springer Series in
OPTICAL SCIENCES

Volume 1

1 **Solid-State Laser Engineering**
By W. Koechner, 5th revised and updated ed. 1999, 472 figs., 55 tabs., XII, 746 pages

Published titles since volume 110

110 **Kramers–Kronig Relations in Optical Materials Research**
By V. Lucarini, J.J. Saarinen, K.-E. Peiponen, E.M. Vartiainen, 2005, 37 figs., X, 162 pages

111 **Semiconductor Lasers**
Stability, Instability and Chaos
By J. Ohtsubo, 2nd edn. 2007, 169 figs., XIII, 475 pages

112 **Photovoltaic Solar Energy Generation**
By A. Goetzberger and V.U. Hoffmann, 2005, 139 figs., XII, 234 pages

113 **Photorefractive Materials and Their Applications 1**
Basic Effects
By P. Günter and J.P. Huignard, 2006, 169 figs., XIV, 421 pages

114 **Photorefractive Materials and Their Applications 2**
Materials
By P. Günter and J.P. Huignard, 2006, 370 figs., XVII, 640 pages

115 **Photorefractive Materials and Their Applications 3**
Applications
By P. Günter and J.P. Huignard, 2007, 316 figs., X, 366 pages

116 **Spatial Filtering Velocimetry**
Fundamentals and Applications
By Y. Aizu and T. Asakura, 2006, 112 figs., XII, 212 pages

117 **Progress in Nano-Electro-Optics V**
Nanophotonic Fabrications, Devices, Systems, and Their Theoretical Bases
By M. Ohtsu (Ed.), 2006, 122 figs., XIV, 188 pages

118 **Mid-infrared Semiconductor Optoelectronics**
By A. Krier (Ed.), 2006, 443 figs., XVIII, 751 pages

119 **Optical Interconnects**
The Silicon Approach
By L. Pavesi and G. Guillot (Eds.), 2006, 265 figs., XXII, 389 pages

120 **Relativistic Nonlinear Electrodynamics**
Interaction of Charged Particles with Strong and Super Strong Laser Fields
By H.K. Avetissian, 2006, 23 figs., XIII, 333 pages

121 **Thermal Processes Using Attosecond Laser Pulses**
When Time Matters
By M. Kozlowski and J. Marciak-Kozlowska, 2006, 46 figs., XII, 217 pages

122 **Modeling and Analysis of Transient Processes in Open Resonant Structures**
New Methods and Techniques
By Y.K. Sirenko, N.P. Yashina, and S. Ström, 2007, 110 figs., XIV, 353 pages

123 **Wavelength Filters in Fibre Optics**
By H. Venghaus (Ed.), 2006, 210 figs., XXIV, 454 pages

124 **Light Scattering by Systems of Particles**
Null-Field Method with Discrete Sources: Theory and Programs
By A. Doicu, T. Wriedt, and Y.A. Eremin, 2006, 123 figs., XIII, 324 pages

Springer Series in

OPTICAL SCIENCES

125 **Electromagnetic and Optical Pulse Propagation 1**
Spectral Representations in Temporally Dispersive Media
By K.E. Oughstun, 2007, 74 figs., XX, 456 pages

126 **Quantum Well Infrared Photodetectors**
Physics and Applications
By H. Schneider and H.C. Liu, 2007, 153 figs., XVI, 250 pages

127 **Integrated Ring Resonators**
The Compendium
By D.G. Rabus, 2007, 243 figs., XVI, 258 pages

128 **High Power Diode Lasers**
Technology and Applications
By F. Bachmann, P. Loosen, and R. Poprawe (Eds.) 2007, 543 figs., VI, 548 pages

129 **Laser Ablation and its Applications**
By C.R. Phipps (Ed.) 2007, 300 figs., XX, 586 pages

130 **Concentrator Photovoltaics**
By A. Luque and V. Andreev (Eds.) 2007, 250 figs., XIII, 345 pages

131 **Surface Plasmon Nanophotonics**
By M.L. Brongersma and P.G. Kik (Eds.) 2007, 147 figs., VII, 271 pages

132 **Ultrafast Optics V**
By S. Watanabe and K. Midorikawa (Eds.) 2007, 339 figs., XXXVII, 562 pages. With CD-ROM

133 **Frontiers in Surface Nanophotonics**
Principles and Applications
By D.L. Andrews and Z. Gaburro (Eds.) 2007, 89 figs., X, 176 pages

134 **Strong Field Laser Physics**
By T. Brabec, 2007, approx. 150 figs., XV, 500 pages

135 **Optical Nonlinearities in Chalcogenide Glasses and their Applications**
By A. Zakery and S.R. Elliott, 2007, 92 figs., IX, 199 pages

136 **Optical Measurement Techniques**
Innovations for Industry and the Life Sciences
By K.E. Peiponen, R. Myllylä and A.V. Priezzhev, 2008, approx. 65 figs., IX, 300 pages

137 **Modern Developments in X-Ray and Neutron Optics**
By A. Erko, M. Idir, T. Krist and A.G. Michette, 2008, approx. 150 figs., XV, 400 pages

138 **Optical Micro-Resonators**
Theory, Fabrication, and Applications
By R. Grover, J. Heebner and T. Ibrahim, 2008, approx. 100 figs., XXII, 330 pages

GPSR Compliance

The European Union's (EU) General Product Safety Regulation (GPSR) is a set of rules that requires consumer products to be safe and our obligations to ensure this.

If you have any concerns about our products, you can contact us on ProductSafety@springernature.com

In case Publisher is established outside the EU, the EU authorized representative is:

Springer Nature Customer Service Center GmbH
Europaplatz 3
69115 Heidelberg, Germany

Batch number: 10755461

Printed by Printforce, the Netherlands